T0189739

Springer Series in Operations Research and Financial Engineering

Series Editors:

Thomas V. Mikosch
Sidney I. Resnick
Stephen M. Robinson

For further volumes:
http://www.springer.com/series/3182

Ludger Rüschendorf

Mathematical Risk Analysis

Dependence, Risk Bounds,
Optimal Allocations and Portfolios

 Springer

Ludger Rüschendorf
Department of Mathematical Stochastics
University of Freiburg
Freiburg
Germany

ISSN 1431-8598
ISBN 978-3-642-43016-9   ISBN 978-3-642-33590-7 (eBook)
DOI 10.1007/978-3-642-33590-7
Springer Heidelberg New York Dordrecht London

JEL Classification: C1, C2, G1

Mathematics Subject Classification (2010): 62P05, 91B30, 91Gxx, 60G70

Printed on acid-free paper

Springer is part of Springer Science+Business Media (www.springer.com)

Preface

This book gives an introduction to basic concepts and methods in mathematical risk analysis, in particular to those parts of risk theory which are of particular relevance in finance and insurance. The description of the influence of dependence in multivariate stochastic models for risk vectors is the main focus of the text.

In the first part we introduce basic probabilistic tools and methods of distributional analysis and describe their use in the modelling of dependence and to derive risk bounds in these models. The second part of this book is concerned with measures of risk with a particular view towards risk measures in financial and insurance context. The final parts are devoted to relevant applications such as optimal risk allocations problems and optimal portfolio problems. Throughout we give short accounts of the basic methods used from stochastic ordering, from duality theory, from extreme value theory, convex analysis, and empirical processes as they are needed in the applications.

The focus is on presentation of the main ideas and methods and on their relevance for practical application. In favour of better readability we give only proofs of those results which do not need too much preparation or which are not too extended. The text is mostly self-contained. We make use of several basic results from stochastic ordering and distributional analysis which can be spotted easily in related textbooks.

The book can be used as a textbook for an advanced undergraduate third year course or for a graduate level mathematical risk analysis course. Good knowledge of basic probability and statistics as well as of general mathematics is a prerequisite. Some more advanced mathematical methods are explained on a non-technical level. The book should also be of interest as a reference text giving a clearly structured and up-to-date treatment of the main concepts and techniques used in this area. It also gives a guide to actual research; it points out relevant present research topics and describes the state of the art. In particular its aim is to give orientation to an interested researcher entering the field and moreover to help to acquire a solid fundament for working in this area.

The notion of dependence raises several basic issues. The first one is the construction of stochastic models of dependent risk vectors broad enough to describe all relevant classes of dependent risks. It is pointed out that fundamental tools for this purpose are the distributional transform, the quantile transform, and their multivariate extensions. These tools give an easy access to the Fréchet class, which is a synonym for the class of all possible dependence models. In particular a simple application of these distributional transforms gives the general form of Sklar's representation theorem and thus the notion of copula. The multivariate quantile transform yields a construction method for random vectors with specified general distribution and is a basic tool for simulation. The multivariate distributional transform on the other side, transforms a random vector to a vector with iid uniformly distributed components. This transform extends a classical result of Rosenblatt (1952) and has some important applications to goodness of fit tests and to identification procedures. Some concrete classes of constructions of multivariate copula models are described by methods like L^2-projections or the pair-copula constructions.

A classical topic in the analysis of risk is the development of sharp risk bounds in dependence models. The historical origins of this question are the Hoeffding–Fréchet bounds which give sharp upper and lower bounds for the covariance and the joint distribution function of two random variables X, Y with distribution functions F, G. These results have been extended to the problem of establishing sharp bounds for general risk functionals w.r.t. Fréchet classes. Important progress on this class of problems was obtained by the development of a corresponding duality theory, which was motivated by this problem of getting bounds for dependence functionals. It turned out that this duality theory in case of a two-fold product space connects up with the Monge–Kantorovich mass-transportation theory which aimed to describe minimal distances or transport costs between two distributions. By means of duality theory several basic sharp dependence bounds could be determined.

As a consequence the notion of comonotonicity is identified as the worst case dependence structure, in case the components of the portfolio are real. These findings were further extended by means of various stochastic ordering results concerning diffusion type orderings (as convex order or stop-loss order) and also concerning dependence orderings (like super-modular or directionally-convex ordering). W.r.t. all convex law invariant risk measures comonotonicity is the worst case dependence structure of the joint portfolio.

An exposition of the representation theory of convex risk measures and its basic properties like continuity properties is given on spaces of L^p-risks. Also extensions to risk measures on portfolio vectors are detailed. These extensions allow one to include for optimal allocation or portfolio problems the important aspect of dependence within the portfolio components. A fundamental question concerning the dependence structure is on the existence and form of a worst case dependence structure – generalizing comonotonicity – for a sample $X_1, \ldots, X_n$ of portfolio vectors. It turns out however that a universally worst case dependence structure does not exist any more in higher dimension. But it is possible to describe worst case dependent portfolios w.r.t. specific multivariate risk measures. Here again a close connection with mass transportation comes into play. The max-correlation risk measures which are defined via mass transportation problems are the building blocks of all law invariant risk measures and thus take the role of the "average value at risk" risk measure in one dimension. Worst case dependence structures then are identified by comonotonicity w.r.t. worst case scenario measures.

In the final two chapters some relevant classes of optimal risk allocation and portfolio optimization problems are dealt with. The risk allocation problems are closely connected with optimal investment problems or minimal demand problems in finance and insurance. We also discuss classical and recent results on optimal (re-)insurance contracts. By combination of stochastic ordering results and results on worst case risk measures some simplified derivation of these optimality results can be given. Optimal portfolios are determined also from the point of view to minimize the sensibility to extremal risk events. These results are based on extreme value theory and supplement the usual finite risk analysis given by extensions of the classical Markowitz theory to the frame of risk measures. The notion of asymptotic portfolio loss order allows us to compare in this respect different stochastic loss models.

The aim of the book is to present relevant methods and tools to deal with the influence of dependence on various problems of risk analysis. It also discusses in detail some relevant applications to optimal risk allocation and optimal portfolio problems in finance and in insurance. The content of this book represents areas of my research over the last 20–30 years. The work on dependence, stochastic ordering, and risk bounds has been combined in more recent years with the new developments on risk measures and related optimization problems. As a result the book is not an encyclopaedic presentation. Several relevant subjects of mathematical risk analysis are not dealt with in this book. The basis of this book are several of my survey papers and oral presentations on dependence, risk bounds, and stochastic orderings, as dealt with in this book. Of particular mention are surveys on the theory of Fréchet bounds, which were presented and published in the series of volumes of the conferences on *Probabilities with given marginals* started by the Rome 1990 conference. The main topics of stochastic orderings in particular dependence orderings as described in this text, the related duality theory, and the distributional transforms go back to my habilitation thesis in 1979 and some related publications in the following years.

The basis and motivation for working in the area of risk measures naturally arose from the fundamental work of Delbaen (2000) and Föllmer and Schied (2011). My particular interest in this area was to combine this theory with the analysis of dependence properties. This aim has also been followed with more focus on applications in risk management in the book of McNeil et al. (2005b) giving a rich source of techniques and methods. In the book of Pflug and Römisch (2007) this theory is combined with statistical and decision theoretic concepts. My more recent interest was also driven by insurance applications and problems of optimal insurance contracts as presented in Kaas et al. (2001) and in more detail in the more recent book of Denuit et al. (2005) which is also focused on the role of stochastic ordering, risk measures, and dependence in insurance. In that book a much broader exposition of basics in these areas is given and applied insurance problems are included in much more detail.

Finally, I would like to express my gratitude and appreciation to all those whom I had the pleasure to work with over the years or who gave some inspiration to the work, either personally or by their work.

Fruitful cooperation with Svetlozar T. Rachev in the 1990s on mass transportation problems and also with Norbert Gaffke, Juan Cuesta-Albertos, Doraiswany Ramachandran, and Michael Ludkovski, enriched my view on the area of considered risk problems over a period of more than 10 years. The part on risk measures and newer developments on risk bounds and optimal allocation and portfolio problems reflects also a lot of work and cooperation with several of my former students. I would like to mention Ludger Uckelmann, Thomas Goll, Maike Kaina, Irina Weber, Christian Burgert, Jan Bergenthum, Swen Kiesel, Victor Wolf, and Georg Mainik. I must also mention recent joint work with Giovanni Puccetti and Paul Embrechts on extended risk bounds.

An essential impetus to start this work came from Damir Filipovic (who suggested I undertake this project). However it took two years to plan this book and for it to come to life based on a sabbatical. Particular thanks are due to Monika Hattenbach for the excellent typing and organizational work. She was supported in some parts by Thomas Lais. Many thanks also to Thomas Mikosch for his comments on parts of the manuscript, to several reviewers for their mostly friendly and encouraging comments and to Catriona Byrne for her kind and competent guidance through the problems of choosing the right book series and her careful organization of the review and production process.

Freiburg, Germany   Ludger Rüschendorf
December 2012

Contents

Part II   Risk Measures and Worst Case Portfolios

Part III   Optimal Risk Allocation

Part IV Optimal Portfolios and Extreme Risks

Part I
Stochastic Dependence and Extremal Risk

In financial and insurance risk management as in many further areas of probabilistic modelling as for example in network analysis there are typically several sources of risks. In insurance there are the different contracts held by an insurance company, in finance the portfolio held by a bank is composed of many risky assets. The various risk components are typically not independent of each other. In consequence it became necessary to develop the proper tools to describe the relevant statistical models for dependent variables and to analyse their properties.

There are two basic tasks to face this situation. The first one is the problem of "dependence modelling". Here we are given a vector of risks $X = (X_1, \ldots, X_n)$ where the components X_i have known distributions P_i. X_i themselves might be one-dimensional or higher dimensional risks. The joint distribution P^X of X then is an element of the "Fréchet class" $\mathcal{M}(P_1, \ldots, P_n)$ of all probability measures P on the product space which have marginals P_i, i.e. $P^{\pi_i} = P_i$, $1 \leq i \leq n$, where π_i are the projections on the i-th component. To describe models for the dependence structure of X is equivalent to describe (parametric) submodels of the Fréchet class $\mathcal{M}(P_1, \ldots, P_n)$. In the case of one-dimensional marginals with distribution functions F_i also the notion $\mathcal{F}(F_1, \ldots, F_n)$ is used for the Fréchet class of corresponding distribution functions. The basic notion of copula introduced by Sklar aims to separate the description of the dependence part of a distribution and the marginal part in the case of one-dimensional marginals. In consequence dependence models are given by specifying the marginals and specifying the corresponding copula models.

The second main subject of describing "dependence" are notions of dependence orderings, corresponding dependence measures and the description of bounds on the possible influence of dependence on certain risk functionals like the risk of the joint portfolio in finance or in insurance. Basic results in this direction are the classical Hoeffding–Fréchet bounds which specify upper and lower bounds for the joint distribution function $F = F_X$ of the risk vector X. In this connection the notion of "comonotonicity" is used to describe the worst case dependence structure for one-dimensional marginal risks.

In Chapter 1 we give an introduction to the basic notion of copulas and some related useful tools for the construction of probability models. We introduce the distributional transform, the quantile transform, and their multivariate variants, the multivariate distributional transform and the multivariate quantile transform, which give a basic construction and simulation method and which extend the classical Rosenblatt transform. We discuss applications to copulas, to the conditional tail expectation, and to the empirical copula process, which is a basic statistical tool for dependence analysis.

In Chapter 2 we describe various problems of generalized Fréchet bounds. The basic tool to deal with these problems is a duality theorem describing the range of influence of dependence on an integral functional by a dual representation. In case $n = 2$ this dual representation is closely related to a corresponding result in mass transportation theory going back in its earliest version to Kantorovich. Based on this duality result sharp bounds on the influence of dependence on various risk functionals can be given. Also the description of optimal couplings is closely related to this duality result. We describe in detail in Chapters 3 and 4 resulting bounds for the value at risk (equivalently for the distribution function) and for the excess of loss of the joint portfolio. The result for the excess of loss explains the universal worst case character of the comonotonic dependence structure. Under restrictions on the class of possible dependence structures one can give strongly improved bounds for the risk functionals. We discuss the restriction to positive dependent risk vectors and the restriction on the dependence structure induced by higher order marginals in Chapter 5. Finally in Chapter 6 the risk bounds can in some cases be made more informative by the method of dependence orderings which allow to describe some structure of the influence of dependence within the Fréchet classes and to compare different models concerning the risk induced by their internal dependence properties.

Chapter 1
Copulas, Sklar's Theorem, and Distributional Transform

In this chapter we introduce some useful tools in order to construct and analyse multivariate distributions. The distributional transform and its inverse the quantile transform allow to deal with general one-dimensional distributions in a similar way as with continuous distributions. A nice and simple application is a short proof of the general Sklar Theorem. We also introduce multivariate extensions, the multivariate distributional transform, and its inverse, the multivariate quantile transform. These extensions are a useful tool for the construction of a random vector with given distribution function respectively allow to build functional models of classes of processes. They are also a basic tool for the simulation of multivariate distributions. We also describe some applications to stochastic ordering, to goodness of fit tests and to a general version of the empirical copula process. We introduce to some common classes of copula models and explain the pair copula construction method as well as a construction method based on projections. In the final part extensions to generalized Fréchet classes with given overlapping multivariate marginals are discussed. The construction of dependence models by projections is extended to the generalized Fréchet class where some higher dimensional marginals are specified.

1.1 Sklar's Theorem and the Distributional Transform

The notion of copula was introduced in Sklar (1959) to decompose an n-dimensional distribution function F into two parts, the marginal distribution functions F_i and the copula C, describing the dependence part of the distribution.

Definition 1.1 (Copula). Let $X = (X_1, \ldots, X_n)$ be a random vector with distribution function F and with marginal distribution functions F_i, $X_i \sim F_i$, $1 \leq i \leq n$. A distribution function C with uniform marginals on $[0, 1]$ is called a "copula" of X if

$$F = C(F_1, \ldots, F_n). \tag{1.1}$$

L. Rüschendorf, *Mathematical Risk Analysis*, Springer Series in Operations Research and Financial Engineering, DOI 10.1007/978-3-642-33590-7_1,
© Springer-Verlag Berlin Heidelberg 2013

By definition the set of copulas is identical to the Fréchet class $\mathcal{F}(\mathcal{U}, \ldots, \mathcal{U})$ of all distribution functions C with uniform marginals distribution functions

$$\mathcal{U}(t) = t, \quad t \in [0, 1]. \tag{1.2}$$

For the case that the marginal distribution functions of F are continuous it is easy to describe a corresponding copula. Define C to be the distribution function of $(F_1(X_1), \ldots, F_n(X_n))$. Since $F_i(X_i) \sim U(0, 1)$ have uniform distribution C is a copula and furthermore we obtain the representation

$$\begin{aligned} C(u_1, \ldots, u_n) &= P(F_1(X_1) \le u_1, \ldots, F_n(X_n) \le u_n) \\ &= P(X_1 \le F_1^{-1}(u_1), \ldots, X_n \le F_n^{-1}(u_n)) \\ &= F_X(F_1^{-1}(u_1), \ldots, F_n^{-1}(u_n)). \end{aligned} \tag{1.3}$$

Here F_i^{-1} denotes the generalized inverse of F_i, the "quantile transform", defined by

$$F_i^{-1}(t) = \inf\{x \in \mathbb{R}^1; \ F_i(x) \ge t\}.$$

C is a copula of F since by definition of C

$$\begin{aligned} F(x_1, \ldots, x_n) &= P(X_1 \le x_1, \ldots, X_n \le x_n) \\ &= P(F_1(X_1) \le F_1(x_1), \ldots, F_n(X_n) \le F_n(x_n)) \\ &= C(F_1(x_1), \ldots, F_n(x_n)). \end{aligned} \tag{1.4}$$

The argument for the construction of the copula is based on the property that for continuous distribution functions F_i, $F_i(X_i)$ is uniformly distributed on $(0, 1)$: $F_i(X_i) \sim U(0, 1)$. There is a simple extension of this transformation which we call "distributional transform."

Definition 1.2 (Distributional transform). Let Y be a real random variable with distribution function F and let V be a random variable independent of Y, such that $V \sim U(0, 1)$, i.e. V is uniformly distributed on $(0, 1)$. The modified distribution function $F(x, \lambda)$ is defined by

$$F(x, \lambda) := P(X < x) + \lambda P(Y = x). \tag{1.5}$$

We call

$$U := F(Y, V) \tag{1.6}$$

the (generalized) "distributional transform" of Y.

For continuous distribution functions F, $F(x, \lambda) = F(x)$ for all λ, and $U = F(Y) \overset{d}{=} U(0, 1)$, $\overset{d}{=}$ denoting equality in distribution. This property is easily extended to the generalized distributional transform (see e.g. Rü[1] (2005)).

Proposition 1.3 (Distributional transform). *Let $U = F(Y, V)$ be the distributional transform of Y as defined in* (1.6). *Then*

$$U \overset{d}{=} U(0, 1) \text{ and } Y = F^{-1}(U) \text{ a.s.} \tag{1.7}$$

An equivalent way to introduce the distributional transform is given by

$$U = F(Y-) + V(F(Y) - F(Y-)), \tag{1.8}$$

where $F(y-)$ denotes the left-hand limit. Thus at any jump point of the distribution function F one uses V to randomize the jump height.

The distributional transform is a useful tool which allows in many respects to deal with general (discontinuous) distributions similar as with continuous distributions. In particular it implies a simple proof of Sklar's Theorem in the general case (see Moore and Spruill (1975) and Rü (1981b, 2005)).

Theorem 1.4 (Sklar's Theorem). *Let $F \in \mathcal{F}(F_1, \ldots, F_n)$ be an n-dimensional distribution function with marginals $F_1, \ldots, F_n$. Then there exists a copula $C \in \mathcal{F}(\mathcal{U}, \ldots, \mathcal{U})$ with uniform marginals such that*

$$F(x_1, \ldots, x_n) = C(F_1(x_1), \ldots, F_n(x_n)). \tag{1.9}$$

Proof. Let $X = (X_1, \ldots, X_n)$ be a random vector on a probability space $(\Omega, \mathfrak{A}, P)$ with distribution function F and let $V \sim U(0, 1)$ be independent of X. Considering the distributional transforms $U_i := F_i(X_i, V)$, $1 \le i \le n$, we have by Proposition 1.3 $U_i \overset{d}{=} U(0, 1)$ and $X_i = F_i^{-1}(U_i)$ a.s., $1 \le i \le n$. Thus defining C to be the distribution function of $U = (U_1, \ldots, U_n)$ we obtain

$$F(x) = P(X \le x) = P(F_i^{-1}(U_i) \le x_i, 1 \le i \le n)$$
$$= P(U_i \le F_i(x_i), 1 \le i \le n) = C(F_1(x_1), \ldots, F_n(x_n)),$$

i.e. C is a copula of F. $\qquad\qquad\qquad\qquad\qquad\qquad\qquad\qquad\qquad\qquad\qquad\square$

Remark 1.5. (a) Copula and dependence. From the construction of the distributional transform it is clear that the distributional transform is not unique in the case when the distribution has discrete parts. Different choices of the

---

[1] Within the whole book the author's name is abbreviated to Rü.

Figure 1.1 Copula of
uniform distribution on line
segments (published in: *Journal of Stat.
Planning Inf.*, 139: 3921–3927, 2009)

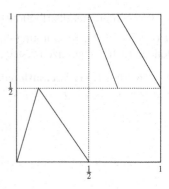

randomizations V at the jumps or in the components, i.e. choosing $U_i =$ $F_i(X_i, V_i)$, may introduce *artificial* local dependence between the components of a random vector on the level of the copula. From the copula alone one does not see whether some local positive or negative dependence is a real one or just comes from the choice of the copula. For dimension $n = 2$ the copula in Figure 1.1 could mean a real switch of local positive and negative dependence for the original distribution, but it might also be an artefact resulting from the randomization in case the marginals are e.g. both two-point distributions while the joint distribution in this case could be even comonotone. Thus the copula information alone is not sufficient to decide all dependence properties.

(b) **Conditional value at risk.** A more recent application of the distributional transform is to risk measures. It is well known that the conditional tail expectation

$$\text{TCE}_\alpha(X) := -E(X \mid X \le q_\alpha), \tag{1.10}$$

where q_α is the lower α-quantile of the risk X, does not define a coherent risk measure except when restricted to continuous distributions. This defect can be overcome by using the distributional transform $U = F(X, V)$ and defining the modified version, which we call conditional value at risk (CVR_α)

$$\text{CVR}_\alpha(X) = -E(X \mid U \le \alpha). \tag{1.11}$$

By some simple calculations (see Burgert and Rü (2006b)) one sees that

$$\text{CVR}_\alpha(X) = -\frac{1}{\alpha}\left[EX1(X < q_\alpha) + q_\alpha(\alpha - P(X < q_\alpha)) \right] = ES_\alpha(X). \tag{1.12}$$

Thus the more natural definition of CVR_α coincides with the well-established "expected shortfall risk measure" $ES_\alpha(X)$ which is a coherent risk measure.

As a consequence the expected shortfall is represented as conditional expectation and our definition in (1.11) of the conditional value at risk seems to be appropriate for this purpose.

(c) **Stochastic ordering.** The construction of copulas based on the distributional transform as in the proof of Sklar's Theorem above has been used in early papers on stochastic ordering. The following typical example of this type of result is from Rü (1981b, Proposition 7).

Proposition 1.6. *Let F_i, G_i be one-dimensional distribution functions with $F_i \leq G_i$ (or equivalently $G_i \leq_{st} F_i$), $1 \leq i \leq n$. Then to any $F \in \mathcal{F}(F_1, \ldots, F_n)$ there exists an element $G \in \mathcal{F}(G_1, \ldots, G_n)$ with $G \leq_{st} F$. Here $\leq_{st}$ denotes the multivariate stochastic ordering w.r.t. increasing functions.*

Proof. Let $X = (X_1, \ldots, X_n) \sim F$ and let $U_i = F_i(X_i, V)$ denote the distributional transforms of the components X_i. Then $U = (U_1, \ldots, U_n)$ is a copula vector of F. Define $Y = (Y_1, \ldots, Y_n)$ as vector of the quantile transforms of the components of U, $Y_i = G_i^{-1}(U_i)$. Then $Y \sim G \in \mathcal{F}(G_1, \ldots, G_n)$ and from the assumption $F_i \leq G_i$ we obtain that $Y \leq X$ pointwise. In consequence $G \leq_{st} F$. □

In particular the above argument shows that $G \leq_{st} F$ if F and G have the same copula. ◇

1.2 Copula Models and Copula Constructions

In order to construct a distributional model for a random vector X it is by Sklar's representation theorem sufficient to specify a copula model fitting to the normalized data. There are a large number of suggestions and principles for the construction of copula models. Classical books on these models for $n = 2$ are Mardia (1962) and Hutchinson and Lai (1990). More recent books for $n \geq 2$ are Joe (1997), Nelsen (2006), Mari and Kotz (2001), Denuit et al. (2005), and McNeil et al. (2005b). A lot of material on copulas can also be found in the conference volumes of the "Probability with given marginals" conferences – i.e. the series of conferences in Rome (1990), Seattle (1993), Prague (1996), Barcelona (2000), Québec (2004), Tartu (2007) and Sao Paulo (2010) as well as in the recent proceedings edited by Jaworski, Durante, Härdle, and Rychlik (2010). A huge basket of models has been developed and some of its properties have been investigated, concerning in particular the following questions:

- What is the "range of dependence" covered by these models and measured with some dependence index. Is the range of dependence wide and flexible enough?
- What models exhibit tail dependence and thus are able to model situations with strong positive dependence in tails?
- Is some parameter available describing the degree of dependence?

- Is there a natural probabilistic representation respectively interpretation of these models describing situations when to apply them?
- Is there a closed form of the copula or a simple simulation algorithm so that goodness of fit test can be applied to evaluate whether they fit the data?

Several questions of this type have been discussed in the nice survey papers of Schweizer (1991) and Durante and Sempi (2010).

1.2.1 Some Classes of Copulas

Copulas, their properties and applications are discussed in detail in the above-mentioned literature. Copulas are not for any class of distributions the suitable standardizations as for example for elliptical distributions. In extreme value theory a standardization by the exponential distribution may be better adapted. For some applications it may be more natural and simpler to model the distributions or their densities directly without referring to copulas. Conceptually and for some investigation of dependence properties like tail-dependence the notion of copula is however a useful tool.

Some classical classes of copulas are the following.

- The "Farlie–Gumbel–Morgenstern (FGM) copula":

$$C_\alpha(u, v) = uv(1 + \alpha(1 - u)(1 - v)), \quad |\alpha| \le 1 \tag{1.13}$$

as well as its generalized versions, "EFGM copulas" are extensions of FGM to the n-dimensional case and given by

$$C(u) = \prod_{i=1}^n u_i \left(1 + \sum_{T \subset \{1,\dots,n\}} \alpha_T \prod_{j \in T}(1 - u_j)\right) \tag{1.14}$$

for suitable $\alpha_T \in \mathbb{R}^1$.
- The "Archimedean copulas"

$$C_\Phi(u_1, \dots, u_n) = \Phi^{-1}(\Phi(u_1) + \cdots + \Phi(u_n)) \tag{1.15}$$

for some generator $\Phi : (0, 1] \to \mathbb{R}_+$ with $\Phi(1) = 0$ such that Φ is completely monotone. In fact C_Φ is an n-dimensional copula if and only if Φ is "n-monotone", i.e. Φ is $(n - 2)$ times differentiable and

$$(-1)^k \Phi^{(k)}(t) \ge 0, \quad 0 \le k \le n - 2, \forall t > 0 \tag{1.16}$$

and $(-1)^{n-2}\Phi^{(n-2)}$ is decreasing and convex (see McNeil and Nešlehová (2010)).

In case $\Phi(t) = \frac{1}{\alpha}(t^{-\alpha} - 1)$ one gets the "Clayton copula"

$$C_\alpha(u) = \left(\sum u_i^{-\alpha} - n + 1 \right)^{-1/\alpha}, \quad \alpha > 0. \tag{1.17}$$

In case $\Phi_\delta(t) = -\frac{1}{\delta}\log(1 - (1 - e^{-\delta})e^{-t})$ one gets the "Frank copula"

$$C_\delta(u) = -\frac{1}{\delta}\log\left(1 + \frac{\prod_{i=1}^n (e^{-\delta u_i} - 1)}{(e^{-\delta} - 1)^{n-1}} \right). \tag{1.18}$$

Φ is by Bernstein's theorem completely monotone if and only if it is n-monotone for all $n \in \mathbb{N}$.

Archimedean copulas are connected with mixing models ("frailty models"). Let

$$F_X(x) = \int \prod_{i=1}^n (G_i(x_i))^\delta dF_\Psi(\delta),$$

where Ψ is a positive real mixing variable. Then with L_Ψ, the Laplace transform of Ψ, we obtain that the i-th marginal is given by

$$F_i(x_i) = \int \exp(\delta \ln G_i(x_i)) dF_\Psi(\delta) = L_\Psi(-\ln G_i(x_i)).$$

As a consequence one gets

$$F_X(x) = L_\Psi\left(\sum_{i=1}^n L_\Psi^{-1}(F_i(x_i)) \right), \tag{1.19}$$

i.e. a representation as in (1.15) holds with the generator $\Phi = L_\Psi$ which by Bernstein's theorem is completely monotone.

Archimedean copulas arise from multivariate distributions that have a stochastic representation of the form

$$X \overset{d}{=} R \cdot U \tag{1.20}$$

where U is uniformly distributed on the unit simplex (1-sphere) in $\mathbb{R}_+^n$, $\{x \in \mathbb{R}_+^n; \sum_{i=1}^n x_i = 1\}$ respectively the simplex in $\mathbb{R}^n$, and R is a scaling random variable with values in $\mathbb{R}_+$ independent of U. Formulae for the relation between the distribution function F_R and the generator Φ are available (see McNeil and Nešlehová (2010)).

- "Elliptical distributions" arise from stochastic representations of the form

$$X = \mu + RAU, \tag{1.21}$$

where U is uniformly distributed on the unit 2-sphere $S_{n-1} = \{x \in \mathbb{R}^n; \sum_{i=1}^n u_i^2 = 1\}$ in $\mathbb{R}^n$. R is a real positive random variable independent from U and A is an $n \times n$-matrix. X has a density of the form

$$f(x) = |\Sigma|^{-1}\Phi((x - \mu)^{\mathsf{T}}\Sigma^{-1}(x - \mu)) \text{ with } \Sigma = AA^{\mathsf{T}}.$$

The distribution is denoted as $\mathcal{E}(\mu, \Sigma, F_R)$. Multivariate normal distributions are elliptical (or elliptically contoured). The copulas of elliptical distributions are in general not available in closed form.

- "Extreme value copulas" are defined as possible limits of max sequences. Let $X_i = (X_{i,1}, \ldots, X_{i,d})$, $1 \leq i \leq n$ (note that here d = dimension but n = sample index) be a sequence of iid random vectors with copula C_F. Let $M_n = \max\{X_1, \ldots, X_n\} = (M_{n,1}, \ldots, M_{n,d})$ be the componentwise max of the X_i, then the copula C_n of M_n is given by

$$C_n(u) = C_F^n(u_1^{1/n}, \ldots, u_d^{1/n}).$$

C is called an "extreme value copula" if for any $n \in \mathbb{N}$ there exist some copula C_F such that

$$C = C_F^n(u_1^{1/n}, \ldots, u_d^{1/n}). \tag{1.22}$$

A well-known fact is:
C is an extreme value copula if and only if C is "max stable", i.e. $C(u) = C^m(u_1^{1/m}, \ldots, u_d^{1/m})$ for all $m \in \mathbb{N}$.

De Haan and Resnick (1977) and Pickands (1981) gave the following representation of extreme value copulas:
C is an extreme value copula if and only if it has a representation of the form

$$C(u) = \exp(-l(-\log u_1, \ldots, -\log u_d)). \tag{1.23}$$

Here the "tail dependence function" $l : (0, \infty)^d \to [0, \infty]$ is given by

$$l(x) = \int_{S_{d-1}} \max_{j \leq d}(w_j x_j) dH(w) \tag{1.24}$$

where H is a measure on the simple S_{d-1} satisfying $\int_{S_{d-1}} w_j dH(w) = 1$, $1 \leq j \leq d$. H is called the "spectral measure" of C.

An example of copulas which are of interest in extreme value theory are "Gumbel copulas" given by

$$C_\delta(u) = \exp\left(-\left(\sum_{i=1}^n (-\log u_i)^\delta\right)^{1/\delta}\right) \tag{1.25}$$

for $\delta \in [1, \infty)$. Gumbel copulas are extreme value copulas. The parameter δ is a dependence parameter. If $1 \leq \delta_1 < \delta_2$, then G_{δ_2} is more strongly dependent (in the sense of supermodular ordering) than G_{δ_1} (see Wei and Hu (2002)).

1.2.2 Copulas and L^2-Projections

A general principle for the construction of dependence models (copulas) was introduced in Rü (1985). It is based on the following representation result. Let $C_n(\lambda^n)$ denote the set of all copulas on $[0, 1]^n$ which are Lebesgue-continuous and thus have Lebesgue-densities f. Let $C_n^s(\lambda^n)$ be the set of all signed measures on $[0, 1]^n$ with uniform marginals and which have Lebesgue-densities f.

For any integrable $f \in L^1([0, 1]^n, \lambda^n)$ and $T \subset \{1, \ldots, n\}$ we define

$$f_T = \int f \prod_{i \in T} dy_i, \tag{1.26}$$

i.e. we integrate over the components in T. We consider f_T as a real function on $[0, 1]^n$ being constant in the T components. The following linear operator $S : L^1(\lambda^n) \to L^1(\lambda^n)$ is of interest,

$$Sf = f - \sum_{\substack{T \subset \{1,\ldots,n\} \\ |T| = n-1}} f_T + (n-1)f_{\{1,\ldots,n\}} \tag{1.27}$$

since it leads to the following representation result.

Theorem 1.7 (Copulas and L^2-projections). *All probability distributions on $[0, 1]^n$ with Lebesgue-density and uniform marginals are of the form $(1 + Sf)\lambda^n$ where $f \in L^1(\lambda^n)$, more precisely*

$$C_n^s(\lambda^n) = \{Q = (1 + Sf)\lambda^n; \ f \in L^1(\lambda^n)\} \tag{1.28}$$

and

$$C_n(\lambda^n) = \{P = (1 + Sf)\lambda^n; \ f \in L^1(\lambda), Sf \geq -1\}. \tag{1.29}$$

Proof. Any $P \in C_n^s(\lambda^n)$ has by the Radon–Nikodým theorem a density of the form $P = (1 + f)\lambda^n$ where $f_T = 0$ for $|T| \geq n - 1$. This implies that $Sf = f$, i.e. $P = (1 + Sf)\lambda^n$. Conversely, let $f \in L^1(\lambda^n)$ and $P = (1 + Sf)\lambda^n$. Then for $T_0 \subset \{1, \ldots, n\}, |T_0| = n - 1$ and any further $T \subset \{1, \ldots, n\}$ with $|T| = n - 1$, $T \neq T_0$ we have $(f_T)_{T_0} = f_{\{1,\ldots,n\}}$ and, therefore,

$$(Sf)_{T_0} = \left(f - f_{T_0} - \sum_{\substack{|T|=n-1 \\ T \neq T_0}} f_T + (n-1)f_{\{1,\ldots,n\}} \right)_{T_0} \tag{1.30}$$

$$= f_{T_0} - f_{T_0} - \sum_{\substack{|T|=n-1 \\ T \neq T_0}} (f_T)_{T_0} + (n-1)f_{\{1,\ldots,n\}} = 0. \tag{1.31}$$

Therefore $Q = (1 + Sf)\lambda^n \in \mathcal{C}_n^s(\lambda^n)$. This implies (1.28). Equation (1.29) is immediate from (1.28). □

The operator S in (1.27) can be considered as an L^2-projection of a (signed) measure $Q = (1 + f)\lambda^n$, $f \in L^1(\lambda^1)$ to a copula $P = (1 + Sf)\lambda^n$ if $Sf \geq -1$. The basic idea of the construction method for dependence (copula) models in Rü (1985) is to describe in a first step the relevant dependence by the density function $1 + f$. In a second step this model $Q = (1 + f)\lambda^n$ is projected to the class of copulas in the form $P = (1 + Sf)\lambda^n$.

The resulting proposed model construction method is the following:

Construction by "projection method": Let f_ϑ, $\vartheta \in \Theta$ be a parametric family of functions in $L^1(\lambda^n)$ (describing the intended dependence structure) such that $Sf_\vartheta \geq -1$, $\vartheta \in \Theta$. Then the proposed model obtained by the projection method is given by

$$\mathcal{P} = \{P_\vartheta = (1 + Sf_\vartheta)\lambda^n; \ \vartheta \in \Theta\}. \tag{1.32}$$

The inherent idea in this construction is the hope that the projection on the correct marginal structure does not change the intended dependence structure too much. Obviously we can use the same procedure as above also for the construction of elements in the Fréchet class $M(P_1, \ldots, P_n)$ which are continuous w.r.t. a product measure $\mu_1 \otimes \cdots \otimes \mu_n$.

This projection idea is also underlying Sklar's representation theorem. Let $G \in \mathcal{F}_n(G_1, \ldots, G_n)$ be a multivariate distribution function with marginals $G_1, \ldots, G_n$ and with copula C. Assume that G is a good description of the dependence structure for a model. At the same time assume that the correct marginals should be $F_1, \ldots, F_n$ instead of $G_1, \ldots, G_n$. Then by Sklar's Theorem the following "nonmetric projection" would seem natural:

$$G \to C \to F = C(F_1, \ldots, F_n). \tag{1.33}$$

By this transition the dependence structure is retained in the copula C. The difficulty of calculation may however prevent use of this transition in examples. Typically it will be easier to calculate (only) densities.

The following examples illustrate the projection method, see Rü (1985) and Mari and Kotz (2001, pp. 73–78).

Example 1.8. Let $f \in L^1(\lambda^n)$, let $\alpha_0 = \inf\{(Sf)(x); \ x \in [0, 1]^n\} > -\infty$ and consider $f_\vartheta(x) = \vartheta f(x)$, $\vartheta \in [0, \frac{1}{|\alpha_0|}]$.

(1) **Generalized FGM-family:** If $f(x) = \prod_{i=1}^n v_i(x_i)$ where $\int v_i(x_i)dx_i = 0$, $1 \leq i \leq n$, then $Sf = f$ and $\mathcal{P}$ gives a "generalized FGM-family" (see Johnson and Kotz (1975) and Kimeldorf and Sampson (1975a,b), and (1.13)).

(2) **Polynomial copula models:** If $f(x) = \prod_{i=1}^n x_i^{m_i}$ is a polynomial, then

$$Sf(x) = \prod_{i=1}^n x_i^{m_i} - \sum_{i=1}^n \left(\prod_{j \neq i} \frac{1}{m_j + 1}\right) x_i^{m_i} + (n - 1)\prod_{i=1}^n \frac{1}{m_i + 1} \tag{1.34}$$

gives a "polynomial copula model" which is the same as the FGM-model generated by $v_i(x_i) = x_i^{m_i} - \frac{1}{m_i+1}$. Linear combinations then yield the general form of polynomial copula models. The restrictions to the parameters are given in Mari and Kotz (2001, p. 74).

This polynomial model has also been discussed in Wei, Fang, and Fang (1998). Several families of polynomial copulas have been suggested as local approximations of other copula families. So the Ali–Mikhail–Haq family is a polynomial approximate in the neighbourhood of the independence copula, the FGM is an approximation of the Plackett family (see Mari and Kotz (2001, p. 75)).

(3) **A copula concentrated strongly near the diagonal:** Let $n = 2$ and consider $f(x, y) = \frac{1}{\sqrt{|x-y|}}$. The function f has singularities at the diagonal, i.e. we expect a large density in the neighbourhood of the diagonal. Then by simple calculation one gets

$$Sf(x, y) = \frac{1}{\sqrt{|x - y|}} - 2\left(\sqrt{x} + \sqrt{1 - x} + \sqrt{y} + \sqrt{1 - y}\right) + \frac{8}{3}. \quad (1.35)$$

$\mathcal{P}$ describes a family of copulas which is concentrated near the diagonal. ϑ is a positive dependence parameter; higher values of ϑ yield higher order positive dependence.

(4) **Diagonal weakly concentrated copula:** For $\Theta = [0, 1]$, $n = 2$ define

$$f_\vartheta(x, y) = -1_{\{|x-y|>\vartheta\}},$$

so $(1 + f_\vartheta)\lambda^2$ is concentrated in a neighbourhood of the diagonal but it is uniformly distributed on $\{(x, y) \in [0, 1]^2 : |x - y| \le \vartheta\}$ in contrast to the singular concentration in Example 1.8, (1.8).

By the projection to the correct marginals we obtain

$$Sf_\vartheta(x, y) = f_\vartheta(x, y) + g_\vartheta(x) + g_\vartheta(y) - (1 - \vartheta)^2, \quad (1.36)$$

where

$$g_\vartheta(x) = (2 - 2\vartheta)1_{\{\vartheta \le x \le 1-\vartheta\}} + (1 - x - y)1_{(0 \le x < \vartheta)} - (x - y)1_{(x>1-\vartheta)}$$

$$\text{for } 0 \le \vartheta \le \frac{1}{2}$$

and $g_\vartheta(x) = (x - \vartheta)1_{(x>\vartheta)} + (\vartheta - x)1_{(x<1-\vartheta)}$, $\frac{1}{2} < \vartheta \le 1$.

The projected probability model $P_\vartheta = (1 + Sf_\vartheta)\lambda^2$ approaches for $\vartheta \to 1$ the product measure while for ϑ small (neglecting *small and large x-values*)

$$1 + (Sf_\vartheta)(x, y) \approx \begin{cases} 2 - 2\vartheta & \text{for } |x - y| < \vartheta, \\ 1 - 2\vartheta & \text{for } |x - y| > \vartheta. \end{cases} \quad (1.37)$$

In order to introduce a stronger dependence effect one can start with $f_{\vartheta,a}(x,y)$ $= af_\vartheta(x,y)$ which for ϑ small and a large centres the distribution near the diagonal.   ◇

1.3   Multivariate Distributional and Quantile Transform

The distributional transform $F(X,V)$ as well as the inverse quantile transform $F^{-1}(V)$ have been generalized to the multivariate case.

Definition 1.9 (Multivariate quantile transform).   Let F be a d-dimensional distribution function and let $V_1, \ldots, V_n$ be iid $U(0,1)$-distributed random variables. Then the "multivariate quantile transform" $Y := \tau_F^{-1}(V)$ is defined recursively as

$$
\begin{aligned}
Y_1 &:= F_1^{-1}(V_1) \\
Y_k &:= F_{k|1,\ldots,k-1}^{-1}(V_k \mid Y_1, \ldots, Y_{k-1}), \quad 2 \le k \le n,
\end{aligned}
\tag{1.38}
$$

where $F_{k|1,\ldots,k-1}$ denote the conditional distribution functions of the k-th component π_k given the first $k-1$ components $\pi_1, \ldots, \pi_{k-1}$.

The multivariate quantile transform is a basic method to construct a random vector Y with specified distribution function F from a given vector $V = (V_1, \ldots, V_n)$ of iid uniformly on $[0,1]$ distributed random variables. By construction the multivariate quantile transform uses only one-dimensional conditional distribution functions.

Theorem 1.10 (Multivariate quantile transform).   *The multivariate quantile transform $Y = \tau_F^{-1}(V)$ is a random vector with distribution function F.*

Proof. In case $n = 2$ and for any $x_1, x_2 \in \mathbb{R}$ we have using the independence of (V_i)

$$
\begin{aligned}
P(Y_1 \le x_1, Y_2 \le x_2) &= \int_{-\infty}^{x_1} P(Y_2 \le x_2 \mid Y_1 = y_1) dF_1(y_1) \\
&= \int_{-\infty}^{x_1} P(F_{2|1}^{-1}(V_2 \mid y_1) \le x_2 \mid Y_1 = y_1) dF_1(y_1) \\
&= \int_{-\infty}^{x_1} P(F_{2|1}^{-1}(V_2 \mid y_1) \le x_2) dF_1(y_1) \\
&= \int_{-\infty}^{x_1} F_{2|1}(x_2 \mid y_1) dF_1(y_1) = F_{12}(x_1, x_2).
\end{aligned}
$$

By induction this argument extends to yield $P(Y_1 \le x_1, \ldots, Y_n \le x_n) = F(x_1, \ldots, x_n)$.   □

The multivariate quantile transform was introduced in O'Brien (1975), Arjas and Lehtonen (1978), and Rü (1981b). The construction Y is an inductive construction of a random vector with specified distribution function F and is called "regression representation" since Y is of the form

$$
\begin{aligned}
Y_1 &= h_1(V_1), \\
Y_2 &= h_2(Y_1, V_2), \\
Y_3 &= h_2(Y_1, Y_2, V_3), \\
&\vdots \\
Y_n &= h_n(Y_1, \ldots, Y_{n-1}, V_n)
\end{aligned}
\tag{1.39}
$$

representing Y_k as function of the past $Y_1, \ldots, Y_{k-1}$ and some innovation V_k. As a consequence this implies the "standard representation"

$$
Y_k = f_k(V_1, \ldots, V_k), \quad 1 \le k \le n.
\tag{1.40}
$$

The multivariate quantile transform and particularly Theorem 1.10 is a basic method for the "simulation of multivariate distributions". In order to construct a random vector with distribution function F one needs to simulate iid $U(0,1)$-random variables and to determine the inverses of the one-dimensional conditional distribution functions $F_{i|x_1,\ldots,x_{i-1}}(x_i)$, which can be done in many examples either analytically or numerically. There is a large literature on (Monte Carlo) simulation of multivariate distributions, which uses essentially the multivariate quantile transform. This transform was introduced first in the above-mentioned papers dealing with representation of stochastic sequences and with stochastic orderings.

Definition 1.11 (Multivariate distributional transform). Let X be an n-dimensional random vector and let $V_1, \ldots, V_n$ be iid $U(0,1)$-distributed random variables, $V = (V_1, \ldots, V_n)$. For $\lambda = (\lambda_1, \ldots, \lambda_n) \in [0,1]^n$ define

$$
\tau_F(x, \lambda) := \big(F_1(x_1, \lambda_1), F_2(x_2, \lambda_2 \mid x_1), \ldots, F_d(x_d, \lambda_d \mid x_1, \ldots, x_{d-1})\big),
\tag{1.41}
$$

where

$$
F_1(x_1, \lambda_1) = P(X_1 < x_1) + \lambda_1 P(X_1 = x_1),
$$

$$
\begin{aligned}
F_k(x_k, \lambda_k \mid x_1, \ldots, x_{k-1}) &= P(X_k < x_k \mid X_j = x_j, j \le k-1) \\
&\quad + \lambda_k P(X_k = x_k \mid X_j = x_j, j \le k-1)
\end{aligned}
$$

are the distributional transforms of the one-dimensional conditional distributions. Finally the "multivariate distributional transform" of X is defined as

$$
U := \tau_F(X, V).
\tag{1.42}
$$

Rosenblatt (1952) introduced this transformation in the special case of absolutely continuous conditional distributions allowing the application of the transformation formula. Therefore this transformation is also called the "Rosenblatt transformation" in this case. For general distributions the multivariate distributional transform (generalized Rosenblatt transform) was introduced in Rü (1981b). The basic property is stated in the following theorem in Rü (1981b).

Theorem 1.12 (Multivariate distributional transform). *Let X be a random vector and let $U = \tau_F(X, V)$ denote its multivariate distributional transform. Then*

(a)
$$U \sim U((0, 1)^d), \tag{1.43}$$

    i.e. the components U_i of U are iid $U(0, 1)$ distributed.

(b) The multivariate quantile transform τ_F^{-1} is inverse to the multivariate distributional transform, i.e.

$$X = \tau_F^{-1}(U) = \tau_F^{-1}(\tau_F(X, V)) \ a.s. \tag{1.44}$$

Proof. $U_1 = F_1(X_1, V_1) \sim U(0, 1)$ by Proposition 1.3. Consider next $U_2 = F_2(X_2, V_2 \mid X_1)$. Conditionally given $X_1 = x_1$ we have again from Proposition 1.3 that

$$U_2 \stackrel{d}{=} F_2(X_2, V_2 \mid x_1) \sim U(0, 1).$$

Furthermore, $P^{U_2 \mid X_1 = x_1} = U(0, 1)$ is independent of x_1 and thus U_2, X_1 are independent. Since $U_1 = F_1(X_1, V_1)$ we get from iterated conditional expectation that also U_2, U_1 are independent. The general case then follows by induction. □

Remark 1.13 (Regression and standard representation).

(a) By combining the multivariate quantile and the multivariate distributional transform one gets for any given stochastic sequence (X_k) a pointwise "standard (innovation) representation"

$$X_k = f_k(U_1, \ldots, U_k) \ a.s., \tag{1.45}$$

respectively a pointwise "regression representation"

$$X_k = f_k(X_1, \ldots, X_{k-1}, U_k) \ a.s., \quad 1 \le k \le n \tag{1.46}$$

with some iid sequences (U_i), $U_i \sim U(0, 1)$. This result was proved first for $n = 2$ in Skorohod (1976). Related functional representations of classes of stochastic models are given in Rü and de Valk (1993). For example in the case of Markov chains the regression representation reduces to the following functional representation of (any) Markov chain:

Corollary 1.14 (Regression representation of Markov chains). *Any Markov chain (X_n) has a representation as a nonlinear regression model*

$$X_k = f_k(X_{k-1}, U_k) \ a.s., \tag{1.47}$$

where (U_k) is an iid sequence of $U(0, 1)$-distributed random variables, U_k is independent of $X_1, \ldots, X_{k-1}$.

(b) The copula transformation

$$X = (X_1, \ldots, X_d) \to U = (U_1, \ldots, U_d), U_i = F_i(X_i, V_i)$$

which transforms a vector X to a copula vector U, where U corresponds to the copula of X, forgets about the marginals but retains essential information on the dependence structure of X. On the contrary the multivariate distributional transform forgets also about the dependence structure. This is an interesting property, when one wants to identify a distribution. These two different properties of the copula transformation and the multivariate distributional transform lead to different kinds of applications. Some of them are described in Section 1.5. ◊

1.4 Pair Copula Construction of Copula Models

Besides the multivariate quantile transform in Theorem 1.10, which is based on the one-dimensional conditional distributions $F_{i|x_1,\ldots,x_{i-1}}$, several further methods to represent a distribution function F in terms of conditional distribution functions have been proposed. Particular interest in the recent literature and in various applications arose from the "pair copula construction (PCC)" method which is based on a series of certain (organized) pairs of variables.

In the original example Joe (1997) used the following pairwise construction of an m-dimensional distribution function F. For $m = 3$, $F = F_{123}$ can be represented as

$$F_{123}(x) = \int_{-\infty}^{x_2} F_{13|z_2}(x_1, x_3) dF_2(z_2) \tag{1.48}$$

where $F_{13|z_2}$ is the conditional distribution function of the pair X_1, X_3 given $X_2 = z_2$. By Sklar's Theorem this can also be written in terms of the conditional copula $C_{13|z_2}$ in the form

$$F_{123}(x) = \int_{-\infty}^{x_2} C_{13|z_2}(F_{1|z_2}(x_1), F_{3|z_2}(x_3)) dF_2(z_2). \tag{1.49}$$

Similarly for general m one obtains recursively the representation

$$F_{1\ldots m}(x) = \int_{-\infty}^{x_2} \cdots \int_{-\infty}^{x_{m-1}} C_{1m|z_2,\ldots,z_{m-1}}\big(F_{1|z_2,\ldots,z_{m-1}}(x_1),$$
$$F_{m|z_2,\ldots,z_{m-1}}(x_m)\big) dF_{2\ldots m-1}(z_2, \ldots, z_{m-1}) \tag{1.50}$$

which is given in terms of pairwise conditional copulas.

In Bedford and Cooke (2001, 2002) and Kurowicka and Cooke (2006) some general classes of graphical organization principles representing multivariate distributions were developed. See also the survey of Czado (2010). Two basic examples of these classes of constructions are C-vines and D-vines (C = canonical, D = drawable).

(a) **D-vines:** The construction of D-vines is based on densities and uses the representation

$$f(x_1, \ldots, x_n) = \prod_{i=2}^{n} f_{i|x_1, \ldots, x_{i-1}}(x_i) f_1(x_1). \tag{1.51}$$

By Sklar's Theorem we have

$$f_{12}(x_1, x_2) = c_{12}(F_1(x_1), F_2(x_2)) f_1(x_1) f_2(x_2), \tag{1.52}$$

where c_{12} is a bivariate copula density. This implies for the conditional density

$$f_{1|x_2}(x_1) = c_{12}(F_1(x_1), F_2(x_2)) f_1(x_1). \tag{1.53}$$

Using (1.53) for the conditional density of (X_1, X_i) given $X_2, \ldots, X_{i-1}$ we obtain by recursion

$$f_{i|x_1, \ldots, x_{i-1}}(x_i) = c_{1,i|2, \ldots, i-1} \cdot f_{i|x_2, \ldots, x_{i-1}}(x_i)$$
$$= \left(\prod_{j=1}^{i-2} c_{j,i|j+1, \ldots, i-1} \right) c_{i-1,i} \cdot f_i(x_i) \tag{1.54}$$

using the conditional copula densities

$$c_{j,i|j+1, \ldots, i-1} = c_{j,i|x_{j+1}, \ldots, x_{i-1}}(F_{j|x_{j+1}, \ldots, x_{i-1}}(x_j), F_{i|x_{j+1}, \ldots, x_{i-1}}(x_i)).$$

As a result we obtain the "D-vine" density decomposition

$$f(x_1, \ldots, x_n) = \left(\prod_{i=2}^{n} \prod_{j=1}^{i-2} c_{ji|j+1, \ldots, i-1} \right) \prod_{i=2}^{n} c_{i-1,i} \prod_{l=1}^{n} f_l(x_l)$$
$$= \left(\prod_{i=1}^{n-1} \prod_{j=1}^{n-i} c_{j,i+j|(j+1, \ldots, j+i-1)} \right) \prod_{l=1}^{n} f_l(x_l). \tag{1.55}$$

The conditional copula densities in (1.55) are evaluated at the conditional distribution functions $F_{j|x_{j+1}, \ldots, x_{j+i-1}}, F_{i+j|x_{j+1}, \ldots, x_{j+i-1}}$.

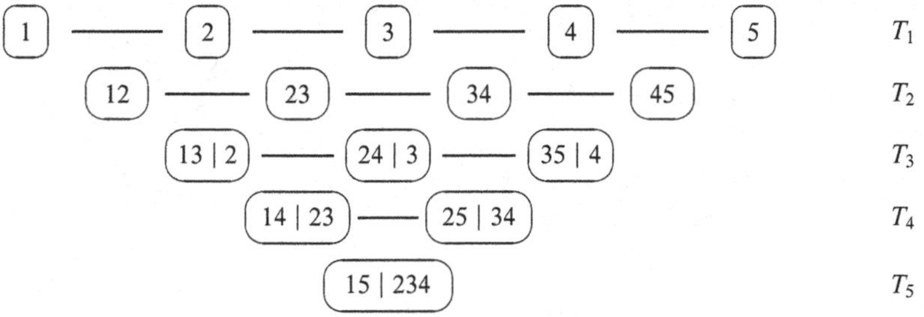

Figure 1.2 D-vine tree for $n = 5$

The D-vine decomposition in (1.55) can be organized iteratively by pairwise copulas using iteratively levels $T_1, \ldots, T_{n-1}$. For $n = 5$ we obtain the representation

$$f(x_1, \ldots, x_5)$$

$$= \prod_{l=1}^{5} f_l(x_l) c_{12} \cdot c_{23} \cdot c_{34} \cdot c_{45} \cdot c_{13|2} \cdot c_{24|3} \cdot c_{35|4} \cdot c_{14|23} \cdot c_{25|34} \cdot c_{15|234}.$$

This is described in the following graphical organization. Each transition step from level T_i to level T_{i+1} involves a (conditional) pair copula (Figure 1.2).

(b) **C-vines:** C-vine decompositions are obtained when applying the representation of the conditional density successively to the conditional distribution of X_{i-1}, X_i given $X_1, \ldots, X_{i-2}$. This gives

$$f_{i|x_1, \ldots, x_{i-1}}(x_i) = c_{i-1,i|x_1, \ldots, x_{i-2}} f_{i|x_1, \ldots, x_{i-1}}(x_i). \tag{1.56}$$

Using (1.56) instead of (1.54) in (1.51) we obtain the C-vine decomposition

$$f(x_1, \ldots, x_n) = f_1(x_1) \prod_{i=2}^{n} \prod_{k=1}^{i-1} c_{i-k,i|1,\ldots,i-k-1} f_k(x_i)$$

$$= \left(\prod_{i=2}^{n} \prod_{k=1}^{k-1} c_{i-k,i|1,\ldots,i-k-1} \right) \prod_{k=1}^{n} f_k(x_k)$$

$$= \left(\prod_{j=1}^{n-1} \prod_{i=1}^{n-j} c_{j,j+i|1,\ldots,j-1} \right) \prod_{k=1}^{n} f_k(x_k). \tag{1.57}$$

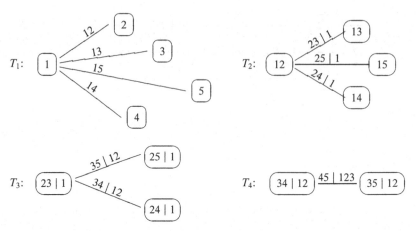

Figure 1.3 C-vine for $n = 5$

For $n = 5$ (1.57) gives a decomposition of the form

$$f(x_1,\ldots,x_5) = \prod_{k=1}^{5} f_k(x_k)c_{12}c_{13}c_{14}c_{15}c_{23|1}c_{24|1}c_{25|1}c_{34|12}c_{35|12}c_{45|123},$$

which is represented by the following graph in Figure 1.3 with levels $T_1,\ldots,T_4$. Again each transition of level T_i to T_{i+1} involves (conditional) pair copulas.

Remark 1.15. (a) PCC and Markov random fields:
More general systems to organize pairwise conditional copulas in order to represent uniquely an n-dimensional distribution function respectively density are described in Bedford and Cooke (2001, 2002) and Kurowicka and Cooke (2006) under the denomination "regular vine". These decompositions are represented graphically as a nested tree. Edges of the tree denote the indices used in the conditional copula densities. The representation of the density in terms of these conditional distributions is an analogue of the Hammersley–Clifford theorem for Markov random field specifications by certain conditional distributions (of the components given their neighbourhoods). Depending on the choice of the neighbourhoods there is some strong similarity between both ways of constructing multivariate (spatial) models.

(b) **Reduction of vine representation:**
Haff, Aas, and Frigessi (2010) consider a reduction of the (great) complexity of regular vine models by assuming that the conditional copula densities $c_{j,i+j|j+1,\ldots,j+i-1}$ respectively copulas $C_{1j|2,\ldots,j-1}$ do not depend on the variables x_k in the condition but they depend only through the conditional distribution functions $F_{j|x_{j+1},\ldots,x_{j+i-1}}$ etc. This reduction seems to yield good approximations in several applications.   ◊

1.5 Applications of the Distributional Transform

1.5.1 Application to Stochastic Ordering

Let $\leq_{\text{st}}$ denote the usual stochastic ordering on $\mathbb{R}^n$, i.e. the integral induced ordering w.r.t. the class $\mathcal{F}_i$ of increasing functions which is defined by

$$X \leq_{\text{st}} Y \text{ if } Ef(X) \leq Ef(Y) \text{ for all } f \in \mathcal{F}_i \tag{1.58}$$

such that the expectations exist.

The following sufficient condition for the multitivariate stochastic order is a direct consequence of Theorem 1.10 on the multivariate quantile transform.

Proposition 1.16. *Let X, Y be n-dimensional vectors with distribution functions F, $G \in \mathcal{F}_n$ and let $(V_i)_{1 \leq i \leq n}$ be iid, $U(0, 1)$-distributed, then*

$$\tau_F^{-1}(V) \leq \tau_G^{-1}(V) \text{ implies that } X \leq_{\text{st}} Y. \tag{1.59}$$

Condition (1.59) is stated in Rü (1981b). It implies various sufficient conditions for stochastic ordering going back to classical results of Veinott (1965), Kalmykov (1962), and Stoyan (1972). The comparison result of Veinott (1965) states

Corollary 1.17 (Comparison w.r.t. stochastic order $\leq_{\text{st}}$). *Let X, Y be n-dimensional random vectors such that $X_1 \leq_{\text{st}} Y_1$ and for $2 \leq i \leq n$*

$$P^{X_i | X_1 = x_1, \ldots, X_{i-1} = x_{i-1}} \leq_{\text{st}} P^{Y_i | Y_1 = y_1, \ldots, Y_{i-1} = y_{i-1}} \tag{1.60}$$

for all $x_j \leq y_j$, $1 \leq j \leq i-1$, then

$$X \leq_{\text{st}} Y.$$

Proof. Condition (1.60) implies (by induction) that $\tau_F^{-1}(V) \leq \tau_G^{-1}(V)$ where $F = F_X$, $G = F_Y$. Since $\tau_F^{-1}(V) \overset{d}{=} X$ and $\tau_G^{-1}(V) \overset{d}{=} Y$ this implies $X \leq_{\text{st}} Y$. □

The standard construction in (1.45) respectively the regression representation in (1.46) however are not applicable in general when $P \leq_{\text{st}} Q$, i.e. they do not in general produce pointwise a.s. constructions $X \sim P$, $Y \sim Q$ such that $X \leq Y$ a.s. The existence of such a coupling is however true under general conditions as follows from the following theorem due to Strassen (1965).

Let $(S, \leq)$ be a Polish space supplied with a closed semiorder "$\leq$", i.e., the set $\{(x, y) \in S \times S; \ x \leq y\}$ is closed w.r.t. the product topology. The closed order $\leq$ induces the "stochastic order" $\leq_{\text{st}}$ on the set $M^1(S)$ of probability measures on S defined by

$$P \leq_{\text{st}} Q \text{ if } \int f dP \leq \int f dQ \tag{1.61}$$

for all integrable increasing real functions $f \in \mathcal{F}_i = \mathcal{F}_i(S, \leq)$.

Theorem 1.18 (Strassen's ordering theorem, Strassen (1965)).
Let $(S, \leq)$ be a Polish space supplied with a closed partial order. Let $P, Q \in M^1(S)$, then

$$P \leq_{\text{st}} Q$$

   $\Leftrightarrow$ *There exist random variables $X \sim P, Y \sim Q$ such that $X \leq Y$ a.s.*

In Section 3.4 we will discuss extensions of this a.s. ordering theorem to the class of "integral induced orderings $\leq_{\mathcal{F}}$" for some function class $\mathcal{F}$. These are defined via

$$P \leq_{\mathcal{F}} Q \text{ if } \int f dP \leq \int f dQ \text{ for all integrable } f \in \mathcal{F}.$$

Early sources for integral induced orderings are Marshall and Olkin (1979), Rü (1979), and Whitt (1986).

The regression and standard constructions are used essentially in various papers and textbooks on stochastic ordering and are closely connected with some notions of stochastic ordering respectively dependence orderings. We state as one example the notion of conditional increasing in sequence (CIS).

Definition 1.19 (Conditional increasing in sequence (CIS)).
A random vector $X = (X_1, \ldots, X_n)$ is called "conditional increasing in sequence" (CIS)if for $2 \leq i \leq n$

$$X_i \uparrow_{\text{st}} (X_1, \ldots, X_{i-1}), \tag{1.62}$$

i.e. the conditional distribution $P_{i|x_1,\ldots,x_{i-1}} = P^{X_i|X_1=x_1,\ldots,X_{i-1}=x_{i-1}}$ is stochastically increasing in $(x_1, \ldots, x_{i-1})$.

The CIS-property of a random vector X is a positive dependence property of X. This property is equivalent to the condition that the standard representation based on the multivariate quantile transform $Y = \tau_F^{-1}(V)$ is an increasing function in V, i.e.

$$Y_k = f_k(V_1, \ldots, V_k) \text{ are increasing in } V, \quad 2 \leq k \leq n. \tag{1.63}$$

Proposition 1.20 (CIS and multivariate quantile transform).
Let X be a random vector with distribution function F, then X is CIS if and only if the construction by the multivariate quantile transformation $Y = \tau_F^{-1}(V) = (f_1(V_1), f_2(V_1, V_2), \ldots, f_n(V_1, \ldots, V_n))$ is monotonically increasing in V.

1.5.2 Optimal Couplings

The multivariate quantile transform is by Theorems 1.10 and 1.12 a basic construction method for random vectors. An extension of this construction leads to interesting connections with the construction of (optimal) couplings of distributions.

Let $h : \mathbb{R}^n \to \mathbb{R}^m$ be a measurable function and let $P \in M^1(\mathbb{R}^n, \mathfrak{B}^n)$ have distribution function F. Let furthermore S, V be random variables on $(\Omega, \mathfrak{A}, R)$ such that the "distributional equation"

$$P^h = R^S,$$

holds, i.e. h and S have the same distributions w.r.t. P respectively R.

Theorem 1.21 (Solution of stochastic equations, Rü (1985) and Rachev and Rü (1991)). *Let S be a random variable on a probability space $(\Omega, \mathfrak{A}, R)$ that satisfies the distributional equation*

$$P^h = R^S.$$

Assume that $V = (V_i)_{1 \le i \le n}$ are further iid $U(0, 1)$-distributed random variables on $(\Omega, \mathfrak{A}, R)$ such that S, V are independent. Then there exists a random variable X on $(\Omega, \mathfrak{A}, R)$ such that

(a) $R^X = P$, i.e. X has distribution function F
and
(b) X is a solution of the "stochastic equation"

$$h \circ X = S \text{ a.s. with respect to } R.$$

Remark 1.22. The situation is described by the following diagram:

If the distributional equation $R^S = P^h$ holds, then there exists a *solution* X with $R^X = P$ solving the stochastic equation $h \circ X = S$ a.s. and thus making the diagram commutative. $\Diamond$

Proof. We denote by

$$F_{i|x_1,\dots,x_{i-1},s} = F_i(\cdot \mid x_1, \dots, x_{i-1}, s) = P^{\pi_i \mid \pi_1 = x_1, \dots, \pi_{i-1} = x_{i-1}, h = s}$$

the conditional distribution function of the i-th projection π_i given $\pi_j = x_j$, $j \le i - 1$ and given $h = s$ and define inductively a random vector X by

$$X_1 = F_1^{-1}(V_1 \mid S), \quad X_2 = F_2^{-1}(V_2 \mid X_1, S), \dots,$$
$$X_n = F_n^{-1}(V_n \mid X_1, \dots, X_{n-1}, S). \tag{1.64}$$

Equation (1.64) is an extension of the multivariate quantile transform. By the independence assumptions we obtain

$$R^{X_1 \mid S = s} = R^{F_1^{-1}(V_1 \mid S) \mid S = s} = R^{F_1^{-1}(V_1 \mid S = s)} = P^{\pi_1 \mid h = s}.$$

Similarly,

$$R^{X_2 \mid X_1 = x_1, S = s} = R^{F_2^{-1}(V_2 \mid X_1, S) \mid X_1 = x_1, S = s}$$
$$= R^{F_2^{-1}(V_2 \mid x_1, s)} = P^{\pi_2 \mid \pi_1 = x_1, h = s}$$

implying

$$R^{(X_1, X_2) \mid S = s} = \int R^{X_2 \mid X_1 = x_1, S = s} dR^{X_1 \mid S = s}$$
$$= \int P^{\pi_2 \mid \pi_1 = x_1, h = s} dP^{\pi_1 \mid h = s}(x_2) = P^{(\pi_1, \pi_2) \mid h = s}.$$

By induction we find $R^{X \mid S = s} = P^{\pi \mid h = s}$ and, therefore, $R^X = P$.

Since almost surely w.r.t. P^h it holds that $P^{\pi \mid h = s}(\{x; \, h(x) = s\}) = 1$ we obtain $R^{X \mid S = s}(\{x; \, h(x) = s\}) = 1[R^S]$ and thus

$$R(\{h(X) = S\}) = \int R^{X \mid S = s}(\{x; \, h(x) = s\}) dR^S(x) = 1. \qquad \square$$

Remark 1.23. Using a measure isomorphism argument Theorem 1.21 on the solutions of stochastic equations extends to Borel spaces E, F, replacing $\mathbb{R}^n$, $\mathbb{R}^m$ (see Rü (1985) and Rachev and Rü (1991)), i.e. let $(E, P) \xrightarrow{h} F$, and $(\Omega, R) \xrightarrow{f} F$ be functions such that the distributional equation

$$P^h = R^f \tag{1.65}$$

holds. If (E, P) is rich enough, i.e. it allows a uniformly $U(0, 1)$-distributed random variable V on (E, P) independent of h, then there exists a random variable $X : E \to F$ with $P^X = R$ such that X solves the stochastic equation

$$h = f \circ X[P]. \tag{1.66}$$

An interesting application of Theorem 1.21 is to the construction of optimal couplings. Let $T : (\mathbb{R}^n, \mathfrak{B}^n) \to (\mathbb{R}^m, \mathfrak{B}^m)$ and for probability measures $P, Q \in \mathcal{M}^1(\mathbb{R}^n, \mathfrak{B}^n)$ define the optimal coupling problem:

$$c_T(P, Q) := \inf \left\{ E \|T(X) - T(Y)\|^2; \ X \stackrel{d}{=} P, Y \stackrel{d}{=} Q \right\}. \qquad (1.67)$$

Equation (1.67) is the optimal coupling of $T(X), T(Y)$ over all possible couplings X, Y of P, Q. Then the following result holds (see Rü (1986)):

Corollary 1.24 (Optimal coupling of T). *Let $P_1 = P^T$, $Q_1 = Q^T$ be the distributions of T under P, Q. Then*

$$c_T(P, Q) = \inf \left\{ E \|T(X) - T(Y)\|^2; \ X \stackrel{d}{=} P, Q \stackrel{d}{=} Q \right\} = \ell_2^2(P_1, Q_1),$$

where $\ell_2(P_1, Q_1)$ is the "minimal ℓ_2-metric" of P_1, Q_1 given by

$$\ell_2(P_1, Q_1) = \inf \left\{ (E \|U - V\|^2)^{1/2}; \ U \stackrel{d}{=} P_1, V \stackrel{d}{=} Q_1 \right\}. \qquad (1.68)$$

In case $m = 1$ it holds that

$$\ell_2^2(P_1, Q_1) = \int_0^1 (F_1^{-1}(u) - G_1^{-1}(u))^2 du, \qquad (1.69)$$

where F_1, G_1 are the distribution functions of P_1, Q_1.

1.5.3 Identification and Goodness of Fit Tests

For the construction of a goodness of fit test for the hypothesis $H_0 : F = F_0$ the multivariate distributional transform allows to construct simple test statistics by checking whether the transformed random vectors $Y_i = \tau_{F_0}(X_i, V^i)$, $1 \le i \le n$, are uniformly distributed on the unit cube $[0, 1]^d$. Standard tests for this purpose are based on Kolmogorov–Smirnov test statistics $T_m = \sup_{t \in [0,1]} |\widehat{F}_m(t) - t|$ on Cramér–von Mises statistics $\int (\widehat{F}_m(t) - t)^2 dt$ or on weighted variants of them. Here $m = dn$ is the sample size and $\widehat{F}_m$ the corresponding empirical distribution function. A detailed discussion of this principle and its practical and theoretical properties is given in Prakasa Rao (1987). A main problem for the practical application of this construction method is the calculation of conditional distribution functions.

This principle of standardization is also useful for various other kinds of identification problems and for statistical tests as for example for the test of the two-sample problem $H_0 : F = G$. For this problem we use the empirical version of the distributional transform based on the pooled sample. We have to check whether the transformed sample is a realization of a $U([0, 1]^d)$-distributed variety.

1.5.4 Empirical Copula Process and Empirical Dependence Function

We consider the problem of testing or describing dependence properties of multivariate distributions based on a sequence of observations. The construction of test statistics is typically based on some classical dependence measures like Kendall's τ or Spearman's ϱ (see Nelsen (2006)) or related dependence functionals. Empirical versions of the dependence functionals can often be represented as functionals of the reduced empirical process, the empirical copula function and the normalized empirical copula process. The distributional transform allows to extend some limit theorems known for the case of continuous distributions to more general distribution classes.

Let $X_j = (X_{j,1}, \ldots, X_{j,k})$, $1 \leq j \leq n$ be k-dimensional random vectors with distribution function $F \in \mathcal{F}(F_1, \ldots, F_k)$. For the statistical analysis of dependence properties of F a useful tool is the "reduced empirical process", which is also called "copula process", and is defined for $t \in [0, 1]^k$ by

$$V_n(t) := \frac{1}{\sqrt{n}} \sum_{j=1}^{n} \left(I\left(U_{j,1} \leq t_1, \ldots, U_{j,k} \leq t_k\right) - C(t) \right). \qquad (1.70)$$

Here $U_j = (U_{j,1}, \ldots, U_{j,k})$ are the copula vectors of X_j, $U_{j,i} = F_i(X_{j,i}, V^j)$, and C is the corresponding copula $C(t) = P(U_j \leq t)$.

The construction of the distributional transforms $U_{j,i}$ is based on knowing the marginal distribution functions F_i. If F_i are not known it is natural to use empirical versions of them. Let

$$\widehat{F}_i(x_i) = \frac{1}{n} \sum_{j=1}^{n} 1_{(-\infty, x_i]}(X_{j,i}) \qquad (1.71)$$

denote the empirical distribution functions of the i-th components of $X_1, \ldots, X_n$. Then in the case of a continuous distribution function F the empirical counterparts of the distributional transforms are

$$\widehat{U}_{j,i} := \widehat{F}_i(X_{j,i}), \quad \widehat{U}_j = \left(\widehat{U}_{j,1}, \ldots, \widehat{U}_{j,k}\right). \qquad (1.72)$$

For continuous distribution function F_i we have that

$$n\widehat{U}_{j,i} = n\widehat{F}_i(X_{j,i}) = R_{j,i}^n \qquad (1.73)$$

are the ranks of $X_{j,i}$ in the n-tuple of i-th components $X_{1,i}, \ldots, X_{n,i}$ of $X_1, \ldots,$ X_n and the ranks $R_{1,i}, \ldots, R_{n,i}$ are a.s. a permutation of $1, \ldots, n$. The "empirical copula function" is then given by

$$\widehat{C}_n(t) = \frac{1}{n} \sum_{j=1}^{n} I\left(\widehat{U}_j \leq t\right), \quad t \in [0, 1]^k. \tag{1.74}$$

$\widehat{C}_n$ is an estimator of the copula function C. $\widehat{C}_n$ induces the "normalized empirical copula process"

$$L_n(t) := \sqrt{n}\left(\widehat{C}_n(t) - C(t)\right)$$
$$= \frac{1}{\sqrt{n}} \sum_{j=1}^{n} \left\{ I(R_{j,1}^n \leq nt_1, \ldots, R_{j,k}^n \leq nt_k) - C(t) \right\}, \quad t \in [0, 1]^k. \tag{1.75}$$

This normalized empirical copula process was introduced in Rü (1974, 1976) under the name *multivariate rank order process*. In that paper more generally the sequential version of the process

$$L_n(s, t) = \frac{1}{\sqrt{n}} \sum_{j=1}^{[ns]} \left\{ I(\widehat{U}_j \leq t) - C(t) \right\}, \quad s \in [0, 1], t \in [0, 1]^k \tag{1.76}$$

was introduced and analysed for nonstationary and mixing random variables.

The empirical copula function $\widehat{C}_n$ was also introduced in Deheuvels (1979) and called "empirical dependence function". Based on limit theory for the reduced empirical process it is shown in Rü (1974, 1976) and also in a series of papers of Deheuvels starting with Deheuvels (1979) that the normalized empirical copula process converges to a Gaussian process. Several nonparametric measures of dependence like Spearman's ϱ or Kendall's τ have corresponding empirical versions which can be represented as functionals of L_n. As a consequence one obtains asymptotic distributions for these test statistics for testing dependence properties.

The distributional transform suggests to consider an extension of the empirical copula process to the case of general distribution functions F. The empirical versions of the $U_{j,i}$ are now defined as

$$\widehat{U}_{j,i} = \tau_{\widehat{F}_i}\left(X_{j,i}, V^j\right) \tag{1.77}$$

which are exactly $U(0, 1)$ distributed. In order to avoid artificial dependence it is natural to let the copula $C_j(t) = P(U_j \leq t), t \in [0, 1]^k$, be based on the same randomization V^j in all components of the j-th random vector such that $C_j(t) = C(t)$, $1 \leq j \leq n$. We define the normalized empirical copula process by

$$L_n(t) = \sqrt{n}\left(\widehat{C}_n(t) - C(t)\right), \quad t \in [0, 1]^k. \tag{1.78}$$

The copula C has bounded nondecreasing partial derivatives a.s. on $[0,1]^k$ (see Nelsen (2006, p. 11)). Now the proof of Theorem 3.3 in Rü (1976) extends to the case of general distributions.

The basic assumption of this theorem is convergence of the reduced sequential empirical process, the sequential version of V_n in (1.70) (defined as in (1.76) for L_n). This assumption has been established for various classes of independent and mixing sequences of random vectors.

(A)   Assume that the reduced sequential process $V_n(s,t)$ converges weakly to an a.s. continuous Gaussian process V_0 in the Skorohod space D_{k+1}.

The additional assumptions on V_0 made in Rü (1976) served there to obtain stronger convergence results or to deal with more general assumptions on the distributions.

Theorem 1.25 (Limit theorem for the normalized empirical copula process, Rü (1976, 2009)). *Under condition (A) the sequential version $L_n(s,t)$ of the normalized empirical copula process converges weakly to the a.s. continuous Gaussian process L_0 given by*

$$L_0(s,t) = V_0(s,t) - s \sum_{i=1}^{n} \frac{\partial C(t)}{\partial t_i} V_0(1,\ldots,1,t_i,\ldots,1). \qquad (1.79)$$

Based on this convergence result asymptotic distributions of test statistics testing dependence properties can be derived as in the continuous case. The proofs are based on representations or approximations of these statistics by functionals of the empirical copula process L_n. For examples of this type see Rü (1974, 1976) and Deheuvels (1979, 1981). For applications to the estimation of dependence functionals and extensions to the empirical tail copula process see Schmidt and Stadtmüller (2006).

1.6 Multivariate and Overlapping Marginals

In this section we consider the case that not only one-dimensional (marginal) distributions of the risk vector X are known. We assume that also for certain subsets J of the components the joint distribution of $(X_j)_{j \in J}$ is known. This is motivated from available empirical information contained in certain historical data sets or from functional knowledge of the random mechanism.

1.6.1 Generalized Fréchet Class

Let $(E_j, \mathfrak{A}_j)$, $1 \le j \le n$ be n measure spaces and let $\mathcal{E} \subset \mathcal{P}(\{1,\ldots,n\})$ be a system of subsets $J \subset \{1,\ldots,n\}$ such that $\cup_{J \in \mathcal{E}} J = \{1,\ldots,n\}$. Let $P_J \in$

Figure 1.4 Multivariate
marginals

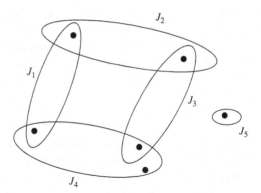

$\mathcal{M}^1(E_J, \mathfrak{A}_J)$, $J \in \mathcal{E}$, be a consistent system of probability distributions on $(E_J, \mathfrak{A}_J) = \bigotimes_{j \in J}(E_j, \mathfrak{A}_j)$. We assume that we know the joint distributions of the components in J for all $J \in \mathcal{E}$.

This assumption is a restriction on the joint dependence structure in the model. In comparison the Fréchet class (with only single marginals fixed) includes the set of all possible dependence structures (Figure 1.4).

Definition 1.26 (Generalized Fréchet class).   To a given consistent system $(P_J)_{J \in \mathcal{E}}$ of probability measures we define the "generalized Fréchet class" $\mathcal{M}_\mathcal{E}$ by

$$\mathcal{M}_\mathcal{E} = \mathcal{M}(P_J, J \in \mathcal{E}) = \left\{P \in \mathcal{M}^1(E, \mathfrak{A}); \ P^{\pi_J} = P_J, J \in \mathcal{E}\right\}, \qquad (1.80)$$

where $(E, \mathfrak{A}) = \bigotimes_{i=1}^n (E_i, \mathfrak{A}_i)$ and π_J are the projections on the components in J.

Using the generalized Fréchet class as a model class for a risk vector X means that the distribution of $X_J = (X_j)_{j \in J}$ is specified to be P_J for all sets $J \in \mathcal{E}$.

In the particular case that $\mathcal{E} = \{\{1\}, \ldots, \{n\}\}$ where $\mathcal{E}$ consists of singletons we get the usual Fréchet class $\mathcal{M}(P_1, \ldots, P_n)$. If $\mathcal{E} = \mathcal{E}^s = \{\{i, i+1\}, 1 \le i \le n-1\}$ we get the "series case", where all neighboured pairwise distributions are known. A system $\mathcal{E}$ is called "decomposable" (or "regular"), if there do not exist cycles in $\mathcal{E}$. The most simple nondecomposable (nonregular) system is given by $\mathcal{E}_3^2 = \{\{1, 2\}, \{2, 3\}, \{1, 3\}\}$ or in more general form by the "pairwise system" $\mathcal{E} = \mathcal{E}_n^2 = \{\{i, j\}; \ 1 \le i < j \le n\}$.

The "marginal problem" is the question whether there exist joint distributions with the given multivariate marginals. A classical result due to Vorobev (1962) and Kellerer (1964) states in the situation of Polish spaces:

Theorem 1.27 (Marginal problem).   *Let $\mathcal{E} \subset \mathcal{P}(\{1, \ldots, n\})$ and $(E_j, \mathfrak{A}_j)$ be Polish spaces. Then the statement*

$$\textit{Consistency of } (P_J) \textit{ implies } \mathcal{M}_\mathcal{E} \ne \phi$$

is equivalent to the condition that $\mathcal{E}$ is decomposable.

Thus in general consistency is not enough to imply the existence of joint distributions. A simple counterexample is the following. Let $\mathcal{E} = \{\{1, 2\}, \{2, 3\}, \{1, 3\}\}$ and let $P_{12} = P_{23} = P_{13}$ be the distribution of the pair $(U, 1 - U)$ where $U \sim U(0, 1)$. Then this system is consistent but $\mathcal{M}_{\mathcal{E}} = \phi$. If there would exist some element $P \in \mathcal{M}_{\mathcal{E}}$ and $X = (X_1, X_2, X_3) \sim P$ then we would get

$$\text{Cov}(X_1, X_3) = \text{Cov}(X_1, 1 - X_2) = -\text{Cov}(X_1, X_2) > 0,$$

a contradiction. Some characterizations of nonemptiness of $\mathcal{M}_{\mathcal{E}}$ are known (see Rü (1991a)), which however are not easy to apply but may serve to produce counterexamples.

Assuming $\mathcal{M}_{\mathcal{E}} \neq \phi$, a natural idea to construct submodels $\mathcal{P} \subset \mathcal{M}_{\mathcal{E}}$ describing the dependence structure in a correct way is the following extension of the projection method discussed in Section 1.2 for the simple marginal case. Let $\{P_{\vartheta}; \; \vartheta \in \Theta\}$ be a parametric class of probability measures on $(E, \mathfrak{A})$ with densities $f_{\vartheta} \sim P_{\vartheta}$ describing the dependence of the components correctly. Then one may try to determine the projections $P_{\vartheta} \rightarrow P'_{\vartheta} \in \mathcal{M}_{\mathcal{E}}$ w.r.t. some suitable distance in order to fit the marginal structure. The hope is that even after projection the dependence structure is essentially not changed (see Section 1.2).

In the case that $\mathcal{M}_{\mathcal{E}} = \mathcal{M}(P_1, \ldots, P_n)$ one can interpret Sklar's Theorem in this sense, i.e. transforming an element $G \in \mathcal{F}(G_1, \ldots, G_n)$ to some $F \in \mathcal{F}(F_1, \ldots, F_n)$ with the correct marginals $F_1, \ldots, F_n$ via the copula C,

$$G \rightarrow C \rightarrow F = C(F_1, \ldots, F_n). \tag{1.81}$$

For the Kullback–Leibler distance the projection is characterized by a density of product form

$$\frac{dP}{dQ} = \prod_{i=1}^{n} f_i.$$

The "iterative proportional fitting algorithm (IPF)" has been shown to converge to the projection (see Rü (1995a)) under some conditions.

In the case of general overlapping marginals a characterization of all L^2-projections (with restriction to the probability measures continuous w.r.t. the product measure) is given in Rü (1985). For the Kullback–Leibler distance a product form of the density

$$\frac{dP}{dQ}(x) = \prod_{J \in \mathcal{E}} f_J(x_J) \tag{1.82}$$

is sufficient for the projection and up to a closedness property also necessary.

In general however a natural extension of Sklar's Theorem giving a construction

$$\mathcal{M}(Q_J, J \in \mathcal{E}) \rightarrow \mathcal{M}(P_J, J \in \mathcal{E}) \tag{1.83}$$

for two marginal systems is still an open question.

There are some particular results on the connection described in (1.83). For $P \in \mathcal{M}^1(\mathbb{R}^n, \mathfrak{B}^n)$ denote $\mathcal{C}(P)$ the set of all copulas of P, then the following relations can be easily seen.

Proposition 1.28. *(a) If $\mathcal{M}_{\mathcal{E}}(P_J, J \in \mathcal{E}) \neq \phi$, then there exist $C_J \in \mathcal{C}(P_J)$, $J \in \mathcal{E}$ such that $\mathcal{M}_{\mathcal{E}}^C := \mathcal{M}_{\mathcal{E}}(C_J, J \in \mathcal{E}) \neq \phi$.*
(b) If C_J are copulas of P_J, $J \in \mathcal{E}$ and $\mathcal{M}_{\mathcal{E}}^C = \mathcal{M}_{\mathcal{E}}(C_J, J \in \mathcal{E}) \neq \phi$, then $\mathcal{M}_{\mathcal{E}} = \mathcal{M}_{\mathcal{E}}(P_J, J \in \mathcal{E}) \neq \phi$.
(c) In general $\mathcal{M}_{\mathcal{E}}(P_J, J \in \mathcal{E}) \neq \phi$ does not imply $\mathcal{M}_{\mathcal{E}}^C(C_J, J \in \mathcal{E}) \neq \phi$ for any choice of copulas $C_J \in \mathcal{C}(P_J)$.

1.6.2 Copulas with Given Independence Structure

In Section 1.2 the projections w.r.t. L^2-distance have been described in the simple marginal case. These results have been extended to the multivariate marginal case in Rü (1985). Let $\mathcal{E}$ be an index class $\mathcal{E} = \{T_1, \ldots, T_k\}$ with $\cup T_j = \{1, \ldots, n\}$ and let $P_{T_1}, \ldots, P_{T_k}$ be a given consistent system of distributions on $[0, 1]^{|T_i|}$, $1 \leq i \leq k$. We assume that all univariate marginals are uniform $U(0, 1)$ and thus the generated Fréchet class $\mathcal{M}_{\mathcal{E}}$ is a subclass of the set of all copulas. As in Section 1.2 we restrict in the following to the Lebesgue-continuous elements in $\mathcal{M}_{\mathcal{E}}$

$$\mathcal{M}_{\mathcal{E}}(\lambda^n) = \{P \in \mathcal{M}_{\mathcal{E}}; \ P \ll \lambda^n\}; \tag{1.84}$$

the signed version of this class we denote by $\mathcal{M}_{\mathcal{E}}^s(\lambda^n)$.

In the first part we consider the special case of distributions with given independence structure, i.e. we assume that

$$P_{T_i} = \lambda^{|T_i|}, \quad 1 \leq i \leq k. \tag{1.85}$$

Thus we consider the class of probability models for a random vector X such that $(X_j)_{j \in T_i}$ are iid $U(0, 1)$-distributed for any $i \leq k$.

To describe the corresponding generalized Fréchet class we need a second linear operator V supplementing the operator S defined in Section 1.2 in (1.27).

Define $V : L^1(\lambda^n) \to L^1(\lambda^n)$ inductively by

$$f_{(1)} := f_{R_1}, f_{(m+1)} := f_{(m)} - (f_{(m)})_{R_{m+1}} \text{ for } m < k \text{ and } V(f) := f_{(k)}, \tag{1.86}$$

where $R_m = T_m^c = \{1, \ldots, n\} \setminus T_m$ and f_R is obtained from f by integrating over the components in R.

Theorem 1.29 (Distributions with given independence structure). *For the independence structure $\mathcal{E}$ given in (1.85) we have the representation of the generalized Fréchet class $\mathcal{M}_n^s(\mathcal{E})$ by*

$$\mathcal{M}_n^s(\mathcal{E}) = \{(1 + V \circ Sf)\lambda^n; \ f \in L^1(\lambda^n)\}$$

respectively

$$M_n(\mathcal{E}) = \{(1 + V \circ Sf)\lambda^n; \ f \in L^1(\lambda^n), 1 + V \circ Sf \geq 0\}.$$

The proof of Theorem 1.29 is similar to that of Theorem 1.7 in Section 1.2 (see Rü (1985)).

Special attention has been given in the literature to the case that $\mathcal{E} = \mathcal{E}_k = \{T \subset \{1, \ldots, n\}; \ |T| = k\}$ i.e. the case that all choices of k-components are independent. In this case a more compact representation of the solutions is possible.

Define for $f \in L^1(\lambda^n)$ and $1 \leq k < n$ inductively linear operators $V_1, \ldots, V_n$ by

$$V_1 f = f, \ V_{k+1} f = V_k f - \sum_{|T|=n-k+1} (V_k f)_T. \qquad (1.87)$$

Call a signed measure P "k-independent" if the distribution of any k-components $(\pi_{i_1}, \ldots, \pi_{i_k})$ is $\lambda^k / [0, 1]^k$.

Theorem 1.30 (k-independent probability measures). *The set of all k-independent (signed) measures has the representation*

$$M_n^s(\mathcal{E}_k) = \{(1 + V_k \circ Sf)\lambda^n; \ f \in L^1(\lambda^n)\}$$

respectively

$$M_n(\mathcal{E}_k) = \{(1 + V_k \circ Sf)\lambda^n; \ f \in L^1(\lambda^n), 1 + V_k \circ Sf \geq 0\}. \qquad (1.88)$$

This result follows by reduction from Theorem 1.29.

Example 1.31 (FGM-distributions). Consider the generalized FGM-distributions defined as $(1 + f)\lambda^n$, where $f(x) = \prod_{i=1}^n v_i(x_i)$ such that $\int v_i(x_i)dx_i = 0$, $1 \leq i \leq n$. If $\lambda^n(\{f = 0\}) < 1$, then $Sf = f$ and furthermore

$$V_1 f = V_2 f = \cdots = V_{n-1} f = f. \qquad (1.89)$$

This implies that the FGM-distribution $(1 + f)\lambda^n$ is $(n-1)$-independent but not n-independent. This observation indicates the lack of strong dependence in higher dimensional FGM-families. Simultaneously, it gives some natural examples of $(n-1)$-independent distributions which are not n-independent. Similarly, $(1+f)\lambda^n$ is k-independent but not $(k + 1)$-independent, where we define f as

$$f(x) = \alpha \sum_{|T|=k} \prod_{j \in T} v_j(x_j), \qquad (1.90)$$

α being a factor such that $1 + f \geq 0$. $\qquad\qquad\qquad\qquad\qquad\qquad \Diamond$

1.6.3 Copulas, Overlapping Marginals, and L^2-Projections

The construction in Theorem 1.29 can be extended to the construction in the general Fréchet class case. Define for $J \subset \{1, \dots, k\}$, $T_J := \bigcap_{j \in J} T_j$, and for $f \in L^1(\lambda^n)$, f_T by integrating over the T components. f_T again is considered as a function on $[0,1]^n$. Let for $T \in \mathcal{E}$, P_T have densities g^T, $P_T = g^T \lambda^{|T|}$ and define

$$h(x) = \sum_{m=1}^{k} (-1)^{m-1} \sum_{\substack{J \subset \{1,\dots,k\} \\ |J|=m}} g^{T_J}(x), \quad x \in [0,1]^n, \tag{1.91}$$

where $g^{T_J}(x) = 0$ if T_J is empty. Then we have the following representation of the general Fréchet class (see Rü (1985)). Define the operator $T_\mathcal{E}$ by

$$T_\mathcal{E} f := h + V \circ Sf. \tag{1.92}$$

Theorem 1.32 (Representation of general Fréchet class). *The class of all (signed) measures in the generalized Fréchet class which are Lebesgue-continuous has the representation*

$$\mathcal{M}_n^s(\mathcal{E}) = \{(T_\mathcal{E} f)\lambda^n; \ f \in L^1(\lambda^n)\}$$

respectively

$$M_n(\mathcal{E}) = \{(T_\mathcal{E} f)\lambda^n; \ f \in L^1(\lambda^n), T_\mathcal{E} f \geq 0\}. \tag{1.93}$$

Proof. In the first step we prove that $h\lambda^n \in \mathcal{M}_n^s(\mathcal{E})$ or, equivalently, that $h_{R_i} = g^{T_i}$, where $R_i = T_i^c$, $1 \leq i \leq k$. Without loss of generality we consider the case $i = 1$. By definition of h we get

$$h = g^{T_1} + \sum_{m=1}^{k} (-1)^{m-1} \sum_{\substack{|J|=m, \\ J \neq \{1\}}} g^{T_J}$$

$$= g^{T_1} + \sum_{m=1}^{k} (-1)^{m-1} \left(\sum_{\substack{|J|=m, \\ 1 \in J, J \neq \{1\}}} g^{T_{J \cup \{1\}}} + \sum_{\substack{|J|=m, \\ 1 \notin J}} g^{T_J} \right)$$

$$= g^{T_1} + \sum_{m=1}^{k} (-1)^{m-1} \left(\sum_{\substack{|J|=m-1, \\ 1 \notin J, J \neq \phi}} g^{T_{J \cup \{1\}}} + \sum_{\substack{|J|=m, \\ 1 \notin J}} g^{T_J} \right)$$

$$= g^{T_1} + \sum_{m=1}^{k} (-1)^{m} \left(\sum_{\substack{|J|=m, \\ 1 \notin J}} g^{T_{J \cup \{1\}}} - \sum_{\substack{|J|=m, \\ 1 \notin J}} g^{T_J} \right)$$

and from the relation $(g^{T_{J \cup \{1\}}})_{R_1} = (g^{T_J})_{R_1}$ we obtain the assertion $h_{R_1^c} = g^{T_1}$.

Let now $P = g\lambda^n \in M_n^s(\mathcal{E})$, then $g = h + (g - h) = h + V \circ S(g - h)$, since $g - h$ is by the first part of this proof a fixpoint of $V \circ S$. Conversely, for $f \in L^1(\lambda^n)$ and $T \in \mathcal{E}$

$$(h + V \circ Sf)_{T^c} = h_{T^c} = g^T$$

by definition of V, i.e. $(h + V \circ Sf)\lambda^n \in M(\mathcal{E})$.   □

Theorem 1.32 allows in certain cases to construct families of probability measures with given multivariate marginals. The idea is to find a function $f \in L^1(\lambda^n)$, such that $V \circ Sf$ is balancing the negative parts of h. Some natural candidates for f are functions which allow an explicit and simple determination of the transform $V \circ Sf$, such as e.g. linear combinations of functions of the type $\prod_{i=1}^n v_i(x_i)$ where $\int v_i(x_i)dx_i = 0$, $1 \le i \le n$. The following is an example of this kind of construction.

Example 1.33. Let $n = 3$, $\mathcal{E} = \{\{1, 2\}, \{2, 3\}, \{1, 3\}\}$.

(a) When the marginal densities are $f_{12}(x_1, x_2) = 1$, $f_{23}(x_1, x_3) = 1 + \left(x_2 - \frac{1}{2}\right)$ $\cdot \left(x_3 - \frac{1}{2}\right)$, $f_{13}(x_1, x_3) = 1 + \left(x_1 - \frac{1}{2}\right)\left(x_3 - \frac{1}{2}\right)$, then

$$h(x_1, x_2, x_3) = \frac{3}{2} + x_1 x_2 + x_2 x_3 - x_3 - \frac{x_1 + x_2}{2}$$

is already a non-negative density with the given marginals.

(b) If $f_{12}(x_1, x_2) = 1 + 3\left(x_1 - \frac{1}{2}\right)\left(x_2 - \frac{1}{2}\right)$, $f_{13}(x_1, x_3) = 1 - 3\left(x_1 - \frac{1}{2}\right)$ $\cdot \left(x_3 - \frac{1}{2}\right)$, and $f_{23}(x_2, x_3) = 1$, then

$$h(x_1, x_2, x_3) = f_{13}(x_1, x_3) + f_{12}(x_1, x_2) - 1$$

and $\min\{h(x_1, x_2, x_3)\} = \frac{1}{2} = h(1, 0, 1) = h(0, 1, 0)$. A function balancing these negative parts is given by

$$f(x_1, x_2, x_3) = -6\left(x_1 - \frac{1}{2}\right)\left(x_2 - \frac{1}{2}\right)\left(x_3 - \frac{1}{2}\right),$$

so that

$$h(x_1, x_2, x_3) + f(x_1, x_2, x_3)$$
$$= 1 - 6\left(x_1 - \frac{1}{2}\right)\left(x_2 - \frac{1}{2}\right)\left(x_3 - \frac{1}{2}\right)$$
$$+ 3\left(x_1 - \frac{1}{2}\right)\left(x_2 - \frac{1}{2}\right) - 3\left(x_1 - \frac{1}{2}\right)\left(x_3 - \frac{1}{2}\right)$$

gives a non-negative density with the given marginals as can easily be seen by discussing the cases $x_1, x_2 \le \frac{1}{2}$, $x_3 \ge \frac{1}{2}$, etc. Instead of the factor 6 in the balancing function, one can use a factor a in an interval around 6, in this way obtaining a parametric class of distributions with given multivariate marginals.

Chapter 2
Fréchet Classes, Risk Bounds, and Duality Theory

Our aim in this chapter is to describe the possible influence of dependence between the components of a risk vector $X = (X_1, \ldots, X_n)$ on the expectation of a real functional $f(X)$, i.e. to determine the range of $Ef(X)$ under all possible dependence structures. When the distribution of X_i is given by P_i, $1 \leq i \leq n$ then the set of all possible dependence structures is given by the Fréchet class $\mathcal{M}(P_1, \ldots, P_n)$. We assume that P_i are probability measures on Polish spaces $(E_i, \mathfrak{A}_i)$, $1 \leq i \leq n$; typically $E_i = \mathbb{R}$ or $E_i = \mathbb{R}^d$. Then the Fréchet class $\mathcal{M}(P_1, \ldots, P_n)$ is defined as

$$\mathcal{M}(P_1, \ldots, P_n) = \left\{ P \in M^1(E, \mathfrak{A}) \mid P^{\pi_i} = P_i, 1 \leq i \leq n \right\} \qquad (2.1)$$

where $(E, \mathfrak{A}) = \bigotimes_{i=1}^{n} (E_i, \mathfrak{A}_i)$ and π_i is the projection of the product space E on the i-th component.

The Fréchet class allows to introduce the basic notion of positive dependence (respectively negative dependence) of an element $P \in \mathcal{M}(P_1, \ldots, P_n)$ w.r.t. some dependence ordering $\leq$ on $M^1(E, \mathfrak{A})$. If

$$\bigotimes_{i=1}^{n} P_i \leq P, \qquad (2.2)$$

then P is called "positive dependent", if $P \leq \bigotimes_{i=1}^{n} P_i$, then P is called negative dependent. Similarly, for a random vector X we call X positive dependent if $X^{\perp} \leq X$, where $X^{\perp}$ is the vector with independent components $X_i^{\perp} \stackrel{d}{=} X_i$ and $\leq$ is the same ordering defined for random vectors. For a measurable function $f : E \to \mathbb{R}$ the "generalized Fréchet bounds" are defined as

L. Rüschendorf, *Mathematical Risk Analysis*, Springer Series in Operations Research and Financial Engineering, DOI 10.1007/978-3-642-33590-7_2, © Springer-Verlag Berlin Heidelberg 2013

$$M(f) := \sup \left\{ \int f dP \mid P \in \mathcal{M}(P_1, \ldots, P_n) \right\}$$

$$m(f) := \inf \left\{ \int f dP \mid P \in \mathcal{M}(P_1, \ldots, P_n) \right\}. \tag{2.3}$$

The open or closed interval $[(m(f), M(f))]$ describes exactly the possible influence of dependence on the expectation $Ef(X)$. The classical results of Hoeffding (1940) and Fréchet (1951) concern the real case $E_i = \mathbb{R}$ and $n = 2$ and state the following sharp bounds:

$$F^-(x_1, x_2) := (F_1(x_1) + F_2(x_2) - 1)_+ \leq F(x_1, x_2) \tag{2.4}$$

$$\leq F^+(x_1, x_2) := \min\{F_1(x_1), F_2(x_2)\}$$

and

$$\iint (F^-(x_1, x_2) - F_1(x_1) F_2(x_2)) dx_1 dx_2$$

$$\leq \mathrm{Cov}(X_1, X_2) \leq \iint (F^+(x_1, x_2) - F_1(x_1) F_2(x_2)) dx_1 dx_2. \tag{2.5}$$

Equation (2.5) is a consequence of (2.4) and the covariance representation

$$\mathrm{Cov}(X_1, X_2) = \iint (F(x, y) - F_1(x) F_2(y)) dx dy. \tag{2.6}$$

The upper and lower Fréchet bounds for the distribution function F in (2.4) yield also the joint distributions having the maximal respectively minimal possible covariance and correlation. Also from the covariance representation it is obvious that the upper and lower bounds in (2.5) are attained only by the upper and lower Fréchet bounds in (2.4), i.e. we have uniqueness.

Remark 2.1 (Fréchet bounds vs. Hoeffding–Fréchet bounds). Due to the historical results stated in Eqs. (2.4) and (2.5) it is also justified to use the denomination "Hoeffding–Fréchet bounds" instead of "Fréchet bounds". The work of Hoeffding uses the results in (2.4) and (2.5) mainly for the introduction of a normalized scale-invariant version of the correlation. The concept, metric structure and risk bounds for the marginal classes $\mathcal{M}(P_1, P_2)$ were however only introduced and analysed in Fréchet's work. Therefore, we use in this book the denomination Fréchet bounds. This was also the reason why this denomination was used in the early survey papers on this subject in Rü (1981a, 1991a,b,1996). There is no doubt that the results of Hoeffding belong to the original sources of the circle of problems concerned with the dependence structure and with the structure of Fréchet classes. The notion Fréchet class and the general class $\mathcal{F}(F_1, \ldots, F_n)$ was introduced in Dall'Aglio (1956) and many of its properties were investigated in the Italian school of probability. This school is a basic source of the development of the theory of Fréchet classes.   ◊

The basic tool to determine generalized Fréchet bounds is a duality theorem, which connects the problem of dependence bounds to linear programming and to mass transportation problems. In Section 2.1 we introduce the basic duality theorem and the connection to mass transportation. Then as a consequence we discuss in Section 2.2 sharp upper bounds for several risk functionals. For several important risk functionals like the excess of loss or convex functionals of the joint portfolio the worst case dependence structure is given by the comonotonic risk vector. This is not true however for example for the value at risk or for the maximal risk. Bounds for the distribution function and the value at risk of the joint portfolio will be the subject of Chapter 4 while details on the excess of loss and the related convex ordering are given in Chapter 3.

2.1   Dual Representation of Generalized Fréchet Bounds

The problem of determining the generalized Fréchet bounds in (2.3) can be considered as an infinite dimensional linear optimization problem. The linear functional $\int f\,dP$ which is linear in f and P has to be maximized respectively minimized over the convex Fréchet class $\mathcal{M} = \mathcal{M}(P_1, \ldots, P_n)$. Based on experience with finite linear programming it is natural to formulate dual problems corresponding to (2.3). Define the dual functionals

$$I(f) := \inf\left\{ \sum_{i=1}^{n} \int f_i\,dP_i;\ f_i \in L^1(P_i), \bigoplus f_i \geq f \right\} \qquad (2.7)$$

where $\bigoplus f_i(x) = \sum_{i=1}^{n} f_i(x_i)$ and

$$i(f) := \sup\left\{ \sum_{i=1}^{n} \int f_i\,dP_i;\ f_i \in L^1(P_i), \bigoplus f_i \leq f \right\}. \qquad (2.8)$$

Since $M(-f) = -m(f)$ and $I(-f) = -i(f)$ we can restrict to the problem, whether "duality holds" for the sup-problem i.e.

$$M(f) = I(f). \qquad (2.9)$$

Depending on the class of functions f considered also several related formulations of duality have been given which for example include additional boundedness or continuity properties.

Remark 2.2 (Fréchet bounds and mass transportation problems). In case $n = 2$ the problem of generalized Fréchet bounds in (2.3) is identical to the mass transportation problem

$$m(c) = \inf\left\{\int c(x, y)d\mu(x, y) \mid \mu \in \mathcal{M}(P_1, P_2)\right\}, \qquad (2.10)$$

where c is a cost function measuring the cost of transportation of a mass distribution P_1 to a mass distribution P_2. Any element $\mu \in \mathcal{M}(P_1, P_2)$ can be interpreted as an admissible mass transfer plan from mass distribution P_1 to P_2. Kantorovich (1942) was the first to formulate this problem motivated by economic assignment type problems and stated a duality result as in (2.8) in case $E_1 = E_2$ are compact metric spaces and c is a continuous cost function. His proof however was valid only for the case where $c(x, y) = d(x, y)$ is a metric. In the metric case the duality result takes a more specific form. This form was given in the paper of Kantorovich and Rubinstein (1957). In this case $m(d) = \ell_1(P_1, P_2)$ defines a metric on $M^1(E_1)$, the set of probability measures on E_1, the minimal ℓ_1-metric. This metric is also called a Wasserstein metric denominated by an influential paper in ergodic theory after a transcript of its author Vasershtein (1969). (For the history of this notion see Rü (2001).)

In the case that the solution of the mass transfer problem is given by a transfer map $x \to T(x)$ such that $P_2 = P_1^T$, and further P_1, P_2 are uniform distributions on some domains $U_1, U_2 \subset \mathbb{R}^k$ and $c(x, y) = \|x - y\|$ is the Euclidean distance the solution of the transport problem (2.10) is also a solution of the classical geometric Monge problem. Kantorovich was not aware of this connection when writing his 1942 paper but noticed it only later in a short note in 1948. The duality Theorem 2.9 became the essential tool for the development of the geometric-analytic transportation problem since the 1990s and led to an enormous range of applications in geometry and analysis. Due to the above connection this research area is now called Monge–Kantorovich theory.

Motivated by the probabilistic Fréchet type problems to describe the influence of dependence on risk functionals a general duality problem as in (2.9) and some related duality results were given Rü (1979, 1981a) and Gaffke and Rü (1981) without being aware at that time of Kantorovich's mass transportation problem. In these papers duality results were established in particular for bounded continuous functions f and also for some classes of bounded and integrable measurable functions. Motivated by these results Kellerer (1984a) proved interesting extensions of the duality theorem by investigating in detail continuity properties of the primal and dual functionals M, I. Further interesting early contributions were given in Rachev (1985), Levin and Milyutin (1979), and Levin (1984) mainly in the context of transportation respectively the related transshipment problem. These and further developments as well as many probabilistic applications were described in Rachev and Rü (1998a,b). The corresponding development of the mass transportation problem and connections to geometry and analysis are given in Villani (2003, 2009). The general Fréchet classes $\mathcal{M}(P_1, \ldots, P_n)$ and related classes restricting the dependence structure like $\mathcal{M}_{\mathcal{E}}$ or positive dependence restrictions do not have a natural interpretation in terms of transportation problems and thus have an importance of their own. ◊

The following is a list of some basic duality results for the generalized transportation problem respectively the problem of generalized Fréchet bounds. We assume that the basic spaces $(E_i, \mathfrak{A}_i)$ are Polish.

Theorem 2.3 (Duality theorem).

(a) *The duality $M(f) = I(f)$ holds for the class $\mathcal{L}_m(E)$ of all lower-majorized product measurable functions, where*

$$\mathcal{L}_m(E) = \left\{ f \in \mathcal{L}(\mathfrak{A}_1 \otimes \cdots \otimes \mathfrak{A}_n); \ \exists f_i \in \mathcal{L}^1(P_i) \text{ such that } f \geq \bigoplus_{i=1}^{n} f_i \right\}. \quad (2.11)$$

(b) *Duality holds on $\overline{F}(E)$,* $\quad\quad\quad\quad\quad\quad\quad\quad\quad\quad\quad\quad\quad\quad\quad\quad\quad\quad$ (2.12)
   where $F(E)$ is the class of upper-semicontinuous functions and the closure is w.r.t. $I(|f - g|)$.
(c) *Existence of an optimal measure for the M-functional holds on $\overline{F}$.* $\quad$ (2.13)
(d) *Existence of minimal dual functions (f_i) for the I-functional holds on $\mathcal{L}_m(E)$.*
(e) *For $P \in \mathcal{M}(P_1, \ldots, P_n)$ and $f_i \in \mathcal{L}^1(P_i)$, $\bigoplus f_i \geq f$ it holds that:*

   P respectively (f_i) are solutions of the primal respectively M, I

$$dual \ problems \ \Leftrightarrow \ f = \bigoplus_{i=1}^{n} f_i[P]. \quad (2.14)$$

Remark 2.4. (a) The duality and existence results were proved in Rü (1979, 1981a) and Gaffke and Rü (1981) for the class C_b of bounded continuous functions and for some classes of bounded measurable and integrable functions. For the class of all bounded measurable functions f duality $\widetilde{M}(f) = I(f)$ was shown to hold true with a modified Fréchet class including all finite additive measures with given marginals P_i. The extension to the general result as stated in Theorem 2.3 is due to Kellerer (1984a). The conditions in Theorem 2.3 are sharp, i.e. there exist counterexamples of the duality and existence results for example for the class $G(E)$ of lower semicontinuous functions or without the lower boundedness condition in (d).
It is easy to see that the Fréchet class $\mathcal{M}(P_1, \ldots, P_n)$ is a tight and closed class of probability measures. Therefore, existence for the M functional follows for $f \in F$ since $P \to \int f dP$ is an upper semicontinuous functional on $\mathcal{M}(P_1, \ldots, P_n)$. Recent developments of the duality theorem are given in Ramachandran and Rü (1995, 2000), Rachev and Rü (1998a,b), Beiglböck and Schachermayer (2011), and Beiglböck et al. (2009).

(b) Condition (2.14) characterizes optimal measures P under the existence condition on the dual problem as e.g. for $f \in \mathcal{L}_m(E)$.

The sufficiency part does not need any conditions, i.e. the equality $f = \bigoplus_{i=1}^{n} f_i[P]$ for some $P \in \mathcal{M}(P_1, \ldots, P_n)$ in (2.14) implies optimality of P and (f_i).

(c) In Ramachandran and Rü (1995, 2000) the duality theorem is extended from Polish spaces to the class of perfect measures spaces. It is shown that in a strong form the duality theorem cannot be extended beyond these spaces.   ◊

Let $E_1 = \cdots = E_n$ and $P_1 = \cdots = P_n = Q$ and let f be a symmetric function on E satisfying any of the conditions in Theorem 2.3 as for example of the form $f(x) = \varphi\left(\sum_{i=1}^{n} x_i\right)$. Then there is a simplified dual representation.

Proposition 2.5. *Let* $P_1 = \cdots = P_n = Q$, *let* f *satisfy any of the conditions in Theorem 2.3 and let* f *be symmetric, i.e.* $f(x_{\sigma(1)}, \ldots, x_{\sigma(1)}) = f(x)$ *for all permutations* σ, *then*

$$M(f) = \inf\left\{ n \int h\, dQ;\ h \in L^1(Q), \sum_{i=1}^{n} h(x_i) \geq f(x) \right\}. \tag{2.15}$$

Proof. For admissible $f_i \in L^1(Q)$ in the definition of $I(f)$, i.e. $\bigoplus f_i \geq f$ we define

$$h(y) := \frac{1}{n} \sum_{i=1}^{n} f_i(y).$$

Then $h \in \mathcal{L}^1(Q)$ and

$$\sum_{i=1}^{n} h(x_i) = \frac{1}{n} \sum_{i=1}^{n} \sum_{j=1}^{n} f_j(x_i)$$

$$= \frac{1}{n} \sum_{i=1}^{n} \sum_{j=1}^{n} f_j\left(x_{\sigma_i(j)}\right)$$

$$\geq \frac{1}{n} \sum_{i=1}^{n} f\left(x_{\sigma_i(1)}, \ldots, x_{\sigma_i(n)}\right)$$

$$= f(x)$$

where $\sigma_i(j) = i + j(\mathrm{mod}\, n)$. So we have

$$n \int h\, dQ = \int \left(\sum_{i=1}^{n} f_i\right) dQ = \sum_{i=1}^{n} \int f_i\, dQ$$

which implies equality in (2.15).   □

An interesting consequence of the duality theorem is the following result due to Strassen (1965) (see also Kellerer (1984a)) concerning the case $n = 2$ and $f_A = 1_A$ for some $A \in \mathfrak{A}_1 \otimes \mathfrak{A}_2$. We use the notation $M(A) := M(f_A)$, $m(A) := m(f_A)$.

Theorem 2.6 (Duality theorem for sets).   *Let* $(E_i, \mathfrak{A}_i)$ $i = 1, 2$ *be Polish spaces and let* $A \in \mathfrak{A}_1 \otimes \mathfrak{A}_2$ *then*

$$M(A) = \inf \left\{ P_1(A_1) + P_2(A_2);\ A_i \in \mathfrak{A}_i, A \subset A_1 \times E_2 \cup E_1 \times A_2 \right\} \quad (2.16)$$

and

$$m(A) = \sup \left\{ (P_1(A_1) + P_2(A_2) - 1)_+;\ A_i \in \mathfrak{A}_i, A \supset A_1 \times A_2 \right\}. \quad (2.17)$$

If A *is closed then the* A_i *can be restricted to the class of closed sets.*

Equations (2.16) and (2.17) say that for $n = 2$ one can restrict in the dual problem to two-valued functions. There are generalizations of the duality theorem to the case of multivariate and overlapping marginals as considered in Section 1.3. These can be imbedded in the following "generalized marginal problem":

Let on a general measure space $(E, \mathfrak{B})$ (not necessarily a product space) $\mathfrak{B}_i \subset \mathfrak{B}$, $1 \leq i \leq n$ be a finite system of sub-σ-algebras with a consistent system of probability measures $P_i \in \mathcal{M}^1(E, \mathfrak{B}_i)$, i.e. $P_i / (\mathfrak{B}_i \cap \mathfrak{B}_j) = P_j / (\mathfrak{B}_i \cap \mathfrak{B}_j)$ for all $i \neq j$. Define the "generalized Fréchet class" (respectively restriction class)

$$\mathcal{M} = \mathcal{M}((P_i)) = \left\{ P \in \mathcal{M}(E, \mathfrak{B});\ P / \mathfrak{B}_i = P_i, 1 \leq i \leq n \right\}. \quad (2.18)$$

We also consider the extended class $\mathrm{ba}(E, \mathfrak{B})$ of finitely additive measures on $(E, \mathfrak{B})$ with restrictions P_i, i.e. such that $P / \mathfrak{B}_i = P_i$.

$$\widetilde{\mathcal{M}} = \widetilde{\mathcal{M}}((P_i)) = \left\{ P \in \mathrm{ba}(E, \mathfrak{B});\ P / \mathfrak{B}_i = P_i, 1 \leq i \leq n \right\}. \quad (2.19)$$

Furthermore, we define

$$F := \left\{ \sum_{i=1}^{n} f_i;\ f_i \in \mathcal{L}^1(P_i, \mathfrak{B}_i), 1 \leq i \leq n \right\} = \bigoplus_{i=1}^{n} \mathcal{L}^1(\mathfrak{B}_i, P_i), \quad (2.20)$$

the direct sum of the $\mathfrak{B}_i$-measurable P_i-integrable real functions on E. F is a vector subspace of the vector space

$$\mathcal{L}^m = \left\{ \varphi \in \mathcal{L}(E, \mathfrak{B});\ \exists f \in F, \varphi \leq f \right\} \quad (2.21)$$

the set of measurable functions on $(E, \mathfrak{B})$ which are majorized by an element of F. By consistency the linear operator

$$T : F \to \mathbb{R}, \quad T \left(\sum_{i=1}^{n} f_i \right) := \sum_{i=1}^{n} \int f_i \, dP_i \quad (2.22)$$

is well defined on F. The following duality theorem is given in Rü (1991b). The method of proof is an extension of the method used in Rü (1979 and 1981a), Gaffke and Rü (1981) for the case $\mathcal{M}(P_1, \ldots, P_n)$. The first part of the following theorem has been established in more general form in Lembcke (1972, b) and Luschgy and Thomsen (1983).

Theorem 2.7 (Generalized duality theorem for $\widetilde{\mathcal{M}}$).

(a) "Marginal problem":

$$\widetilde{\mathcal{M}} \neq \phi \text{ if and only if } T \geq 0 \tag{2.23}$$

i.e. $f \in F$, $f \geq 0$ implies that $Tf \geq 0$.
(b) "Duality": For $\varphi \in \mathcal{L}^m$ we have

$$\widetilde{\mathcal{M}}(\varphi) := \sup\left\{\int \varphi \, dP; \; P \in \widetilde{\mathcal{M}}\right\} = I(\varphi) := \inf\left\{Th; \; h \in F, \varphi \leq h\right\}. \tag{2.24}$$

(c) If $I(\varphi) > -\infty$, then there exists a $P \in \widetilde{\mathcal{M}}$ with $\widetilde{\mathcal{M}}(\varphi) = \int \varphi \, dP$.

Proof.   (a) The direction "$\Rightarrow$" is obvious. For the converse direction note that I is sublinear on $\mathcal{L}^m$ and $If = Tf$ for $f \in F$. If S is a linear functional on $\mathcal{L}^m$, $S \leq I$ then for $f \in F$, $f \geq 0$ it holds that:

$$-Sf = S(-f) \leq I(-f) = \inf\{Th; \; -f \leq h, h \in F\} \leq T0 = 0.$$

Thus $S \geq 0$ and $S/F = T$.

By the Hahn–Banach Theorem there exists an extension S of T to $\mathcal{L}^m$ such that $S \leq U$. Riesz's representation theorem ensures the existence of an element $P \in ba(X, \mathcal{B})$ representing S. Since $S/F = T$ if follows that $P \in \widetilde{\mathcal{M}}$ and thus $\widetilde{\mathcal{M}} \neq \phi$.

(b, c) A corollary to the Hahn–Banach Theorem is the existence of an extension S with $S\varphi = I\varphi$ if $I\varphi > -\infty$. The element $P \in \widetilde{\mathcal{M}}$ representing S then yields the conclusions (b), (c) in case $I\varphi > -\infty$.
If $I\varphi = -\infty$, then also $\widetilde{\mathcal{M}}(\varphi) = -\infty$ and so (b) holds true in general.   □

Thus the duality theorem is true for a general class of functions φ if one admits finitely additive measures. Under the following assumptions (A1) and (A2) we get as a consequence a duality theorem for $M(\varphi)$.

(A1) $(E_i, \mathcal{B}_i, P_i)$ are compactly approximable, i.e. there exist compact set systems $\mathcal{E}_i \subset \mathcal{B}_i$ such that $P_i(B_i) = \sup\{P_i(E_i); \; E_i \subset B_i, E_i \in \mathcal{E}_i\}$, $1 \leq i \leq n$.
(A2) $(E, \mathcal{B})$ is a topological space with Borel σ-algebra $\mathcal{B}$ and $\mathfrak{R} = \mathfrak{R}(\bigcup_{i=1}^{n} \mathcal{B}_i)$, the algebra generated by $\mathcal{B}_i$, contains a countable basis of the topology.

Conditions (A1) and (A2) are satisfied in particular in the case of Polish spaces. Let $\mathcal{L}^1(E, \mathfrak{R}, P)$ denote the set of P integrable functions, where P is considered as a finitely additive measure on the algebra $\mathfrak{R}$ (cf. Dunford and Schwartz (1967,

Definition 17, p. 112)). Then we have a generalized duality theorem for the M-functional (see Rü (1991b)).

Theorem 2.8 (Generalized duality theorem for M).

(a) If (A1) holds true, then

$$M \neq \phi \text{ if and only if } T \geq 0. \tag{2.25}$$

Furthermore, duality $M(\varphi) = I(\varphi)$ holds for all $\varphi \in \bigcap_{P \in \mathcal{M}} \mathcal{L}^1(E, \mathcal{R}, P)$.

b) If (A1) and (A2) hold true, then duality $M(\varphi) = I(\varphi)$ holds for all $\varphi \in C_b(E)$.

Proof. Condition (A1) implies that any $P \in \widetilde{M}$ is compactly approximable on $\mathfrak{R}$. This allows us to use extension results ensuring that for $P \in \widetilde{M}$ and $\varphi \in \mathcal{L}^1(\mathfrak{R}, P)$ (respectively $\varphi \in C_b(E)$ if (A2) holds) there exists an element $P \in M$ such that $\int \varphi \, dP = \int \varphi \, d\widetilde{P}$. For details see Rü (1981a, 1991b). $\square$

There are two basic classes of applications of the generalized duality results. One of them concerns the case where $T_i : (E, \mathfrak{B}) \to (E_i, \mathfrak{B}_i)$ are measurable functions and $\mathfrak{B}_i = \sigma(T_i) \subset \mathfrak{B}$ are the induced σ-algebras. In this case $P_i(T_i \in B_i) = P_i^{T_i}(B_i)$ are uniquely given by the image distributions $P_i^{T_i}$. Thus the generalized Fréchet class $\mathcal{M}$ describes the class of all probability measures P on $(E, \mathfrak{B})$ such that

$$P^{T_i} = \mu_i := P_i^{T_i}, \quad 1 \leq i \leq n \tag{2.26}$$

i.e. we obtain the class of all probability measures with fixed distributions of certain functions $T_1, \ldots, T_n$.

A second interesting application is to the multivariate marginal problem with possibly overlapping marginals (see Chapter 6). On the product space $(E, \mathfrak{A}) = \bigotimes_{i=1}^n (E_i, \mathfrak{A})$ let for a given marginal system $(P_J)_{J \in \mathcal{E}}$, where $\mathcal{E} \subset \mathcal{P}(\{1, \ldots, n\})$, $\mathcal{M}_{\mathcal{E}} = \mathcal{M}_{\mathcal{E}}((P_J)_{J \in \mathcal{E}})$ denote the corresponding generalized Fréchet class. By $\mathfrak{B}_J$ we denote the sub-σ-algebra induced by $\mathfrak{A}_J = \bigotimes_{j \in J} \mathfrak{A}_j$ in $(E, \mathfrak{A})$, $J \in \mathcal{E}$.

Theorem 2.9. *If (A1) and (A2) hold, then for*

$$\varphi \in \bigcap_{P \in \mathcal{M}} \mathcal{L}^1(E, \mathfrak{R}, P) \cup C_b(E) \cup F(E)$$

duality holds, i.e.

$$\mathcal{M}_{\mathcal{E}}(\varphi) := \sup \left\{ \int \varphi \, dP; \ P \in \mathcal{M}_{\mathcal{E}} \right\}$$

$$= I_{\mathcal{E}}(\varphi) := \inf \left\{ \sum_{J \in \mathcal{E}} \int f_J \, dP_J; \ f_J \in \mathcal{L}^1(P_J), J \in \mathcal{E}, \bigoplus_{J \in \mathcal{E}} f_J \geq \varphi \right\}. \tag{2.27}$$

For the case $\varphi \in \bigcap_{P \in \mathcal{M}_\mathcal{E}} \mathcal{L}^1(E, \mathfrak{R}, P) \cup C_b(E)$ Theorem 2.9 is a corollary of Theorem 2.8 (see Rü (1984)). The case of upper-semicontinuous $\varphi \in F(E)$ is due to Kellerer (1988). An example in that paper shows that for lower-semicontinuous functions the duality theorem does not hold in general.

Remark 2.10 (Additional restrictions). Related existence and duality results under additional restrictions on the measures can be proved in a similar way. Let in the situation of Theorem 2.7 μ be a finite measure on $(E, \mathfrak{B})$ and define

$$\widetilde{\mathcal{M}}_\mu := \{P \in ba(E, \mathfrak{B}); \; P/\mathfrak{B}_i = P_i, 1 \leq i \leq n \text{ and } P \leq \mu\}, \qquad (2.28)$$

the marginal class with the additional restriction $P \leq \mu$. If the dual operator I is replaced by

$$I_\mu(\varphi) := \inf\left\{I(\varphi_0) + \int h_+ \, d\mu; \; \varphi_0 + h \geq \varphi\right\} \qquad (2.29)$$

then the existence and duality results

$$\widetilde{M}_\mu(\varphi) = I_\mu(\varphi) \quad \text{respectively } M_\mu(\varphi) = I_\mu(\varphi) \qquad (2.30)$$

can be established as in Theorems 2.7 and 2.8. Extensions to more general classes of restrictions are considered in Ramachandran and Rü (1997). ◊

For generalized marginal systems $(P_J)_{J \in \mathcal{E}}$ as in Theorem 2.9 there is an interesting extension of the marginal problem. We assume that $\Lambda \subset \mathcal{M}^1(E, \mathfrak{A})$ is a specified subset of the class of all probability measures on $(E, \mathfrak{A})$ and ask the question, whether there exists an element in Λ which has the marginals $P_J, J \in \mathcal{E}$. If for example

$$\Lambda = \left\{P \in \mathcal{M}^1(E, \mathfrak{A}); \; P\left(\sum_{i=1}^n x_i \geq c\right) \geq 1 - \alpha\right\} \qquad (2.31)$$

then the above question can be translated in the language of risk bounds: Does there exist in the case of measures with (known) marginals $(P_J)_{J \in \mathcal{E}}$ a measure P with value at risk of the joint portfolio at level α

$$\text{VaR}_\alpha\left(\sum_{i=1}^n X_i\right) \geq c. \qquad (2.32)$$

In option price theory we can use Λ to describe the restriction that option prices of certain derivatives $Y_1, \ldots, Y_m$ are known (from the market) i.e.

$$\Lambda = \left\{P \in \mathcal{M}^1(E, \mathfrak{A}); \; P^{Y_i} = \mu_i, 1 \leq i \leq m\right\} \qquad (2.33)$$

(considering E as our basic probability space here). Also under this restriction interesting questions on the worst case value at risk of the joint portfolio $\sum_{i=1}^{n} X_i$ or on the maximal price of some derivative $T(X)$ can be put in this framework (see for example Piterbarg (2011)).

For the following result we consider $M^1(E, \mathfrak{A})$ as a subset of $(C_b(E))^*$, the dual of $C_b(E)$). We assume:

(A)   (a) Λ is a convex and relatively closed subset of $\mathcal{M}_1(E, \mathfrak{A})$ w.r.t. weak $*$-topology on $\mathcal{M}^1(E, \mathfrak{A})$,

     (b) $\bigcup_{J \in \mathcal{E}} J = \{1, \ldots, n\}$.

We denote $F = \{\bigoplus_{J \in \mathcal{E}} f_J;\ f_J \in C_b(E_J, \mathfrak{A}_J)\}$. The following general existence result is due to Gaffke and Rü (1984, Theorem 2).

Theorem 2.11 (Existence with additional constraints). *If assumption (A) holds, there exists an element* $P \in M^1(E;\ \mathfrak{A})$ *with marginals* $(P_J)_{J \in \mathcal{E}}$ *and such that* $P \in \Lambda$ *i.e.* $M_{\mathcal{E}} \cap \Lambda \neq \phi$ *if and only if*

$$\inf \left\{ \int \bigoplus_{J \in \mathcal{E}} f_J\, d\mu;\ \mu \in \Lambda \right\} \leq \sum_{J \in \mathcal{E}} \int f_J\, dP_J,\ \text{for all } \bigoplus_{J \in \mathcal{E}} f_J \in F. \quad (2.34)$$

Typically condition (2.34) will be difficult to verify, but it may serve to construct counterexamples and bounds.

Several different types of restrictions such as the assumption that the distribution function F of P is in $\mathcal{F}(F_1, \ldots, F_n)$ and for some subset S in $E = \mathbb{R}^n$ the bounds $F(x) \leq F_0(x),\ x \in S$ respectively $F(x) \geq F_0(x),\ x \in S$ holds true for some distribution function F_0 are considered in Rachev and Rü (1994). Applications of this type of restrictions to option pricing are given in Tankov (2011).

2.2 Fréchet Bounds Comonotonicity and Extremal Risk

As a consequence of the duality results in the previous section we obtain a series of interesting applications to describe the maximal influence of dependence on a risk functional. Some interesting risk functionals that we consider in this and in the following sections are the following:

$$\varphi(x) = \left(\sum_{i=1}^{n} x_i - K \right)_+ \to \text{excess of loss}$$

$$\varphi_t(x) = \mathbb{1}_{[t,\infty)} \left(\sum_{i=1}^{n} x_i \right) \to \text{value of risk}$$

$$\psi(x) = \max_{i \leq n} x_i \to \text{maximal risk}.$$

The maximal expectation of these functions w.r.t. all possible dependence structures describes the worst case risk. For the function φ this yields the worst case excess of loss of the joint portfolio. The worst case value at risk is given by the inverse of the expectation of φ_t, $\mathrm{VaR}_\alpha(P) = (E_P\varphi_t)^{-1}(\alpha)$. Finally ψ is related to the question of worst case maximal risk.

The following bounds for risk functionals extend in particular the classical Hoeffding–Fréchet bounds. These characterize the class of all distribution functions F with marginal distribution functions $F_1, \ldots, F_n$ by lower and upper bounds F_-, F_+ for F.

Theorem 2.12 (Classical Hoeffding–Fréchet bounds). *Let $F \in \mathcal{F}(F_1, \ldots, F_n)$ be an n-dimensional distribution function with marginal distribution functions $F_1, \ldots, F_n$. Then*

(a) $F_-(x) := \left(\sum_{i=1}^{n} F_i(x_i) - (n-1) \right)_+ \le F(x) \le F_+(x) := \min_{1 \le i \le n} F_i(x_i).$ (2.35)

   F_-, F_+ *are called lower respectively upper Fréchet bounds (respectively Hoeffding–Fréchet bounds).*
(b) F_+ *is an n-dimensional distribution function with marginals $F_1, \ldots, F_n$, i.e. $F_+ \in \mathcal{F}(F_1, \ldots, F_n)$. For $n = 2$ also $F_- \in \mathcal{F}(F_1, F_2)$.*
(c) *Let F be a distribution function on $\mathbb{R}^n$. Then*

$$F \in \mathcal{F}(F_1, \ldots, F_n) \Leftrightarrow F_- \le F \le F_+.$$

In particular there exists for any n a largest distribution function with marginals F_i namely the upper Hoeffding–Fréchet bound F_+. For $n = 2$ there exists also a smallest distribution function with marginals F_i, the lower Hoeffding–Fréchet bound F_-. The upper and lower Hoeffding–Fréchet bounds have an intuitive representation in terms of random variables.

Definition 2.13 (Comonotonicity, countermonotonicity). Let $F_1, \ldots, F_n$ be one-dimensional distribution functions and let $U \sim U(0, 1)$ be a random variable uniformly distributed on $(0, 1)$.

(a) The random vector

$$X^c := \left(F_1^{-1}(U), \ldots, F_n^{-1}(U) \right),$$ (2.36)

   with $F_i^{-1} = \inf\{y; F_i(y) \ge x\}$ the generalized inverse of F_i, is called "comonotonic" risk vector.
(b) For $n = 2$

$$X_c := \left(F_1^{-1}(U), F_2^{-1}(1-U) \right)$$ (2.37)

   is called "countermonotonic" risk vector.

Theorem 2.14 (Fréchet bounds and comonotonicity).
Let $X = (X_1, \ldots, X_n)$ *be a risk vector defined on* $(\Omega, \mathfrak{A}, P)$ *with distribution function* $F \in \mathcal{F}(F_1, \ldots, F_n)$.

(a) The following are equivalent:

1. $F = F_+$, *F is the upper Fréchet bound.*
2. *F is the distribution function of a comonotonic risk vector* X^c.
3. *There exists a real random variable Z and monotonically nondecreasing functions*

$$f_1, \ldots, f_n \text{ such that } X \stackrel{d}{=} (f_1(Z), \ldots, f_n(Z)). \tag{2.38}$$

4. *The* X_i, $1 \le i \le n$, *are similarly ordered, i.e. for all* $i \ne j$ *it holds that:*

$$\big(X_i(\omega) \le X_i(\omega') \Rightarrow X_j(\omega) \le X_j(\omega')\big) \text{ a.s.} \tag{2.39}$$

with respect to the product measure $P \otimes P$ *on* Ω^2.

(b) For $n = 2$ *the following are equivalent:*

1. $F = F_-$.
2. *F is the distribution function of a countermonotonic risk vector* X_c.
3. *There exists a real random variable Z such that*

$$X \stackrel{d}{=} (f_1(Z), f_2(Z))$$

with f_1 *monotonically nondecreasing and* f_2 *monotonically nonincreasing.*
4. X_1 *and* X_2 *are oppositely ordered, i.e. a.s. with respect to* $P \otimes P$ *it holds that:*

$$X_1(\omega) \le X_1(\omega') \Rightarrow X_2(\omega) \ge X_2(\omega').$$

Theorem 2.14 is a long-known result. It is not easy to state the original sources. For a proof see for example Cuesta-Albertos, Rü, and TueroDiaz (1993). Theorem 2.14 implies that for any $n \ge 2$ the Fréchet bounds F_+ and for $n = 2$ also the lower Fréchet bounds F_- are sharp bounds. It was proved in Dall'Aglio (1972) that the lower Fréchet bounds F_- only in exceptional cases are distribution functions and these cases are explicitly described in that paper. Sharpness of the lower Hoeffding–Fréchet bounds holds true however in general as follows from the following generalized version of Fréchet bounds.

Theorem 2.15 (Sharpness of Fréchet bounds, Rü (1980)). *Let* $(E_i, \mathfrak{A}_i)$, $1 \le i \le n$, *be Polish spaces,* $P_i \in \mathcal{M}^1(E_i, \mathfrak{A}_i)$ *and* $A_i \in \mathfrak{A}_i$, $1 \le i \le n$. *Then for any* $P \in \mathcal{M}(P_1, \ldots, P_n)$ *it holds that:*

$$\left(\sum_{i=1}^n P_i(A_i) - (n-1)\right)_+ \le P(A_1 \times \cdots \times A_n) \le \min\{P_i(A_i); \ 1 \le i \le n\} \tag{2.40}$$

and the upper and lower bounds in (2.40) are sharp.

Proof. The statement of the theorem is equivalent to

$$M(A_1 \times \cdots \times A_n) = \min(P_i(A_i))$$

$$m(A_1 \times \cdots \times A_n) = \left(\sum_{i=1}^{n} P_i(A_i) - (n-1) \right)_+. \qquad (2.41)$$

The bounds in (2.41) are a consequence of the duality theorem (Theorem 2.3). It is by some reduction arguments not difficult to show that

$$I(A_1 \times \cdots \times A_n) = \min(P_i(A_i))$$

$$\text{and} \quad i(A_1 \times \cdots \times A_n) = \left(\sum_{i=1}^{n} P_i(A_i) - (n-1) \right)_+.$$

An alternative proof is by induction. For $n = 2$ it follows from Strassen's theorem (Theorem 2.6) that

$$M(A_1 \times A_2) = \inf \left\{ P_1(B_1) + P_2(B_2); \ A_1 \times A_2 \subset B_1 \times E_2 \cup E_1 \times B_2 \right\}$$

$$= \min \left\{ P_1(A_1), P_2(A_2) \right\}.$$

For the induction step note that

$$\mathcal{M}(P_1, \dots, P_{n+1}) = \bigcup_{Q \in \mathcal{M}(P_1, \dots, P_n)} M(Q, P_{n+1}). \qquad (2.42)$$

Therefore, by the induction hypothesis we get

$$M(A_1 \times \cdots \times A_n) = \sup_{Q \in \mathcal{M}(P_1, \dots, P_n)} \sup_{P \in M(Q, P_{n+1})} P(A_1 \times \cdots \times A_n \times A_{n+1})$$

$$= \sup_{Q \in \mathcal{M}(P_1, \dots, P_n)} \min \left\{ Q(A_1 \times \cdots \times A_n), P_{n+1}(A_{n+1}) \right\}$$

$$= \min \left\{ M(A_1 \times \cdots \times A_n), P_{n+1}(A_{n+1}) \right\}$$

$$= \min \left\{ P_i(A_i); \ 1 \le i \le n+1 \right\}.$$

The proof of the lower bounds is similar.   □

Remark 2.16. (a) In the case that $E_i = \mathbb{R}^1$, $A_i = (-\infty, x_i]$ Theorem 2.15 implies sharpness of the classical lower Hoeffding–Fréchet bounds in (2.35) for any $n \ge 2$. This sharpness result of the lower Hoeffding–Fréchet bounds was established first in the above-mentioned paper Rü (1980). Note that since $\mathcal{M}(P_1, \dots, P_n)$ is weakly compact the upper Fréchet bounds in (2.40) are attained when $A_i = \overline{A_i}$ are closed or when $P_i(\partial A_i) = P_i(\overline{A_i} \setminus \mathring{A_i}) = 0$, $1 \le i \le n$. A similar result also holds for the lower Fréchet bound.

(b) **Classical Fréchet problem**

The following result, known as the "classical Fréchet problem", concerns the question whether to a given finite measure μ on the product space $(E_1 \times E_2, \mathfrak{A}_1 \otimes \mathfrak{A}_2)$ there exists an element P in the Fréchet class, $P \in \mathcal{M}(P_1, P_2)$, such that $P \leq \mu$. The answer was given in Fréchet (1951) and Kellerer (1964).

Theorem 2.17 (Classical Fréchet problem). *Let $(E_i, \mathfrak{A}_i)$ be Polish spaces. There exists a $P \in \mathcal{M}(P_1, P_2)$ such that $P \leq \mu$ if and only if*

$$\mu(A_1 \times A_2) \geq (P_1(A_1) + P_2(A_2) - 1)_+. \tag{2.43}$$

$\Diamond$

Thus existence holds true for $n = 2$ if and only if μ is on product sets larger than the lower Fréchet bound. This result indicated sharpness of the lower Fréchet bound in general and was a main motivation in Rü (1980) to establish Theorem 2.15. Note however that the generalization of Theorem 2.17 to the case $n \geq 2$ is no longer true (see Gaffke and Rü (1984)).

As a corollary of Theorem 2.15 the following sharp bounds are obtained in marginal classes.

Corollary 2.18 (Multivariate marginals). *Let $F_i \in \mathcal{F}_{k_i}$ be k_i-dimensional distribution functions, $1 \leq i \leq n$. Then for any $x_i \in \mathbb{R}^{k_i}$ and $F \in \mathcal{F}(F_1, \ldots, F_n)$ holds*

$$\left(\sum_{i=1}^{n} F_i(x_i) - (n-1) \right)_+ \leq F(x) \leq \min_{1 \leq i \leq n} F_i(x_i) \tag{2.44}$$

and the bounds are sharp.

Corollary 2.19. *Let $F_i \in \mathcal{F}$ be one-dimensional distribution functions, $1 \leq i \leq n$, and let X_i be random variables with $X_i \sim F_i$. Then*

(a) Maximal concentration

$$\sum_{i=1}^{n} (F_i(b_i) - F_i(a_i)) \leq P(X_i \in (a_i, b_i], 1 \leq i \leq n)$$

$$\leq \min \{ F_i(b_i) - F_i(a_i); \ 1 \leq i \leq n \}. \tag{2.45}$$

(b) Maximal risk

$$H^-(t) := \left(\sum_{i=1}^{n} F_i(t) - (n-1) \right)_+$$

$$\leq P \left(\max_{1 \leq i \leq n} X_i \leq t \right) \leq H^+(t) := \min \{ F_i(t); \ 1 \leq i \leq n \} \tag{2.46}$$

and the bounds in (a) and (b) are sharp.

We next consider from our three specific risk functionals the "maximal risk functional". Let $X^c = (F_1^{-1}(U), \ldots, F_n^{-1}(U))$ be a comonotonic vector, then by Theorem 2.14 the distribution function of $M_n(X^c)$ the maximum of the components of X^c is given by

$$
\begin{aligned}
P(M_n(X^c) \le t) &= P(X_1^c \le t, \ldots, X_n^c \le t) \\
&= F_+(t, \ldots, t) = \min\{F_i(t); \ 1 \le i \le n\} \\
&= H^+(t).
\end{aligned}
$$

Thus for the comonotonic vector X^c the distribution function of the maximum of the components is largest possible. Lai and Robbins (1976, 1978) have constructed for continuous F_i recursively a random vector $\widetilde{X}$ with marginal distribution functions F_i such that the distribution function of $M_n(\widetilde{X})$ is given by H_- (for general F_i see Rü (1980)). This random vector $\widetilde{X}$ is called a "maximally dependent" random vector. It can be constructed by iteratively determining $\widetilde{X}_j$ as countermonotonic to M_{j-1} the maximum of the first $(j-1)$ random variables $\widetilde{X}_i$. The comonotonic random vector has the largest distribution function of the maximum, the maximally dependent vector has the smallest distribution function of the maximum. This implies w.r.t. the stochastic order $\le_{\mathrm{st}}$ the following result.

Corollary 2.20 (Maximal risk, Lai and Robbins (1976, 1978), Rü (1980)). *For any random vector X with marginal distribution functions F_i it holds that:*

$$
M_n(X^c) \le_{\mathrm{st}} M_n(X) \le_{\mathrm{st}} M_n(\widetilde{X}), \tag{2.47}
$$

where X^c respectively $\widetilde{X}$ are the comonotonic respectively the maximally dependent risk vector.

The comonotonic risk vector X^c has the smallest maximal risk. The largest maximal risk is realized by the maximally dependent risk vector $\widetilde{X}$. An interesting result of Lai and Robbins (1978) shows that the maximum of the maximally dependent risk vector $M_n(\widetilde{X})$ in the iid case is close (in an asymptotic sense) to $M_n(X^\perp)$, where $X^\perp$ is the vector with independent components, $X_i^\perp \sim F_i$. Thus the independent case is close to the worst case in the sense of maximal risk. In particular it holds under an integrability condition that

$$
EM_n(X^\perp) \sim EM_n(\widetilde{X}) \sim F^{-1}\left(1 - \frac{1}{n}\right) \tag{2.48}
$$

for iid sequences $F = F_1 = F_2 = \ldots$ where $a_n = F^{-1}\left(1 - \frac{1}{n}\right)$ is the usual normalization for the maximum law (for a more precise formulation see Lai and Robbins (1978)).

The simplest way to derive the upper bound in (2.47) is the following typical duality argument: Note that in case $F_1 = \cdots = F_n$

$$M_n(X) = \max_{i \leq n} X_i \leq v + \sum_{i=1}^{n} (X_i - v)_+ . \tag{2.49}$$

Equality holds in (2.49) if and only if for some "splitting point" v^* the sets $\{X_i \geq v^*\}$ are pairwise disjoint and $\bigcup_{i=1}^{n} \{X_i \geq v^*\} = \Omega$. The maximally dependent random vector is constructed such that there is a splitting point v^* as above. In the inhomogeneous case one can use a generalized version of (2.49)

$$M_n(X) \leq \sum_{i=1}^{n} v_i + \sum_{i=1}^{n} (X_i - v_i)_+ \tag{2.50}$$

where again we need pairwise disjointness of $\{X_i \geq v_i\}$ and $\bigcup_{i=1}^{n} \{X_i \geq v_i\} = \Omega$. This property holds true for the maximally dependent vector $\widetilde{X}$.

Chapter 3
Convex Order, Excess of Loss, and Comonotonicity

In this chapter we derive that comonotonicity is the worst case dependence structure concerning convex order of the joint portfolio $\sum_{i=1}^{n} X_i$ or equivalently of the excess of loss. Two different approaches to this result are given. The first approach due to Meilijson and Nadas (1979) is based on a simple duality argument and uses just a monotonicity property of the inverse distribution function. The second approach in Rü (1983) is based on classical results on rearrangements and Schur ordering going back to Hardy et al. (1929). The connection is obtained by a rearrangement representation of the Fréchet class.

3.1 Convex Order and Comonotonicity

Comonotonicity specifies the strongest possible form of positive dependence between n random variables $X_1, \ldots, X_n$ concerning the pointwise ordering of distribution functions (see Theorem 2.12). Concerning the evaluation of risk bounds a fundamental result says that the comonotonic dependence structure is also the worst case dependence concerning the excess of loss of the joint portfolio $\sum_{i=1}^{n} X_i$. This result can equivalently be formulated in terms of the convex order $\leq_{\mathrm{cx}}$ respectively the increasing convex order $\leq_{\mathrm{icx}}$.

Definition 3.1 (Convex ordering). For two real random variables X, Y we define $X \leq_{\mathrm{cx}} Y$ "X is smaller than Y in convex order" if

$$E\varphi(X) \leq E\varphi(Y) \tag{3.1}$$

for all convex functions φ such that the integrals exist.

Similarly, $X \leq_{\mathrm{icx}} Y$ "X is smaller in increasing convex order than Y" if (3.1) holds for all increasing integrable convex functions.

L. Rüschendorf, *Mathematical Risk Analysis*, Springer Series in Operations Research and Financial Engineering, DOI 10.1007/978-3-642-33590-7_3,
© Springer-Verlag Berlin Heidelberg 2013

Remark 3.2. Convex ordering can be described by means of the "excess of loss function" defined for a random variable X by

$$\pi_X(t) := E(X - t)_+, \quad t \in \mathbb{R}. \tag{3.2}$$

By partial integration one gets

$$\pi_X(t) = \int_t^\infty \overline{F}(s)\,ds, \tag{3.3}$$

where $\overline{F}(s) = 1 - F(s)$ is the survival function of F. Excess of loss functions $\pi = \pi_X$ are decreasing and convex and satisfy

$$\lim_{t \to \infty} \pi(t) = 0 \text{ and } \lim_{t \to -\infty} \pi(t) \text{ exists and is equal to } EX. \tag{3.4}$$

Conversely any decreasing convex function π satisfying (3.4) is an excess of loss function of a random variable Z with distribution function

$$F_Z(t) = 1 + \pi'_+(t), \tag{3.5}$$

$\pi'_+(t)$ being the right derivative at t (see Müller and Stoyan (2002, Theorem 1.5.10)). Equations (3.3) and (3.5) imply in particular that

$$X \leq_{\text{st}} Y \text{ if and only if } \pi_Y - \pi_X \text{ is decreasing.} \tag{3.6}$$

$\Diamond$

Some basic properties of the convex order are collected in the following theorem.

Theorem 3.3 (Convex ordering criteria). *Let X, Y be real, integrable random variables with distribution functions F, G, then:*

(a) $X \leq_{\text{icx}} Y$ if and only if $\pi_X \leq \pi_Y$.
(b) If $X \leq_{\text{icx}} Y$ and $EX = EY$, then $X \leq_{\text{cx}} Y$.
*(c) **Cut criterion** (Karlin and Novikoff (1963))*
     If for some $t_0 \in \mathbb{R}$ $F(t) \leq G(t)$ for $t < t_0$ and $F(t) \geq G(t)$ for $t > t_0$ and if $EX = EY$, then

$$X \leq_{\text{cx}} Y.$$

(d) $X \leq_{\text{icx}} Y$, if and only if there exists a random variable Z such that

$$X \leq_{\text{st}} Z \leq_{\text{cx}} Y.$$

Proof. (a) The proof follows from an approximation argument using the fact that on bounded domains increasing convex functions can be uniformly approximated by positive linear combinations of angle functions $(x - t_i)_+$.

(b) Follows directly from (a).

(c) From the representation of the excess of loss functions π_X, π_Y in (3.2) we obtain

$$\pi_Y(t) - \pi_X(t) = \int_t^\infty (\overline{G}(x) - \overline{F}(x))dx = \int_t^\infty (F(x) - G(x))dx. \quad (3.7)$$

For $t \geq t_0$ it holds that by the cut condition $\pi_Y(t) - \pi_X(t) \geq 0$. Furthermore, $\lim_{t \to -\infty} (\pi_Y(t) - \pi_X(t)) = EY - EX = 0$ and thus for $t < t_0$

$$\pi_Y(t) - \pi_X(t) = (EY - EX) + \int_{-\infty}^t (G(x) - F(x))dx \geq 0$$

again by the cut criterion. Therefore, (a) and (b) imply $X \leq_{cx} Y$.

(d) Defining $\pi(t) = \max\{\pi_X(t), EY - t\}$ one finds that π is convex, decreasing,

$$\lim_{t \to \infty} \pi(t) = 0 \text{ and } \lim_{t \to -\infty} (\pi(t) + t) = EY. \quad (3.8)$$

Therefore by (3.4) and (3.5) there exists a random variable Z such that π is the excess of loss function of Z, i.e. $\pi = \pi_Z$. Since $\pi_Z(t) - \pi_X(t)$ is decreasing to zero as $t \to \infty$ we get by (3.6) that

$$X \leq_{st} Z.$$

Further $\pi_Z(t) \leq \pi_Y(t)$, $t \in \mathbb{R}^1$ and thus as a consequence of (a), (b) and (3.8) we obtain

$$Z \leq_{cx} Y. \qquad \square$$

Remark 3.4. (a) As a consequence of Theorem 3.3 the increasing convex order $\leq_{icx}$ is equivalent to the "stop-loss order" defined by the comparison of the excess of losses

$$E(X - t)_+ \leq E(Y - t)_+, \quad t \in \mathbb{R}.$$

Obviously $X \leq_{icx} Y$ is equivalent to $Y \leq_{dcv} X$, the ordering w.r.t. decreasing concave functions. This ordering is called in the economics literature "second order stochastic dominance" and denoted as

$$Y \leq_{ssd} X.$$

(b) The cut criterion in (c) can be formulated by saying that $S(G - F) = 1$ with sign sequence $(+, -)$. Here $S(G - F)$ counts the sign changes of $G - F$. In case of existing densities f, g we obtain a sufficient criterion by the condition $S(g - f) = 2$ with sign sequence $(+, -, +)$.

(c) The proof of Theorem 3.3(c) implies more generally:
If $EX \leq EY$ and the cut criterion holds, then

$$X \leq_{icx} Y. \quad (3.9)$$

(d) For several further properties on convex ordering we refer to Shaked and
    Shanthikumar (1994) or to Müller and Stoyan (2002).   ◊

The following is "the" basic result concerning risk bounds. It describes the role
of the comonotonic vector as a worst case distribution w.r.t. the excess of loss
risk functional. We describe two approaches to this result. The first approach due
to Meilijson and Nadas (1979) is based on a simple duality argument. The worst
case excess of loss is described as a semilinear program. The second approach
in Rü (1983) relates this convex ordering result to some early developments on
rearrangements and Schur order in analysis.

**Theorem 3.5 (Comonotonic vector and convex order, Meilijson and Nadas
(1979)).** *Let $(X_1, \ldots, X_n)$ be a random vector with marginal distribution functions
F_i, $1 \leq i \leq n$. Let X^c be a corresponding comonotonic vector with $X_i^c \sim F_i$.
Then:*

(a)

$$\sum_{i=1}^{n} X_i \leq_{\mathrm{cx}} \sum_{i=1}^{n} X_i^c, \tag{3.10}$$

*i.e. the portfolio of the comonotonic vector is the worst case portfolio w.r.t. the
convex order.*

(b)

$$E\left(\sum_{i=1}^{n} X_i - t\right)_+ \leq E\left(\sum_{i=1}^{n} X_i^c - t\right)_+ = \Psi_+(t) \tag{3.11}$$

where

$$\Psi_+(t) := \inf_{v=(v_1,\ldots,v_n)} \left\{ \left(\sum_{i=1}^{n} v_i - t\right)_+ + \sum_{i=1}^{n} E(X_i - v_i)_+ \right\}$$

$$= \inf_{\sum_{i=1}^{n} v_i = t} \sum_{i=1}^{n} E(X_i - v_i)_+.$$

Proof. (b) For any vector v holds

$$\left(\sum_{i=1}^{n} X_i - t\right)_+ \leq \sum_{i=1}^{n} (X_i - v_i)_+ + \left(\sum_{i=1}^{n} v_i - t\right)_+. \tag{3.12}$$

This implies that

$$E\left(\sum_{i=1}^{n} X_i - t\right)_+ \leq \Psi_+(t). \tag{3.13}$$

Consider the comonotonic vector $X^c = (F_1^{-1}(U), \ldots, F_n^{-1}(U))$ and define
$u_0 \in (0, 1)$ such that

$$\sum_{i=1}^{n} F_i^{-1}(u_0-) \le t \le \sum_{i=1}^{n} F_i^{-1}(u_0). \qquad (3.14)$$

Let

$$v_i \in [F_i^{-1}(u_0-), F_i^{-1}(u_0)] \text{ such that } \sum v_i = t \qquad (3.15)$$

then for any $u \in (0, 1)$ either $u \le u_0$ and then $F_i^{-1}(u) \le v_i$, $1 \le i \le n$ or $u \ge u_0$ and then $F_i^{-1}(u) \ge v_i$, $1 \le i \le n$.

This geometric property of the monotone functions F_i^{-1} implies that

$$\left(\sum_{i=1}^{n} F_i^{-1}(u) - t \right)_+ = \left(\sum_{i=1}^{n} (F_i^{-1}(u) - v_i) \right)_+$$

$$= \sum_{i=1}^{n} (F_i^{-1}(u) - v_i)_+. \qquad (3.16)$$

Thus for the comonotonic vector X^c and for v_i as chosen in (3.15) we have by (3.13)

$$E \left(\sum_{i=1}^{n} F_i^{-1}(U) - t \right)_+ = \sum_{i=1}^{n} E(F_i^{-1}(U) - v_i)_+$$

$$= \Psi_+(t). \qquad (3.17)$$

Therefore, for this choice of v_i equality holds in (3.12). In particular the proof shows that one can restrict the optimization in (3.11) to vectors v with $\sum_{i=1}^{n} v_i = t$.

(a) follows from (b) and the convex ordering criterion in Theorem 3.3. □

The interesting equality in (3.16) will be of use for several ordering properties considered later in this book.

3.2 Schur Order and Rearrangements

For the second approach to the basic excess of loss Theorem 3.5 we make use of the notion of Schur order $\prec_S$ which is an ordering of the dispersiveness of two random vectors. The Schur order on $\mathbb{R}^n$ was introduced in analysis by Schur, Muirhead, Hardy, Littlewood, and Pólya and Karamata around 1930. An extensive presentation of the stochastic Schur order can be found in Marshall and Olkin (1979) and Shaked and Shanthikumar (1994).

Definition 3.6 (Schur order on $\mathbb{R}^n$). For $a, b \in \mathbb{R}^n$ define

$$a \preceq_S b \text{ if } \sum_{i=1}^{k} a_{(i)} \le \sum_{i=1}^{k} b_{(i)}, \quad 1 \le k \le n \qquad (3.18)$$

where $a_{(1)} \geq \cdots \geq a_{(n)}$ are the components arranged in decreasing order. Further, define

$$a \prec_S b \text{ "a is smaller than b in Schur order" if}$$

$$a \preceq_S b \text{ and if } \sum_{i=1}^n a_i = \sum_{i=1}^n b_i. \tag{3.19}$$

The Schur order $\prec_S$ is also called "majorization order" and $\preceq_S$ "weak majorization order".

Smaller vectors in Schur order $\prec_S$ have less variation in their components. The vectors with identical components are the minimal vectors w.r.t. $\prec_S$. There is a precise way to describe this order by averaging processes which is based on doubly stochastic matrices.

Definition 3.7 (Doubly stochastic matrix). An $n \times n$ matrix $\Pi = (\pi_{i,j}) \in \mathbb{R}_+^{n \times n}$ is called a "doubly stochastic matrix" if for all $i, j \leq n$

$$\sum_{l=1}^n \pi_{i,l} = \sum_{k=1}^n \pi_{k,j} = 1.$$

Let $\mathcal{D}_n$ denote the class of all doubly stochastic matrices of order n.

If $a = \Pi b$ for some $b \in \mathbb{R}^n$ and $\Pi \in \mathcal{D}_n$, then the components of a are convex combinations of the components of b and it is easy to see that $a \prec_S b$. Also by the definition of convexity it is immediate that for any convex function $f : \mathbb{R}^1 \to \mathbb{R}^1$, $a \prec_S b$ implies that

$$\sum_{i=1}^n f(a_i) \leq \sum_{i=1}^n f(b_i). \tag{3.20}$$

In fact by classical results of Hardy et al. (1929) and Karamata (1932) Schur ordering can be characterized by these properties.

Theorem 3.8 (Characterization of Schur order). *For any $a, b \in \mathbb{R}^n$ the following properties are equivalent:*

(a) $a \prec_S b$.
(b) There exists some $\Pi \in \mathcal{D}_n$ such that

$$\Pi b = a. \tag{3.21}$$

(c) For all convex functions $f : \mathbb{R}^1 \to \mathbb{R}^1$ it holds that:

$$\sum_{i=1}^n f(a_i) \leq \sum_{i=1}^n f(b_i). \tag{3.22}$$

Definition 3.9 (Schur convex function). A function $f : \mathbb{R}^n \to \mathbb{R}$ is called "Schur convex" if it is increasing w.r.t. the Schur order $\prec_S$, i.e.

$$a \prec_S b \text{ implies } f(a) \le f(b). \tag{3.23}$$

Remark 3.10. (a) Theorem 3.8 implies Schur convexity of any function h of the form $h(x) = \sum_{i=1}^n f(x_i)$ with f convex. Conversely, assume that $h(x) = \sum_{i=1}^n f(x_i)$ is increasing w.r.t. $\prec_S$. Then define for $a \in \mathbb{R}^n$ the vector b by $b = \left(\frac{1}{n}\sum_{i=1}^n a_i\right) \cdot \mathbb{1} = \left(\frac{1}{n}\sum_{i=1}^n a_i, \ldots, \frac{1}{n}\sum_{i=1}^n a_i\right)$.
Then $b \prec_S a$ and thus $h(b) \le h(a)$. This is equivalent to convexity of f:

$$f\left(\frac{1}{n}\sum_{i=1}^n a_i\right) \le \frac{1}{n}\sum_{i=1}^n f(a_i). \tag{3.24}$$

Thus $h(x) = \sum_{i=1}^n f(x_i)$ is Schur convex if and only if f is convex.
(b) From (3.21) it follows by simple arguments that symmetric convex functions f, as for example $f(x) = \sum_{i=1}^n f(x_i)$ with f convex, are Schur convex. ◇

The integral order on the class of probability measures $M^1(\mathbb{R}^n, \mathcal{B}^n)$ w.r.t. the class of Schur convex functions is called "Schur convex order" on $M^1(\mathbb{R}^n, \mathcal{B}^n)$ or stochastic Schur order. Define for $P, Q \in M^1(\mathbb{R}^n, \mathcal{B}^n)$

$$P \prec_S Q :\Leftrightarrow \int f dP \le \int f dQ \text{ for all Schur convex, integrable } f. \tag{3.25}$$

As a consequence of Strassen's ordering theorem (see Theorem 1.18) we obtain

Corollary 3.11 (Schur ordering on $M^1(\mathbb{R}^n, \mathcal{B}^n)$). *For any $P, Q \in M^1(\mathbb{R}^n, \mathcal{B}^n)$ the following are equivalent:*

$$P \prec_S Q$$

⟺ *There exist random vectors $X \sim P, Y \sim Q$ such that*

$$X \prec_S Y \text{ a.s.}$$

⟺ *There exist a random vector $Y \sim Q$ and a random doubly stochastic matrix $\Pi \in \mathcal{D}_n$ such that*

$$\Pi Y \sim P.$$

An extension of the Schur order on $\mathbb{R}^n$ can be given to measurable functions $f, g : (0, 1) \to \mathbb{R}^1$. It is based on the notion of rearrangements of functions.

Definition 3.12 (Rearrangements). Let $f, g : (0, 1) \to \mathbb{R}^1$ be measurable functions. Then g is called a "rearrangement" of f if for all $t \in \mathbb{R}^1$

$$\lambda(\{f \le t\}) = \lambda(\{g \le t\}), \tag{3.26}$$

where λ is the Lebesgue measure on $(0, 1)$. We use the notation $f \sim_r g$, when f is a rearrangement of g.

In probabilistic terms f, g are rearrangements of each other if they have the same distribution on the probability space $((0, 1), (0, 1)\mathfrak{B}^1; \lambda)$. Rearrangements of functions were introduced in Hardy et al. (1929). They have important applications in many parts of analysis and were studied intensively in Luxemburg (1967), Chong and Rice (1971), Day (1972) and Chong (1974). In generalization to the discrete case the following basic results are not difficult to establish and are given in these papers.

Theorem 3.13 (Rearrangement inequalities). *Let* $f, g : (0, 1) \to \mathbb{R}^1$ *be measurable.*

(a) There exist (essentially w.r.t. λ unique) increasing and decreasing rearrangements f^*, f_* of f.

(b) If $fg \in L^1(\lambda)$ then

$$\int f^* g_* \, d\lambda \le \int fg \, d\lambda \le \int f^* g^* \, d\lambda. \qquad (3.27)$$

(c) $\int f^* g_* \, d\lambda = \int f_* g^* \, d\lambda$ and $\int f^* g^* \, d\lambda = \int f_* g_* \, d\lambda$.

Considering f as a random variable on $((0, 1), (0, 1)\mathfrak{B}^1, \lambda)$ the increasing and decreasing rearrangements f^*, f_* are identified in the following proposition.

Proposition 3.14 (Rearrangements and distribution function). *Let* $f : (0, 1) \to \mathbb{R}^1$ *be a measurable function. Then the essentially unique increasing and decreasing rearrangements of f are given by*

$$f^*(u) = F^{-1}(u), \quad f_*(u) = F^{-1}(1 - u), \qquad (3.28)$$

where $F^{-1}(u) = \inf\{t \in \mathbb{R}^1; \ F(t) \ge u\}$, $u \in (0, 1)$ is the generalized inverse of the distribution function F of λ^f.

Remark 3.15. Considering f, g in Theorem 3.13 as random variables on $(0, 1)$, inequality (3.27) and Proposition 3.14 imply the Hoeffding–Fréchet bounds for random variables defined on $((0, 1), (0, 1)\mathfrak{B}^1, \lambda)$. $\lozenge$

Rearrangement inequalities are closely connected with a generalization of the Schur order $\prec_S$ on $\mathbb{R}^n$ (see Definition 3.6) to integrable functions in $L^1(\lambda)$.

Definition 3.16 (Schur order on $L^1(\lambda)$). For $f, g : (0, 1) \to \mathbb{R}^1$, $f, g \in L^1(\lambda)$ we define

$$f \preceq_S g \text{ if } \int_x^1 f^*(t) d\lambda(t) \le \int_x^1 g^*(t) d\lambda(t) \text{ for all } x \in (0, 1), \qquad (3.29)$$

$$f \prec_S g \text{ if } f \preceq_S g \text{ and } \int_0^1 f(t) d\lambda(t) = \int_0^1 g(t) d\lambda(t). \qquad (3.30)$$

The partial order $\prec_S$ on $L^1(\lambda)$ is called the "Schur order" on $L^1(\lambda)$.

Remark 3.17. (a) Equivalent definitions of the Schur order $f \prec g$ are obtained by

$$\int_0^1 f(t)d\lambda(t) = \int_0^1 g(t)d\lambda(t)$$

and

$$\int_0^x f^*(t)d\lambda(t) \geq \int_0^x g^*(t)d\lambda(t), \quad \forall x \in (0,1) \tag{3.31}$$

or

$$\int_x^1 f_*(t)d\lambda(t) \geq \int_x^1 g_*(t)d\lambda(t), \quad \forall x \in (0,1). \tag{3.32}$$

(b) Characterizations and properties of the "continuous" Schur order were intensively discussed in Ryff (1965), Luxemburg (1967), and Chong and Rice (1971).

◇

The following Schur ordering properties of monotone rearrangements are easily established from the definition (see Day (1972, 6.1, 6.2)).

Proposition 3.18. *Let* $f, g \in L^1(\lambda)$, *then*
(a) $f^* + g_* \prec_S f + g \prec_S f^* + g^*$
(b) $f^* - g^* \prec_S f - g \prec_S f^* - g_*$
(c) $f^* g_* \preceq_S \quad fg \quad \preceq_S f^* g^* \qquad$ *if* $fg \in L^1(\lambda)$.

Also the notion of doubly stochastic matrix has an analogue in the infinite dimensional case based on the Schur order $\prec_S$.

Definition 3.19 (Doubly stochastic operator). A linear operator $T : L^1(\lambda) \to L^1(\lambda)$ is called a "doubly stochastic operator" if for any $f \in L^1(\lambda)$

$$Tf \prec_S f.$$

Typical examples of doubly stochastic operators are conditional expectations w.r.t. some sub-σ-algebras $\mathcal{C} \subset (0,1)\mathfrak{B}^1$, $Tf = E(f \mid \mathcal{C})$. The conditional expectation is taken w.r.t. the Lebesgue measure on $(0,1)$. The characterization corresponding to (3.21) is due to Ryff (1965).

Proposition 3.20. *For* $f, g \in L^1(\lambda)$ *it holds that:* $f \prec_S g$ *if and only if there exists a doubly stochastic operator* T *on* $L^1(\lambda)$ *such that*

$$f = Tg. \tag{3.33}$$

The most basic connection of Schur ordering to convex functions and convex ordering corresponding to (3.22) is given by the following theorem due to Hardy, Littlewood, and Pólya (1952) and Chong (1974).

Theorem 3.21 (Hardy–Littlewood–Pólya Theorem). *For* $f, g \in L^1(\lambda)$ *the following equivalences hold:*

(a) $f \preceq_s g$ *if and only if*

$$\int \varphi \circ f \, d\lambda \leq \int \varphi \circ g \, d\lambda \qquad (3.34)$$

for all convex, nondecreasing, integrable $\varphi : \mathbb{R}^1 \to \mathbb{R}^1$.
(b) $f \prec_s g$ *if and only if (3.34) holds for all convex, integrable $\varphi : \mathbb{R}^1 \to \mathbb{R}^1$.*

Proof. By the definition of rearrangements condition (3.34) is equivalent to the condition

$$\int \varphi \circ F^{-1}(u) d\lambda(u) \leq \int \varphi \circ G^{-1}(u) d\lambda(u),$$

where F, G are the distribution functions of λ^f respectively λ^g. With $U \sim U(0,1)$ the above inequalities are equivalent to the convex ordering

$$F^{-1}(U) \leq_{cx} G^{-1}(U). \qquad (3.35)$$

Furthermore this condition is by Theorem 3.3 equivalent to

$$\int \left(F^{-1}(u) - x\right)_+ d\lambda(u) \leq \int \left(G^{-1}(u) - x\right)_+ d\lambda(u) \qquad (3.36)$$

for all $x \in \mathbb{R}^1$.

If (3.36) holds true, then choosing $x = G^{-1}(\beta)$ we obtain

$$\int_\beta^1 F^{-1}(u) du - (1-\beta)G^{-1}(\beta) \leq \int_\beta^1 \left(F^{-1}(u) - G^{-1}(\beta)\right)_+ du$$

$$\leq \int_\beta^1 \left(G^{-1}(u) - G^{-1}(\beta)\right)_+ du$$

$$= \int_\beta^1 G^{-1}(u) du - (1-\beta)G^{-1}(\beta) \qquad (3.37)$$

and as a consequence for any $\beta \in (0,1)$

$$\int_\beta^1 F^{-1}(u) du \leq \int_\beta^1 G^{-1}(u) du.$$

This implies by Proposition 3.14 $f \preceq_s g$.

Conversely, if $f \preceq_s g$ then for any $u \in \mathbb{R}^1$ we choose some $\alpha \in (0,1]$ such that $F^{-1}(\alpha) \leq u \leq F^{-1}(\alpha+)$ and obtain from partial integration

$$E(F^{-1}(U) - u)_+ = \int_u^\infty \lambda(F^{-1} \geq x) dx = \int_\alpha^1 (F^{-1}(t) - u) dt, \qquad (3.38)$$

where U is $U(0, 1)$-distributed. This implies that

$$E(F^{-1}(U) - u)_+ \leq \int_\alpha^1 (G^{-1}(t) - u)dt$$

$$\leq \int_0^1 (G^{-1}(t) - u)_+ dt$$

$$= E(G^{-1}(U) - u)_+.$$

Thus by Theorem 3.3 the convex ordering condition (3.34) follows. $\qquad\square$

Remark 3.22. The Hardy–Littlewood–Pólya Theorem implies that Schur ordering $f \prec_s g$ is equivalent to the convex ordering

$$f(U) \leq_{cx} g(U) \text{ for } U \sim U(0, 1).$$ $\qquad\Diamond$

3.3 Rearrangements and Excess of Loss

The connection of rearrangements to the notion of Fréchet classes is given in the following theorem stated in Rü (1983, Lemma 1), which is based on a measure theoretic result of Rohlin (1952).

Theorem 3.23 (Rearrangements and Fréchet class). *Let $P_1, \ldots, P_n \in M^1(\mathbb{R}^1, \mathfrak{B}^1)$ be n probability measures on $\mathbb{R}^1$ with corresponding distribution functions $F_1, \ldots, F_n$. Also let U be a $U(0, 1)$-distributed random variable on a probability space $(\Omega, \mathfrak{A}, P)$, then*

$$\mathcal{M}(P_1, \ldots, P_n) = \left\{ P^{(f_1(U), \ldots, f_n(U))}; \ f_i \sim_r F_i^{-1}, 1 \leq i \leq n \right\}. \qquad (3.39)$$

Proof. If f_i is a rearrangement of F_i^{-1}, then

$$P^{f_i(U)} = \lambda^{f_i} = \lambda^{F_i^{-1}} = P_i, \quad 1 \leq i \leq n.$$

So the right-hand side of (3.39) is contained in the left-hand side.

A theorem of Rohlin (1952) (cf. also Parthasarathy (1967), and Whitt (1976, Lemma 2.7)) on the isomorphism of measure spaces implies that each $Q \in \mathcal{M}^1(R^n, B^n)$ has a representation $Q = \lambda^{(f_1, \ldots, f_n)}$, where $f_i = (0, 1) \to R^1$ are measurable. For $Q \in \mathcal{M}(P_1, \ldots, P_n)$

$$\lambda^{f_i} = P_i = \lambda^{F_i^{-1}}, \quad 1 \leq i \leq n,$$

i.e. f_i is a rearrangement of F_i^{-1}, $1 \leq i \leq n$. $\qquad\square$

As a consequence of Theorem 3.23 we obtain that the problem of determining generalized Fréchet bounds can be equivalently formulated as a problem to determine optimal rearrangements (see Rü (1983)).

Theorem 3.24 (Rearrangements and Fréchet bounds). *Let* $P_1, \ldots, P_n \in M^1(\mathbb{R}^1, \mathfrak{B}^1)$ *be probability measures with distribution functions* $F_1, \ldots, F_n$, *then for any* $f : \mathbb{R}^n \to \mathbb{R}$ *measurable the generalized Fréchet bounds as defined in* (2.3) *are equivalently characterized as optimal rearrangements*

$$M(\varphi) = \sup \left\{ \int \varphi(f_1(t), \ldots, f_n(t)) d\lambda(t); \ f_i \sim_r F_i^{-1}, 1 \leq i \leq n \right\} \qquad (3.40)$$

and

$$m(\varphi) = \inf \left\{ \int \varphi(f_1(t), \ldots, f_n(t)) d\lambda(t); \ f_i \sim_r F_i^{-1}, 1 \leq i \leq n \right\}. \qquad (3.41)$$

Remark 3.25 (Rearrangement inequalities, Hoeffding–Fréchet bounds). As a consequence of the measure theoretic representation in Theorems 3.23 and 3.24 the Hardy et al. (1952) inequality for functions in (3.27) implies the classical Hoeffding–Fréchet bounds

$$EF^{-1}(U)G^{-1}(1 - U) \leq EXY \leq EF^{-1}(U)G^{-1}(U) \qquad (3.42)$$

for any random variables X, Y with distribution functions F, G. Here U is a uniformly on $(0, 1)$ distributed random variable $U \sim U(0, 1)$ (see also Remark 3.15). Some authors denote inequality (3.42) as the "Hardy–Littlewood inequality" which however does not seem to be completely justified. The papers of Hoeffding from 1940 and of Fréchet (1951) are the early sources of this result. The basic measure theoretic result in Rohlin making the transfer from rearrangements to general random variables as in Theorem 3.23 possible is only from 1952.

Whitt (1976) gave two proofs of (3.42) based on rearrangements. One proof is based on reducing (3.42) to the finite discrete case and using there the discrete rearrangement inequality of Hardy et al. (1952). A second proof used the continuous time Hardy et al. (1952) rearrangement result and reduced (3.42) by a similar measure theoretic result as above to the function case. ◇

The Hardy–Littlewood–Pólya Theorem 3.21 together with the rearrangement representation in Theorem 3.23 and the convexity criterion in Theorem 3.3 imply as a corollary the following characterization of the convex ordering $\leq_{cx}$.

Corollary 3.26 (Convex order and Schur order). *Let* X, Y *be real, integrable random variables with distribution functions* F, G *and with* $EX = EY$, *then*

$$X \leq_{cx} Y \Leftrightarrow F^{-1} \prec_S G^{-1}$$

$$\Leftrightarrow E(X - x)_+ \leq E(Y - x)_+, \quad \forall x \in \mathbb{R}^1.$$

Remark 3.27 (Lorenz curve, income distribution). For a random variable X with distribution function F, describing the income distribution of a population, the function $L : (0, 1) \rightarrow \mathbb{R}^1$ defined by

$$L(t) = \frac{1}{EX} \int_0^t F^{-1}(u) du \qquad (3.43)$$

is called a "Lorenz curve".

For a non-negative income distribution $L(t)$ represents the proportion of the poorest $100t$ percent of the population. For empirical populations this curve was introduced by the economist Lorenz as a measure of inequality in the income distribution. Corollary 3.26 says that comparison of two Lorenz curves (in case of equal means) is equivalent to comparison in convex order. For a detailed study of Lorenz curves see Arnold (1987). $\Diamond$

Similarly to the equivalence in Corollary 3.26 and the representation Theorem 3.23, Proposition 3.18 implies the following convex ordering results for $X + Y$, $X - Y$, and $X \cdot Y$.

Corollary 3.28. *Let X, Y be real integrable random variables with distribution functions F, G, then*

(a) $\qquad F^{-1}(U) + G^{-1}(1 - U) \leq_{cx} X + Y \leq_{cx} F^{-1}(U) + G^{-1}(U).$ $\qquad (3.44)$

(b) $\qquad F^{-1}(U) - G^{-1}(U) \leq_{cx} X - Y \leq_{cx} F^{-1}(U) - G^{-1}(1 - U).$ $\qquad (3.45)$

(c) If $X \cdot Y$ are integrable, then

$$F^{-1}(U) \cdot G^{-1}(1 - U) \leq_{icx} X \cdot Y \leq_{icx} F^{-1}(U) \cdot G^{-1}(U). \qquad (3.46)$$

As result of the above discussion we next obtain an alternative proof of the basic worst case interpretation of the comonotonic vector in Theorem 3.5 (cf. Rü (1983)).

Corollary 3.29 (Comonotonic vector and convex order). *Let $X_1, \ldots, X_n$ be integrable, real random variables with distribution functions $F_1, \ldots, F_n$, then*

$$\sum_{i=1}^n X_i \leq_{cx} \sum_{i=1}^n F_i^{-1}(U) = \left(\sum_{i=1}^n F_i \right)^{-1}(U). \qquad (3.47)$$

Proof. By the rearrangement representation of the Fréchet class in Theorem 3.23 and the Hardy–Littlewood Theorem 3.21 it is sufficient to prove that for all rearrangements f_i of F_i^{-1}, $1 \leq i \leq n$ it holds that

$$\sum_{i=1}^n f_i \prec_S \sum_{i=1}^n F_i^{-1}. \qquad (3.48)$$

But (3.48) follows from Proposition 3.18 by induction. For the induction step we get

$$\sum_{i=1}^{n+1} f_i = \left(\sum_{i=1}^{n} f_i\right) + f_{n+1} \prec_S F^{-1}_{\sum_{i=1}^{n} f_i} + F^{-1}_{n+1}.$$

By the induction hypothesis $F^{-1}_{\sum_{i=1}^{n} f_i} \prec_S \sum_{i=1}^{n} F^{-1}_i$ and, therefore, we get from the definition of the Schur order

$$\sum_{i=1}^{n+1} f_i \prec_S \sum_{i=1}^{n} F^{-1}_i + F^{-1}_{n+1}.$$

The last equality in (3.47) is direct from the definition and monotonicity of the generalized inverses.   □

3.4 Integral Orders and $\prec_{\mathcal{F}}$-Diffusions

In this section we give extensions of Strassen's ordering theorem (Theorem 1.18) to some general class of integral orders, not necessarily defined by the class of increasing functions w.r.t. a given partial order like the usual partial order $\leq$ or the Schur order $\prec_S$ on $\mathbb{R}^n$.

Definition 3.30 (Integral order $\prec_{\mathcal{F}}$). Let $\mathcal{F}$ be a class of real functions on a measure space $(E, \mathfrak{A})$. The integral order $\prec_{\mathcal{F}}$ is defined on the set of probability measures $\mathcal{M}^1(E, \mathfrak{A})$ by

$$P \prec_{\mathcal{F}} Q \text{ if } \int f \, dP \leq \int f \, dQ \qquad (3.49)$$

for all integrable $f \in \mathcal{F}$.

Typically the same order $\prec_{\mathcal{F}}$ may have various different classes of generators which generate the same order. For some purposes it is of interest to have a small generating class and for other purposes to have a large (maximal) generating class. A maximal generating class exists. It is a convex class and is called $\mathcal{R}_{\mathcal{F}}$. For example for the stochastic order $\leq_{st}$ on $\mathbb{R}^1$ a small generating class is $\mathcal{F}_0 = \{1_{[t,\infty)}; t \in \mathbb{R}^1\}$, a large (maximal) generating class is $\mathcal{F}_i$, the set of all increasing functions on $\mathbb{R}^1$.

The integral order $\prec_{\mathcal{F}}$ defines a partial order on $\mathcal{M}^1(E, \mathfrak{A})$. As particular cases we obtain for $\mathcal{F} = \mathcal{F}_i$ the set of increasing functions on $(E, \leq)$ the "stochastic (increasing) order $\leq_{st}$", for $\mathcal{F} = \mathcal{F}_{cx}$ the class of convex functions the convex order $\leq_{cx}$, for $\mathcal{F} = \mathcal{F}_{icx} := \mathcal{F}_{cx} \cap \mathcal{F}_i$ the increasing convex order $\leq_{icx}$, and so on. Later on we will be particularly interested in some function classes which define different types of dependence orders.

The extension of the ordering theorem is based on certain Markov kernels called $\prec_{\mathcal{F}}$-diffusions.

Definition 3.31. Let $\mathcal{F} \subset \mathcal{L}(E, \mathfrak{A})$ be a class of real measurable functions on $(E, \mathfrak{A})$. A Markov kernel K on E is called an "$\mathcal{F}$-diffusion", if

$$\varepsilon_x \prec_{\mathcal{F}} K(x, \cdot) \text{ for all } x \in E. \tag{3.50}$$

A $\prec_{\mathcal{F}}$-diffusion kernel K "diffuses" locally in any point x mass with respect to $\prec_{\mathcal{F}}$. If for example $\prec_{\mathcal{F}} = \leq_{\mathrm{st}}$ is the stochastic order, then K is an $\mathcal{F}$-diffusion if

$$K(x, \{y \in E; \ x \leq y\}) = 1. \tag{3.51}$$

In fact in general we have to take in (3.51) the outer measure.

The composition KP for $P \in \mathcal{M}^1(E, \mathfrak{A})$ is defined by

$$KP(A) = \int K(x, A) P(dx).$$

Proposition 3.32. *Let K be a $\prec_{\mathcal{F}}$-diffusion, then*

$$P \prec_{\mathcal{F}} KP \text{ for all } P \in \mathcal{M}^1(E, \mathfrak{A}). \tag{3.52}$$

Proof. The proof is direct from the definition. For $f \in \mathcal{F}$ it holds that

$$\int f dKP = \int \left(\int f(y) K(x, dy) \right) dP(x) \geq \int f(x) dP(x),$$

and thus $P \prec_{\mathcal{F}} KP$. $\qquad\qquad\qquad\qquad\qquad\qquad\qquad\qquad\qquad \square$

The following representation result gives under certain conditions on $\mathcal{F}$ a converse of Proposition 3.32. It characterizes integral stochastic orders by corresponding $\mathcal{F}$-diffusions. Results of this type in the case of stochastic order of the convex and convex increasing order were obtained in the classical papers of Blackwell, Stein, Sherman, Cartier, in Meyer (1966, Theorem 53), and Strassen (1965). Based on Strassen's kernel representation theorem (Strassen 1965, Theorem 3) some further classes of examples were given in Rü (1980) and in Mosler and Scarsini (1991a). The following extended version of this representation result was given for part (b) in Müller and Stoyan (2002) and for part (a) in Rü (2005). We define for $f \in C_b(E)$, where $(E, \mathfrak{A})$ is a Polish space,

$$h_f(x) := \sup \left\{ \int f \, dP; \ \varepsilon_x \prec_{\mathcal{F}} P \right\}. \tag{3.53}$$

Let $\mathcal{F}^0$ denote the convex, maximal generator of the order dual to $\prec_{\mathcal{F}}$ such that

$$P \prec_{\mathcal{F}} Q \text{ is equivalent to } Q \prec_{\mathcal{F}^0} P. \tag{3.54}$$

Let $\overline{\mathcal{F}^0}^{P,Q}$ be the closure of $\mathcal{F}^0$ w.r.t. $\mathcal{L}^1(P)$ and $\mathcal{L}^1(Q)$, i.e.

$$\overline{\mathcal{F}^0}^{P,Q} = \mathcal{L}^1(\mathcal{F}^0, P) \cap \mathcal{L}^1(\mathcal{F}^0, Q).$$

Theorem 3.33 (Integral order representation). *Let $(E, \mathfrak{A})$ be a Polish space, $\mathcal{F} \subset C_b(E)$ with dual convex cone $\mathcal{F}^0$ and let $P, Q \in \mathcal{M}^1(E, \mathfrak{A})$. Under either of the assumptions*

(a) $$h_f \in \overline{\mathcal{F}^0}^{P,Q} \text{ for } f \in C_b(E) \tag{3.55}$$

or

(b) $$f, g \in \mathcal{R}_{\mathcal{F}} \Rightarrow \max(f, g) \in \mathcal{R}_{\mathcal{F}} \tag{3.56}$$

it holds that: $P \prec_{\mathcal{F}} Q$ if and only if there exists an $\mathcal{F}$-diffusion K such that

$$Q = KP. \tag{3.57}$$

Proof. (a) Define $\Pi_x := \{P \in \mathcal{M}^1(E, \mathfrak{A}); \ \varepsilon_x \prec_{\mathcal{F}} P\}$. Then Π_x is convex, weakly closed, and

$$h_f(x) = \sup \left\{ \int f \, dP; \ P \in \Pi_x \right\}.$$

For $f \in C_b(E)$ it holds that $f(x) \leq h_f(x)$. Since by assumption $h_f \in \overline{\mathcal{F}^0}^{P,Q}$ we obtain

$$\int f \, dQ \leq \int h_f \, dQ \leq \int h_f \, dP. \tag{3.58}$$

This implies by Strassen's kernel representation theorem (Strassen 1965, Theorem 3) the existence of a kernel K on E with $Q = KP$ and $K(x, \cdot) \in \Pi_x$ for all $x \in E$. By definition K therefore is an $\mathcal{F}$-diffusion and (3.57) holds true.

(b) Under condition (b) this representation result goes back in the compact case to Meyer (1966, Theorem 53). The extension to Polish spaces is given in Müller and Stoyan (2002, Theorem 2.6.1). $\qquad\qquad\qquad\qquad\qquad\qquad\qquad\qquad\qquad\quad\square$

Remark 3.34 (Integral representation in the unbounded case). Theorem 3.33 is formulated for generating classes $\mathcal{F} \subset C_b(E)$. For unbounded generating function classes $\mathcal{F}$ let $b : E \to [1, \infty)$ be a measurable weighting function and consider the function classes $B_b = \{f : (E, \mathfrak{A}) \to (\mathbb{R}^1, \mathfrak{B}^1); \ |f| \leq b\}$ and $\mathcal{M}_b^1 = \mathcal{M}_b^1(E, \mathfrak{A})$ the set of all probability measures Q on $(E, \mathfrak{A})$ such that b is Q-integrable. Then based on the duality pair $(B_b, \mathcal{M}_b^1)$ the integral representation theorem extends to the case that $\mathcal{F} \subset B_b$. Here the condition $h_f \in \overline{\mathcal{F}^0}$ has to be understood as closure w.r.t. the weak topology $\sigma(B_b, \mathcal{M}_b)$ induced on B_b by the signed measures integrating b, i.e. $\int |b| \, d|\mu| < \infty$. Under condition (b) this extension is stated in Müller and Stoyan (2002, Theorem 2.6.1). $\Diamond$

Example 3.35. (a) Stochastic order. In the case of the stochastic order $\leq_{st}$ on a Polish space E Theorem 3.33 implies the order representation result of Strassen (Theorem 1.18):
$P \leq_{st} Q$ if and only if there exist random variables $X \sim P$, $Y \sim Q$ on some probability space $(\Omega, \mathfrak{A}, R)$ such that

$$X \leq Y \, a.s.$$

Just let (X, Y) be any pair of random variables with distribution given by the product kernel $P \times K$ on the product space $E \times E$, defined by

$$P \times K(A_1 \times A_2) = \int_{A_1} K(x, A_2) P(dx). \qquad (3.59)$$

In particular we get the representation of the Schur order $\prec_S$ and the usual stochastic order $\leq_{st}$ on $\mathbb{R}^n$.

(b) **Convex and increasing convex order.** As a corollary we obtain in the case of the convex order $\leq_{cx}$ respectively the increasing convex order $\leq_{icx}$ on $\mathbb{R}^n$ induced by the class of all convex respectively increasing convex functions the following classical result due to Strassen (1965).

Corollary 3.36. *For $P, Q \in \mathcal{M}^1(\mathbb{R}^n, \mathfrak{B}^n)$ it holds that:*
$P \leq_{cx} Q$ (respectively $P \leq_{icx} Q$) if and only if there exist random vectors $X \sim P$, $Y \sim Q$ such that

$$E(Y \mid X) = X \quad (\text{respectively } E(Y \mid X) \geq X). \qquad (3.60)$$

Proof. In this case we take the weight function $b(x) = 1 + \|x\|$. Then the convex order $\leq_{cx}$ on M_b^1 is generated by the class of convex functions in B_b, i.e. by $\mathcal{F} = \mathcal{F}_{cx} \cap B_b$. A kernel K is an $\mathcal{F}$-diffusion if and only if $x = EK(x, \cdot)$, $\forall x \in E$. For the proof note that if $x = EK(x, \cdot)$ and $f \in \mathcal{F}$, then by Jensen's inequality

$$f(x) = f\left(\int y K(x, da)\right) \leq \int f(y) K(x, dy)$$

and thus

$$\varepsilon_x \leq_{cx} K(x, \cdot).$$

The converse direction follows by taking linear functions.
$\mathcal{F}$ is a maximal generator of $\leq_{cx}$ (in B_b) and $f, g \in \mathcal{F} \Rightarrow \max(f, g) \in \mathcal{F}$. Therefore, (3.60) follows from the nonbounded version of Theorem 3.33 (see Remark 3.34).

Alternatively, we can also use condition (3.55). For $f \in B_b$, $h_f(x) = \sup\{\int f\, dP; P \in \Pi_x\}$ is easily seen to be concave and in B_b and thus $h_f \in \overline{\mathcal{F}^0}$, i.e. h_f lies in the closure of the dual cone $\mathcal{F}^0$ w.r.t. $\sigma(B_b, M_b)$. □

(c) **Symmetric convex order** $\leq_{\mathrm{sym,cx}}$. Let $\mathcal{F} = \mathcal{F}^{\mathrm{sym,cx}}$ be the set of all symmetric convex functions on $\mathbb{R}^n$ generating the symmetric convex order $\leq_{\mathrm{sym,cx}}$. Then by similar arguments we get (see Rü (1981b))

Corollary 3.37. *For $P, Q \in M^1(\mathbb{R}^n, \mathfrak{B}^n)$ it holds that: $P \leq_{\mathrm{sym,cx}} Q$ if and only if there exist random vectors $X \sim P$, $Y \sim Q$ such that*

$$X \prec_S E(Y_{(\,)} \mid X), \tag{3.61}$$

where $\prec_S$ is the Schur order and $Y_{(\,)}$ is the ordered version of vector Y.

Proof. Here one finds that a kernel K is an $\mathcal{F}$-diffusion ($\mathcal{F} = \mathcal{F}_{\mathrm{sym,cx}} \cap B_b$) if and only if $x \prec_S \int y_{(\,)}, K(x, dy)$ where $\prec_S$ is the Schur order and $y_{(\,)}$ is the ordered vector y. The further argument is similar to that of Corollary 3.36. □

It is interesting to note that Condition (3.61) is weaker than the Schur ordering condition $X \prec_S Y$. In particular $\leq_{\mathrm{sym,cx}}$ and $\prec_S$ are different partial orders on $M^1(\mathbb{R}^n, \mathcal{B}^n)$.

(d) **Norm increasing order.**
If $\mathcal{F} = \mathcal{F}^{\|\;\|}$ is the class of norm increasing functions $f(x) = g(\|x\|)$ in $C_b(\mathbb{R}^n)$, i.e. $\|\;\|$ is any norm on $\mathbb{R}^n$ and g is continuous, bounded increasing, then we get

Corollary 3.38. *For $P, Q \in M^1(\mathbb{R}^n, \mathfrak{B}^1)$ if holds that $P \prec_{\mathcal{F}^{\|\;\|}} Q$ if and only if there exist random vectors $X \sim P$, $Y \sim Q$ such that*

$$\|X\| \leq \|Y\| \quad a.s. \tag{3.62}$$

Proof. Note that $\varepsilon_x \prec_{\mathcal{F}^{\|\;\|}} P$ if and only if P has support in $\{y; \|y\| \geq \|x\|\}$. Further for any $f \in C_b$ it holds that

$$h_f(x) = \sup\left\{\int f\, dP;\ \varepsilon_x \prec_{\mathcal{F}^{\|\;\|}}\right\}.$$

So $h_f \in \overline{\mathcal{F}^0}^{P,Q}$ and we obtain (3.62) from the kernel representation result in Theorem 3.33. □

◇

Chapter 4
Bounds for the Distribution Function and Value at Risk of the Joint Portfolio

An important problem in quantitative risk measurement in finance and insurance is to determine (sharp) bounds for the distribution function of the joint portfolio $\sum_{i=1}^{n} X_i$ of a risk vector $X = (X_1, \ldots, X_n)$ where the marginal distribution functions $F_i \sim X_i$ are known but the dependence between the components is unspecified. The same problem is also of interest for more general aggregations of risk given by functions $\psi(X_1, \ldots, X_n)$. For a real random variable $Y \sim H$ with distribution function H the (insurance version of the) "value at risk" $\mathrm{VaR}_\alpha(Y)$ is defined as the α-quantile of the distribution function

$$\mathrm{VaR}_\alpha(Y) = H^{-1}(\alpha), \qquad 0 \le \alpha \le 1. \tag{4.1}$$

Thus by inversion (sharp) bounds for distribution functions imply also (sharp) bounds for the value at risk. Let

$$M_n(t) := \sup \left\{ P\left(\sum_{i=1}^{n} X_i \le t \right); \ X_i \sim F_i,\ 1 \le i \le n \right\}$$

$$m_n(t) := \inf \left\{ P\left(\sum_{i=1}^{n} X_i < t \right); \ X_i \sim F_i,\ 1 \le i \le n \right\}. \tag{4.2}$$

Then

$$1 - m_n(t) = \sup \left\{ P\left(\sum_{i=1}^{n} X_i \ge t \right); \ X_i \sim F_i,\ 1 \le i \le n \right\} =: M_n^+(t)$$

L. Rüschendorf, *Mathematical Risk Analysis*, Springer Series in Operations Research and Financial Engineering, DOI 10.1007/978-3-642-33590-7_4,
© Springer-Verlag Berlin Heidelberg 2013

and thus one obtains the sharp bounds

$$\text{VaR}_\alpha \left(\sum_{i=1}^n X_i \right) \leq M_n^{-1}(\alpha)$$

$$\text{respectively } \text{VaR}_\alpha \left(\sum_{i=1}^n X_i \right) \leq (1 - m_n)^{-1}(1 - \alpha). \qquad (4.3)$$

The basic reason for using the $\leq$ sign for $M_n(t)$ and the $<$ sign for $m_n(t)$ is that with this definition the bounds are attained. For the comonotonic risk vector $X^c = (F_1^{-1}(U), \dots, F_n^{-1}(U))$ the value at risk is additive (see (3.47))

$$\text{VaR}_\alpha \left(\sum_{i=1}^n F_i^{-1}(U) \right) = \sum_{i=1}^n \text{VaR}_\alpha \left(F_i^{-1}(U) \right) \qquad (4.4)$$

and thus the value at risk is easy to calculate for comonotone portfolios. The value at risk VaR_α is also additive on certain distribution classes as on the class of elliptical distributions (see McNeil et al. (2005b)). But the comonotonic dependence structure is not the worst case dependence structure and thus does not give bounds for the distribution function of the sum respectively the value at risk.

4.1   Standard Bounds

So-called "standard bounds" for $M_n(t)$, $m_n(t)$ have been obtained in the literature by simple arguments based on Bonferoni type inequalities (see Sklar (1973), Moynihan et al. (1978), Denuit et al. (1999), and Rü (2005)).

Proposition 4.1 (Standard bounds). *Let $X = (X_1, \dots, X_n)$ be a random vector with marginal distribution functions $F_1, \dots, F_n$. Then for any $t \in \mathbb{R}^1$ it holds that:*

$$\left(\bigvee_{i=1}^n F_i(t) - (n - 1) \right)_+ \leq P \left(\sum_{i=1}^n X_i \leq t \right)$$

$$\leq \min \left(\bigwedge_{i=1}^n F_i(t), 1 \right) \qquad (4.5)$$

where $\bigwedge_{i=1}^n F_i(t) := \inf \left\{ \sum_{i=1}^n F_i(u_i); \sum_{i=1}^n u_i = t \right\}$ is the "infimal convolution" of the (F_i) and $\bigvee_{i=1}^n F_i(t) := \sup \left\{ \sum_{i=1}^n F_i(u_i); \sum_{i=1}^n u_i = t \right\}$ is the "supremal convolution" of the (F_i).

Proof. For any $u_1, \ldots, u_n$ with $\sum_{i=1}^{n} u_i = t$ holds

$$P\left(\sum_{i=1}^{n} X_i \leq t\right) \leq P\left(\bigcup_{i=1}^{n}\{X_i \leq u_i\}\right)$$

$$\leq \sum_{i=1}^{n} F_i(u_i) \qquad (4.6)$$

which implies the upper bound in (4.5). Similarly, using the Fréchet lower bound in (2.34) we obtain

$$P\left(\sum_{i=1}^{n} X_i \leq t\right) \geq P\left(X_1 \leq u_1, \ldots, X_n \leq u_n\right)$$

$$\geq \left(\sum_{i=1}^{n} F_i(u_i) - (n-1)\right)_+. \qquad (4.7)$$

$\square$

In general the standard bounds in Proposition 4.1 are not sharp and can be improved. In the case $n = 2$ it was however proved independently in Makarov (1981) and Rü (1982) that the standard bounds are sharp.

Define for general n

$$A_n(t) := \left\{(x_1, \ldots, x_n); \sum_{i=1}^{n} x_i \leq t\right\},$$

$$A_n^+(t) := \left\{(x_1, \ldots, x_n); \sum_{i=1}^{n} x_i < t\right\}, \quad t \in \mathbb{R}^1$$

and let $(F_1 \wedge F_2)^-(t) = \inf\{F_1(x-) + F_2(t-x); x \in \mathbb{R}^1\}$ denote the left continuous version of $F_1 \wedge F_2$; similarly $(F_1 \vee F_2)^-(t)$ the left continuous version of $F_1 \vee F_2$.

Theorem 4.2 (Sharpness of standard bounds, $n = 2$). *If X_i have distribution functions F_i, $i = 1, 2$ then*

$$P(X_1 + X_2 \leq t) \leq M_2(t) = (F_1 \wedge F_2)^-(t) \qquad (4.8)$$

and

$$P(X_1 + X_2 < t) \geq m_2(t) = \left((F_1 \vee F_2)^-(t) - 1\right)_+. \qquad (4.9)$$

Proof. The proof given in Rü (1982) is based on Strassen's Theorem (cf. Theorem 2.6) which implies

$$M_2(t) = M(A_2(t))$$
$$= 1 - \sup \left\{ P_2(U) - P_1(\pi_1(A_2(t) \cap \mathbb{R}^1 \times U)); \ U \subset \mathbb{R}^1 \text{open} \right\}, \quad (4.10)$$

where π_1 is the projection on the first component. Since

$$\pi_1(A_2(t) \cap (\mathbb{R}^1 \times U)) = \left\{ x \in \mathbb{R}^1; \ \exists y \in U, x + y \leq t \right\} = (-\infty, t - \inf U)$$

we can restrict to open intervals $U = (x, \infty)$ and obtain

$$M_2(t) = 1 - \sup \left\{ P_2((x, \infty)) - P_1((-\infty, t - x)); \ x \in \mathbb{R}^1 \right\}$$
$$= \inf \left\{ F_2(x) + F_1((t - x)-); \ x \in \mathbb{R}^1 \right\}$$
$$= (F_1 \wedge F_2)^-(t)$$

the left continuous version of the infimal convolution $F_1 \wedge F_2(t)$.
  Similarly

$$M((A_2^+(t))^c) = 2 - \sup \left\{ F_1(x-) + F_2(t - x); \ x \in \mathbb{R}^1 \right\}$$
$$= 2 - (F_1 \vee F_2)^-(t)$$

which implies that

$$m_2(t) = m(A_2(t)) = (F_1 \vee F_2(t) - 1)_+ . \qquad \square$$

Remark 4.3. Alternatively the sharpness of the standard bounds in Theorem 4.2 can also be derived directly from the duality theorem (Theorem 2.3). The proof of Makarov (1981) uses direct arguments on the copulas. A similar proof is given in the paper of Frank et al. (1987). The upper and lower bounds in (4.8) and (4.9) are attained pointwise for any t. But they are not attained uniformly. There is no worst case dependence structure w.r.t. the value at risk independent of α. ◇

To construct pointwise solutions we use the following lemma.

Lemma 4.4. *Define* $h(t) = (F_1 \wedge F_2)^-(t)$ *and* $g(t) = F_1^{-1} \vee F_2^{-1}(t)$, $0 \leq t \leq 1$. *Then*

(a) g, h *are monotonically nondecreasing,*
(b) $g \circ h \leq t, 0 \leq t \leq 1$.

Proof. (a) follows by definition. For (b) let $s = F_1(u)$, then using that $F_i^{-1} \circ F_i(x) \leq x$ we obtain

$$
\begin{aligned}
F_1^{-1}(s) + F_2^{-1}(h(t) - s) &= F_1^{-1} \circ F_1(u) + F_2^{-1}(h(t) - F_1(u)) \\
&\leq u + F_2^{-1}(F_2(t - u)) \\
&\leq u + (t - u) \\
&= t.
\end{aligned}
$$

This implies (b) since for the sup in the definition of $F_1^{-1} \vee F_2^{-1}(t)$ it is enough to consider real numbers t of the form $F_1(u)$. $\square$

Let λ be the Lebesgue measure on $[0, 1]$ and define random variables

$$
Y_1(s) = F_1^{-1}(s), \quad Y_2(s) = F_2^{-1}(\varphi(s)), \quad s \in [0, 1] \tag{4.11}
$$

with $\varphi(s) = h(t) - s, 0 \leq s \leq h(t)$ and $\varphi(s) = s, h(t) \leq s \leq 1$. Then the random variables Y_1, Y_2 are a worst case pair concerning the distribution function of the sum at point t (see Rü (1983, Proposition 2)).

Proposition 4.5 (Worst case pairs). *The random variables Y_1, Y_2 defined in (4.11) satisfy*

(a) $Y_1 \sim F_1, Y_2 \sim F_2$
(b) $P(Y_1 + Y_2 \leq t) = M_2(t) = (F_1 \wedge F_2)^-(t)$.

Proof. The Lebesgue measure λ is invariant w.r.t. φ i.e. $\lambda^\varphi = \lambda$. Therefore $\lambda^{Y_i} = \lambda^{F_i^{-1} \circ \varphi} = \varphi^{F_i^{-1}}$ and thus $Y_i \sim F_i, i = 1, 2$. From Lemma 4.4 and definition of Y_i it follows that

$$
\lambda(\{Y_1 + Y_2 \leq t\}) \geq h(t) = (F_1 \wedge F_2)^-(t).
$$

This implies part (b) and moreover

$$
h(t) = \lambda(\{Y_1 + Y_2 \leq t\}) = F_1^{-1} \vee F_2^{-1}(t) = g(t). \tag{4.12}
$$

$\square$

The solution Y_1, Y_2 for given t is constructed by using the antitonic rearrangements of $F_1^{-1} \, F_2^{-1}$ on the interval $[h(t), 1]$. The maximal probability $\alpha^* = F_1 \wedge F_2(t)$ can be characterized in the following way. Let $f \sim_r h$ denote that f is a rearrangement of h.

Corollary 4.6. $\alpha^* = M_2(t) = 1 - \inf\{\alpha \in [0, 1]$ *there exist* $f^\alpha \sim_r F_j^{-1}$ *on* $[0, \alpha]$, *such that* $f_1^\alpha(s) + f_2^\alpha(s) = t, \forall s \in [0, \alpha]\}$.

This structure characterizes also worst case solutions in the general case $n \geq 2$ (see Rü (1983) and Puccetti and Rü (2012a)).

Theorem 4.7 (Structure of worst case solutions). *For all $t \in \mathbb{R}^1$ and distribution functions $F_1, \ldots, F_n$ it holds that*

$$M_n(t) = 1 - \inf\Big\{\alpha \in [0,1]; \text{ there exist } f_j^\alpha \sim_{\mathrm{r}} F_j^{-1} \text{ on } [0,\alpha],$$
$$0 \le j \le n, \text{ such that } \sum_{j=1}^n f_j^\alpha \le t \text{ on } [0,\alpha]\Big\} \quad (4.13)$$

and similarly

$$m_n(t) = 1 - \sup\Big\{\alpha \in [0,1]; \text{ there exist } f_j^\alpha \sim_{\mathrm{r}} F_j^{-1}/[\alpha,1]$$
$$\text{such that } \sum_{j=1}^n f_j^\alpha \ge t \text{ on } [\alpha,1]\Big\}. \quad (4.14)$$

Proof. If $f_j^\alpha \sim_{\mathrm{r}} F_j^{-1}/[0,\alpha]$ then f_j^α can be extended to rearrangements F_j^{-1} on $[0,1]$. Since $f_j^\alpha \overset{d}{=} F_j$ we obtain $M_n(t) \ge P(f_1^\alpha + \cdots + f_n^\alpha) \le t$ which implies the $\ge$ inequality in (4.13).

For the $\le$ inequality, let $f_j^* \sim_{\mathrm{r}} F_j^{-1}$ be solutions of (4.13) and define

$$A := \Big\{u \in [0,1]; \ \sum_{j=1}^n f_j^*(u) \le t\Big\}.$$

Then by the rearrangement characterization of the Fréchet class $\lambda(A) = M_n(t)$. With $\alpha = 1 - M_n(t)$ there exists a λ-preserving transformation $\varphi : [0,1] \to [0,1]$ such that $A = \varphi([0,\alpha])$. Therefore, we can assume w.l.g. that $A = [0,\alpha]$. Moreover there exist $\varphi_j : [0,1] \to [0,1]$, $\varphi_j \sim_{\mathrm{r}} F_j^{-1}$, $1 \le j \le n$, such that $f_j^\alpha = \varphi_j/[0,\alpha] \sim_{\mathrm{r}} F_j^{-1}/[0,\alpha]$ and $f_j^\alpha(u) \le f_j^*(u)$, $u \in [0,\alpha]$. Define for example

$$A_j^* := \Big\{u \in [0,\alpha]; \ f_j^*(u) \le F_j^{-1} \le F_j^{-1}(\alpha)\Big\}$$

and

$$f_j^\alpha/_{[\alpha,1]} := f_j/_{A_j^*} + F_j^{-1}/_{[0,\alpha]\backslash A_j^*}.$$

Then we can use an extension of $f_j^\alpha/_{[\alpha,1]}$ to $[0,1]$ such that $f_j^\alpha \sim_{\mathrm{r}} F_j^{-1}$. This construction implies using that $A = [0,\alpha]$ that $\sum_{j=1}^n f_j^\alpha(u) \le \sum_{j=1}^n f_j^*(u) \le t$. In consequence we get the $\le$ inequality in (4.13).

The proof of (4.14) is similar. $\qquad \square$

By Theorem 4.7 the problem to get sharp bounds on the distribution function of the sum has been reduced to a rearrangement problem. This rearrangement formulation is useful to obtain a fast algorithm to approximate the sharp bounds numerically (see Puccetti and Rü (2012b) and Embrechts et al. (2012)). The proposed algorithm works well for general inhomogeneous portfolios in high dimensions (up to dimension $d \leq 600$).

In some examples based on the duality theorem in Proposition 2.5 the sharp bounds $M_n(t)$, $m_n(t)$ have been evaluated (see Rü (1982, 1983)).

Example 4.8. (a) If $P_i = U(0, 1)$, $1 \leq i \leq n$, then

$$M_n(t) = \frac{2}{n}t, \quad 0 \leq t \leq \frac{n}{2}. \tag{4.15}$$

The solution of the dual problem in Proposition 2.5 is given by the linear function

$$f(x) = \begin{cases} \dfrac{2}{n} - tx, & \text{if } 0 \leq x \leq \dfrac{2}{t \cdot n}, \\ 0, & \text{otherwise.} \end{cases} \tag{4.16}$$

The worst case solution is constructed similarly to the case $n = 2$ by a rearrangement of the structure as in Theorem 4.7. It has the property of being mixing on the interval $[0, \alpha^*]$ i.e. $\sum_{j=1}^{n} f_j^{\alpha}(u) = t$ for $u \in [0, \alpha]$. In particular in the case $n = 3$ we obtain

$$M_3(t) = \begin{cases} \dfrac{2}{3}t, & 0 \leq t \leq \dfrac{3}{2}, \\ 1, & t > \dfrac{3}{2}, \end{cases} \qquad m_3(t) = \begin{cases} \dfrac{2}{3}t - 1, & 0 \leq t \leq 3, \\ 1, & t \geq 3. \end{cases}$$

The standard bounds from Proposition 4.1 in this case are much more crude and are given by

$$\min\left(1, \bigwedge_{i=1}^{3} F_i(t)\right) = \min(1, t) \quad \text{and} \quad \left(\bigvee_{i=1}^{3} F_i(t) - 2\right)_+ = (t - 2)_+.$$

(b) Bernoulli distribution: If $P_i = B(1, p)$, $1 \leq i \leq n$ are Bernoulli distributions, then for $k \leq np$

$$M_n(k) = \frac{n}{n - k}(1 - p). \tag{4.17}$$

Again the argument is similar as for (a). ◊

For distributions with a decreasing density on $[0, 1]$ Wang and Wang (2011) have established the following sharp result under a moderate moment condition.

Theorem 4.9 (Monotone density with bounded support). *Let the distribution function F have a decreasing density on its support* $[0, 1]$ *with mean* μ *and let*

$$E(Y \mid Y \geq t) \geq t + \frac{1-t}{n} \quad \text{for any } t \in [0, 1] \tag{4.18}$$

and $Y \sim F$. *Let* $G = F^{-1}$ *and define* $\psi(t) = E(Y \mid Y \geq G(t))$, $t \in [0, 1)$, *then*

$$m_n(s) = \begin{cases} 0, & s \leq n\mu, \\ \psi^-\left(\dfrac{s}{n}\right), & n\mu < s < n, \\ 1, & s \geq n, \end{cases}$$

where ψ^- *is the left-continuous generalized inverse of* ψ.

Proof. By the moderate mean condition (4.18) F is completely mixable with index n i.e. there exist $X_i \sim F$ such that $\sum_{i=1}^n X_i = n\mu$ (see Corollary 2.9 of Wang and Wang (2011)). This implies that $m_n(s) = 0$ for $s \leq n\mu$. For $n\mu < s < n$ by the conditional moment method (cf. Theorem 4.12) it holds that

$$m_n(s) \geq \psi\left(\frac{s}{n}\right). \tag{4.19}$$

Furthermore, with $a = \psi^-\left(\frac{s}{n}\right)$ and $V \sim U([a, 1])$, $G(V)$ has also a decreasing density and is therefore by the result mentioned above mixing on $[a, 1]$, i.e. there exist $V_i \sim U([a, 1])$ such that $\sum_{i=1}^n G(V_i) = n\psi(a) = s$. Defining $Y_i = G(U)\mathbb{1}_{(U \leq a)} + G(V_i)\mathbb{1}_{(U>a)}$ with $U \sim U(0, 1)$ and U independent of V_i we obtain $Y_i \sim F$ and

$$P\left(\sum_{i=1}^n Y_i < s\right) = P(U \leq a) = a.$$

As a consequence the lower bound in (4.19) is attained. $\qquad \square$

Remark 4.10. (a) The structure of the optimal coupling is of the form given in Theorem 4.7. It is of the same form as in the example of uniform distributions (see Example 4.8). The solution is mixing on the largest possible interval $[\alpha, 1]$.
(b) The result is transferred directly to the case of general bounded support. For unbounded support with decreasing density a related result was given in Wang et al. (2011). $\qquad \diamond$

The standard bounds in Proposition 4.1 have been extended to general monotonically nondecreasing aggregation functions $\psi(x_1, \ldots, x_n)$ under slightly varying regularity conditions in Williamson and Downs (1990) for $n \geq 2$, in Embrechts et al. (2003), Rü (2005), and Embrechts and Puccetti (2006b). For a multivariate distribution function H on $\mathbb{R}^n$ denote by $\overline{H}$ the corresponding survival function

$\overline{H}(x) = P_H([x, \infty))$. Furthermore, for componentwise monotonically nondecreasing aggregation functions ψ denote by

$$A_{\psi}^{+}(t) := \left\{u = (u_1, \ldots, u_n) \in \mathbb{R}^n; u \text{ is a maximal point in } \{\psi \leq t\}\right\}$$

and the ψ-inf- respectively ψ-sup-convolutions:

$$\bigwedge_{\psi} F_i(t) := \inf \left\{ \sum_{i=1}^{n} F_i(u_i); \ u \in A_{\psi}^{+}(t) \right\} \tag{4.20}$$

respectively

$$\bigvee_{\psi} F_i(t) := \sup \left\{ \sum_{i=1}^{n} F_i(u_i); \ u \in A_{\psi}^{+}(t) \right\}. \tag{4.21}$$

Theorem 4.11 (Standard bounds for monotone aggregation functions). *Let* $X = (X_1, \ldots, X_n)$ *be a random vector with distribution function* $F \in \mathcal{F}(F_1, \ldots, F_n)$ *and let* ψ *be monotonically nondecreasing. Then*

$$\left(\bigvee_{\psi} F_i(t) - (n-1) \right)_{+} \leq P\left(\psi(X) \leq t\right) \leq \bigwedge_{\psi} F_i(t). \tag{4.22}$$

Proof. For any $u \in A_{\psi}^{+}$ it holds that using the maximality of u

$$P\left(\psi(X) \leq t\right) \leq P\left(\bigcup_{i=1}^{n} \{X_i \leq u_i\} \right) \leq \sum_{i=1}^{n} F_i(u_i).$$

This implies the upper bound. Furthermore,

$$P\left(\psi(X) \leq t\right) \geq P\left(X_1 \leq u_1, \ldots, X_n \leq u_n\right) \geq \left(\sum_{i=1}^{n} F_i(u_i) - (n-1) \right)_{+}$$

by the lower Fréchet bounds.   □

The standard bounds on the distribution function respectively the tail risk of the joint portfolio can be essentially improved when addition information on the dependence structure of X is available (see Chapter 5).

4.2 Conditional Moment Method

There is a useful alternative to the standard bounds which is from the computational point of view often preferable compared with the inf-convolution calculations necessary for the standard bounds. This method depends only on the first conditional moments of the marginals. It is however restricted to the case of the joint portfolio aggregation $\psi(x) = \sum_{i=1}^{n} x_i$. The method was developed in the case of homogeneous portfolios with monotone densities on $[0, 1]$ in Wang and Wang (2011) and then extended to general inhomogeneous distributions $F_1, \ldots, F_n$ in Puccetti and Rü (2012b) and Wang et al. (2011).

Let $X_i \sim F_i$, $G_i = F_i^{-1}$ be the generalized inverse of F_i, $G = \sum_{i=1}^{n} G_i$ and assume that $\mu_i = E[X_i]$ exists. For $a \in [0, 1]$, define $\Psi(a)$ as the sum of the conditional first moments, given $X_i \geq G_i(a)$,

$$\Psi(a) = \frac{1}{1-a} \int_a^1 G(t) dt = \sum_{i=1}^{n} E[X_i \mid X_i \geq G_i(a)].$$

The function Ψ is determined by the conditional first moments of the marginal distributions F_i. Obviously, Ψ is monotonically nondecreasing and $\Psi(0) = \mu = \sum_{i=1}^{n} \mu_i$.

Theorem 4.12 (Method of conditional moments). *Let $X_i \sim F_i$ have first moments μ_i, $1 \leq i \leq n$. Then, for $s \geq \mu$, we have*

$$M_n^+(s) = \sup \left\{ P\left(\sum_{i=1}^{n} X_i \geq s \right); \ X_i \sim F_i, 1 \leq i \leq n \right\} \leq 1 - \Psi^-(s) \quad (4.23)$$

where $\Psi^-(s) = \sup \{t \in [0, 1]; \Psi(t) \leq s\}$ is the left-continuous generalized inverse of Ψ.

Proof. With $X_i \sim F_i$ and $S = \sum_{i=1}^{n} X_i$, we have

$$\mu = \sum_{i=1}^{n} \mu_i = E[S] \geq E\left[S 1_{\{S < s\}}\right] + s P\left(S \geq s\right)$$

$$= \int_0^{P(S<s)} G(t) dt + s P\left(S \geq s\right) = \mu - \int_{P(S<s)}^1 G(t) dt + s P\left(S \geq s\right). \quad (4.24)$$

If $P(S \geq s) > 0$, this implies that $\Psi\big(P(S < s)\big) \geq s$ and thus

$$P\left(S < s\right) \geq \Psi^-(s).$$

As a consequence, we obtain

$$P\left(S \geq s\right) \leq 1 - \Psi^-(s). \qquad \square$$

Remark 4.13. (a) The conditional bound in (4.23) is sharp if and only if the estimate in (4.24) is an equality, that is if for the optimal coupling it holds true that $\{S \geq s\} = \{S = s\}$ a.s. This means, by Theorem 4.7, that the corresponding optimal rearrangements f_i^α on $[\alpha, 1]$ satisfy

$$\sum_{i=1}^{n} f_i^\alpha(u) = s \text{ for all } u \in [\alpha, 1],$$

with $1 - \alpha = M(s)$. In Wang and Wang (2011), this property is called *mixing* on the interval $[\alpha, 1]$. In that paper, it is established that mixing holds true in the homogeneous case of monotone densities on $[0, 1]$ under the moderate moment condition (4.18).

(b) For unbounded domains, the bound (4.23) typically fails to be sharp. To be a good bound it is indeed necessary that

$$\sum_{i=1}^{n} E\left[X_i \mid X_i \geq G_i^{-1}(\alpha)\right] \approx s.$$

(c) The method to get upper bounds for $M(s)$ implies directly also a lower bound for $P(S > s)$. Denote by H the conditional moment function associated to the random variable $-X_i$, we obtain

$$P(S > s) = 1 - P((-S) \geq (-s)) \geq H^{-1}(-s). \qquad \diamond$$

The method of conditional moments in Theorem 4.12 can also be extended to give good bounds in the case of unbounded domains. For $s \geq \mu$ and $t \in [0, 1]$, define the function $H_t(t_1)$ as a conditional expected moment function on the interval $[t, t_1]$

$$H_t(t_1) = \frac{1}{t_1 - t} \int_t^{t_1} G(u)du = E[G(U_{[t,t_1]})], \text{ for } t_1 \geq t.$$

Here $U_{[t,t_1]}$ denotes a random variable uniformly distributed on the interval $[t, t_1]$. The function H_t is increasing in t_1 and decreasing in t. Let $H_t(1) \geq s$ and $G(t) \leq s$. This allows to define $t_1(t) = H_t^{-1}(s)$. If we assume continuity of the F_i, then we get that the conditional expectation on $[t, t_1]$ is identical to s

$$H_t(t_1(t)) = s. \qquad (4.25)$$

Instead of continuity it is enough in the following to postulate existence of $t_1(t)$ such that the sum of the conditional expectations of the rearrangements on $[t, t_1(t)]$ is equal to s. Next, we define the optimal choice of such t's as

$$t_0 = t_0(s) = \inf\left\{t; \ \left(G_i \mid_{[t,t_1(t)]}\right), 1 \leq i \leq n, \text{ are mixing with value } s\right\},$$

that is t_0 is the infimum of all the t's such that there exist rearrangements $f_i^t \sim_r G_i \mid_{[t,t_1(t)]}$ which satisfy

$$\sum_{i=1}^{n} f_i^t = E\left[\sum_{i=1}^{n} G_i \mid_{[t,t_1(t)]}\right] = s. \qquad (4.26)$$

Under the "mixing assumption" (4.26), there exist some $t \in [0, 1]$ such that $t_1(t) > t$ and the restricted distributions $\left(G_i \mid_{[t,t_1(t)]}\right)$ are mixing. Therefore, as indicated above, we get random variables $\tilde{V}_i \sim U_{[t,t_1(t)]}$ such that $\sum_{i=1}^{n} G_i(\tilde{V}_i) = s$. As a consequence, we also get some random variables $\tilde{\tilde{V}}_i \sim U_{[t_1(t),1]}$ with $\sum_{i=1}^{n} G_i(\tilde{\tilde{V}}_i) \geq s$. Finally, this implies the existence of random variables $V_i \sim U_{[t,1]}$ such that $\sum_{i=1}^{n} G_i(V_i) \geq s$. As a result, we can state the following theorem.

Theorem 4.14 (Extended conditional moment method). *Let $X_i \sim F_i$, $1 \leq i \leq n$ be random variables and assume that the mixing condition (4.26) holds. Then, for $s \geq \mu$ we obtain the lower bound*

$$\inf\left\{ P\left(\sum_{i=1}^{n} X_i < s\right); \ X_i \sim F_i\right\} \geq t_0(s). \qquad (4.27)$$

Remark 4.15. The conditional mixing condition (4.26) is satisfied in the homogeneous case $F_i = F$, $1 \leq i \leq n$ with density f, if f is decreasing on $[t, t_1(t)]$ and if, furthermore, the moderate moment condition

$$s/n \geq \left(1 - \frac{1}{n}\right) G(t) + \frac{1}{n} G(t_1(t))$$

holds; see Corollary 2.9 in Wang and Wang (2011). In the particular case where F is concentrated on $[0, 1]$ (or on a bounded domain) and $t_1(t_0) = 1$ our Theorem 4.12 implies that $t_0(s)$ is in fact a sharp bound under the moderate moment condition (see Theorem 4.9).

An alternative case is obtained when the density of f is symmetric unimodal on some admissible interval $[t, t_1(t)]$ satisfying $E \sum_{i=1}^{n} G_i \mid_{[t,t_1(t)]} = s$. In this case, the mixing condition is satisfied (see Rü and Uckelmann (2002)). Theorem 4.14 gives good bounds also in the case of unbounded domains. It also shows that the problem of establishing further mixing criteria is of interest.   ◊

4.3 Dual Bounds

As noted above the standard bounds in Proposition 4.1 are in general for $n \geq 3$ not sharp. Based on the duality theorems (Theorem 2.3 and Proposition 2.5) improved bounds (called "dual bounds") have been given in Embrechts and Puccetti (2006a,b, 2010) and Puccetti and Rü (2012a).

These dual bounds are obtained as a consequence of the following dual representation of $M_n^+(t)$ and $m_n^+(t)$; which is a specialization of upper and lower Fréchet bounds for the tail risk $P(\sum X_i \geq s)$ in Theorem 2.3.

Theorem 4.16. *The problems* (4.28a) *and* (4.28b) *have the following dual counterparts:*

$$M_n^+(s) = \inf\left\{ \sum_{i=1}^n \int g_i \, dF_i; \; g_i \text{ bounded}, \right.$$

$$\left. 1 \leq i \leq n \text{ with } \sum_{i=1}^n g_i(x_i) \geq 1_{[s,+\infty)}\left(\sum_{i=1}^n X_i\right)\right\}, \qquad (4.28a)$$

$$m_n^+(s) = \sup\left\{ \sum_{i=1}^n \int f_i \, dF_i; \; f_i \text{ bounded}, \right.$$

$$\left. 1 \leq i \leq n \text{ with } \sum_{i=1}^n f_i(x_i) \leq 1_{[s,+\infty)}\left(\sum_{i=1}^n X_i\right)\right\}. \qquad (4.28b)$$

While the dual representations in (4.28) are difficult to evaluate in general, they allow to establish good bounds obtained by choosing admissible piecewise linear dual functions in the dual problem. The resulting bounds are called "dual bounds" (see Embrechts and Puccetti (2006b)).

Theorem 4.17 (Dual bounds). *Let $X_i \sim F_i$ and $\overline{F}_i = 1 - F_i$ be the survival function of F_i. Then, for any $s \in \mathbb{R}$, we have*

$$M_n^+(s) \leq D(s) = \inf_{u \in \overline{\mathcal{U}}(s)} \min\left\{ \frac{\sum_{i=1}^n \int_{u_i}^{s-\sum_{j \neq i} u_j} \overline{F}_i(t)dt}{s - \sum_{i=1}^n u_i}, 1 \right\}, \qquad (4.29a)$$

$$m_n^+(s) \geq d(s) = \sup_{u \in \underline{\mathcal{U}}(s)} \max\left\{ \frac{\sum_{i=1}^n \int_{u_i}^{s-\sum_{j \neq i} u_j} \overline{F}_i(t)dt}{s - \sum_{i=1}^n u_i} - d + 1, 0 \right\}, \qquad (4.29b)$$

where

$$\overline{\mathcal{U}}(s) = \left\{ u \in \mathbb{R}^n; \; \sum_{i=1}^n u_i < s \right\} \text{ and } \underline{\mathcal{U}}(s) = \left\{ u \in \mathbb{R}^n; \; \sum_{i=1}^n u_i > s \right\}.$$

Proof. We give a sketch of the proof of (4.29a). A detailed proof has been given in Theorem 3.2 in Embrechts and Puccetti (2006a) for the case of bounds on $P(S < s)$. The bound in (4.29a) is obtained by proving that the linear functions $g_i : \mathbb{R} \to \mathbb{R}$, $1 \leq i \leq n$, defined as

$$g_i(x_i) = \begin{cases} 0, & \text{if } x_i \leq u_i, \\ \dfrac{x_i - u_i}{s - \sum_{i=1}^{n} u_i}, & \text{if } u_i < x_i \leq s - \sum_{j \neq i} u_j, \\ 1, & \text{otherwise,} \end{cases}$$

are admissible. This can be seen by direct calculation. Substituting these functions into the dual problem in (4.28a) and taking the infimum over all $u \in \overline{\mathcal{U}}(s)$ yields the result.

The proof for (4.29b) is analogous and based on (4.28b). In order to obtain (4.29b), it is sufficient to prove that the functions f_i, $1 \leq i \leq d$, defined as

$$f_i(x_i) = \begin{cases} 1, & \text{if } x_i \leq s - \sum_{j \neq i} u_j, \\ \dfrac{u_i - x_i}{\sum_{i=1}^{n} u_i - s}, & \text{if } s - \sum_{j \neq i} u_j < x_i \leq u_i, \\ 0, & \text{otherwise,} \end{cases}$$

are an admissible choice in (4.28b), and to take the infimum over all $u \in \underline{\mathcal{U}}(s)$.   □

In the above proof, if we choose a vector $u \in \mathcal{U}(s)$, that is $\sum_{i=1}^{n} u_i = s$, the piecewise-linear dual admissible choices become piecewise-constant and thus yield a standard bound. As a consequence, the dual bounds improve the corresponding standard bounds.

The dual bounds turn out in examples to be a strong improvement of the standard bounds. Calculation of the dual bounds requires however to solve an n-dimensional optimization problem which typically will be possible only for small values of n say $n \leq 5$. The situation changes in the homogeneous case where $F_i = F$, $1 \leq i \leq n$. Then we obtain a considerable simplified expression and a one-dimensional problem which can be solved in any dimension (see Embrechts and Puccetti (2006a, Theorem 4.2)).

Theorem 4.18 (Dual bound, homogeneous case). *Let* $F_1 = \cdots = F_n =: F$ *be distribution functions on* $\mathbb{R}_+$. *Then for any* $s \geq 0$ *it holds that*

$$P\left(\sum_{i=1}^{n} X_i \geq s\right) \geq 1 - n \inf_{s \in [0, s/n]} \frac{\int_r^{s-(n-1)r} \overline{F}(x)dx}{t - nr}. \tag{4.30}$$

Theorem 4.18 gives an upper bound for

$$M_n^+(s) = \sup\left\{ P\left(\sup_{1 \leq i \leq n} X_i \geq s \right); \ X_i \sim F_i, 1 \leq i \leq n \right\}. \tag{4.31}$$

This type of bounds has been shown to improve essentially the standard bounds. In the case of decreasing or increasing densities on $[0, b]$ they coincide with the sharp bounds given in Theorem 4.9.

The dual bounds in (4.30) have been proved in Puccetti and Rü (2011) to be sharp in the homogeneous case under some general distributional assumptions. The starting point of proving sharpness is the following simplified version of the duality theorem (cf. Proposition 2.5).

Theorem 4.19 (Duality theorem, homogeneous case). *In the homogeneous case* $F_1 = \cdots = F_n = F$ *it holds that:*

(a) Problem (4.31) has the following dual representation:

$$M(s) := M_n^+(s) = \inf\left\{n\int g\,dF;\ g \in \mathcal{D}(s)\right\}, \qquad (4.32)$$

where

$$\mathcal{D}(s) = \left\{g : \mathbb{R} \to \mathbb{R};\ g \text{ bounded}, \sum_{i=1}^{n} g(x_i) \geq 1_{[s,\infty)} \text{ for } x_1,\ldots,x_n \in \mathbb{R}\right\}.$$

An optimal dual solution $g \in \mathcal{D}(s)$ *exists.*

(b) A random vector X^* *with distribution function* $F_{X^*} \in \mathfrak{F}(F,\ldots,F)$ *is a solution of* $M_n^+(s) = P(\sum X_i^* \geq s)$ *if and only if there exists an admissible function* $g^* \in \mathcal{D}(s)$ *such that*

$$P\left(\sum_{i=1}^{n} g^*(X_i^*) = 1_{[s,\infty)}\left(\sum_{i=1}^{n} X_i^*\right)\right) = 1. \qquad (4.33)$$

(c) The sup in (4.31) is attained and any solution X^* *of (4.32) such that* $M_n^+(s) = P(\sum_{i=1}^{n} X_i^* \geq s)$ *is called an "optimal coupling".*

As in Theorems 4.17 and 4.18 we introduce for $t < \frac{s}{n}$ piecewise linear functions g_t by

$$g_a(x) := \begin{cases} 0, & \text{if } x < t, \\ \dfrac{x-t}{s-nt}, & \text{if } a \leq x \leq s - (n-1)t, \\ 1, & \text{otherwise.} \end{cases} \qquad (4.34)$$

g_t are admissible, i.e. $g_t \in \mathcal{D}(s)$ and they induce the dual bound $D(s)$

$$M(s) \leq D(s) = \inf_{t<s/n}\left(n\int g_t\,dF\right) = n \inf_{t<s/n} \min\left\{\frac{\int_t^{s-(n-1)t} \overline{F}(x)dx}{s-nt}, 1\right\}. \qquad (4.35)$$

Our aim is to prove that equality holds in (4.35) under some general conditions. To this aim a basic tool is the concept of mixability.

Definition 4.20 (n-mixability). A distribution (function) F on $\mathbb{R}$ is called "n-mixable" if there exist n random variables $X_1, \ldots, X_n$ identically distributed with distribution function F such that

$$P(X_1 + \cdots + X_n = n\mu) = 1, \tag{4.36}$$

for some $\mu \in \mathbb{R}$. Any such μ is called a centre of F and any vector $(X_1, \ldots, X_n)$ satisfying (4.36) with $X_i \sim F$, $1 \leq i \leq n$ is called an "n-complete mix".

If F is mixable and has finite mean, then its centre is unique and equal to its mean. Mixability was introduced in early work on variance reduction in Monte-Carlo simulation. Obviously an n-mix $X_1, \ldots, X_n$ minimizes the variance of the arithmetic mean $\frac{1}{n} \sum_{i=1}^{n} X_i$ as an estimator of the mean $\mu = \int x \, dF(x)$.

Remark 4.21. The following statements hold:

(a) The convex sum of n-mixable distributions with centre μ is n-mixable with centre μ.
(b) Any linear transformation $L(x) = mx + q$ of an n-mixable distribution with centre μ is n-mixable with centre $m\mu + q$.
(c) The binomial distribution $B(n, p/q)$, $p, q \in \mathbb{N}$ is q-mixable.
(d) Suppose F is a distribution on the real interval $[a, b]$, $a = F^{-1}(0)$ and $b = F^{-1}(1)$, having mean μ. A necessary condition for F to be n-mixable is that

$$a + \tfrac{b-a}{n} \leq \mu \leq b - \tfrac{b-a}{n}. \tag{4.37}$$

(e) If F is continuous with a monotone density on $[a, b]$, then condition (4.37) is sufficient for F to be n-mixable.   ◇

The structure of optimal couplings is described in the following proposition (see Rü 1982, Proposition 3).

Proposition 4.22. *For any marginal distribution function F there exists an optimal coupling X^* with distribution function $F_{X^*} \in \mathfrak{F}(F, \ldots, F)$ such that $M(s) = P\left(\sum_{j=1}^{n} X_j^* \geq s\right)$ and for any such X^* we have for all $i \leq n$*

$$\{X_i^* > F^{-1}(1 - M(s))\} \subset \left\{ \sum_{j=1}^{n} X_j^* \geq s \right\} \subset \{X_i^* \geq F^{-1}(1 - M(s))\} \ \text{a.s.}$$

$$\tag{4.38}$$

In case F is continuous, one gets that

$$\left\{ \sum_{j=1}^{n} X_j^* \geq s \right\} = \{X_i^* \geq F^{-1}(1 - M(s))\} \ \text{a.s.}$$

Theorem 4.19 and Proposition 4.22 allow to restrict to admissible dual functions $g \geq 0$ such that for an optimal coupling X^* it holds that

$$P\big(g(X_i^*) = 0 \mid X_i^* < a^*\big) = 1, \quad 1 \leq i \leq n, \tag{4.39}$$

where $a^* = F^{-1}(1 - M(s))$.

As a consequence we get the following characterization of optimal couplings by a mixability property.

Theorem 4.23. *Let* $a^* = F^{-1}(1 - M(s))$. *A random vector* X^* *with distribution function* $F_{X^*} \in \mathfrak{F}(F_1, \ldots, F_n)$ *is a solution of* $M(s) = P(\sum_{j=1}^n X_j^* \geq s)$ *if and only if there exists an admissible function* $g^* \in \mathcal{D}(s)$ *such that the conditional distribution of*

$$\big(g_j^*(X_j^*) \mid X_i^* \geq a^*\big), \quad 1 \leq i \leq n.$$

is n-mixable with centre $\mu = 1/n$.

Proof. Assume that X^* and g^* are an optimal coupling, respectively, an optimal dual function. By (4.39) we can assume that any $g^* \in \mathcal{D}(s)$ is zero below the threshold a^*. Using (4.33) and (4.38) we obtain that $g^* \in \mathcal{D}$ is an optimal choice for (4.32) if and only if

$$P\left(\sum_{j=1}^n g_j^*(X_j^*) = 1 \,\Big|\, \sum_{j=1}^n X_j^* \geq s\right) = P\left(\sum_{j=1}^n g_j^*(X_j^*) = 1 \mid X_i^* \geq a^*\right) = 1,$$
$$\tag{4.40}$$

for $1 \leq i \leq n$, i.e. the conditional distribution of

$$\big(g_j^*(X_j^*) \mid X_i^* \geq a^*\big), \quad 1 \leq i \leq n$$

is n-mixable with centre $\mu = 1/n$. □

We prove the sharpness of the dual bound $D(s)$ in two steps. First we state the complete mixability of the dual function $g_t(X_1)$ for a suitable choice of the parameter $t = a$ above a certain threshold a^*. Then we show that $a^* = M^{-1}(s)$, hence obtaining the optimality of g_a.

For the realization of these steps we need the following three conditions:

(A1) Attainment condition: There exists some $a < s/n$ such that

$$D(s) = \inf_{t < s/n} \frac{n \int_t^{s-(n-1)t} \overline{F}(x)dx}{s - nt} = \frac{n \int_a^b \overline{F}(x)dx}{b - a},$$

where $b = s - (n-1)a$ and $a^* = F^{-1}(1 - D(s)) \leq a$.

(A2) Mixability condition: The conditional distribution of $(X_1 \mid X_1 \geq a^*)$ is n-mixable on (a, b).

(A3) Ordering condition: For all $y \geq b$ it holds that

$$(n - 1)(F(y) - F(b)) \leq F(a) - F\left(\frac{s - y}{n - 1}\right).$$

Remark 4.24. The ordering condition (A3) can be formulated in terms of stochastic ordering $\leq_{st}$. Define the distribution functions F_1 and F_2

$$F_1(y) = \frac{\overline{F}\left(\frac{s-y}{n-1}\right) - \overline{F}(a)}{\overline{F}(a^*) - \overline{F}(a)} \quad \text{and} \quad F_2(y) = \frac{\overline{F}(b) - \overline{F}(y)}{\overline{F}(b)}, \quad \text{for } y \geq b.$$

Then condition (A3) is equivalent to

$$F_1 \leq_{st} F_2. \tag{4.41}$$

If F has a density f, then by a well-known condition stochastic ordering $F_1 \leq_{st} F_2$ is implied by the "monotone likelihood ratio" criterion for their densities stating that f_2/f_1 is increasing. In our case this amounts to

$$\frac{f(y)}{f\left(\frac{s-y}{n-1}\right)} \quad \text{is increasing in } y, \text{ for } y \geq b. \tag{4.42}$$

This condition can be checked in examples.   ◊

The following main result about sharpness of the dual bound was established in Puccetti and Rü (2011).

Theorem 4.25 (Sharpness of the dual bound). *Under the attainment condition (A1), the mixing condition (A2) and the ordering condition (A3), the dual bound is sharp, that is*

$$M(s) = D(s) = \inf_{t < s/n} \frac{n \int_t^{s-(n-1)t} \overline{F}(x)dx}{s - nt} = \frac{n \int_a^b \overline{F}(x)dx}{b - a}.$$

Proof. We give the main idea of the proof. For details we refer to Puccetti and Rü (2011). First order conditions for the optimization in the attainment assumption (A1) imply that

$$\frac{n \int_a^b \overline{F}(x)dx}{b - a} = \overline{F}(a) + (n - 1)\overline{F}(b), \tag{4.43}$$

where $b = s - (n - 1)a$ and $a^* = F^{-1}(1 - D(s)) \leq a$. Therefore, a^* satisfies

$$\overline{F}(a^*) = D(s) = \frac{n \int_a^b \overline{F}(x)dx}{b - a} = \overline{F}(a) + (n - 1)\overline{F}(b). \tag{4.44}$$

Let $Y_{a^*} \stackrel{d}{=} (X_1 \mid X_1 \geq a^*)$ and let H be the distribution function of $g_a(Y_{a^*})$. Then the mixing condition (A2) implies using linearity of g_a that H is n-mixable. Since $g_a \in \mathcal{D}(s)$ it follows that

$$M(s) \leq D(s) = \frac{n \int_a^b \overline{F}(x)dx}{b - a} = \overline{F}(a) + (n - 1)\overline{F}(b).$$

The main step of the proof is to show the existence of a coupling X^* with $F_{X^*} \in \mathcal{F}(F, \ldots, F)$ such that

$$P\left(\sum_{i=1}^n X_i^* \geq s \right) = \overline{F}(a^*). \tag{4.45}$$

The construction of X^* is based on the following coupling steps using the mixing condition (A2):

(a) When one X_i^* lies in the interval (a, b), then all X_j^* lie in (a, b) and

$$P\left(\sum_{j=1}^n X_j^* = s \mid X_i^* \in (a, b) \right) = 1.$$

(b) For all $1 \leq i \leq n$ and $X_i^* \geq b$ all X_j^*, $j \neq i$ are coupled in a countermonotonic way to X_i^*:

$$P\left(X_j^* = F_{a^*}^{-1}((n - 1)\overline{F}_{a^*}(X_i^*)) \mid X_i^* \geq b \right) = 1.$$

(c) If any X_i^* is $< a^*$, then all X_j^* are $\leq a^*$ and otherwise undetermined.

By the ordering assumption (A3) we then conclude that the equality in (4.45) holds true. The above arguments imply that

$$M(s) = \overline{F}(a^*)$$

and, furthermore, the conditional distribution of $(g_a(X_1) \mid X_1 > a^*)$ is n-mixing with centre $\mu = 1/n$. Thus by Theorem 4.23 g_a is a solution of the dual problem in (4.32) and

$$M(s) = n \int g_a dF = \frac{n \int_a^b \overline{F}(x)dx}{b - a} = \inf_{t < s/n} \frac{n \int_t^{s - (n-1)t} \overline{F}(x)dx}{s - dt} = D(s). \quad \square$$

Remark 4.26. (a) Monotone densities: Continuous distribution functions F having a positive, decreasing density f on (a^*, ∞) satisfy assumptions (A2) and

(A3) (see Puccetti and Rü (2011)). In consequence Theorem 4.25 implies the result of Wang and Wang (2011) in Theorem 4.9.

(b) **Monotonicity in the tail:** As a consequence of (4.26), sharpness of the dual bound $D(s)$ can be stated, for s large enough, for all continuous, unbounded distribution functions which have an ultimately decreasing density. This is particularly useful in applications of quantitative risk management, where sharp bounds $M(s)$ are typically calculated for high thresholds s and positive, unbounded and continuous distribution functions F.

(c) **Concave densities:** Continuous distribution functions F having a concave density f on the interval (a, b) satisfy the mixing assumption (A2). This result follows from Theorem 4.3 in Puccetti et al. (2012). In order to obtain sharpness of the dual bound $D(s)$ for these distributions, conditions (A1) and (A3) have to be checked numerically.   ◊

Chapter 5
Restrictions on the Dependence Structure

In this chapter we consider some types of restrictions on the dependence structure which reduce the Fréchet class of all possible dependence structures as considered in the previous chapters. In consequence this leads to better bounds on the distribution function and on the tail probability of the aggregate risk when this information is available. One type of restriction we discuss is the restriction to some types of positive dependent distributions. This restriction leads to essential improvements of the bounds, when available. A second type of restriction on the dependence structure is considered in the case that certain higher dimensional joint distributions are known. Various techniques have been introduced in the literature to use this information for improvement on the distributional bounds. A few further results are given in the literature where some additional moment conditions as for example correlations are used to derive improved bounds.

5.1 Restriction to Positive Dependent Risk Vectors

A natural condition which is available in several applications is the condition that the components of the risk vector X have some positive dependence structure. This positive dependence structure can be combined with the knowledge of the marginal distributions $F_1, \ldots, F_n$ and leads to improved bounds for the distribution function, respectively the tail risk. A simple way to describe positive dependence is by the notion of "positive upper orthant dependence" (PUOD), respectively "positive lower orthant dependence" (PLOD). A random vector X is said to be PUOD if

$$\overline{F}_X(x) = P\,(X \geq x) \geq \prod_{i=1}^{n} P(X_i \geq x_i) = \prod_{i=1}^{n} \overline{F}_i(x_i), \text{ for all } x \in \mathbb{R}^n. \quad (5.1)$$

L. Rüschendorf, *Mathematical Risk Analysis*, Springer Series in Operations Research and Financial Engineering, DOI 10.1007/978-3-642-33590-7_5,

X is said to be PLOD if

$$F_X(x) = P\,(X \le x) \ge \prod_{i=1}^{n} P(X_i \le x_i) = \prod_{i=1}^{n} F_i(x_i), \text{ for all } x \in \mathbb{R}^n. \quad (5.2)$$

Moreover, X is called "positive orthant dependent" (POD) if X is both PUOD and PLOD. These notions belong to the earliest notions of positive dependence in probability. They were introduced in Lehmann (1966) and supplement and substitute the more classical notion of positive linear correlation.

Conditions (5.1) and (5.2) can be interpreted by saying that the distribution function F_X, respectively the survival function $\overline{F}_X$, is dominated by the corresponding distribution function, respectively the corresponding tail function, of the product measure. We can consider them as partial ordering relations $\le_{\text{uo}}$, $\le_{\text{lo}}$, $\le_c$ on the class of probability measures on $\mathbb{R}^n$ where $\le_c$ combines $\le_{\text{uo}}$ and $\le_{\text{lo}}$ and is called "concordance order". In this sense, we can consider more general positive dependence bounds by assuming that $F_X \le G$ respectively $G \le F_X$ where $\le$ is any of the above dependence orderings and G is any other distribution on $\mathbb{R}^d$.

The following typically strong improvements of the standard bounds of the joint portfolio under the additional positive dependence restrictions have been given in several similar forms in the literature; see Williamson and Downs (1990), Embrechts et al. (2003), Rü (2005), and Embrechts and Puccetti (2006b).

Recall that $\mathcal{U}(s) = \{u = (u_1, \dots, u_n) \in \mathbb{R}^n; \sum_{i=1}^{n} u_i = s\}$. For an n-dimensional distribution function G, we define the generalized G-infimal and G-supremal convolutions as

$$\bigwedge G(s) = \inf_{u \in \mathcal{U}(s)} G(u) \text{ and } \bigvee G(s) = \sup_{u \in \mathcal{U}(s)} G(u).$$

In case $G(u) = \prod_{i=1}^{n} F_i(u_i)$ is the distribution function of the product measure, we obtain the infimal and supremal convolutions of $\prod F_i$

$$\bigwedge G(s) = \bigwedge \left(\prod_{i=1}^{n} F_i \right)(s), \quad \bigvee G(s) = \bigvee \left(\prod_{i=1}^{n} F_i \right)(s).$$

Theorem 5.1 (Positive dependence restriction). *Let X be an n-dimensional risk vector having marginals F_i, $1 \le i \le n$ and let G be a d-dimensional distribution function.*

(a) If $G \le_{\text{lo}} F_X$, then

$$P \left(\sum_{i=1}^{n} X_i \le s \right) \ge \bigvee G(s); \quad (5.3)$$

(b) If $G \le_{\text{uo}} F_X$, then

$$P \left(\sum_{i=1}^{n} X_i < s \right) \le 1 - \bigvee \overline{G}(s); \quad (5.4)$$

(c) *If X is POD then*

$$\bigvee\left(\prod_{i=1}^{n} F_i\right)(s) \leq P\left(\sum_{i=1}^{n} X_i \leq s\right),$$

$$P\left(\sum_{i=1}^{n} X_i < s\right) \leq 1 - \bigvee\left(\prod_{i=1}^{n} \overline{F}_i\right)(s). \qquad (5.5)$$

Proof. (a) If $G \leq_{\mathrm{lo}} F_X$, that is $G(x) \leq F_X(x)$ for all $x \in \mathbb{R}^n$, then, for all $u \in \mathcal{U}(s)$, we have

$$P\left(\sum_{i=1}^{n} X_i \leq s\right) \geq P(X_1 \leq u_1, \ldots, X_n \leq u_n) = F_X(u) \geq G(u),$$

which implies (5.3).

(b) If $G \leq_{\mathrm{uo}} F_X$, that is $\overline{F}_X(x) \geq \overline{G}(x)$ for all $x \in \mathbb{R}^n$, then, for all $u \in \mathcal{U}(s)$, we have

$$P\left(\sum_{i=1}^{n} X_i < s\right) = 1 - P\left(\sum_{i=1}^{n} X_i \geq s\right)$$

$$\leq 1 - P(X_1 \geq u_1, \ldots, X_n \geq u_n)$$

$$= 1 - \overline{F}_X(u) \leq 1 - \overline{G}(u),$$

which implies (5.4).

(c) is a consequence of (a) and (b) when $G = \prod F_i$ is the product measure. □

Remark 5.2. (a) Similarly to the standard bounds (4.4), the bounds in Theorem 5.1 are in general sharp only when $d = 2$. However, they substantially improve the corresponding standard bounds (4.4), in the case of restriction to the positive dependence scenarios (a), (b) and (c). Under the positive dependence assumption in (c) we get the lower bound

$$\bigvee\left(\prod_{i=1}^{n} F_i\right)(s) = \sup\left\{\prod_{i=1}^{n} F_i(u_i); \ \sum u_i = s\right\} \qquad (5.6)$$

and the upper bound

$$1 - \sup\left\{\prod \overline{F}_i(u_i); \ \sum u_i = s\right\}, \qquad (5.7)$$

which are in fact enormous improvements on the standard bounds. They should be most relevant in applications, where the positive dependence assumption is satisfied. For the relevance of positive dependence assumptions see Remark 5.23.

(b) As is clear from the proof of Theorem 5.1, the functions G and $\overline{G}$ bounding the distribution F and, respectively the survival function $\overline{F}$, do not need in general to be distribution functions. It is enough to assume that G and $\overline{G}$ are nondecreasing functions. Using the Hoeffding–Fréchet bounds, we can generally assume that

$$\max\left\{\sum_{i=1}^{n} F_i(x_i) - (d-1), 0\right\} \leq G(x)$$

$$\text{and } \overline{G}(x) \geq \max\left\{\sum_{i=1}^{n}\overline{F}_i(x_i) - (d-1), 0\right\}.$$

(c) In the case of trivial upper and lower bounds $G, \overline{G}$ the bounds in (5.3), (5.4), and (5.5) reduce to the standard bounds in Proposition 4.1.   ◇

The improved bounds based on positive dependence restrictions extend in a similar way also to monotonically nondecreasing aggregation functions $\psi(x)$ (cf. Theorem 4.11). Let

$$A_\psi^+(t) = \left\{u = (u_1, \ldots, u_n); \ u \text{ is a maximal point in } \mathbb{R}^n \text{ with } \psi(u) \leq t\right\}$$

and

$$A_\psi^-(t) = \left\{u \in \mathbb{R}^n; \ \psi(u) \geq t\right\},$$

then we get the following bounds.

Theorem 5.3. *Let X be an n-dimensional risk vector with marginals $F_1, \ldots, F_n$. Let G be an n-dimensional distribution function and let $\psi = \psi(X)$ be a monotonically nondecreasing aggregation function, then:*

(a) If $G \leq_{\mathrm{lo}} F_X$, then

$$P\bigl(\psi(X) \leq s\bigr) \geq \bigvee_\psi^+ G(s) := \sup_{u \in A_\psi^+(s)} G(u); \tag{5.8}$$

(b) If $G \leq_{\mathrm{uo}} F_X$, then

$$P\bigl(\psi(X) < s\bigr) \leq 1 - \bigvee_\psi^- \overline{G}(s) := 1 - \sup_{u \in A_\psi^-(s)} \overline{G}(u); \tag{5.9}$$

(c) If X is POD, then

$$\bigvee_\psi^+ \left(\prod F_i\right)(s) \leq P\bigl(\psi(X) \leq s\bigr) \tag{5.10}$$

and

$$P\bigl(\psi(X) < s\bigr) \leq 1 - \bigvee_\psi^- \overline{G}(s). \tag{5.11}$$

5.2 Higher Order Marginals

In this section we consider the case that not only the one-dimensional marginal distributions of the risk vector $X = (X_1, \ldots, X_n)$ are known but it is assumed that for a class $\mathcal{C} \subset \mathcal{P}(1, \ldots, n)$ of subsets of $\{1, \ldots, n\}$ the marginal distributions P_J of $X_J = (X_j)_{j \in J}$, $J \in \mathcal{C}$ are known. As a consequence we get as a model the generalized Fréchet class $\mathcal{M}_{\mathcal{C}} = \mathcal{M}(P_J; J \in \mathcal{C})$ of all probability measures on the product space $E = E_1 \times \cdots \times E_n$, where $(E_i, \mathfrak{A}_i)$ are given Borel spaces with $X_i \in E_i$ and typically $E_i = \mathbb{R}$, such that the distribution of the components in J is P_J, $P^{\pi_J} = P_J$ for all $J \in \mathcal{C}$. Here $\pi_J : E \to \prod_{j \in J} E_j$ denotes the projection on the J-components. We generally assume that $\bigcup_{J \in \mathcal{C}} J = \{1, \ldots, n\}$. This implies that

$$\mathcal{M}_{\mathcal{C}} \subset \mathcal{M}(P_1, \ldots, P_n). \tag{5.12}$$

$\mathcal{M}_{\mathcal{C}}$ is a subclass of the simple Fréchet class which describes the class of all possible dependence structures. The knowledge of the higher dimensional marginals restricts the possible dependence structure and thus leads to improved bounds for risk functions. In order that the notion $\mathcal{M}_{\mathcal{C}}$ of generalized Fréchet class makes sense we have to assume "consistency" of the marginal system, i.e. for $J_1, J_2 \in \mathcal{C}$ such that $J_1 \cap J_2 \neq \phi$ holds

$$\pi_{J_1 \cap J_2} P_{J_1} = \pi_{J_1 \cap J_2} P_{J_2}. \tag{5.13}$$

Consistency of P_J, $J \in \mathcal{C}$ is a necessary condition to guarantee that the generalized Fréchet class $\mathcal{M}_{\mathcal{C}} \neq \phi$. It has been shown in Vorobev (1962) and Kellerer (1964) that when $\mathcal{C}$ is regular (i.e. it does not contain cycles) consistency is also sufficient to imply nonemptiness (see also Rü (1991a)). When $\mathcal{C}$ is nonregular as for example in the case $\mathcal{C} = \{\{1, 2\}, \{2, 3\}, \{1, 3\}\}$ the generalized Fréchet class may be empty, even with consistent marginals. Some relevant particular classes of marginal systems include

- The "simple system" $\mathcal{C}_n = \{\{1\}, \ldots, \{n\}\}$ defining the regular Fréchet class $\mathcal{M}(P_1, \ldots, P_n)$
- The "star-like system" $\mathcal{C}^* = \{\{1, j\}, 2 \le j \le n\}$ defining the regular Fréchet class $\mathcal{M}(P_{1,2}, \ldots, P_{1,n})$
- The "series system" $\mathcal{C}^s = \{\{i, i+1\}, 1 \le i \le n-1\}$ defining the regular Fréchet class $\mathcal{M}(P_{1,2}, \ldots, P_{n-1,n})$
- The "pairwise system" $\mathcal{C}^p = \{\{i, j\}, i \le j\}$ defining the nonregular Fréchet class $\mathcal{M}(P_{i,j}; 1 \le i \le j \le n)$.

Our aim is to find good bounds for the generalized Fréchet bounds

$$M_{\mathcal{C}}(\varphi) = \sup \left\{ \int \varphi \, dP; \ P \in \mathcal{M}_{\mathcal{C}} \right\},$$

$$m_{\mathcal{C}}(\varphi) = \inf \left\{ \int \varphi \, dP; \ P \in \mathcal{M}_{\mathcal{C}} \right\}. \tag{5.14}$$

96 5 Restrictions on the Dependence Structure

Relation (5.12) implies that

$$M_C(\varphi) \le M(\varphi), \qquad m_C(\varphi) \ge m(\varphi) \tag{5.15}$$

where $M(\varphi)$ and $m(\varphi)$ are the Fréchet bounds for the simple system $\mathcal{M} = \mathcal{M}(P_1, \ldots, P_n)$. For certain classes of functions a dual representation of the Fréchet bounds holds true (see Theorem 2.9)

$$M_C(\varphi) = U_C(\varphi) := \inf \left\{ \sum_{J \in C} \int f_J dP_J; \ \sum_{J \in C} f_j \circ \pi_J \ge \varphi \right\} \tag{5.16}$$

and

$$m_C(\varphi) = I_C(\varphi) := \sup \left\{ \sum_{J \in C} \int f_J dP_J; \ \sum_{J \in C} f_j \circ \pi_J \le \varphi \right\}. \tag{5.17}$$

Typically the dual problems in (5.16) and (5.17) are too difficult to be solvable. The main techniques to solve generalized Fréchet problems with multivariate marginals are reductions of these problems to simpler marginal systems like simple marginal systems. In some cases this can be done easily. Let for example $E_i = \mathbb{R}$, $1 \le i \le n$, and $C = \{J_1, \ldots, J_d\}$ be a class of non-overlapping multivariate marginals $P_J \in M^1(\mathbb{R}^J, \mathcal{B}^J)$, $J \in C$, i.e. $J_i \cap J_j = \emptyset$ for all $i \ne j$. Let $\varphi(X) = \Psi(\sum_{i=1}^n X_i)$ be a risk functional of the joint portfolio $\sum_{i=1}^n X_i$. Define $Y_i = \sum_{j \in J_i} X_j$ and let $Q_i \sim Y_i$ be the distribution of Y_i. Then we obtain the following obvious exact reduction result reducing the Fréchet problem for multivariate non-overlapping marginals to a simple marginal problem on the real line.

Proposition 5.4 (Non-overlapping marginals). *Let $C = \{J_1, \ldots, J_d\}$ be a class of non-overlapping multivariate marginals and $\varphi(X) = \psi(\sum_{i=1}^n X_i)$ be a risk functional. Then*

$$M_C(\varphi) = M_{\mathcal{H}}(\widetilde{\psi}), \tag{5.18}$$

where $\mathcal{H} = \mathcal{H}(Q_1, \ldots, Q_d)$ is the simple Fréchet class generated by the marginals Q_i and $\widetilde{\psi}(y_1, \ldots, y_d) = \psi\left(\sum_{i=1}^d y_i\right)$.

Two types of reduction techniques have been introduced in Rü (1991b), the conditioning method and a reduction method to Bonferroni bounds. These methods are applicable for certain marginal systems. A reduction technique for general marginal systems – restricted however to risk functionals of the form $\varphi\left(\sum_{i=1}^n X_i\right)$ – was given in Puccetti and Rü (2012b). This technique in general does not lead to sharp bounds but allows to obtain reasonably good bounds for any system of marginal information whether overlapping or non-overlapping.

5.2.1   A Reduction Principle and Bonferroni Type Bounds

We consider the generalized Fréchet class $\mathcal{M}_\mathcal{C}$ as introduced above and assume that $h_i : E_i \to W_i$ are mappings into Borel spaces $(W_i, \mathfrak{B}_i)$, $1 \leq i \leq n$. Define $h = (h_1, \ldots, h_n) : E \to W$ where $W = \prod W_i$ and $h(x) = (h_i(x_i))_{1 \leq i \leq n}$ and define $h_J := (h_j)_{j \in J}$. Then we have the following reduction principle (see Rü (1991a))

Proposition 5.5 (Reduction principle). *Let* $\mathcal{M}_\mathcal{C}^h := \{P^h; P \in \mathcal{M}_\mathcal{C}\}$ *denote the class of induced distributions. Then*

(a) $$\mathcal{M}_\mathcal{C}^h = \mathcal{M}(P_J; J \in \mathcal{C})^h \subset \mathcal{M}(P_J^{h_J}; J \in \mathcal{C}) \tag{5.19}$$

and for $\varphi \in \mathcal{L}^1(\mathcal{M}_\mathcal{C}^h) = \bigcap_{Q \in \mathcal{M}_\mathcal{C}^h} \mathcal{L}^1(Q)$ *holds*

$$M_\mathcal{C}(\varphi \circ h) \leq \sup \left\{ \int \varphi \, dQ; \ Q \in \mathcal{M}\left(P_J^{h_J}, \ J \in \mathcal{C}\right) \right\},$$

$$m_\mathcal{C}(\varphi \circ h) \geq \inf \left\{ \int \varphi \, dQ; \ Q \in \mathcal{M}\left(P_J^{h_J}, \ J \in \mathcal{C}\right) \right\}. \tag{5.20}$$

(b) If $\mathcal{C} = \{\{1\}, \ldots, \{n\}\}$ *is a simple marginal system then*

$$\mathcal{M}(P_1, \ldots, P_n)^h = \mathcal{M}\left(P_1^{h_1}, \ldots, P_n^{h_n}\right). \tag{5.21}$$

Proof. The proof of (a) is obvious. For the proof of b) let $\mu \in \mathcal{M}\left(P_1^{h_1}, \ldots, P_n^{h_n}\right)$ and let $(\Omega, \mathfrak{A}, R)$ be a non-atomic probability space and let $U : (\Omega, \mathfrak{A}) \to (W, \mathfrak{B})$ be such that $R^U = \mu$. Then

$$P_j^{h_j} = R^{U_j}$$

and by the result on the solution of stochastic equations (see Theorem 1.21) there exist random variables $X_j : (\Omega, \mathfrak{A}) \to (E_j, \mathfrak{B}_j)$ such that

$$h_j \circ X_j = U_j \text{ a.s. with respect to } R$$

and furthermore, $R^{X_j} = P_j$, $1 \leq i \leq n$. As a consequence the vector $X = (X_1, \ldots, X_n) : (E, \mathfrak{A}) \to (W, \mathfrak{B})$ has the distribution $R^X \in \mathcal{M}(P_1, \ldots, P_n) = \mathcal{M}$ and

$$h(X) = U[R].$$

This implies that $\mathcal{M}\left(P_1^{h_1}, \ldots, P_n^{h_n}\right) \subset \mathcal{M}\left(P_1, \ldots, P_n\right)^h.$ $\qquad \square$

So in the case of simple marginals there is a reduction of the problem of Fréchet bounds in (5.20) for functions of the type $\varphi \circ h$ to corresponding Fréchet bounds for the function φ w.r.t. the simple Fréchet class $\mathcal{M}(P_1^{h_1}, \ldots, P_n^{h_n})$. This reduction principle can be extended to general regular marginal systems (Rü (1991b)).

Theorem 5.6 (Generalized reduction principle). *Let M_C be a regular marginal system on E, then*

$$\mathcal{M}_C^h = \mathcal{M}\left(P_J^{h_J};\ J \in \mathcal{C}\right). \tag{5.22}$$

Proof. The proof is given for simple marginal systems in Proposition 5.5. It can be given constructively also for star-like systems $\mathcal{C}^*$ and for series systems $\mathcal{C}^s$. It is shown in Rü (1991b) that the case of general regular marginal systems can be reduced to these cases by induction.   □

Remark 5.7. For general marginal systems $\mathcal{C}$ the reduction theorem does not hold true. It is easy to construct examples where $\mathcal{M}_C^h = \phi$ but where

$$\mathcal{M}\left(P_j^{h_J};\ J \in \mathcal{C}\right) \neq \phi. \qquad\qquad \lozenge$$

An interesting application of the reduction principle implies the identity of (certain) sharp Fréchet bounds with the corresponding Bonferroni type bounds. Let $h_i := 1_{A_i}$, $A_i \in \mathfrak{A}_i$, $1 \le i \le n$, so $h_i : E_i \to \{0, 1\}$. Then with $B_i := E_1 \times \cdots \times A_i \times \cdots \times E_n \in \mathfrak{B}$ we obtain for $P \in \mathcal{M}_C = \mathcal{M}\left(P_J;\ J \in \mathcal{C}\right)$

$$P\left(A_1 \times \cdots \times A_n\right) = P\left(\bigcap_{j=1}^n B_j\right). \tag{5.23}$$

We denote by

$$p_J := P\,(B_J) = P_J\,(A_J), \tag{5.24}$$

where $A_J = \prod_{j \in J} A_j$, $B_J := \bigcap_{j=1}^n B_j$, $J \in \mathcal{C}$, the probability of the intersections B_J. The Bonferroni bounds state bounds for the probability of the intersection $P\left(\bigcap_{j=1}^n B_j\right)$ given the marginal probabilities $p_J = P(B_J)$, $J \in \mathcal{C}$. Let

$$M_C^B\left(\bigcap_{j=1}^n B_j\right) = \sup\left\{Q\left(\bigcap_{j=1}^n B_j\right);\ Q \in M^1(W, \mathfrak{B}), Q(B_J) = p_J, J \in \mathcal{C}\right\}$$
$$\tag{5.25}$$

$$m_C^B\left(\bigcap_{j=1}^n B_j\right) = \inf\left\{Q\left(\bigcap_{j=1}^n B_j\right);\ Q \in M^1(W, \mathfrak{B}), Q(B_J) = p_J, J \in \mathcal{C}\right\}$$
$$\tag{5.26}$$

denote the sharp upper and lower Bonferroni bounds. Note that in general knowledge of $p_J = P(\bigcap_{j \in J} B_j) = P(h_j = 1, j \in J)$ does not specify the distribution of h_J. Therefore, (even in the decomposable case) in general the Bonferroni bounds only give upper bounds for the Fréchet bounds. Let $\varphi : \{0, 1\}^n \to \mathbb{R}$, then we get by duality theory on $M^1(\{0, 1\}^n, \mathcal{P}(\{0, 1\}^n))$

$$M_C(\varphi \circ h) \leq M_C^B(\varphi) = U_C^B(\varphi)$$

$$:= \inf \left\{ \sum_{J \in C} \alpha_J p_J; \ \alpha_J \in \mathbb{R}, \ \sum_{J \in C} \alpha_J 1_{B_J} \geq \varphi \right\} \tag{5.27}$$

which reduces for the special case $\varphi(y) = \prod y_i = \prod_{j=1}^n h_j$ to

$$M_C(A_1 \times \cdots \times A_n) \leq M_C^B(\varphi) = U_C^B(\varphi)$$

$$= \inf \left\{ \sum_{J \in C} \alpha_J p_J; \ \sum_{J \in C} \alpha_J \geq 1, \ \sum_{\substack{J \in C \\ J \subset C}} \alpha_J \geq 0, \forall C \subset \{1, \ldots, n\} \right\} \tag{5.28}$$

(cf. Hailperin (1965)). If the marginal system C is founded, i.e.

$$J_1 \in C \text{ and } J_2 \subset J_1, \text{ then } J_2 \in C \tag{5.29}$$

then the distribution of h_J is specified by the knowledge of p_J, $J \in C$. Therefore, we can apply the reduction theorem in the decomposable case.

Theorem 5.8 (Fréchet bounds and Bonferroni bounds). *Let C be a regular, founded marginal system and let $\mathcal{M}_C = \mathcal{M}(P_J, J \in C)$ be the corresponding Fréchet class. For $A_i \in \mathfrak{A}_i$ define $B_i = E_1 \times \cdots \times A_i \times \cdots \times E_n$ and let $p_J = P_J(A_J)$, $J \in C$ be the associated Bonferroni system. Then the sharp Fréchet bounds coincide with the sharp Bonferroni bounds, i.e.*

$$\mathcal{M}_C\big(A_1 \times \cdots \times A_n\big) = \mathcal{M}_C^B\left(\bigcap_{j=1}^n B_j \right) \tag{5.30}$$

and

$$m_C\big(A_1 \times \cdots \times A_n\big) = m_C^B\left(\bigcap_{j=1}^n B_j \right). \tag{5.31}$$

Proof. For the proof we apply the reduction theorem, Theorem 5.6, to $\varphi(y) = \prod_{i=1}^n y_i$, $h_i(x_i) = 1_{A_i}(x_i)$. Then

$$\mathcal{M}_C\big(A_1 \times \cdots \times A_n\big) = \mathcal{M}_C(\varphi \circ h) = \widetilde{M}(\varphi),$$

where $\widetilde{\mathcal{M}} = \mathcal{M}\big(P_J^{h_J}; \ J \in C\big) \subset \mathcal{M}^1(\{0,1\}^n, \mathcal{P}(\{0,1\}^n))$. For $Q \in \widetilde{\mathcal{M}}$ knowledge of $p_J = Q(B_J)$, $J \in C$ implies knowledge of $P_J^{h_J}$ on $\{0,1\}^J$, $J \in C$, as C is founded. Therefore, we conclude

$$\widetilde{\mathcal{M}}(\varphi) = \sup \left\{ \int \varphi dQ; \ Q \in \mathcal{M}\left(P_J^{h_J}; \ J \in C \right) \right\}$$

$$= \sup \left\{ Q\left(\bigcap_{i=1}^{n} B_i \right); \quad Q(B_J) = p_J, J \in \mathcal{C} \right\}$$

$$= \mathcal{M}_{\mathcal{C}}^{B} \left(\bigcap_{i=1}^{n} B_i \right). \qquad \qquad \square$$

We get in particular in the case of simple marginals the equality of the Fréchet bounds with the simple Bonferroni bounds which was established in Proposition 5.5 (see Rü (1981a)) as a consequence of duality theory.

Corollary 5.9 (Sharpness of Fréchet bounds). *Let* $A_i \in \mathfrak{A}_i$, $P_i \in M^1(E_i, \mathfrak{A}_i)$ *and let* $p_i := P_i(A_i)$, $i \leq i \leq n$. *Then with* $M = M(P_1, \dots, P_n)$ *it holds that*

$$M\left(A_1 \times \cdots \times A_n\right) = \min_{1 \leq i \leq n} p_i \qquad (5.32)$$

$$m\left(A_1 \times \cdots \times A_n\right) = \left(\sum_{i=1}^{n} p_i - (n-1) \right)_{+}. \qquad (5.33)$$

So the knowledge of the complete distributions P_i does not help to improve the Bonferroni bounds. As a consequence of results in Rüger (1981) for Bonferroni type bounds and the reduction principle above we get for simple marginal systems (see Rü (1991b)),

Proposition 5.10. *Let* $P_i \in M^1(E_i, \mathfrak{A}_i)$, $1 \leq i \leq n$, $A_i \in \mathfrak{A}_i$ *and define* $p_i = P_i(A_i)$ *and*

$$L_k := \bigcup_{\substack{J \subset \{1,\dots,n\} \\ |J|=k}} \bigcap_{j \in J} \{\pi_j \in A_j\} \subset E$$

the event that for at least k *components* $x_j \in A_j$ *holds true. Then the sharp Fréchet bounds are given by*

$$M(L_k) = b_k := \min\left(1, \frac{1}{k-r} \sum_{i=1}^{n-r} p_{(i)}, \quad 0 \leq r \leq k-1 \right) \qquad (5.34)$$

and

$$m(L_k) = a_k := \max\left(0, \frac{\sum_{i=r+1}^{n} p_{(i)} - (k-1)}{n-r-(k-1)}, \quad 0 \leq r \leq n-k \right) \qquad (5.35)$$

where $p_{(1)} \leq \cdots \leq p_{(n)}$.

Remark 5.11 (Order statistics). As an application we consider real random variables $X_1, \ldots, X_n$ and we denote by $X_{(1)} \leq X_{(2)} \leq \cdots \leq X_{(n)}$ the order statistics (in increasing order). Then with $A_i = \{X_i \geq t\}$ we get

$$L_k = \{X_{(k)} \geq t\}, \quad 1 \leq k \leq n. \tag{5.36}$$

As a consequence (5.34) and (5.35) give sharp Fréchet bounds for the tails of order statistics.   ◊

We next give some consequences of the reduction principle in connection with Bonferroni bounds of higher order. For results on Bonferroni bounds we refer to Galambos (1978) and Tong (1980).

Theorem 5.12 (Higher order bounds). *Let $A_i \in \mathfrak{A}_i$, $1 \leq i \leq n$ and consider a marginal system C. Then*

(a) $M_C(A_1 \times \cdots \times A_n) \leq \min\limits_{J \in C} P_J(A_J).$ (5.37)

(b) If $C = C^k := \{J \subset \{1, \ldots, n\}; \ |J| = k\}$, then

$$M_C(A_1 \times \cdots \times A_n) \leq 1 + \sum_{j=1}^{k} (-1)^j q_j \quad \text{for } k \in 2\mathbb{N}$$

and (5.38)

$$m_C(A_1 \times \cdots \times A_n) \geq 1 + \sum_{j=1}^{k} (-1)^j q_j \quad \text{for } k \in 2\mathbb{N} - 1$$

where $q_j := \sum\limits_{|J|=j} P_J\left(\bigcap\limits_{j \in J} A_j^c\right).$

(c) $m_C(A_1 \times \cdots \times A_n) \geq \begin{cases} \sum\limits_{j=1}^{n-1} (-1)^{j+1} r_j - 1, & n \in 2\mathbb{N} \\[2mm] \sum\limits_{j=1}^{n-1} (-1)^{j+1} r_j - \min\limits_{|J|=n-1} P_J(A_J), & n \in 2\mathbb{N} - 1 \end{cases}$ (5.39)

where $r_J = \sum\limits_{|J|=j} P_J(A_J).$

(d) If $C = C^2$ is the system of pairwise marginals then with $q_i = P\left(A_i^c\right)$, $q_{ij} = P_{ij}\left(A_i^c \times A_j^c\right)$ it holds that

$$M_C(A_1 \times \cdots \times A_n) \leq 1 - \sum_{i=1}^{n} q_i + \sum_{i<j}^{n} q_{ij} \tag{5.40}$$

and

$$m_C(A_1 \times \cdots \times A_n) \geq 1 - \sum_{i=1}^{n} q_i + \sup_{\tau} \sum_{(i,j)\in\tau} q_{ij} \qquad (5.41)$$

where the supremum is on the set of all spanning trees of the complete graph with n nodes $1, \ldots, n$.

Proof. (a) Is obvious from Proposition 5.5 and (5.19).

(b) Follows from (5.19) and the usual Bonferroni bounds of order up to k.

(c) Is a slight modification of Poincaré's Theorem (see Warmuth (1988)).

(d) Let $P \in M_C$, $\varphi \circ h = \max_{\{1\leq i\leq n\}} 1_{\{\pi_i \in A_i^c\}} = 1_{\bigcup_{i=1}^{n}\{\pi_i \in A_i^c\}}$, then

$$P(A_1 \times \cdots \times A_n) = 1 - P\left(\bigcup_{i=1}^{n}\{\pi_i \in A_i^c\}\right) = 1 - \int \varphi \circ h \, dP.$$

From (5.19) we obtain

$$M_C(\varphi \circ h) \leq M_C^B(\varphi) = U_C^B(\varphi)$$

$$= \inf\left\{\sum \alpha_i q_i - \sum \alpha_{ij} q_{ij}; \quad \forall J \subset \{1,\ldots,n\}, \sum_{i\in J}\alpha_i - \sum_{\substack{i<j \\ i,j\in J}}\alpha_{ij} \geq 1\right\}$$

$$\leq \inf\left\{\sum \alpha_i q_i - \sum \alpha_{ij} q_{ij}; \quad \alpha_{ij} \geq 0, \sum_{i\in J}\alpha_i - \sum_{\substack{i<j \\ i,j\in J}}\alpha_{ij} \geq 1, \forall J \in \mathcal{C}\right\}$$

$$= \sum q_i - \sup_{\tau}\sum_{(i,j)\in\tau} q_{ij}$$

(cf. Maurer (1983, Proposition 1) and Hunter (1976)). This implies the lower bound for $m_C(A_1 \times \cdots \times A_n)$. A direct proof of (d) follows also from the inequality

$$1_{\bigcup_{i=1}^{n} B_i} \leq \sum_{i=1}^{n} 1_{B_i} - \sum_{(i,j)\in\tau} 1_{B_i\cap B_j} \qquad (5.42)$$

holding true for all spanning trees τ, where a spanning tree is a system of $n-1$ edges (i, j) containing each of the n nodes of the graph. Relation (5.41) implies that

$$P\left(A_1 \times \cdots \times A_n\right) \geq 1 - \sum_{i=1}^{n} q_i + \sup_{\tau} \sum_{(i,j)\in\tau} q_{ij}$$

and thus the lower bound (5.41) for $m_C\left(A_1 \times \cdots \times A_n\right)$. The upper bound for $M_C\left(A_1 \times \cdots \times A_n\right)$ in (5.40) is obtained by applying the standard upper Bonferroni inequality of second order.   □

Remark 5.13. (a) For $E_j = R^1$, $A_j = (\infty, x_j]$, $P \in M_C$ follows from (5.37) for the distribution function

$$F_P(x) \leq \min_{J \in C} F_J(x_J), \quad F_J = F_{P_J}. \tag{5.43}$$

(5.43) was stated in the case $C = I^k = \{J \subset \{1,\ldots,n\}; \ |J| = k\}$ in Warmuth (1988). The right-hand side in (5.43) is however (in contrast to the statement in Theorem 2.3 of that paper) not a distribution function with marginals P_J and, therefore, the bounds in (5.43) are in general not sharp. Similarly also the corresponding lower bounds for F_P deriving from Theorem 5.12c are not sharp. Various further bounds like Chung–Erdős bounds can be applied in specific applications.

(b) In the case of a series system $C = \{(i, i+1); \ 1 \leq i \leq n-1\}$ the statement in (d) implies the following lower bounds for m_C

$$m_C\left(A_1, \ldots, A_n\right) \geq 1 - \sum_{i=1}^{n} q_i + \sum_{i=1}^{n-1} q_{ii+1}. \tag{5.44}$$

More generally if C is founded one can restrict in the duality representations (5.27) and (5.28) to $\alpha_J \in \{-1, 0, 1\}$. The minimal admissible $\alpha = (\alpha_J)_{J \in C}$ are related to some interesting combinatorial configurations.   ◊

5.2.2   The Conditioning Method

The proof of the reduction principle in Section 5.2.1 is based on the property that general regular structures can be reduced to some simple configurations, namely the simple marginals $\mathcal{M}(P_1, \ldots, P_n)$, the series system $\mathcal{M}(P_{1,2}, P_{2,3}, \ldots, P_{n-1,n})$, and the star-like system $\mathcal{M}(P_{1j}; \ 2 \leq j \leq n)$. In this section we derive (sharp) bounds for these basic structures by the method of conditioning and then apply this method also to some nonregular structures. The results in this section are mostly due to Rü (1991b). Some similar results can be found in Joe (1997).

For the basic series structure $\mathcal{C} = \{\{1,2\},\{2,3\}\}$ we can give a reduction of the Fréchet problem to the simple marginal case by conditioning on the second component. Let

$$P_{1|x_2} := P_{12}^{\pi_1|\pi_2=x_2} \tag{5.45}$$

denote the conditional distribution of the first component given the second component is x_2.

Proposition 5.14 (Basic series system). *For $A \in \mathfrak{A}_1 \otimes \mathfrak{A}_2 \otimes \mathfrak{A}_3$ and $\mathcal{C} = \mathcal{C}_3^s = \{\{1,2\},\{2,3\}\}$ we have*

$$M_{\mathcal{C}}(A) = \int M_{1,3|x_2}(A_{x_2})dP_2(x_2)$$

and $\tag{5.46}$

$$m_{\mathcal{C}}(A) = \int m_{1,3|x_2}(A_{x_2})dP_2(x_2)$$

where $M_{1,3|x_2}$, $m_{1,3|x_2}$ are the sharp upper and lower Fréchet bounds w.r.t. the simple marginal class $M(P_{1|x_2}, P_{3|x_2})$ and where A_{x_2} is the x_2-section of A.

Proof. For any $P \in M_{\mathcal{C}}$ we have by conditioning under x_2

$$P(A) = \int P_{1,3|x_2}(A_{x_2})dP_2(x_2),$$

where

$$P_{13|x_2} = P^{(\pi_1,\pi_3)|\pi_2=x_2} \in M(P_{1|x_2}, P_{3|x_2}).$$

This implies that

$$M_{\mathcal{C}}(A) \le \int M_{1,3|x_2}(A_{x_2})dP_2(x_2)$$

and

$$m_{\mathcal{C}}(A) \ge \int m_{1,3|x_2}(A_{x_2})dP_2(x_2).$$

Let on the other hand $x_2 \to P_{13|x_2}(A_{x_2})$ be a maximum point of $x_2 \to M_{1,3|x_2}(A_{x_2})$ which can be chosen as Markov kernel w.r.t. the completed σ-algebras (see Rü (1985)). Then $P^* := P_{13|x_2}^* \times P_2(dx_2) \in M_{\mathcal{C}}$ and, therefore,

$$M_C(B) = \int P^*_{13|x_2}(B_{x_2})dP_2(x_2) = P^*(B). \qquad \square$$

Remark 5.15. For $A \in \mathfrak{A}_1 \otimes \mathfrak{A}_2$ the upper and lower bounds under the integral in (5.46) have an *explicit* representation by the Strassen theorem (see Theorem 2.6)

$$M_{1,3|x_2}(A) = U_{13|x_2}(A)$$
$$:= \inf\{P_{1|x_2}(A_1) + P_{3|x_2}(A_2); \ A \subset A_1 \times E_3 \cup E_1 \times A_3\}$$

and

$$m_{1,3|x_2}(A) = L_{13|x_2}(A)$$
$$:= \sup\{P_{1|x_2}(A_1) + P_{3|x_2}(A_2) - 1; \ A \supset A_1 \times A_2\}.$$

Similarly to Proposition 5.14, one obtains for functions $\varphi = \varphi(x_1, x_2, x_3)$

$$M_C(\varphi) = \int U_{13|x_2}\varphi_{x_2}dP_2(x_2) \tag{5.47}$$

and

$$m_C(\varphi) = \int L_{13|x_2}\varphi_{x_2}dP_2(x_2) \tag{5.48}$$

where φ_{x_2} denotes the x_2 section of φ. $\qquad \diamondsuit$

For the star-like system $C^s = \{\{1, j\}; \ 2 \le j \le n\}$ we obtain similarly to Proposition 5.14 for the basic series case an exact reduction to the simple marginal case.

Proposition 5.16 (Reduction of star-like systems). *For any integrable function* $\varphi : (E, \mathfrak{A}) \to (\mathbb{R}^1, \mathfrak{B}^1)$ *it holds that*

$$M_{C^s}(\varphi) = \int M_{2,\dots,n|x_1}\varphi_{x_1}dP_1(x_1) \tag{5.49}$$

and

$$m_{C^s}(\varphi) = \int m_{2,\dots,n|x_1}\varphi_{x_1}dP_1(x_1). \tag{5.50}$$

For the simplest nonregular structure, a circle of length 3, we obtain as a consequence the following reduction to a simple marginal structure.

Proposition 5.17 (Pairwise marginals, $\mathcal{C} = C_3^p = \{\{1,2\}\{2,3\}\{1,3\}\}$). *For the pairwise marginal structure $\mathcal{C} = C_3^p$ and $\varphi = \varphi(x_1, x_2, x_3)$ we have the following bounds:*

$$M_{\mathcal{C}}(\varphi) \leq \min \left\{ \int M_{23|x_1}(\varphi_{x_1}) dP_1(x_1), \right.$$

$$\left. \int M_{13|x_2}(\varphi_{x_2}) dP_2(x_2), \int M_{12|x_3}(\varphi_{x_3}) dP_3(x_3) \right\} \quad (5.51)$$

and

$$m_{\mathcal{C}}(\varphi) \geq \max \left\{ \int m_{23|x_1}(\varphi_{x_1}) dP_1(x_1), \right.$$

$$\left. \int m_{13|x_2}(\varphi_{x_2}) dP_2(x_2), \int m_{12|x_3}(\varphi_{x_3}) dP_3(x_3) \right\}. \quad (5.52)$$

Proof. The proof follows from the representation

$$\mathcal{M}_{\mathcal{C}} = \mathcal{M}(P_{12}, P_{13}, P_{23}) = \mathcal{M}(P_{12}, P_{23}) \cap \mathcal{M}(P_{13}, P_{23}) \cap \mathcal{M}(P_{12}, P_{13})$$

and the reduction of the basic series system in Proposition 5.14.   □

Remark 5.18 (Higher order Fréchet bounds). For $A = A_1 \times A_2 \times A_3$ and $x_1 \in A_1$ it holds that:

$$M_{23|x_1}(A_1) = U_{23|x_1}(A_2 \times A_3) = \min(P_{2|x_1}(A_2), P_{3|x_1}(A_3))$$

and, therefore, we get from (5.51)

$$M_{C_3^p}(A) \leq \min \left\{ \int_{A_1} U_{23|x_1}(A_2 \times A_3) dP_1(x_1), \right.$$

$$\left. \int_{A_2} U_{13|x_2}(A_1 \times A_3) dP_2(x_2), \int_{A_3} U_{13|x_3}(A_1 \times A_2) dP_3(x_3) \right\},$$

$$(5.53)$$

which is an improvement of the Bonferroni bounds $\min\{P_{12}(A_1 \times A_2), P_{13}(A_1 \times A_3), P_{23}(A_2 \times A_3)\}$ in (5.37). Similarly

$$m_{23|x_1}(A_{x_1}) = L_{23|x_1}(A_2 \times A_3) = \left(P_{2|x_1}(A_2) + P_{3|x_1}(A_3) - 1\right)_+ \quad (5.54)$$

and

$$m_{C_3^p}(A) \geq \max \left\{ \int_{A_1} L_{23|x_1}(A_2 \times A_3) dP_1(x_1), \right.$$

$$\int_{A_2} L_{13|x_2}(A_1 \times A_3) dP_2(x_2),$$

$$\left. \int_{A_3} L_{12|x_3}(A_1 \times A_2) dP_3(x_3) \right\}. \tag{5.55}$$

In the case of $E_i = \mathbb{R}^1$ and $A_i = (-\infty, x_i]$ we get the following collection of higher order Fréchet bounds on the distribution function $F = F(x_1, x_2, x_3)$. From (5.53) we have the conditional upper bound

$$F(x) \leq \min \left\{ \int_{-\infty}^{x_1} \min(F_{2|y_1}(y_2), F_{3|y_1}(y_2)) dF_1(y_1), \right.$$

$$\int_{-\infty}^{x_3} \min(F_{1|y_3}(y_1), F_{2|y_3}(y_2)) dF_3(y_3),$$

$$\left. \int_{-\infty}^{x_2} \min(F_{1|y_2}(y_1), F_{3|y_2}(y_3)) dF_2(y_2) \right\}$$

$$\leq \min(F_{12}(x_1, x_2), F_{13}(x_1, x_3), F_{23}(x_2, x_3)). \tag{5.56}$$

With $F_{ij} = F_{ij}(x_i, x_j)$, $F_i = F_i(x_i)$ and $\overline{F}_{ij} = \overline{F}_{ij}(x_i, x_j)$, $\overline{F}_i = \overline{F}_i(x_i)$ we obtain from Theorem 5.12 the Bonferroni upper bound for $n \geq 3$

$$F(x) \leq 1 - \sum_i \overline{F}_i + \sum_{i<j} \overline{F}_{ij}. \tag{5.57}$$

The identity $\overline{F}(x) = 1 - \sum_i F_i + \sum_{i<j} F_{ij} - F(x)$ for $n = 3$ implies

$$F(x) \leq 1 - \sum_i F_i + \sum_{i<j} F_{ij}. \tag{5.58}$$

Finally from (5.41) we have the lower bound

$$F(x) \geq 1 - \sum_i \overline{F}_i + \max\{\overline{F}_{12} + \overline{F}_{23}, \overline{F}_{13} + \overline{F}_{12}, \overline{F}_{13} + \overline{F}_{23}\}. \tag{5.59}$$

$\Diamond$

5.2.3   Reduction Bounds for the Joint Portfolio in General Marginal Systems

In this section we consider the case that the components X_i of the risk vector X are real and we are interested in the distribution function or the tail of the joint portfolio $\sum_{i=1}^{n} X_i$. A reduction technique for these risk bounds has been given in the series case in Embrechts and Puccetti (2010) and extended to general marginal systems in Puccetti and Rü (2012b).

Let $\mathcal{C} = \{J_1, \ldots, J_m\}$ be any marginal system with corresponding Fréchet class $\mathcal{M}_\mathcal{C} = \{P_J; \ J \in \mathcal{C}\}$. Let $\eta_i := |J_i|$ denote the number of components in J_i, $1 \le i \le m$. We associate to the risk vector X with distribution $P^X \in \mathcal{M}_\mathcal{C}$ random variables Y_r defined by

$$Y_r = \sum_{i \in J_r} \frac{X_i}{\eta_i}, \quad r = 1, \ldots, m. \tag{5.60}$$

Let Q_r be the distribution of Y_r and let $\mathcal{H}$ denote the simple Fréchet class

$$\mathcal{H} = \mathcal{M}(Q_1, \ldots, Q_m).$$

Then we obtain the following reduction result reducing the Fréchet problems $M_\mathcal{C}$, $m_\mathcal{C}$ for the general Fréchet class $\mathcal{M}_\mathcal{C}$ to the Fréchet problems $M_\mathcal{H}$, $m_\mathcal{H}$ for the simple Fréchet class $\mathcal{H}$.

$$M_\mathcal{H}(s) = \sup\left\{ P\left(\sum_{i=1}^{m} Y_i \ge s \right); \ P^Y \in \mathcal{H} \right\}$$

$$m_\mathcal{H}(s) = \inf\left\{ P\left(\sum_{i=1}^{m} Y_i \ge s \right); \ P^Y \in \mathcal{H} \right\}. \tag{5.61}$$

Proposition 5.19 (Reduction to simple Fréchet class). *Let* $\mathcal{C} = \{J_1, \ldots, J_m\}$ *be a consistent marginal system. Then we obtain:*

$$M_\mathcal{C}(s) \le M_\mathcal{H}(s) \quad and \quad m_\mathcal{C}(s) \ge m_\mathcal{H}(s). \tag{5.62}$$

Proof. Since the distribution F_X of the vector X belongs to $\mathcal{M}_\mathcal{C} = M(P_J, J \in \mathcal{C})$, the distribution H_Y of Y belongs to the simple Fréchet class $\mathfrak{F}(H_1, \ldots, H_m)$. Using that

$$\sum_{i=1}^{d} X_i = \sum_{r=1}^{n} \sum_{i \in J_r} \frac{X_i}{\eta_i} = \sum_{r=1}^{m} Y_r, \tag{5.63}$$

we obtain

$$\begin{aligned} M_\mathcal{C}(s) &= \sup\{P(X_1 + \cdots + X_n \ge s); \ F_X \in \mathfrak{F}_\mathcal{C}\} \\ &= \sup\{P(Y_1 + \cdots + Y_m \ge s); \ F_X \in \mathfrak{F}_\mathcal{C}\} \\ &\le \sup\{P(Y_1 + \cdots + Y_m \ge s); \ F_Y \in \mathfrak{H}\} = M_\mathfrak{H}(s), \end{aligned}$$

and similarly

$$m_C(s) = \inf\{P(X_1 + \cdots + X_n \geq s); \ F_X \in \mathfrak{F}_C\}$$
$$= \inf\{P(Y_1 + \cdots + Y_m \geq s); \ F_X \in \mathfrak{F}_C\}$$
$$\geq \inf\{P(Y_1 + \cdots + Y_m \geq s); \ F_Y \in \mathfrak{H}\} = m_{\mathfrak{H}}(s). \qquad \square$$

As a consequence of the standard and dual bounds given in Chapter 4, we obtain the following standard and dual bounds for general marginal systems.

Theorem 5.20 (Reduced standard bounds). *Let P_J, $J \in C$, with $C = \{J_1, \ldots, J_m\}$, be a consistent system of marginals generating a nonempty Fréchet class $\mathcal{M}_C$. Let $X = (X_1, \ldots, X_d)$ be a random vector with distribution P^X in $\mathcal{M}_C$. For a given threshold $s \in \mathbb{R}$, define the set $\mathcal{U}(s) \subset \mathbb{R}^n$ as*

$$\mathcal{U}(s) = \left\{ u = (u_1, \ldots, u_n) \in \mathbb{R}^n; \ \sum_{r=1}^{n} u_r = s \right\}.$$

Then, we have

$$M_C(s) \leq M_{\mathfrak{H}}(s) \leq S_{\mathfrak{H}}(s) = \inf_{u \in \mathcal{U}(s)} \left\{ \min\left\{ \overline{H}_1(u_1^-) + \sum_{r=2}^{m} \overline{H}_r(u_r), 1 \right\} \right\},$$

$$m_C(s) \geq m_{\mathfrak{H}}(s) \geq s_{\mathfrak{H}}(s) = \sup_{u \in \mathcal{U}(s)} \left\{ \max\left\{ \sum_{r=1}^{n} \overline{H}_r(u_r^-) - n + 1, 0 \right\} \right\} \quad (5.64)$$

where H_i, $\overline{H}_i$ are the distribution respectively survival function of Y_i.

Based on Theorem 5.20 we obtain the following reduced dual bounds.

Theorem 5.21 (Reduced dual bounds). *Under the assumptions of Theorem 5.20, we have*

$$M_C(s) \leq M_{\mathfrak{H}}(s) \leq D_{\mathfrak{H}}(s) = \inf_{u \in \overline{\mathcal{U}}(s)} \min\left\{ \frac{\sum_{r=1}^{n} \int_{u_r}^{s - \sum_{j \neq r} u_j} \overline{H}_r(t)dt}{s - \sum_{r=1}^{n} u_r}, 1 \right\},$$

$$m_C(s) \geq m_{\mathfrak{H}}(s) \geq d_{\mathfrak{H}}(s)$$

$$= \sup_{u \in \underline{\mathcal{U}}(s)} \max\left\{ \frac{\sum_{r=1}^{n} \int_{u_r}^{s - \sum_{j \neq r} u_j} \overline{H}_r(t)dt}{s - \sum_{r=1}^{n} u_r} - n + 1, 0 \right\}, \quad (5.65)$$

where

$$\overline{\mathcal{U}}(s) = \left\{ u \in \mathbb{R}^n; \ \sum_{r=1}^n u_r < s \right\} \ and \ \underline{\mathcal{U}}(s) = \left\{ u \in \mathbb{R}^n; \ \sum_{r=1}^n u_r > s \right\}.$$

In case $\mathcal{C} = \{\{1\}, \ldots, \{d\}\}$ is the system of simple marginals, in the above theorems we have $m = n$ and $H_i = F_i, 1 \leq i \leq d$. Thus, the reduced bounds (5.64) and (5.65) are equivalent forms of the standard, respectively dual bounds given in Chapter 4. Moreover, since the standard bounds are particular cases of the dual ones, we have that

$$D_{\mathfrak{H}}(s) \leq S_{\mathfrak{H}}(s) \ and \ d_{\mathfrak{H}}(s) \geq s_{\mathfrak{H}}(s), \ for \ all \ s \in \mathbb{R}.$$

It is possible that the two inequalities in (5.62) are strict, implying that

$$s_{\mathfrak{H}}(s) \leq d_{\mathfrak{H}}(s) \leq m_{\mathfrak{H}}(s) < m_{\mathcal{C}}(s), \tag{5.66}$$

$$M_{\mathcal{C}}(s) < M_{\mathfrak{H}}(s) \leq D_{\mathfrak{H}}(s) \leq S_{\mathfrak{H}}(s). \tag{5.67}$$

In order to reduce the gap between the sharp bounds $m_{\mathcal{C}}(s)$ and $M_{\mathcal{C}}(s)$ and their reduced standard and dual counterparts, one can parameterize the Fréchet class $\mathfrak{H}$ as follows. Let $\alpha = \{\alpha_i^r \in [0,1]; \ i = 1, \ldots, n, \ r = 1, \ldots, m\}$ be a set of weighting parameters such that

$$\alpha_i^r > 0 \ if \ and \ only \ if \ i \in J_r, \ and \ \sum_{r=1}^m \alpha_i^r = 1. \tag{5.68}$$

The main idea is to choose for the reduction to simple marginal classes an optimal system of weights putting different weights on the components within their groups J_i. Then, for $1 \leq r \leq m$, define the random variables Y_r^α as

$$Y_r^\alpha = \sum_{i=1}^n \alpha_i^r X_i.$$

Hence, we have

$$\sum_{r=1}^m Y_r^\alpha = \sum_{r=1}^m \sum_{i=1}^n \alpha_i^r X_i = \sum_{i=1}^n \sum_{r=1}^m \alpha_i^r X_i = \sum_{i=1}^n X_i \sum_{r=1}^m \alpha_i^r = \sum_{i=1}^n X_i. \tag{5.69}$$

We denote by H_r^α the distribution of the random variable Y_r^α. Since $\alpha_i^r > 0$ if and only if $i \in J_r$, the random variable Y_r^α depends only on the X_i whose index i appears within the same J_r. Considering also that $P^X \in \mathcal{M}(P_J; \ J \in \mathcal{C})$, the marginals of the random vector $Y^\alpha = (Y_1^\alpha, \ldots, Y_m^\alpha)$ turn out to be fixed.

Therefore, the joint distribution H_Y^α of Y^α belongs to the simple Fréchet class $\mathfrak{H}^\alpha = \mathcal{M}(H_1^\alpha, \ldots, H_n^\alpha)$ and we obtain a reduction result analogous to Theorem 5.19.

Theorem 5.22 (Weighting scheme bounds). *Let $\mathcal{C} = \{J_1, \ldots, J_m\}$ be an arbitrary consistent marginal system with corresponding Fréchet class $\mathcal{M}_\mathcal{C} = \mathcal{M}(F_J; J \in \mathcal{C})$. For α satisfying conditions (5.68), define the simple Fréchet bounds*

$$M_{\mathfrak{H}}^\alpha(s) = \sup\{P(Y_1^\alpha + \cdots + Y_n^\alpha \geq s); \ P_Y \in \mathfrak{H}^\alpha\},$$
$$m_{\mathfrak{H}}^\alpha(s) = \inf\{P(Y_1^\alpha + \cdots + Y_n^\alpha \geq s); \ P_Y \in \mathfrak{H}^\alpha\}.$$

Then, for any $s \in \mathbb{R}$, we have

$$M_\mathcal{C}(s) \leq M_{\mathfrak{H}}^\alpha(s) \ \text{ and } \ m_\mathcal{C}(s) \geq m_{\mathcal{C}_n}^\alpha(s). \tag{5.70}$$

By applying Theorems 5.20 and 5.21 to the Fréchet class $\mathfrak{H}^\alpha$, we define the reduced standard bounds $S_{\mathfrak{H}}^\alpha(s)$, $s_{\mathfrak{H}}^\alpha(s)$, and the reduced dual bounds $D_{\mathfrak{H}}^\alpha(s)$, $d_{\mathfrak{H}}^\alpha(s)$. These bounds are obtained by replacing, in Eqs. (5.64) and (5.65), the marginals H_r by the parameterized marginals H_r^α. Note that, choosing $\alpha_i^r = 1/\eta_i$, $i \in J_r, 1 \leq r \leq n$, these parameterized reduced standard and dual bounds coincide with the ones given in (5.64) and (5.65).

To summarize, for any fixed set of parameters $\alpha_i^r \in [0, 1]$ satisfying (5.68), the following inequalities hold:

$$s_{\mathfrak{H}}^\alpha(s) \leq d_{\mathfrak{H}}^\alpha(s) \leq m_\mathcal{C}(s),$$
$$M_\mathcal{C}(s) \leq D_{\mathfrak{H}}^\alpha(s) \leq S_{\mathfrak{H}}^\alpha(s).$$

At this point, one can look for the α yielding the best possible parameterized dual bound $d^*(s)$ and $D^*(s)$, defined as

$$d^*(s) = \sup_{\alpha \in \mathfrak{A}} d_{\mathfrak{H}}^\alpha(s) \quad \text{and} \quad D^*(s) = \inf_{\alpha \in \mathfrak{A}} D_{\mathfrak{H}}^\alpha(s), \tag{5.71}$$

where $\mathfrak{A} \subset \mathbb{R}^n \times \mathbb{R}^d$ denotes the set of admissible choice of the parameters α.

Remark 5.23 (Rearrangement algorithm and relevance of dependence assumptions). As mentioned in Section 4.1 a numerical rearrangement algorithm (RA) for the approximation of sharp upper and lower bounds for the tail probabilities $P(\sum_{i=1}^n X_i \geq s)$ of general inhomogeneous portfolios was introduced in Puccetti and Rü (2012b). RA allows to approximate the sharp bounds up to dimension $n = 40$.

In Embrechts, Puccetti, and Rü (2012) this rearrangement algorithm was modified and extended to the approximation of sharp upper and lower bounds of the value at risk $\mathrm{VaR}_\alpha(\sum_{i=1}^n X_i)$. This modification allows to deal with high dimensional portfolios up to about dimension $n = 600$.

In this paper it was observed in several examples that positive dependence information as dealt with in Section 5.1 does not improve the upper bounds on the value at risk essentially. This statement is not in contrast with the finding in Section 5.1 that this positive dependence information leads to a strong improvement of the standard bounds. On the other hand it is shown there that higher order marginal information may lead to an essential improvement of the upper bound for the value of risk.   ◊

Chapter 6
Dependence Orderings of Risk Vectors and Portfolios

There are several dependence orderings $\prec$ for risk vectors X, Y available which allow to infer conclusions of the type $X \prec Y$, i.e. Y is stronger positive dependent than X, implies that Y bears more risk than X. Of particular interest is the case where X, Y have identical marginals, i.e. belong to the same Fréchet class. While the results on the Fréchet bounds in the previous chapters identify the worst case dependence structures as for example the comonotonic distribution, when considering the convex risk of the joint portfolio (see Theorem 3.5) we want to describe in this chapter in more detail the dependence structure within the Fréchet class. If $\prec$ is a positive dependence order, then we call X "positive dependent" (w.r.t. $\prec$) if

$$X^{\perp} \prec X \tag{6.1}$$

where $X^{\perp}$ has independent components and $X_i^{\perp}$ have the same distribution as X_i, $X_i^{\perp} \sim X_i$.

6.1 Positive Orthant Dependence and Supermodular Ordering

A classical notion of a positive dependence ordering is based on the positive orthant ordering going back to Lehmann (1966).

Definition 6.1 (Orthant ordering). For two random vectors X, Y in $\mathbb{R}^n$ we define

(a) $X \leq_{\text{uo}} Y$ – "upper orthant order" – if $\overline{F}_X(x) \leq \overline{F}_Y(x)$, $\forall x \in \mathbb{R}^n$

(b) $X \leq_{\text{lo}} Y$ – "lower orthant order" – if $F_X(x) \leq F_Y(x)$, $\forall x \in \mathbb{R}^n$

(c) $X \leq_{\text{c}} Y$ – "concordance order" – if $X \leq_{\text{uo}} Y$ and $X \leq_{\text{lo}} Y$

(d) X is called "positive upper orthant dependent" (PUOD) if $X^{\perp} \leq_{\text{uo}} X$. Similarly we define PLOD and POD by $X^{\perp} \leq_{\text{lo}} X$ respectively $X^{\perp} \leq_{\text{c}} X$.

L. Rüschendorf, *Mathematical Risk Analysis*, Springer Series in Operations Research and Financial Engineering, DOI 10.1007/978-3-642-33590-7_6,
© Springer-Verlag Berlin Heidelberg 2013

We also use the notation $P \leq_{uo} Q$ for the distributions P, Q respectively $F \leq_{uo} G$ for the distribution functions F, G of X, Y. Obviously for $n = 2$ and assuming identical marginals $X_i \stackrel{d}{=} Y_i$ it holds that

$$X \leq_{lo} Y \Leftrightarrow X \leq_{uo} Y \Leftrightarrow X \leq_c Y. \tag{6.2}$$

For $n \geq 3$ these orders are different.

The orthant orders are weaker than the stochastic order $\leq_{st}$ on $\mathbb{R}^n$. In fact they are a combination of a dependence order and a stochastic order of the marginals.

Proposition 6.2. *If $F \in \mathcal{F}(F_1, \ldots, F_n)$ and $G = \mathcal{F}(G_1, \ldots, G_n)$ and $F \leq_{lo} G$, then $F_i \leq_{st} G_i$, $1 \leq i \leq n$ and for $X \sim F$, $Y \sim G$ it holds that*

$$\mathrm{Cov}(X_i, X_j) \leq \mathrm{Cov}(Y_i, Y_j). \tag{6.3}$$

Proof. Equation (6.3) is a consequence of the Hoeffding representation of the covariance

$$\mathrm{Cov}(X_i, X_j) = \int \int (F_{X_i, X_j}(x, y) - F_i(x)F_j(y))dxdy. \tag{6.4}$$

The ordering relation $F_i \leq_{st} G_i$ directly follows from the definition of the $\leq_{lo}$ ordering:

$$F(x) \leq G(x) \text{ for all } x \quad \text{and in consequence} \quad F_i(x_i) \leq G_i(x_i) \text{ for all } x_i. \quad \square$$

Remark 6.3 (Consistency with dependence measures, Pearson's ϱ_p, Kendall's τ and Spearman's ϱ_s).

(a) **Pearson's correlation coefficient:** For random vectors X, Y with distribution functions F, G and identical marginal distribution functions $F_i = G_i$, $1 \leq i \leq n$, (6.3) implies that ordering of the "Pearson correlation coefficient"

$$\varrho_p(X_i, X_j) = \frac{\mathrm{Cov}(X_i, X_j)}{\sqrt{\mathrm{Var}(X_i)\,\mathrm{Var}(X_i)}} \tag{6.5}$$

is implied by the orthant orders:

$$X \leq_{lo} Y \Rightarrow \varrho_p(X_i, X_j) \leq \varrho_p(Y_i, Y_j). \tag{6.6}$$

ϱ_p is the classical measure of (linear) dependence. It has the disadvantage of not being scale invariant and it needs for its definition a moment condition, which is not satisfied in applications with heavy tail distributions.

(b) **Kendall's τ:** There are alternative dependence measures which do not have these disadvantages. Let $X = (X_1, X_2)$ have copula C and copula vector $U = (U_1, U_2)$, then Kendall's τ dependence measure is defined as

$$\tau(X_1, X_2) = P\big((X_1 - X_1')(X_2 - X_2') > 0\big) - P\big((X_1 - X_1')(X_2 - X_2') < 0\big) \quad (6.7)$$

where $X' = (X_1', X_2')$ is independent of X and $X' \sim X$.

Thus τ is the difference of the probability of concordance of X with X' and the probability of discordance of X with X'. In the case of continuous marginal distribution functions F_1, F_2 Kendall's τ can be represented as

$$
\begin{aligned}
\tau(X_1, X_2) &= 2P\big((X_1 - X_1')(X_2 - X_2') > 0\big) - 1 \\
&= 4P\big(X_1 \le X_1', X_2 \le X_2'\big) - 1 \\
&= 4 \int\int F(x, y) dF(x, y) - 1 \\
&= 4 \int\int C(u, v) dC(u, v) - 1 \\
&= 4EC(U_1, U_2) - 1,
\end{aligned}
\qquad (6.8)
$$

where $U \sim C$. Thus τ depends only on the copula C. Note however that in the case of discontinuous marginal distribution functions the last expression depends on the choice of copula C. In the case of ties between the components one has to decide how to evaluate them.

(c) **Spearman's rank correlation coefficient ϱ_S:** Spearman's rank correlation coefficient is defined in the case of continuous marginals as

$$
\begin{aligned}
\varrho_S(X_1, X_2) &= \varrho_P\big(F_1(X_1), F_2(X_2)\big) \\
&= \varrho_P(U_1, U_2) \\
&= \frac{EU_1 U_2 - 1/4}{1/12} \\
&= 12EU_1 U_2 - 3
\end{aligned}
\qquad (6.9)
$$

where $U = (U_1, U_2)$ is a copula vector of X.

ϱ_S can be represented by the concordance probability w.r.t. $X^\perp = (X_1^\perp, X_2^\perp)$, a vector with independent components, and $X_i^\perp \sim X_i$

$$
\begin{aligned}
\varrho_S(X_1, X_2) = 3\big[&P\big((X_1 - X_1^\perp)(X_2 - X_2^\perp) > 0\big) \\
&- P\big((X_1 - X_1^\perp)(X_2 - X_2^\perp) < 0\big)\big].
\end{aligned}
\qquad (6.10)
$$

The representation in (6.9) shows that ϱ_S is consistent with the orthant orderings.

More precisely the above given representations imply in the case that $X = (X_1, X_2)$, $Y = (Y_1, Y_2)$ have identically distributed marginals $X_i \sim Y_i$ then $X \leq_{\text{uo}} Y$ implies that

$$\varrho_P(X_1, X_2) \leq \varrho_P(Y_1, Y_2),$$
$$\varrho_S(X_1, X_2) \leq \varrho_S(Y_1, Y_2), \tag{6.11}$$
$$\text{and} \quad \tau(X_1, X_2) \leq \tau(Y_1, Y_2)$$

i.e. the orthant orders $\leq_{\text{uo}}$, $\leq_{\text{lo}}$, $\leq_c$ are "dependence order". $\Diamond$

The orthant orders $\leq_{\text{uo}}$, $\leq_{\text{lo}}$, $\leq_c$ are generated by the function classes of indicators $\{1_{[x,\infty)};\ x \in \mathbb{R}^n\}$, $\{1_{(-\infty,x]};\ x \in \mathbb{R}^n\}$ respectively the union of both. They are also generated by functions of the product form $\prod_{i=1}^n f_i(x_i)$, where $f_i \geq 0$ are increasing in case of $\leq_{\text{uo}}$ respectively by $f_i \geq 0$ decreasing in case of $\leq_{\text{lo}}$ (see Bergmann (1978)). In case of $\leq_{\text{uo}}$ it holds that:

Proposition 6.4. *Let X, Y be n-dimensional random vectors. Then*

$$X \leq_{\text{uo}} Y \Leftrightarrow E \prod_{i=1}^n f_i(X_i) \leq E \prod_{i=1}^n f_i(Y_i) \tag{6.12}$$

for all increasing $f_i \geq 0$.

A similar result holds for $\leq_{\text{lo}}$. In the case of identical marginals of X, Y, i.e. $X_i \sim Y_i$ one gets directly from the definition:

$$X \leq_{\text{uo}} Y \Leftrightarrow (f_1(X_1), \ldots, f_n(X_n)) \leq_{\text{uo}} (f_1(Y_1), \ldots, f_n(Y_n)) \tag{6.13}$$

for all increasing functions f_i.

In particular PUOD is equivalent to the positive dependence concept of "weak association" in Rü (1981c). In the class of PUOD random vectors X the independent case is characterized by some mixed moment conditions (see Lehmann (1966) for $n = 2$, Jogdeo (1970) for $n = 3$, and Rü (1981d) for $n \geq 3$).

Proposition 6.5. *Let $X = (X_1, \ldots, X_n)$ be POD and assume that $E \prod_{j \in T} X_j$ exists for all $T \subset \{1, \ldots, n\}$. If $E \prod_{j \in T} X_j = \prod_{j \in T} EX_j$ for all $T \subset \{1, \ldots, n\}$ then $X_1, \ldots, X_n$ are stochastically independent.*

A maximal generating class of the upper orthant order $\leq_{\text{uo}}$ is given by the class of Δ-monotone functions, introduced in Rü (1980).

Definition 6.6 (Δ-monotone functions). For functions $f : \mathbb{R}^n \to \mathbb{R}^1$ define the difference operator Δ_i^ε, $\varepsilon > 0$, $1 \leq i \leq n$ by $\Delta_i^\varepsilon f(x) = f(x + \varepsilon e_i) - f(x)$ where e_i is the i-th unit vector.

f is called "Δ-monotone" if for any subset $J = \{i_1, \ldots, i_k\} \subset \{1, \ldots, n\}$ and any $\varepsilon_1, \ldots, \varepsilon_k > 0$ it holds that

$$\Delta_{i_1}^{\varepsilon_1} \ldots \Delta_{i_k}^{\varepsilon_k} f(x) \geq 0. \tag{6.14}$$

Let $\mathcal{F}_\Delta$ denote the class of all Δ-monotone functions on $\mathbb{R}^n$.

Remark 6.7. All multivariate distribution functions are Δ-monotone. Differentiable functions f are Δ-monotone if and only if

$$\frac{\partial^k f}{\partial x_{i_1}, \ldots, \partial x_{i_k}} \geq 0 \quad \text{for all } k \leq n \text{ and } i_1 < \cdots < i_k. \tag{6.15}$$

$$\diamond$$

The class of $\mathcal{F}_\Delta$ of Δ-monotone functions is convex and closed under pointwise convergence. Conversely, right continuous Δ-monotone functions can be approximated by positive linear combinations of indicator functions $1_{[x,\infty)}$. This leads to the following interesting characterization result. We denote by $\leq_\Delta$ the integral order generated by $\mathcal{F}_\Delta$, i.e.

$$\leq_\Delta = \leq_{\mathcal{F}_\Delta}.$$

Theorem 6.8 (Δ-monotone functions, Rü (1980)). *For random vectors X, Y in $\mathbb{R}^n$ it holds that:*

$$X \leq_{\text{uo}} Y \Leftrightarrow X \leq_\Delta Y. \tag{6.16}$$

Remark 6.9 (Lower orthant order).

(a) Similarly, the lower orthant order $\leq_{\text{lo}}$ is generated by the class of functions

$$\mathcal{F}_\Delta^- = \{h : \mathbb{R}^n \to \mathbb{R}^1; \exists f \in \mathcal{F}^\Delta \text{ such that } h(x) = f(-x)\}.$$

So the elements $h \in \mathcal{F}_\Delta^-$ are decreasing and satisfy

$$(-1)^k \Delta_{i_1}^{\varepsilon_1} \cdots \Delta_{i_k}^{\varepsilon_k} h(x) \geq 0 \quad \text{for all } x \in \mathbb{R}^n \tag{6.17}$$

respectively

$$(-1)^k \frac{\partial^k}{\partial x_{i_1} \cdots \partial x_{i_k}} h(x) \geq 0 \quad \text{for all } k \leq n \text{ and } i_1 < \cdots < i_k. \tag{6.18}$$

In consequence we obtain a characterization of the concordance order $\leq_c$ in terms of $\mathcal{F}_\Delta \cup \mathcal{F}_\Delta^-$.

(b) The characterization in (6.16) implies in particular Proposition 6.5. It also implies sharpened versions of several classical integral inequalities (see Rü (1980)). $\diamond$

Δ-monotone functions have a nice conservation property which leads to the following result.

Corollary 6.10. *If X, Y are random vectors in $\mathbb{R}^2$ such that $X \leq_{uo} Y$, then for any $f \in \mathcal{F}_\Delta$ it holds that:*

$$f(X) \leq_{icx} f(Y), \tag{6.19}$$

where $\leq_{icx}$ denotes the ordering w.r.t. increasing convex functions.

Proof. It is enough by general reduction, respectively smoothing results (see Müller and Stoyan (2002)) to consider two times differentiable, increasing convex functions g, i.e. $g' \geq 0$, $g'' \geq 0$, and to assume that $f \in \mathcal{F}_\Delta$ is differentiable, i.e. $\frac{\partial f}{\partial x_i} \geq 0$, $\frac{\partial^2 f}{\partial x_1 \partial x_2} \geq 0$. As a consequence we obtain that $g \circ f \in \mathcal{F}_\Delta$ checking the differentiability condition in (6.18). This implies $Eg \circ f(X) \leq Eg \circ f(Y)$ and thus (6.19). □

The following conclusion (see Rü (1980) and Baccelli and Makowski (1989)) is a direct consequence of the definition of POD.

Corollary 6.11. *If X, Y are n-dimensional random vectors, then*

(a)
$$X \leq_{uo} Y \Rightarrow \min_{i \leq n} X_i \leq_{st} \min_{i \leq n} Y_i \text{ and } \max_{i \leq n} X_i \geq_{st} \max_{1 \leq n} Y_i. \tag{6.20}$$

(b) If X is PUOD, then

$$\min_{i \leq n} X_i \leq_{st} \min_{i \leq n} X_i^{\perp} \text{ and } \max_{i \leq n} X_i \geq_{st} \max_{i \leq n} X_i^{\perp}.$$

An interesting strengthening of the dependence orders $\leq_{uo}$, $\leq_\Delta$, $\leq_{lo}$, $\leq_c$ is the ordering by the class of supermodular functions.

Definition 6.12 (Supermodular functions). $f : \mathbb{R}^n \to \mathbb{R}^1$ is called " supermodular" if for all $1 \leq i < j \leq n$ and $\varepsilon, \delta > 0$ it holds that

$$\Delta_i^\varepsilon \Delta_j^\delta f(x) \geq 0 \quad \text{for all } x \in \mathbb{R}^n. \tag{6.21}$$

Let $\mathcal{F}_{sm}$ denote the class of supermodular functions and let $\leq_{sm}$ denote the corresponding integral order, the "supermodular ordering".

Remark 6.13. (a) By definition $\mathcal{F}_\Delta \subset \mathcal{F}_{sm}$. Thus comparison w.r.t. the supermodular order $\leq_{sm}$ is stronger than comparison w.r.t. the orthant orders. Supermodular functions have been introduced in the literature also under the name "L-superadditive" (L for lattice, see Marshall and Olkin (1979)) as quasi-monotone functions and as functions with increasing differences. In comparison with the Δ-monotone functions only second order differences are assumed to be monotone.

(b) Differentiable functions f are supermodular if and only if

$$\frac{\partial^2 f}{\partial x_i \partial x_j}(x) \geq 0, \quad \forall i < j \text{ and } \forall x \in \mathbb{R}^n. \tag{6.22}$$

Again for the comparison $\leq_{sm}$ it is enough to restrict to (infinite) differentiable supermodular functions.   ◇

A basic comparison result for the supermodular order $\leq_{sm}$ is the following comparison theorem. This theorem was established first in Tchen (1980). The proof was given by Tchen by discrete approximation and reduction in the discrete case to a classical discrete rearrangement theorem of Hardy et al. (1952). The theorem was stated independently (but later) in Rü (1983).[1] Here the proof was reduced by the rearrangement theorem (see Theorem 3.23) to a theorem of Lorentz (1953) on the rearrangement of real functions. Therefore, the theorem is also called the "Lorentz Theorem" in this paper.

Theorem 6.14 (Supermodular order and comonotonicity). *Let X be an n-dimensional random vector with marginal distribution functions $F_1, \ldots, F_n$, then*

$$X \leq_{sm} X^c = (F_1^{-1}(U), \ldots, F_n^{-1}(U)), \tag{6.23}$$

where $U \sim U(0, 1)$.

It is of interest that for $n = 2$ the supermodular order $\leq_{sm}$ when restricted to a Fréchet class $\mathcal{F}(F_1, F_2)$ is identical to the orthant orderings $\leq_{lo}$, $\leq_{uo}$. This classical important result is due to Cambanis et al. (1976).

Theorem 6.15. *Let $X = (X_1, X_2)$, $Y = (Y_1, Y_2)$ have identical marginal distribution functions F_1, F_2. Then it holds that:*

(a)
$$X \leq_{sm} Y \Leftrightarrow X \leq_{lo} Y \Leftrightarrow X \leq_{uo} Y; \tag{6.24}$$

(b)
$$X_c \leq_{sm} X \leq_{sm} X^c \tag{6.25}$$

where $X_c = (F_1^{-1}(U), F_2^{-1}(1 - U))$ is the countermonotonic vector.

Proof. (a) A proof of (a) is obtained by reduction to the differential case and using then a partial integration formula of the form

$$E\varphi(X) - E\varphi(Y) = \int (F_Y(x, y) - F_X(x, y))\mu_\varphi(dx, dy),$$

where $\varphi \in \mathcal{F}_{sm}$ and μ_φ is the measure associated with φ (see Rachev and Rü (1998a, Theorem 3.1.2)).
(b) follows from (a) and the classical Hoeffding–Fréchet bounds.   □

Corollary 6.16. *If X, Y are random vectors in $\mathbb{R}^n$ such that $X \leq_{sm} Y$ and $h \in \mathcal{F}_{sm}$ is monotone, then*
$$h(X) \leq_{icx} h(Y), \tag{6.26}$$

where $\leq_{icx}$ is the increasing, convex order.

---

[1]This paper was submitted in 1980 but only published in 1983.

Proof. For $\varphi \in \mathcal{F}_{\mathrm{icx}}$ we have to verify that $f = \varphi \circ h \in \mathcal{F}_{\mathrm{sm}}$. This problem can be reduced to the differentiable case and then follows from the relation

$$\frac{\partial^2}{\partial x_i \partial x_j} \varphi \circ h = \varphi'' \circ h \frac{\partial h}{\partial x_i} \frac{\partial h}{\partial x_j} + \varphi' \circ h \frac{\partial^2 h}{\partial x_i \partial x_j} \quad \text{for } i \neq j. \qquad \square$$

Remark 6.17 (Excess of loss).

(a) As a particular consequence we obtain from Theorem 6.14 and Corollary 6.16 the basic convex comparison result for the excess of loss of the joint portfolio in Theorem 3.5, stating that

$$\sum_{i=1}^{n} X_i \leq_{\mathrm{cx}} \sum_{i=1}^{n} X_i^c.$$

The comonotonic vector X^c is the worst case joint portfolio. For $n = 2$ the lower bound w.r.t. convex order is given by the countermonotonic vector $X_c = (F_1^{-1}(U), F_2^{-1}(1 - U)) = (X_1^{cm}, X_2^{cm})$

$$\sum_{i=1}^{2} X_i^{cm} \leq_{\mathrm{cx}} \sum_{i=1}^{2} X_i. \tag{6.27}$$

(b) For $P, Q \in M^1(\mathbb{R}^n, \mathfrak{B}^n)$ with identical $(n-1)$-dimensional marginal distributions one obtains that the (weak) upper orthant ordering $\leq_{\mathrm{uo}}$ implies supermodular ordering $\leq_{\mathrm{sm}}$,

$$P \leq_{\mathrm{uo}} Q \Rightarrow P \leq_{\mathrm{sm}} Q \tag{6.28}$$

(see Tchen (1980) and Rü (1980, Theorem 3b)). $\qquad \Diamond$

6.2 Association, Conditional Increasing Vectors, and Positive Supermodular Dependence

There are several well-established (positive) dependence notions and orders for random vectors. In this section we introduce some of them and describe relations to the supermodular ordering $\leq_{\mathrm{sm}}$. Our general aim in this section is to establish some dependence and ordering notions which imply the possibility to compare a broad class of risk functionals as given by the supermodular functions.

In the following definition we consider various notions of association.

Definition 6.18 (Positive dependence notions).   A random vector $X =$ $(X_1, \ldots, X_n)$ is called

(a) "associated" if $Ef(X)g(X) \geq Ef(X)Eg(X)$ for all increasing functions f, g.
(b) "weakly associated" if $E \prod_{i=1}^n f_i(X_i) \geq \prod_{i=1}^n Ef_i(X_i)$ for all $f_i \geq 0$, f_i increasing.
(c) $X \leq_{\mathrm{wcs}} Y$, X is smaller than Y in the "weakly conditional increasing in sequence order" if for all $t \in \mathbb{R}^1$, $i \leq n-1$ and f increasing

$$\mathrm{Cov}(1(X_i > t), f(X_{(i+1)})) \leq \mathrm{Cov}(1(Y_i > t), f(Y_{(i+1)})), \qquad (6.29)$$

where $X_{(i+1)} = (X_{i+1}, \ldots, X_n)$ respectively $Y_{(i+1)} = (Y_{i+1}, \ldots, Y_n)$.
(d) X is called "weakly associated in sequence (WAS)" if

$$X^{\perp} \leq_{\mathrm{wcs}} X.$$

(e) X is said to be "positive supermodular dependent (PSMD)" if

$$X^{\perp} \leq_{\mathrm{sm}} X.$$

Remark 6.19. The notion "association" goes back to early work of Esary et al. (1967). It is a strong positive dependence notion. The notion of "weak association in sequence" and the related $\leq_{\mathrm{wcs}}$ ordering was introduced in Rü (2004). We remark, that

$$X \text{ is } WAS \Leftrightarrow P^{X_{(i+1)}|X_i > t} \geq_{\mathrm{st}} P^{X_{(i+1)}} \quad \forall i \leq n-1, t \in \mathbb{R}^1, \qquad (6.30)$$

i.e. the conditional distribution of the future $X_{(i+1)}$ given $X_i > t$ is stochastically larger than the distribution of $X_{(i+1)}$. ◇

The last remark gives a connection to a second type of dependence orders defined via conditional increasing distributions.

We define that $X \uparrow_{\mathrm{st}} Y$, X is "stochastically increasing" in Y if the conditional distribution $P^{X|Y=y}$ is increasing in y with respect to $\leq_{\mathrm{st}}$.

Definition 6.20 (Conditional increasing property).   A random vector $X =$ $(X_1, \ldots, X_n)$ is called

(a) "conditional increasing (CI)" if for all $i \leq n$

$$X_i \uparrow_{\mathrm{st}} X_J \quad \forall J \subset \{1, \ldots, n\}$$

(b) "conditional increasing in sequence (CIS)" if

$$X_i \uparrow_{\mathrm{st}} (X_1, \ldots, X_{i-1}), \quad 2 \leq i \leq n$$

(c) "positive dependent through stochastic ordering (PDS)" if

$$(X_i, i \neq j) \uparrow_{\text{st}} X_j \quad \forall 1 \le j \le n.$$

Remark 6.21. (a) The condition that X is CIS is equivalent to the condition that the multivariate distributional transform τ_F (see Definition 1.11) is monotone increasing. This observation implies by induction the conclusion that X is associated using the regression representation of X in (1.46), i.e. X CIS implies that X is associated.

(b) The CIS condition says that "the present state depends positively on the past" while the WAS condition says that the "future depends positively on the present".

(c) **Relations between dependence notions:** The following implications between the various positive dependence notions have been established (see for reference Müller and Stoyan (2002, p. 146)):

$$\text{CI} \Rightarrow \text{CIS} \Rightarrow \text{Association} \Rightarrow WA \Rightarrow \text{WAS} \qquad (6.31)$$

and

$$\text{PDS} \Rightarrow \text{WAS}. \qquad (6.32)$$

(d) **MTP$_2$ property:** For a vector X such that P^X has a density f w.r.t. a product measure $\mu = \mu_1 \times \cdots \times \mu_n$, where typically μ is the Lebesgue measure or the counting measure, we define:

X (or f) is "multivariate totally positive of order 2 (MTP$_2$)" if

$$f(x)f(y) \le f(x \wedge y)f(x \vee y) \qquad (6.33)$$

or equivalently if

$$\ln f \text{ is supermodular.} \qquad (6.34)$$

The MTP$_2$ property is typically relatively easy to verify. Karlin and Rinott (1980) proved

$$X \text{ is MTP}_2 \Rightarrow X \text{ is } CI. \qquad (6.35)$$

$$\Diamond$$

The following result states that the $\le_{\text{wcs}}$ ordering implies the supermodular ordering $\le_{\text{sm}}$ (see Rü (2004, Theorem 2.2)).

Theorem 6.22. *Let X, Y be n-dimensional random vectors with identical marginal distributions $P_i = Q_i$, $1 \le i \le n$. Then $X \le_{\text{wcs}} Y$ implies that $X \le_{\text{sm}} Y$.*

Proof. Let $X^\perp$, $Y^\perp$ be random vectors with independent components and $X_i^\perp \overset{d}{=} P_i$, $Y_i^\perp \overset{d}{=} Q_i$, $1 \le i \le n$. Then it is sufficient to compare expectations of functions $f \in \mathcal{F}_{\text{sm}}$ which are bounded and twice differentiable (see e.g. the approximation argument in Christofides and Vaggelatou (2004, Proposition 1)). Denote $g(t, x_{(2)}) := \frac{\partial f}{\partial x_1}(t, x_{(2)})$ and note that $g(t, \cdot)$ is increasing since $f \in \mathcal{F}_{\text{sm}}$. The proof is given by induction in n. We obtain using the simple representation formula of Christofides and Vaggelatou (2004)

$$E(f(X) - f(X_1^\perp, X_{(2)})) = \int \text{Cov}(1(X_1 > t), g(t, X_{(2)}))dt \qquad (6.36)$$

$$\leq \int \text{Cov}(1(Y_1 > t), g(t, Y_{(2)}))dt$$

$$= E(f(Y) - f(Y_1^\perp, Y_{(2)})),$$

where the inequality follows from $X \leq_{\text{wcs}} Y$ and the assumption of identical marginals $X_1 \sim Y_1$. This implies that $Ef(X) \leq Ef(Y) + A_{n-1}$, where $A_{n-1} := Ef(X_1^\perp, X_{(2)}) - Ef(Y_1^\perp, Y_{(2)})$. The function $\int f(x_1, \cdot) \, dP_1(x_1)$ is a supermodular function in $(n-1)$ arguments. Therefore, using that $X_1^\perp \sim Y_1^\perp$, we obtain from induction $A_{n-1} \leq 0$ and thus $Ef(X) \leq Ef(Y)$.   □

As a corollary we obtain

Corollary 6.23 (Positive supermodular dependence criterion).
For a risk vector X it holds that: Weak association of X in sequence implies positive supermodular dependence, i.e.

$$X \text{ WAS } \Rightarrow X \text{ PSMD}.$$

This conclusion gives an important sufficient condition for PSMD.

Remark 6.24. (a) The positive supermodular dependence criterion in Corollary 6.23 is due to Christofides and Vaggelatou (2004) under the stronger assumption of weak association. Under the even stronger CIS-condition this conclusion is due to Meester and Shanthikumar (1999).

(b) Since by Corollary 6.16 $X \leq_{\text{sm}} Y$ implies that $h(X) \leq_{\text{icx}} h(Y)$ for all monotonically nondecreasing, supermodular functions h one obtains as a consequence of Theorem 6.22 a convex comparison result for a general class of risk-functionals of the random vectors.

Corollary 6.25. *If X is weakly associated in sequence (WAS) then:*

$$h(X^\perp) \leq_{\text{icx}} h(X),$$

i.e. $h(X)$ has higher increasing convex risk than $h(X^\perp)$ for any monotonically nondecreasing supermodular function h.

Thus the positive dependence notion WAS implies a higher risk compared with independent portfolios for the general class of increasing supermodular functionals. Denuit et al. (2001) have established this conclusion for the special case of the combined portfolio $X_1 + \cdots + X_n$ for an associated random vector X, i.e. if X is associated, then

$$\sum_{i=1}^{n} X_i^{\perp} \leq_{\text{icx}} \sum_{i=1}^{n} X_i. \tag{6.37}$$

Christofides and Vaggelatou (2004) have stated this interesting conclusion for weakly associated random vectors X.

(c) In the case $n = 2$ holds for random vectors X, Y with identical marginals, $X_i \sim Y_i$,

$$X \leq_{\text{wcs}} Y \text{ if and only if } X \leq_{\text{uo}} Y. \tag{6.38}$$

In this case the result of Theorem 6.22 goes back to the classical paper of Cambanis et al. (1976) (see Theorem 6.15).

(d) **Positive dependence of normal vectors:** In the case of a multivariate normal distribution the following characterizations hold:
Let $X \sim N(\mu, \Sigma)$ where $\Sigma = (\sigma_{ij})$, then:

(d_1) $$X \text{ is POD} \Leftrightarrow X \text{ is associated}$$
$$\Leftrightarrow \sigma_{ij} \geq 0. \tag{6.39}$$

(d_2) $$X \text{ is MTP}_2 \Leftrightarrow X \text{ is } CI$$
$$\Leftrightarrow \Sigma^{-1} \text{is an } M\text{-matrix, i.e. its offdiagonal} \tag{6.40}$$
$$\text{elements are non-positive.}$$

Equation (6.39) is due to Pitt (1982). In (6.40) the first equivalence is due to Barlow and Proschan (1975), the second equivalence is given in Rü (1982).   ◊

6.3   Directionally Convex Order

The strong positive dependence order $\leq_{\text{sm}}$ and its related notion of PSMD are restricted to comparing two elements of the same Fréchet class, i.e. with the same marginal distributions. For the comparison of risk vectors X, Y it is also of interest to compare vectors where the marginals satisfy $X_i \leq_{\text{cx}} Y_i$ respectively $X_i \leq_{\text{icx}} Y_i$. There may exist two sources of the increase of risk. One source is the convex increase of the marginals. The second source is the increase of positive dependence.

6.3.1   Basic Properties of the Directionally Convex Order

A suitable class of functions reflecting both sources of possible risk are the directionally convex functions.

Definition 6.26 (Directionally convex functions and order).

(a) A function $f : \mathbb{R}^n \to \mathbb{R}^1$ is called "directionally convex" if for all $1 \le i \le j \le n$ and all $\varepsilon, \delta > 0$

$$\Delta_i^\varepsilon \Delta_j^\delta f(x) \ge 0 \quad \text{for all } x \in \mathbb{R}^n.$$

Let $\mathcal{F}_{\text{dcx}}$ denote the class of all directionally convex functions.

(b) Let X, Y be random vectors in $\mathbb{R}^n$, then

$$X \le_{\text{dcx}} Y, \quad X \text{ is smaller than } Y \text{ in the directionally convex order},$$

if $Ef(X) \le Ef(Y)$ for all integrable $f \in \mathcal{F}_{\text{dcx}}$, i.e. $\le_{\text{dcx}}$ is the induced integral order.

Remark 6.27. (a) For a twice differentiable function f it holds that

$$f \in \mathcal{F}_{\text{dcx}} \Leftrightarrow \frac{\partial^2 f}{\partial x_i \partial x_j}(x) \ge 0, \quad \forall 1 \le i \le j \le n. \tag{6.41}$$

In comparison to supermodularity directional convex functions are supermodular and additionally convex in each component x_i. $\le_{\text{dcx}}$ is generated by the class of all infinite differentiable directionally convex functions with asymptotically linear growth, i.e. $f(x) = O(\|x\|)$, as $\|x\| \to \infty$ (see Müller and Stoyan (2002)).

(b) For any $g : R^1 \to \mathbb{R}^1$ convex, the function $f(x) = g(\sum_{i=1}^n \alpha_i x_i)$ with $\alpha_i \ge 0$ is directionally convex. As a consequence this implies:

$$X \le_{\text{dcx}} Y \Rightarrow \sum_{i=1}^n \alpha_i X_i \le_{\text{cx}} \sum_{i=1}^n \alpha_i Y_i \quad \text{for all } \alpha_i \ge 0. \tag{6.42}$$

The right-hand side in (6.42) is denoted "positive linear convex order", i.e. $X \le_{\text{plcx}} Y$. $\diamondsuit$

The following result has been stated in Rü (1983) as the Ky Fan–Lorentz Theorem. Its proof is based on the rearrangement representation of the Fréchet class (Theorem 3.23) combined with a theorem in Fan and Lorentz (1954, Theorem 1) on a rearrangement inequality for functions.

Theorem 6.28 (Ky Fan–Lorentz Theorem, Rü (1983)). *Let $F_1, \ldots, F_n$ and $G_1, \ldots, G_n$ be distribution functions such that $F_i \le_{\text{cx}} G_i$, $1 \le i \le n$. Denote by*

$$X^c := (F_i^{-1}(U))_{1 \le i \le n}, \quad Y^c := (G_i^{-1}(U))_{1 \le i \le n}$$

the corresponding comonotone vectors, then

$$X^c \le_{\text{dcx}} Y^c.$$

An extension of Theorem 6.28 is due to Müller and Scarsini (2001). It postulates identical CI-copulas for the vectors X, Y.

Theorem 6.29 (Comparison criterion for $\leq_{\text{dcx}}$, Müller and Scarsini (2001)).
Let X, Y be vectors in $\mathbb{R}^n$ which have a common conditionally increasing copula C. If $X_i \leq_{\text{cx}} Y_i$, $1 \leq i \leq n$, then

$$X \leq_{\text{dcx}} Y.$$

For the proof Müller and Scarsini (2001) reduce the result via the standard representation of X, Y and the CI-property of C to the statement of the Ky Fan–Lorentz Theorem. For general copulas the statement of Theorem 6.29 is not true as the following simple example in Müller and Scarsini (2001) shows.

Example 6.30. For an integrable random variable W define $X = (W, -W)$, $Y = (W, -EW)$. Then $Y_i \leq_{\text{cx}} X_i$, $1 \leq i \leq 2$ but $X_1 + X_2 = 0$ while $Y_1 + Y_2 = W - ES$ and thus $X_1 + X_2 \leq_{\text{cx}} Y_1 + Y_2$. ◇

6.3.2 Further Criteria for $\leq_{\text{dcx}}$

In order to state extensions of the a comparison result in Theorem 6.29 to the case of not necessarily identical and conditional increasing copulas we need a condition which ensures increase in the dependence structure and a condition for the increase in the convex ordering of the marginals. Both properties are inherently contained in the wcs-ordering.

Theorem 6.31 ($\leq_{\text{dcx}}$ ordering for different copulas, Rü (2004)).
Let X, Y be random vectors with marginals P_i, Q_i such that $P_i \leq_{\text{cx}} Q_i$, $1 \leq i \leq n$. Then

$$X \leq_{\text{wcs}} Y \text{ implies } X \leq_{\text{dcx}} Y.$$

Proof. Let $f \in \mathcal{F}_{\text{dcx}}$ be twice differentiable and let $g := \frac{\partial}{\partial x_1} f$, then as in the proof of Theorem 6.22 by the wcs-ordering and using monotonicity of $g(t, \cdot)$ we obtain

$$Ef(X) - Ef(X_1^{\perp}, X_{(2)}) = \int \text{Cov}(1(X_1 > t), g(t, X_{(2)}))dt$$

$$\leq \int \text{Cov}(1(Y_1 > t), g(t, Y_{(2)}))dt$$

$$= Ef(Y) - Ef(Y_1^{\perp}, Y_{(2)}).$$

Since $X_1^{\perp} \leq_{\text{cx}} Y_1^{\perp}$ and $f(\cdot, y_{(2)})$ is convex this implies

$$Ef(X) \le Ef(Y) + Ef(X_1^\perp, X_{(2)}) - Ef(Y_1^\perp, Y_{(2)}) \tag{6.43}$$
$$\le Ef(Y) + Ef(Y_1^\perp, X_{(2)}) - Ef(Y_1^\perp, Y_{(2)})$$
$$= Ef(Y) + A_{n-1}$$

with $A_{n-1} := Ef(Y_1^\perp, X_{(2)}) - Ef(Y_1^\perp, Y_{(2)})$. Now using that $f(y_1^\perp, \cdot) \in \mathcal{F}_{dcx}$ we conclude by induction that $A_{n-1} \le 0$ and thus the result follows. $\square$

Remark 6.32. For the conclusion that X is smaller than Y w.r.t. the increasing directionally convex order $X \le_{idcx} Y$ one can weaken the wcs-ordering to the $\le_{iwcs}$ ordering of X and Y defined as in (6.29) but restricted to non-negative, monotonically nondecreasing f. Then similar to Theorem 6.31 we obtain

$$P_i \le_{icx} Q_i, 1 \le i \le n \text{ and } X \le_{iwcs} Y \text{ implies that } X \le_{idcx} Y. \tag{6.44}$$
$$\diamond$$

The wcs-ordering condition simplifies if only one marginal increases convexly.

Corollary 6.33. Let $X = (X_1, X_{(2)})$, $Y = (Y_1, X_{(2)})$ and assume

$$\operatorname{Cov}(1(X_1 > t), f(X_{(2)})) \ge \operatorname{Cov}(1(Y_1 > t), f(X_{(2)})), \quad \forall t \tag{6.45}$$

for all nondecreasing bounded f.

$$\text{If } X_1 \le_{cx} Y_1, \text{ then } X \le_{dcx} Y.$$

Proof. In the proof of Theorem 6.31 one can omit the induction step after formula (6.43) if only one marginal increases convexly. $\square$

In the case $n = 2$ Theorem 6.31 yields a simple sufficient condition for comparing risks in terms of the survival functions $\overline{F}(u, v) = P(X_1 \ge u, X_2 \ge v)$, $\overline{G}(u, v) = P(Y_1 \ge u, Y_2 \ge v)$. We formulate also a variant of this result for increasing convex risks.

Corollary 6.34. Let $X = (X_1, X_2)$, $Y = (Y_1, Y_2)$. Assume that for the survival functions $\bar{F}, \bar{G}$ of X and Y it holds that

$$\bar{F}(u, v) - \bar{F}_1(u)\bar{F}_2(v) \le \bar{G}(u, v) - \bar{G}_1(u)\bar{G}_2(v). \tag{6.46}$$

(a) If $X_i \le_{cx} Y_i$, $i = 1, 2$, then $X \le_{dcx} Y$.

(b) If $X_i \le_{icx} Y_i$, $i = 1, 2$, then $X \le_{idcx} Y$.

Here $\le_{idcx}$ denotes the ordering w.r.t. the class $\mathcal{F}_{idcx}$ of increasing directionally convex functions.

Proof. The wcs-ordering condition is in the case $n = 2$ identical to the dependence ordering in (6.46) of the survival functions. $\square$

A sufficient condition for the wcs-ordering is stated in the following proposition.

Proposition 6.35. *Let X, Y be random vectors with the same conditionally increasing copula and let $X_i \leq_{\text{cx}} Y_i, 1 \leq i \leq n$, then $X \leq_{\text{wcs}} Y$.*

Proof. In the first step we assume that X, Y differ only in the first component in distribution, i.e.

$$X_1 \leq_{\text{cx}} Y_1 \text{ and } X_{(2)} \overset{d}{=} Y_{(2)}.$$

We have to establish

$$\text{Cov}(f_i(X_i), f(X_{(i+1)})) \leq \text{Cov}(f_i(Y_i), f(Y_{(i+1)})), \quad 1 \leq i \leq n,$$

for f_i, f monotonically nondecreasing. Without loss of generality we consider the case $i = 1$. Since $X_{(2)} \overset{d}{=} Y_{(2)}$ we also may assume that $X_{(2)} = Y_{(2)}$ and $Ef(X_{(2)}) = 0$. Using the standard representation for the joint copula of X, Y we obtain

$$X_1 = F_1^{-1}(V_1), \qquad X_i = F_i^{-1} \circ f_i(V_1, \dots, V_i),$$
$$Y_1 = G_1^{-1}(V_1), \qquad Y_i = X_i, \qquad\qquad i \geq 2.$$

This representation implies that conditionally given $\Theta = (V_2, \dots, V_n) = \vartheta$.
$f_1(X_1) = h_1(V_1), f(X_{(2)}) = h_2(V_1)$ and $f_1(Y_1) = g_1(V_1)$ where h_1, g_1 are monotonically nondecreasing, $g_1 = g_1(\cdot, \vartheta)$. Thus conditionally given $\Theta = \vartheta$ by the Ky Fan–Lorentz Theorem (Theorem 6.28)

$$(f_1(X_1), f(X_{(2)})) \leq_{\text{dcx}} (f_1(Y_1), f(Y_{(2)}))$$

and, therefore,

$$\text{Cov}(f_1(X_1), f(X_{(2)})) = Ef_1(X_1)f(X_{(2)}) \leq \text{Cov}(f_1(Y_1), f(Y_{(2)})).$$

If X, Y differ only in the i-th component, then using a suitable permutation π and X_π, Y_π, we reduce this case to the case $i = 1$. This is possible since X_π, Y_π are conditional increasing in sequence (CIS). Finally we obtain the general case by induction. $\qquad\square$

Proposition 6.35 implies that the directionally convex ordering result in Theorem 6.29 is a consequence of the ordering result in Theorem 6.31.

6.3.3 Directionally Convex Order in Functional Models

In the final part of this section we describe several functional models which allow to establish the directionally convex ordering condition $\leq_{\text{dcx}}$ or the corresponding increasing variant $\leq_{\text{idcx}}$. The following results are taken from Rü (2004).

Some comparable results have been given in Shaked and Tong (1985), Bäuerle (1997), and Bäuerle and Müller (1998).

We start with a simple mixture model.

Theorem 6.36 (Mixture model). *Let V_1, Θ be independent random vectors, V_1 real, and let $X_i = h_i(V_1, \Theta)$, $Y_i = g_i(V_1, \Theta)$, $1 \le i \le n$, where $h_i(\cdot, \vartheta)$ and $g_i(\cdot, \vartheta)$ are monotonically nondecreasing. If for all ϑ*

$$h_i(V_1, \vartheta) \le_{cx} g_i(V_1, \vartheta), \quad 1 \le i \le n \tag{6.47}$$

then $X \le_{dcx} Y$.

Proof. By assumption, conditionally given $\Theta = \vartheta$, $X \mid \vartheta$ and $Y \mid \vartheta$ are comonotone vectors. Therefore, by the Ky Fan–Lorentz Theorem componentwise convex ordering of the marginals as in (6.47) implies $X \mid \vartheta \le_{dcx} Y \mid \vartheta$. Therefore, by mixing we obtain $X \le_{dcx} Y$.   □

One can interpret the representation $X_i = h_i(V_1, \Theta)$, $Y_i = g_i(V_1, \Theta)$ as a model with functional dependence on an internal factor V_1 and an external factor Θ common to both. Both models depend stochastically increasing on V_1 while for any common external factor ϑ the second model has more risk than the first model.

A functional type representation of X, Y as in Theorem 6.36 can be obtained by the regression construction respectively the standard construction of a multivariate distribution with distribution function F given by $X = (h_1(V_1), h_2(V_1, V_2), \ldots, h_n(V_1, \ldots, V_n))$ where (V_i) are uniform, $U(0, 1)$-distributed (see (1.45)) considering $\Theta = (V_2, \ldots, V_n)$. This construction is also used in Müller and Scarsini (2001) for the proof of Theorem 6.29.

We next consider a more complex functional model. Let (U_i), $((V_i), V)$ be independent sequences where V_i, V are one-dimensional. Based on these we define functional models $X = (X_1, \ldots, X_n), Y = (Y_i), Z = (Z_i)$ and $W = (W_i)$ by

$$
\begin{aligned}
X_i &= g_i(U_i, V_i), & Y_i &= g_i(U_i, V), \\
Z_i &= \widetilde{g}_i(U_i, V_i), & W_i &= \widetilde{g}_i(U_i, V).
\end{aligned}
\tag{6.48}
$$

Here (U_i) is a sequence of basic stochastic variables, while V_i, V are random external sources whose influence is described by functions $g_i, \widetilde{g}_i$. The model for X depends on various different sources V_i while the model for Y depends only on one external random source V.

We introduce the following assumptions:

(A1) (U_i) are independent.

(A2) $V_i \stackrel{d}{=} V, 1 \le i \le n$.

(A3) $g_i(u, \cdot), \widetilde{g}_i(u, \cdot)$ are monotonically nondecreasing.

Let $\le_{ccx}$ denote the componentwise convex order, i.e. the integral order induced by the class $\mathcal{F}_{ccx}$ of componentwise convex functions $f(x_1, \ldots, x_n)$. The first part of the following theorem is stated in Bäuerle (1997, Theorem 3.1) for independent (U_i).

Theorem 6.37. *Under (A2), (A3) it holds that*

(a) $$X \leq_{\mathrm{sm}} Y, \qquad Z \leq_{\mathrm{sm}} W. \tag{6.49}$$

(b) If additionally $g_i(u_i, V) \leq_{\mathrm{cx}} \widetilde{g}_i(u_i, V)$, *for all* u_i, *then*

$$Y \leq_{\mathrm{dcx}} W. \tag{6.50}$$

Proof. (a) For $\varphi \in \mathcal{F}_{\mathrm{sm}}$ it holds that by the Lorentz Theorem (Theorem 6.28)

$$
\begin{aligned}
E\varphi(X) &= E_U E(\varphi(X) \mid U_1 = u_1, \dots, U_n = u_n) \\
&= E_U E\varphi(g_1(u_1, V_1), \dots, g_n(u_n, V_n)) \\
&\leq E_U E\varphi(g_1(u_1, V), \dots, g_n(u_n, V)) \\
&= E\varphi(g_1(U_1, V), \dots, g_n(U_n, V)) = E\varphi(Y).
\end{aligned}
$$

Here E_U denote the marginal expectation w.r.t. the random vector U. The proof of $Z \leq_{\mathrm{sm}} W$ is similar.

(b) The proof of $X \leq_{\mathrm{dcx}} W$ follows from (a) and the Ky Fan–Lorentz Theorem by a conditioning argument on the U_i. □

We next compare the above introduced vectors X, Y, Z, W under the assumption that the marginals increase convexly.

Theorem 6.38. *Under conditions (A1), (A2), (A3) the following holds: If for all* v, $\widetilde{g}_i(U_i, v) \leq_{\mathrm{cx}} g_i(U_i, v)$ *then*

$$Z \leq_{\mathrm{ccx}} X, \quad W \leq_{\mathrm{ccx}} Y, \quad and \ Z \leq_{\mathrm{dcx}} Y. \tag{6.51}$$

Proof. By Theorem 6.37 it holds that $X \leq_{\mathrm{sm}} Y$. Further for any componentwise convex function φ it holds that, when conditioning under $V_i = v_i$ and using the assumption on $g_i, \widetilde{g}_i$:

$$
\begin{aligned}
E\varphi(X) &= E_V E\varphi\big(g_1(U_1, v_1), \dots, g_n(U_n, v_n)\big) \\
&\geq E_V E\varphi\big(\widetilde{g}_1(U_1, v_1), \dots, g_n(U_n, v_n)\big) \\
&= E\varphi(Z).
\end{aligned}
$$

Thus we get $Z \leq_{\mathrm{ccx}} X$.

E_V is the marginal expectation w.r.t. V. Here at the inequality we used that by (A1)

$$\big(g_1(U_1, v_1), \dots, g_n(U_n, v_n)\big) \leq_{\mathrm{ccx}} \big(\widetilde{g}_1(U_1, v_1), \dots, \widetilde{g}_n(U_n, v_n)\big).$$

So we get

$$Z \leq_{\mathrm{ccx}} X \leq_{\mathrm{sm}} Y,$$

implying

$$Z \leq_{\text{dcx}} Y.$$

The inequality $W \leq_{\text{ccx}} Y$ is similar.   □

Remark 6.39. The random vectors Y, Z and X, W which are compared w.r.t. $\leq_{\text{dcx}}$ in Theorem 6.38 do not have the same dependence structure (copula). The vectors X, Y, Z, W considered in Theorem 6.38 are not necessarily positive dependent. Since we do not assume independence of the (V_i), any $F \in \mathcal{F}_n$ can be represented in the form $(g_i(U_i, V_i))$, with g_i satisfying (A3). Thus the comparison results of Theorems 6.37 and 6.38 concern a large class of models.   ◇

6.4 Dependence Orderings in Models with Multivariate Marginals

We consider random vectors $X = (X_1, \ldots, X_n)$, where X_i are k_i-dimensional random vectors with distribution functions F_i and corresponding probability measures $P_1, \ldots, P_n$. The distribution function $F = F_X \in \mathcal{F}(F_1, \ldots, F_n)$ has (known) multivariate marginal distribution functions $F_i \sim P_i$. In comparison to the case of one-dimensional marginals the multivariate marginals case has not been considered a lot in the literature. It seems however to be natural that for several applications one can determine the joint distribution of some subgroups and would like to control the influence of dependence between the subgroups. Some general results and principles for these kind of problems have been discussed in Section 3.2. In the following we consider the analogue of the classical Fréchet bounds and obtain as a consequence some bounds on the integrals of Δ-monotone functions. Let $k = \sum_{i=1}^{n} k_i$ be the dimension of X. We state again the results on the sharp Fréchet bounds (see Corollary 2.18 and Theorem 2.12). The results in this section are based on Rü (2004).

Theorem 6.40 (Fréchet bounds for multivariate marginals).

(a) If $n = 2$, $A \in \mathcal{B}^k$ is closed, and $\pi_1(x_1, x_2) = x_1$ is the first projection then

$$M(A) := \sup \{ P(A); P \in M(P_1, P_2) \} \qquad (6.52)$$

$$= 1 - \sup \{ P_2(O) - P_1 \left(\pi_1 \left(A \cap (\mathbb{R}^{k_1} \times O) \right) \right); O \subset \mathbb{R}^{k_2} \ open \}.$$

(b) $$F_-(x) := \left(\sum_{i=1}^{n} F_i(x_i) - (n-1) \right)_+ \leq F(x_1, \ldots, x_n) \qquad (6.53)$$

$$\leq \min_{i \leq n} F_i(x_i) =: F_+(x)$$

$$\overline{F}_-(x) := \left(\sum_{i=1}^{n} \overline{F}_i(x_i) - (n-1) \right)_+ \leq \overline{F}(x_1, \ldots, x_n) \tag{6.54}$$

$$\leq \min_{i \leq n} \overline{F}_i(x_i) =: \overline{F}_+(x)$$

where $\overline{F}_i(x_i) = P(X_i \geq x_i)$, $\overline{F}(x_1, \ldots, x_n) = P(X_i \geq x_i)$, $1 \leq i \leq n$ are the (multivariate) survival functions. The bounds in (6.53) and (6.54) are sharp.

Remark 6.41. (a) In the case of one-dimensional marginals (6.53) and (6.54) are the classical Fréchet bounds. F_+, the upper Fréchet bound, is a distribution function while the lower Fréchet bound F_- is a distribution function only in exceptional cases with large jumps (for details see Dall'Aglio (1972)).

(b) "Bounds for tails of $\Psi(X_1, X_2)$:" Part (a) of Theorem 6.40 allows to obtain sharp upper and lower bounds for tail probabilities $P(\Psi(X_1, X_2) \geq t)$ of general functionals $\Psi(X_1, X_2)$. One obtains simpler bounds for monotonically nondecreasing functions Ψ:

If Ψ is monotonically nondecreasing upper-semicontinuous and $A_\Psi^- := \{(u, v) \text{ minimal in } \mathbb{R}^2; \Psi(u, v) \geq t\}$, then

$$P(\Psi(X_1, X_2) \geq t) \leq \inf_{(u,v) \in A_\Psi^-(t)} (F_1(u) + F_2(v)). \tag{6.55}$$

In analogy to the case of combined risks in the real case where $\Psi(x, y) = x + y$ the bounds in (6.55) are called the infimal (respectively supremal) Ψ-convolution of P_1, P_2 (see Theorem 5.3).   ◊

In the one-dimensional case the comonotone random vectors $(F_1^{-1}(U), \ldots, F_n^{-1}(U))$ attain the upper Fréchet bound (i.e. have distribution function F_+) and are by Theorem 3.5 the riskiest random vectors. In the multivariate case typically there will not exist "comonotone vectors" $(X_1, \ldots, X_n)$ with $X_i \sim F_i$ in the sense that $(X_1, \ldots, X_n) = (f_1(U), \ldots, f_n(U))$ with nondecreasing $f_i : [0, 1] \to \mathbb{R}^{k_i}$. Even in the case that $k_i = k_1$, $1 \leq i \leq n$, and $F_1 = F_2 = \cdots = F_n$ the "*natural comonotone vector*" with identical components $X = (X_1, X_1, \ldots, X_1)$ will not attain the sharp upper Fréchet bound $F_+(x)$ and thus does not yield the riskiest portfolio distribution.

Proposition 6.42 (Comonotone random vector and Fréchet bounds).

(a) *In general the upper and lower Fréchet bounds for $k_i \geq 2$ do not define distribution functions.*

(b) *If $F_1 = F_2 = \cdots = F_n$ is a k_1-dimensional distribution function, and $X_1 \sim F_1$, then the "comonotone" random vector $X = (X_1, \ldots, X_1)$ has distribution function*

$$F(x) = F_1(x_1 \wedge \cdots \wedge x_n) \leq F_+(x) = \min_{1 \leq i \leq n} F_1(x_i), \quad x_i \in \mathbb{R}^k. \tag{6.56}$$

In general there is strict inequality in (6.56).

Proof. (a) The reason for (a) is the following. Let w.l.o.g. $n = 2$, $k_1 = k_2 = 2$. Assume that for $G, H \in \mathcal{F}_2$ with one-dimensional marginals G_1, G_2, H_1, H_2 the lower Fréchet bound $F_- = F_-(G, H)$ is a four-dimensional distribution function. Then for $X \sim F_-$ we conclude from (6.53) that

$$(X_1, X_3) \sim F_-(G_1, H_1), \qquad (X_1, X_4) \sim F_-(G_1, H_2),$$
$$(X_2, X_3) \sim F_-(G_2, H_1), \text{ and } (X_2, X_4) \sim F_-(G_2, H_2).$$

This however would imply strong positive correlation of (X_3, X_4) and of (X_1, X_2) which is not $F_-(G_1, H_1)$ according to our assumption. Except for some special cases with big jumps (as described in Dall'Aglio's classical 1972 paper) we would obtain that $(X_1, X_2) \sim F_+(G_1, G_2)$ and $(X_3, X_4) \sim F_+(H_1, H_2)$. Similarly, the argument for the upper Fréchet bound yields the same contradiction.

(b) follows directly from the definition. Only for $k_1 = 1$ equality holds in (6.56) in general. □

Remark 6.43. The discussion in the proof of (a) in Proposition 6.42 shows that for $n \geq 3$ up to some exceptional cases (with big jumps) only for $G = F_+(G_1, \ldots, G_n)$ and $H = F_+(H_1, \ldots, H_n)$, where G_i, H_i are one-dimensional distribution functions, the Fréchet bounds are distribution functions. This holds in particular true if G_i, H_i are continuous distribution functions. ◊

For Δ-monotone functions f one can conclude from (5.2) and (5.3) upper and lower bounds for the integrals.

Theorem 6.44. Let $F \in \mathcal{F}(F_1, \ldots, F_n)$ and let $f : \mathbb{R}^n \to \mathbb{R}^1$ be Δ-monotone and P_F-integrable. Assume that for $1 \leq i \leq n$, $\lim_{x_i \to -\infty} f(x_1, \ldots, x_i, \ldots, x_n) = 0$ for all $x_1, \ldots, x_{i-1}, x_{i+1}, \ldots, x_n$. Then

$$\int \overline{F}_-(x) df(x) \leq \int f \, dF \leq \int \overline{F}_+(x) df(x). \tag{6.57}$$

Proof. The proof follows from the Fréchet bounds in (5.2) and (5.3) applied to the partial integration formula in Rü (1980, proof of Theorem 3). By this formula one obtains for any $a \in \mathbb{R}^k$

$$\int_{[a,\infty)} \Delta_a^x f \, dF = \int_{[a,\infty)} \overline{F}(x) df(x). \tag{6.58}$$

For $a \to -\infty$ the integrals converge by our assumption on f to yield

$$\int f \, dF = \int \overline{F}(x) df(x) \tag{6.59}$$

and so (6.57) follows from the Fréchet bounds in (6.54). □

Remark 6.45. The bounds in (6.57) for the integrals are not sharp in general, since F_+, F_- are not distribution functions.   ◊

Example 6.46 (Antimonotone and comonotone variates).   In the following examples we investigate some *natural* multivariate extensions of anticomonotone and comonotone variates for various examples of functions φ. In particular we also give an example where the natural "comonotone" random vector yields the lowest risk for a directionally convex function φ. The various examples in this section show that in case of multivariate marginals there does not exist a general worst case distribution like in the case of one-dimensional marginals.

We consider as an example the case $k_i = 2$, $1 \leq i \leq n$, and $F_i = F_-(G_i, H_i)$, $1 \leq i \leq n$, i.e. for a uniform random variable U, $(G_i^{-1}(U), H_i^{-1}(1-U)) \sim F_i$, $1 \leq i \leq n$. As alternative random vectors with distribution functions in $\mathcal{F}(F_1, \ldots, F_n)$ we consider three vectors

$$
\begin{aligned}
W_+ &= ((G_1^{-1}(U), H_1^{-1}(1-U)), \ldots, (G_n^{-1}(U), H_n^{-1}(1-U))), \\
Z &= ((G_1^{-1}(U_1), H_1^{-1}(1-U_1)), \ldots, (G_n^{-1}(U_n), H_n^{-1}(1-U_n))), \quad (6.60)
\end{aligned}
$$

and

$$
W_- = ((G_1^{-1}(U), H_1^{-1}(1-U)), (G_2^{-1}(1-U), H_2^{-1}(U))) \quad \text{if } n = 2,
$$

where (U_i) are independent uniform. W_+ is a "generalized comonotone" vector, Z the independent vector and W_- a generalized "anticomonotone vector" with marginals F_i. Then for $x = (x_1, \ldots, x_n)$, $x_i = (y_i, z_i)$ we obtain as distribution functions of W_+ and Z

$$
\begin{aligned}
F_{W_+}(x) &= (\min G_i(y_i) + \min H_i(z_i) - 1)_+, \\
F_Z(x) &= \prod_{i=1}^{n} (G_i(y_i) + H_i(z_i) - 1)_+. \quad (6.61)
\end{aligned}
$$

Further the Fréchet bounds with marginals F_i are given by

$$
\begin{aligned}
F_+(x) &= \min(G_i(y_i) + H_i(z_i) - 1)_+, \quad \text{and} \\
F_-(x) &= \left(\sum_{i=1}^{n} (G_i(y_i) + H_i(z_i) - 1)_+ - (n-1) \right)_+. \quad (6.62)
\end{aligned}
$$

F_{W_+}, F_Z are not uniformly comparable with each other.

To demonstrate that in dimension ≥ 2 the "comonotone" vector W_+ is not genuinely the worst case dependence structure we consider at first the directionally convex function $\varphi_1(x) = (\sum y_i + \sum z_i)^2$ and the special case where $G_i = G$, $H_i = H$, both with expectations zero. Then

$$E\varphi_1(W_+) = \text{Var}\left(\sum_{i=1}^{n} G^{-1}(U) + H^{-1}(1-U)\right)$$

$$= n^2 \text{Var}(G^{-1}(U) + H^{-1}(1-U)). \qquad (6.63)$$

Similarly,

$$E\varphi_1(Z) = \text{Var}\left(\sum_{i=1}^{n}(G^{-1}(U_i) + H^{-1}(1-U_i))\right)$$

$$= n\,\text{Var}(G^{-1}(U) + H^{-1}(1-U)). \qquad (6.64)$$

So $E\varphi_1(W_+) = nE\varphi_1(Z)$. The "comonotone vector" $W_+ = (X_1, X_1, \dots, X_1)$, $X_1 = (G^{-1}(U), H^{-1}(1-U))$ induces in this case a much higher variance of the sum than the independent vector $Z = (X_1, \dots, X_n)$ where X_i are independent, $X_i \sim X_1$. In fact in this case we obtain from the classical Fréchet bounds that

$$E\varphi_1(W_+) \ge E\varphi_1(X) \qquad (6.65)$$

for all X with marginals $F_i = F_-(G, H)$, i.e. the risk measured by φ_1 is maximal for the "comonotone" vector W_+ (this is true for all marginals F_i).

For $\varphi_2(x) = \max y_i + \max z_i$, the sum of the maximal risks in the first and second components, the situation is different: $E\varphi(W_+) = EG^{-1}(U) + EH^{-1}(1-U)$ is the smallest possible value. From extreme value theory it is known that the value of the independent vector Z, $E\varphi_2(Z)$, is of the order $a_n = G^{-1}(1 - \frac{1}{n}) + H^{-1}$ $(1 - \frac{1}{n})$ under the corresponding domain of attraction conditions. It was shown in Lai and Robbins (1976) that this is close to the maximal possible value attained by *maximally dependent* random variables (see our discussion in Section 2.2). The fact that $E\varphi_2(W_+) \le E\varphi_2(Z)$ is to be expected in this case since φ_2 is Δ-antitone.

Finally consider $\varphi_3(x) = \sum_{i=1}^{n-1}(y_i z_{i+1} + y_{i+1} z_i)^2$ and let G, H have support in $\mathbb{R}_+$. Then φ_3 is directionally convex (so a proper measure of risk). For the "comonotone" vector W_+ we obtain

$$E\varphi_3(W_+) = (n-1)E(G^{-1}(U)H^{-1}(1-U) + H^{-1}(U)G^{-1}(1-U))^2. \quad (6.66)$$

In this case the "comonotone" vector W_+ yields the smallest possible risk. The reason is the strong negative dependence between the components of the marginal distributions. The largest possible value is attained in case $n = 2$ by the "anticomonotone" vector

$$W_- = ((G^{-1}(U), H^{-1}(1-U)), (G^{-1}(1-U), H^{-1}(U))). \qquad (6.67)$$

Some further discussion of possible multivariate notions of "comonotonicity" has been given in Puccetti and Scarsini (2010). Our discussion in this section shows that in the case of multivariate marginals there does not exist a notion of "comonotonicity" describing in a similar universal way as in dimension 1 a worst case dependence structure.

Part II
Risk Measures and Worst Case Portfolios

The purpose of a risk measure is to specify the " riskiness" of an insurance contract X or of a portfolio held by a financial institution, by an index $\phi(X)$. In insurance the risks are typically positive and we use the notation $\Psi(X)$ for risk measures which are monotone in the usual sense; $X \leq Y$ implies that $\Psi(X) \leq \Psi(Y)$. In finance, the losses are typically negative and $X \leq Y$, therefore, implies that the risk of X is larger than the risk of Y, i.e. $\varrho(X) \geq \rho(Y)$ for the risk measure denoted by ρ in this context.

Risk measures have a long tradition in insurance going back to the 1970s (see Bühlmann (1970)). They serve to calculate premiums of insurance contracts and the discussion of the "premium calculation principles" is a basic part of risk management in insurance. In mathematical finance risk measures were introduced in the late 1990s and their relevance for handling financial risk has become a predominant subject.

For an introduction to risk measures we refer in particular to Gerber (1979), Goovaerts et al. (1984), Kaas et al. (2001), and Denuit et al. (2005) for the actuarial side and for the financial risk theory to Föllmer and Schied (2011), Delbaen (2002), and Pflug and Römisch (2007). The main subject in this chapter is the investigation of risk measures for risk vectors and portfolios. Besides the risk coming from the components X_i of the risk vector X we are in particular interested in describing the risk coming from possible dependence between the components. Risk measures should evaluate this additional "dependence risk" in a correct way. This postulate is not easy to formulate and some early postulates in risk theory turned out in the more recent literature to need essential modification and specification.

One prominent example of this type of changing view is the postulate on risk measures that "diversification of risk should be encouraged by the risk measure ρ", i.e. the postulate of "subadditivity"

$$\varrho \left(\sum_{i=1}^{n} X_i \right) \leq \sum_{i=1}^{n} \varrho(X_i) \tag{II.1}$$

or the related weaker variant of "convexity"

$$\varrho\left(\sum_{i=1}^{n} \alpha_i X_i\right) \leq \sum_{i=1}^{n} \alpha_i \rho_i(X_i). \tag{II.2}$$

This kind of "economic axiom" was used to argue against the classical value at risk VaR_α risk measure which is neither subadditive nor convex and to propose to use instead the expected shortfall risk measure $ES_\alpha(x)$ defined (in the insurance version) as excess of loss

$$ES_\alpha(X) = E(X - \mathrm{VaR}_\alpha(X))_+ \tag{II.3}$$

or some variant of it. The expected shortfall measures the magnitude of loss above the "default level" $\mathrm{VaR}_\alpha(X)$.

Diversification however is in general not a useful and justified postulate. If we take for example a portfolio $X = (X_1, \ldots, X_n)$ of heavy tailed components X_i say X_i are symmetric α-stable, i.e. the characteristic function of X_i is given by

$$\varphi_{X_i}(t) = e^{-c|t|^\alpha}$$

with some constant $c > 0$. If we assume that (X_i) are independent, then the joint portfolio $\sum_{i=1}^{n} X_i$ is distributed as $n^{1/\alpha} X_1$ or equivalently for the diversified portfolio $\frac{1}{n} \sum_{i=1}^{n} X_i$ it holds that

$$\frac{1}{n} \sum_{i=1}^{n} X_i \overset{d}{=} n^{1/\alpha - 1} X_1. \tag{II.4}$$

For the proof of (II.4) note that

$$\varphi_{\sum_{i=1}^{n} X_i}(t) = (\varphi_{X_1}(t))^n = e^{-cn|t|^\alpha} = \varphi_{n^{1/\alpha} X_1}(t).$$

In consequence in the practically relevant case $\alpha < 1$ the diversified portfolio is by a magnitude $n^{1/\alpha - 1}$ worse than the undiversified. Any risk measure that would be convex on this model would evaluate the risk in a wrong way. For $\alpha > 1$ we have existence of first moments and a positive diversification effect arises in (II.4). Useful risk measures should reflect this effect. For bounded risks this effect does not appear but for non-integrable risks the axioms of convexity or subadditivity do not make sense in general. One principal problem of risk measures is that the risk distribution (respectively loss distribution) which gives the complete picture of the risk is difficult (or even impossible) to represent by an index $\varrho(X)$ or $\Psi(X)$.

To demonstrate the importance of describing the risk coming from dependence in the portfolio we consider the following simple example.

Example II.1 (Dependence effect in mixture models).   Let $X_i = \Theta Y_i + (1 - \Theta)$ Z_i, $1 \leq i \leq 10^5$ be a portfolio of 10^5 contracts, where (Y_i), (Z_i) are iid, independent of Θ and

$$Y_i \sim \mathcal{B}\left(1, \frac{1}{100}\right), \quad Z_i \sim \mathcal{B}\left(1, \frac{1}{1000}\right), \quad \Theta \sim \mathcal{B}\left(1, \frac{1}{100}\right)$$

are binomial distributed. Θ introduces a 'small' positive dependence in terms of correlation between the components of the X_i. What is the influence of this small positive dependence on the risk of the joint portfolio $S_n = \sum_{i=1}^{n} X_i$?

Note that $X_i \sim \mathcal{B}(1, \frac{1}{1000})$. Let (W_i), $1 \leq i \leq 10^5$ be a related iid portfolio $W_i \sim \mathcal{B}(1, \frac{1}{1000})$ and denote $T_n = \sum_{i=1}^{n} W_i$ the risk of the joint iid portfolio. Then from the central limit theorem we get an approximation by the normal distribution

$$T_n \sim N(100, 100).$$

The distribution of the related mixture portfolio S_n is approximatively given by a mixture distribution

$$S_n \sim \frac{1}{100} N(1000, 1000) + \frac{99}{100} N(100, 100).$$

In consequence the iid portfolio (W_i) is safe w.r.t. the excess of loss with retention limit $t = 150$ (which is the 5σ domain) $E(T_n - 150)_+$ is in the range of 10^{-8}. But $E(S_n - 150)_+ \sim 8.5$ is of considerable magnitude. Note that $P(S_n > 150) \sim \frac{1}{100}$. Thus there exists a considerable risk probability for the mixture model, which is induced by the small positive dependence in the sample.   ◊

After a short introduction to examples and properties of risk measures of real risks we consider their robust representation and continuity properties on L^p-spaces in Chapter 7. Right from the beginning we allow for unbounded risks, since the typical models of risks are not bounded. In Chapter 8 we concentrate on the structure, examples, and properties of risk measures for portfolio vectors. It turns out that their representation is closely connected with mass transportation problems. This connection was first described in Rü (2006). In particular one finds that the role of the expected shortfall and more generally of the spectral risk measures is taken in the multivariate case by the "max correlation risk measures" which are defined by a mass transportation problem.

In contrast to the real case there is no longer a useful notion of comonotonicity which describes a "universal" worst case dependence structure in the case of risk vectors. Instead we obtain depending on a risk measure ϱ (or Ψ) a worst case dependence structure, which one could call in analogy to the one-dimensional case "ϱ-comonotone dependence structure". A general characterization of this worst case dependence structure was given in Rü (2012a). For the special case of max-correlation risk measures see also Ekeland et al. (2010). We also describe the worst case diversification effect measured by

$$\inf \left\{ \sum_{i=1}^{n} \varrho(X_i) - \varrho \left(\sum_{i=1}^{n} X_i \right) \right\} \qquad \text{(II.5)}$$

respectively

$$\inf \left\{ \frac{1}{n} \sum_{i=1}^{n} \varrho(X_i) - \varrho \left(\frac{1}{n} \sum_{i=1}^{n} X_i \right) \right\} , \qquad \text{(II.6)}$$

the inf being over all possible dependence structures. The max correlation risk measures are (up to some regularity condition) the only risk measures where the worst case diversification effect is zero.

Chapter 7
Risk Measures for Real Risks

Measuring the risk of a financial position is a complex process which is connected with several features in the financial or the insurance market. One postulate is that measures of risk should include the aspect of securization of risk, i.e. the possibility to transfer risks by hedging actions or by buying or selling derivatives to the market. The price of the hedging action or of the derivatives is one part of the risk of a position. The remaining part of the risk has to be evaluated based on the underlying probability model and on the preference of the risk taker. Thus measuring of risk is connected with probabilistic modelling, with pricing, and with preferences – this triad is called "the three P's" in Lo (1999).

Let C be the set of available hedging actions including addition of capital, trading actions in a basic process X like $\int_o^T \xi_s dX_s$ or buying derivatives. Let π be a relevant price system on C and let finally $\mathcal{A}$ describe the set of acceptable positions which is related to the preference system of the risk taker. Then a reasonable risk measure is given by

$$\varrho(X) = \inf\{\pi(Y);\ Y \in C, X + Y \in \mathcal{A}\}. \tag{7.1}$$

$\varrho(X)$ is the minimal price of a hedging action which, added to the risk, makes it acceptable. In case that the hedging actions are restricted to adding capital, i.e. $C = \mathbb{R}$ we have no pricing problem and get

$$\varrho_{\mathcal{A}}(X) = \inf\{m \in \mathbb{R};\ X + m \in \mathcal{A}\}. \tag{7.2}$$

$\varrho_{\mathcal{A}}$ is the minimal amount of capital which added to the risk makes it acceptable. The hedging point of view for constructing risk measures in (7.1) was introduced in its general form in Scandolo (2004) and Frittelli and Scandolo (2006). It is still a challenge to put this view into an integrated risk management system.

Generally risk measures ϱ respectively Ψ for real risks are defined on a class $\mathcal{X}$ of financial or insurance risk positions quantifying the risk of X by some number $\varrho(X)$ respectively $\Psi(X)$. Any risk measure has a natural domain $\mathcal{X} \subset L^0$ of definition. Here $L^0 = L^0(\Omega, \mathfrak{A}, P)$ is the class of all measurable elements on some probability

L. Rüschendorf, *Mathematical Risk Analysis*, Springer Series in Operations Research and Financial Engineering, DOI 10.1007/978-3-642-33590-7_7,
© Springer-Verlag Berlin Heidelberg 2013

space $(\Omega, \mathfrak{A}, P)$ which we assume generally to be non-atomic. $\mathfrak{A}$ is assumed generally to be countably generated. The value at risk $\text{VaR}_\alpha(X)$ for example is a risk measure naturally defined on L^0, while the tail conditional expectation (finance version) $TC_\alpha(X) = E(-X \mid X \leq q_\alpha(X))$ with $q_\alpha(X) = F_X^{-1}(\alpha) = -\text{VaR}_\alpha(X)$ is naturally defined on $\mathcal{X} = \{X \in L^0; \ X^- \in L^1\}$. The classical domain of risk measures in finance has been the class of essentially bounded risks $\mathcal{X} = L^\infty$ and is presented in detail in Delbaen (2002) and Föllmer and Schied (2011). In insurance unbounded domains were predominant from the beginning. Since typically risks are modelled also in finance by unbounded random variables as e.g. normal, exponential Lévy or stable distributed random variables, the literature on the more theoretical representation and continuity properties of risk measures has been extended to L^p or L^p-type spaces like Orlicz spaces (see e.g. Cheridito and Li (2009), Kaina and Rü (2009), and Filipović and Svindland (2012)) concentrating in particular on the class of monetary respectively convex risk measures.

7.1   Some Classes of Risk Measures for Real Variables

Risk functionals ϱ, Ψ are defined on some domain $\mathcal{X} \subset L^0$ of random variables (risks) with values in $\mathbb{R} \cup \{\infty\}$.

7.1.1   Basic Properties of Risk Measures

Before discussing some classes of risk functionals we introduce some properties of interest for risk functionals. We use the notation ϱ for the finance version of risk measures and Ψ for the insurance version (with the simpler monotonicity condition). We use only one of these when there is no difference. We assume throughout that the classes $\mathcal{X}$ are closed under sums and include constants.

1. **Monotonicity:** For $X, Y \in \mathcal{X}$:

$$X \leq Y [P] \text{ implies } \varrho(Y) \leq \varrho(X) \tag{7.3}$$

(respectively $\Psi(X) \leq \Psi(Y)$).

2. **Subadditivity:** $\qquad\qquad\qquad \varrho(X + Y) \leq \varrho(X) + \varrho(Y). \tag{7.4}$

3. **Positive homogeneity:** $\qquad\qquad \varrho(\lambda X) = \lambda \varrho(X) \text{ for } \lambda > 0. \tag{7.5}$

4. **Cash invariance (equivariance, translation equivariance):**

$$\varrho(X + a) = \varrho(X) - a, \quad \forall a \in \mathbb{R} \tag{7.6}$$

(respectively $\Psi(X + a) = a + \Psi(x)$).

5. **Convexity:** For $X, Y \in \mathcal{X}, \alpha \in (0,1)$ it holds that

$$\varrho(\alpha X + (1 - \alpha)Y) \le \alpha\varrho(X) + (1 - \alpha)\varrho(Y). \tag{7.7}$$

6. **Law invariance (objectivity, law dependence):**

$$X \overset{d}{=} Y \text{ implies } \varrho(X) = \varrho(Y). \tag{7.8}$$

7. **Comonotone additivity:**

$$\varrho(X + Y) = \varrho(X) + \varrho(Y) \text{ for } X, Y \text{ comonotone.} \tag{7.9}$$

Remark 7.1. (a) **Continuity:** Of interest for the application of risk measures are also various continuity properties which may vary in dependence of the domain $\mathcal{X}$ (see the discussion of the case $\mathcal{X} = L^p$ in Section 7.2). The classical continuity in distribution $X_n \overset{d}{\to} X$, i.e.

$$F_{X_n}(x) \to F_X(x) \text{ for all continuity points } x \text{ of } F_X \tag{7.10}$$

implies $\varrho(X_n) \to \varrho(X)$ typically is too restrictive. For example the expected shortfall would not satisfy this condition.
(b) **Equivariance:** Cash invariance of ϱ implies the important properties

$$\varrho(X + \varrho(X)) = 0 \tag{7.11}$$

and

$$\varrho(a) = \varrho(0) - a. \tag{7.12}$$

Equation (7.11) implies that $\varrho(X)$ is just the amount of "cash" which added to the risk X makes it riskless. The notion "equivariance" for functionals Ψ satisfying $\Psi(X + a) = \Psi(X) + a$ is common in statistics and is a proper notion of this property. It is natural to postulate the normalization condition $\varrho(0) = 0$ or $\varrho(a) = -a$ for all $a \in \mathbb{R}^1$.
(c) **Law invariance:** If a risk functional is law invariant then $\varrho(X)$ depends only on the distribution of X, i.e. we can consider ϱ as a functional on the class of distributions or distribution functions

$$\varrho(X) = \varrho(P^X) = \varrho(F_X). \tag{7.13}$$

Thus a better notation of this property would be "law dependence" instead of "law invariance". In Pflug and Römisch (2007) this property is called adequately "version independence". While the classical examples of risk measures are law dependent typical hedging based risk measures as in (7.1) will not be law invariant but depend on the random variables X themselves.

Law invariance is a useful property since it allows that the risk measure can be
estimated from empirical data. Let $\widehat{F}_n$ be the empirical distribution function of
an iid sample $X_1, \ldots, X_n$, then $\widehat{F}_n \to F$ a.s. by the law of large numbers. If ϱ
is continuous w.r.t. this convergence then

$$\widehat{\varrho}_n := \varrho(\widehat{F}_n) \to \varrho(F) \text{ a.s.}$$

Moreover by the central limit theorem precise error bounds can be given. For
details on this see Tasche (2002) and Belomestny and Krätschmer (2012).

(d) **Subadditivity, convexity, and comonotone additivity:** Subadditivity of a risk
functional ϱ corresponds to the idea that diversification reduces risk (see the
discussion in the introduction). For $X_1, \ldots, X_n$ the "diversification effect" of a
subadditive risk function ϱ

$$D((X_i)) := \sum_{i=1}^n \varrho(X_i) - \varrho\left(\sum_{i=1}^n X_i\right) \tag{7.14}$$

is non-negative. Similarly for convex risk measures one can consider the
(average) diversification

$$D^a((X_i)) = \frac{1}{n}\sum_{i=1}^n \varrho(X_i) - \varrho\left(\frac{1}{n}\sum_{i=1}^n X_i\right). \tag{7.15}$$

Subadditive and convex risk measures make sense only for classes $\mathcal{X}$ of integrable random variables as follows from generalized strong laws as explained
in the introduction for independent α-stable random variables (see Nešlehová
et al. (2006b) and Mainik and Rü (2012)). Substitutes for these properties
are postulates of consistency with some natural orderings, which combine
stochastic (increasing) ordering and variability orderings.

There are also some arguments in favour of additivity of risk measures on
some subclasses. For example additivity on the class of independent risks is
motivated by the classical Bienaymé formula for the variance. An example is

$$\Psi(X) = EX + \lambda \operatorname{Var}(X). \tag{7.16}$$

The notion of comonotone additivity generalizes the additivity on constants
$\varrho(\sum_{i=1}^n a_i) = \sum_{i=1}^n \varrho(a_i)$, $a_i \in \mathbb{R}$ and extends to

$$\varrho\left(\sum_{i=1}^n X_i\right) = \sum_{i=1}^n \varrho(X_i) \tag{7.17}$$

for comonotone $X_1, \ldots, X_n$. $\Diamond$

Some classes of risk measures have attracted particular attention in the more recent literature.

Definition 7.2. (a) A risk functional $\varrho : \mathcal{X} \to \mathbb{R} \cup \{\infty\}$ (respectively Ψ) is called

(1) "risk measure" if it is monotone
(2) "monetary risk measure" if it is monotone and cash invariant
(3) "convex risk measure" if it is monotone, cash invariant, and convex
(4) "coherent risk measure" if it is monotone, positive homogeneous, and subadditive.

(b) If ϱ is a risk measure on $\mathcal{X}$ then $\mathcal{A}_\varrho := \{X \in \mathcal{X}; \; \varrho(X) \le 0\}$ is called "acceptance set" of ϱ.

Remark 7.3 (Acceptance sets and risk measures).

(a) Let ϱ be a monetary risk measure $\varrho \ne \infty$ and let $\mathcal{A} = \mathcal{A}_\varrho$ be its acceptance set, then

(1) $\mathcal{A}$ is monotone
(2) ϱ is convex if and only if $\mathcal{A}$ is convex
(3) $\varrho(X) = \begin{cases} \infty & \text{if } X + a \notin \mathcal{A} \text{ for all } a \in \mathbb{R}, \\ \inf\{a \in \mathbb{R}; \; X + a \in \mathcal{A}\} & \text{else,} \end{cases}$
(4) ϱ is coherent if and only if $\mathcal{A}$ is a convex cone.

(b) Conversely, if $\mathcal{A} \subset \mathcal{X}$, $\mathcal{A} \ne \phi$ is convex and monotone such that

$$\inf\{a \in \mathbb{R}; \; a + Y \in \mathcal{A}\} < \infty \text{ for all } Y \in \mathcal{X},$$

then $\mathcal{A}$ induces a convex risk measure $\varrho_\mathcal{A}$ by

$$\varrho_\mathcal{A}(X) := \inf\{a \in \mathbb{R}; \; X + a \in \mathcal{A}\}, \quad X \in \mathcal{X} \tag{7.18}$$

and $\mathcal{A} \subset \mathcal{A}_{\varrho_\mathcal{A}}$.

As a consequence risk measures can be introduced equivalently by the "preference" of the risk taker which is described by the class $\mathcal{A}$ of acceptable positions.

(c) Similarly a (insurance) risk measure Ψ can be defined by a set $\mathcal{A}$ of acceptable risks

$$\Psi_\mathcal{A}(X) = \inf\{a \in \mathbb{R}^1; \; X - a \in \mathcal{A}\}. \tag{7.19}$$

In insurance risk measures were discussed as premium calculation principles. Classical, i.e. law invariant coherent and convex risk measures, were discussed in Bühlmann (1970), Gerber (1979), and Goovaerts et al. (1984), and many others. Version dependent (not law invariant) convex risk measures were first introduced in Deprez and Gerber (1985). In mathematical finance coherent and convex risk measures were introduced in Artzner et al. (1999), Delbaen (2000), Föllmer and Schied (2011), and Frittelli and Rosazza Gianin (2002), and many other later papers. In comparison in mathematical finance the cash invariance property and in particular the robust (dual) representation of risk measures have been of dominant interest.   ◊

7.1.2   Examples of Risk Measures

In this subsection we collect some classes of interesting risk measures.

Value at Risk

The "value at risk" risk measure is a law invariant risk measure defined on $X = L^0$ in terms of the quantiles of the distribution of a random variable X with distribution function F. For $\lambda \in (0, 1)$ a real number q is called a "λ-quantile" if

$$F(q) = P(X \le q) \ge \lambda \ge P(X < q). \tag{7.20}$$

The λ-quantiles form an interval $[q_X^-(\lambda), q_X^+(\lambda)]$ where

$$q_X^-(\lambda) = \sup\{x; \; P(X < x) < \lambda\}$$
$$= \inf\{x; \; F(x) \ge \lambda\} = F^{-1}(\lambda) \tag{7.21}$$

$$\text{and} \quad q_X^+(\lambda) = \inf\{x; \; F(x) > \lambda\} = \sup\{x; \; P(X < x) \le \lambda\}.$$

The lower λ-quantile $q_X^-(\lambda)$ is identical to the generalized inverse distribution function $F^{-1}(\lambda)$ as used throughout this text.

The insurance risk version of the "value at risk" at level λ is defined as

$$\text{VaR}_\lambda(X) = F^{-1}(\lambda); \tag{7.22}$$

i.e. as the smallest λ-quantile of the distribution of X. The finance risk version of the "value at risk" is defined as

$$\text{V@R}_\lambda(X) = -q_X^+(\lambda), \tag{7.23}$$

i.e. as minus the largest λ-quantile. (We follow here the notation in Föllmer and Schied (2011)).

The value at risk $\text{VaR}_\lambda(X)$ defined in (7.22) is the smallest number q such that a higher damage arises only with probability $\le 1 - \lambda$. Note that in insurance risk one is interested in right tails, i.e. in λ close to one, while in finance one is interested in left tails of X, i.e. in λ close to zero. Switching from X to $-X$ in definition (7.23) we have

$$V@R_\lambda(X) = -q_X^+(\lambda) = q_{-X}^-(1 - \lambda)$$
$$= VaR_{1-\lambda}(-X) \tag{7.24}$$

and thus both definitions are consistent. In some parts of the literature also the $(1 - \lambda)$-quantile $F^{-1}(1 - \lambda)$ is denominated as value at risk $VaR_\lambda(X)$ at level λ. Some properties of the value at risk are given in the following proposition. By relation (7.24) we only consider one of the versions of value at risk.

Proposition 7.4 (Properties of the value at risk).

(a) VaR_λ is a monotone, law and cash invariant, positive homogeneous, comonotone additive risk measure
(b) VaR_λ is continuous w.r.t. convergence in distribution
(c) In general VaR_λ is neither homogeneous nor convex
(d) For integrable X $VaR_\lambda(X)$ is the minimal solution α^* of the optimization problem

$$E(X - \alpha^*)_+ + (1 - \lambda)\alpha^* = \min_{\alpha \in \mathbb{R}} \tag{7.25}$$

(e) On $\mathcal{X} = L^\infty$ it holds that

$$VaR_\lambda(X) = \min \left\{ \Psi(X); \ \Psi \text{ is a convex risk measure on } L^\infty, \right.$$
$$\left. \text{continuous from below, } \Psi \geq VaR_\lambda(X) \right\}. \tag{7.26}$$

Continuity from below is defined as $X_n \uparrow X$ pointwise implies $\Psi(X_n) \uparrow \Psi(X)$.

Proof. (a), (b), (c) For a comonotonic vector $X^c = (X_1^c, \ldots, X_n^c)$ it holds that

$$VaR_\lambda \left(\sum_{i=1}^n X_i^c \right) = F_{\sum_{i=1}^n X_i^c}^{-1}(\lambda) = \sum_{i=1}^n F_{X_i^c}^{-1}(\lambda)$$
$$= \sum_{i=1}^n VaR_\lambda(X_i^c) \quad \text{(see (3.47))}. \tag{7.27}$$

The other properties are easy to establish.
(d) The function $h(t) := E(X - t)_+ + (1 - \lambda)t = \pi_X(t) + (1 - \lambda)t$ are defined with stop-loss function $\pi_X(t)$ given by

$$\pi_X(t) = E(X - t)_+ = \int_t^\infty \overline{F}(x)dx.$$

For $x < VaR_\lambda(X)$ it holds that $\overline{F}(x) > 1 - \lambda$, while for $x \geq VaR_\lambda(X)$ it holds that $\overline{F}(x) \leq 1 - \lambda$. This implies that $VaR_\lambda(X)$ is the smallest minimum point of h, i.e. the smallest solution of (7.25).

(e) A proof is given in Proposition 4.42 in Föllmer and Schied (2011).   □

Remark 7.5. (a) The fact that for fixed λ the $\text{VaR}_\lambda(X)$ is not coherent respectively convex and thus contradicts the "diversification principle" was the starting point of Artzner et al. (1999) of their axiomatic approach to risk measures and led to the suggestion of the expected shortfall risk measure.

(b) From a practical point of view it is useful to consider the value at risk $\text{VaR}_\lambda(X)$ for several levels $\lambda < \lambda_1 < \cdots < \lambda_k < 1$. In this way not only does one get information about the quantile $F_Y^{-1}(\lambda)$ which is with probability $1 - \lambda$ not surpassed but one also gets information about the magnitude of loss above $F_X^{-1}(\lambda)$.   ◊

Tail Conditional Expectation and Expected Shortfall

The tail conditional expectation and the expected shortfall give in addition to information on the probability of a high risk also information on the magnitude of this tail risk. Again there are a finance and an insurance version. The finance version of the tail conditional expectation is given by

$$TCE_\alpha(X) = -E(X \mid X \le q_\alpha), \tag{7.28}$$

where q_α is any α-quantile of X. Correspondingly, the insurance version is

$$TCE_\alpha^+(X) = E(X \mid X \ge q_\alpha). \tag{7.29}$$

The "tail expectation" at level α

$$TE_\alpha^+(X) = E(X - \text{VaR}_\alpha(X))_+ = \pi_X(\text{VaR}_\alpha(X)) \tag{7.30}$$

is the stop-loss premium with retention limit $\text{VaR}_\alpha(X)$. This index has been used in early economics literature. Similarly $TE_\alpha(X) = E(q_\alpha(X) - X)_+$ is the tail expectation in the finance context.

There is a slight modification of TCE_α called "expected shortfall"

$$ES_\alpha(X) = CTE_\alpha(X) = -\frac{1}{\alpha}\left(EX1_{\{X \le q_\alpha^+\}} + q_\alpha^+(\alpha - P(X \le q_\alpha^-))\right) \tag{7.31}$$

and the related insurance version

$$ES_\alpha^+(X) = CTE_\alpha^+(X)$$
$$= \frac{1}{\alpha}\left(EX1_{\{X \ge F^{-1}(\alpha)\}} + F^{-1}(\alpha)(P(X \ge F^{-1}(\alpha)) - (1 - \alpha))\right).$$

For continuous distribution functions $TCE_\alpha = CTE_\alpha = ES_\alpha$ and $TCE_\alpha^+ = CTE_\alpha^+ = ES_\alpha^+$. The TCE_α is only subadditive on the class of continuous distributions while the expected shortfall $ES_\alpha = CTE_\alpha$ is subadditive in general. One can also interpret $ES_\alpha(X) = CTE_\alpha(X)$ as a conditional expectation conditioned on the distributional transform $U = F(X, V)$ (see (1.12))

$$ES_\alpha(X) = CTE_\alpha(X) = -E(X \mid U \le \alpha) \qquad (7.32)$$

and similarly

$$ES_\alpha^+(X) = CTE_\alpha^+(X) = E(X \mid U \ge \alpha). \qquad (7.33)$$

From limit theory of order statistics one gets that the natural estimator

$$\widehat{S}_{n,\alpha} = -\frac{1}{n\alpha} \sum_{i=1}^{[n\alpha]} X_{(i)} \to CTE_\alpha(X) \text{ a.s.} \qquad (7.34)$$

Also one gets directly from the definition that $ES_\alpha(X) = CTE_\alpha(X)$ coincides with the "average value at risk"

$$ES_\alpha(X) = CTE_\alpha(X) = \text{AV@R}_\alpha(X) = -\frac{1}{\alpha} \int_0^\alpha q_\beta^+(X)d\beta \qquad (7.35)$$

respectively

$$ES_\alpha^+(X) = CTE_\alpha^+(X) = \text{AVaR}_\alpha(X) = \frac{1}{\alpha} \int_{1-\alpha}^1 F^{-1}(u)du.$$

For the numerical calculation it is of interest that the average value at risk can be obtained as solution of the following optimization problem

$$\text{AV@R}_\alpha(X) = \max \left\{ x - \frac{1}{\alpha} E(Y - x)_-; \; x \in \mathbb{R} \right\}, \qquad (7.36)$$

(see Rockafellar and Uryasev (2000)).

Finally there is an interesting dual representation of the average value at risk given by

$$\text{AV@R}_\alpha(X) = -\min \left\{ EXZ; \; 0 \le Z \le \frac{1}{\alpha}, EZ = 1 \right\}. \qquad (7.37)$$

Representation (7.37) in particular implies convexity of AV@R_α and also implies that AV@R_α is a coherent risk measure.

Spectral Risk Measures, Distortion Risk Measures

A richer class of risk measures are the spectral risk measures also called weighted
V@R measures and the distortion risk measures. These classes were introduced in
the insurance literature (see Wang (1996) and Wang et al. (1997)). "Spectral risk
measures" are defined as

$$\varrho_\mu(X) = \int_{[0,1]} \varrho_\lambda(X)\mu(d\lambda), \tag{7.38}$$

where $\varrho_\lambda = $ AV@R_λ (respectively AVaR$_\lambda$) and μ is a probability measure on $[0,1]$
weighting the tails.

ϱ_μ can be decomposed in a regular part $\int_{(0,1]} \varrho_\lambda(X)$ and the nonregular part
$\mu(\{0\})\varrho_0(X) = -\mu(\{0\})$ ess inf X, which makes sense on L^∞ but misses the
"usual" continuity properties (see the discussion in Section 7.2).

Distortion risk measures are defined as

$$\varrho_g(X) = \int_{-\infty}^0 g(F(x))dx + \int_0^\infty (g(F(x)) - 1)dx, \tag{7.39}$$

where $g : [0,1] \to [0,1]$ is an increasing, concave function with $g(0) = 0, g(1) = 1$,
called "distortion".

The corresponding insurance version is given by

$$\Psi_g(X) = -\int_{-\infty}^0 (1 - g(\overline{F}(x)))dx + \int_0^\infty g(\overline{F}(x))dx \tag{7.40}$$

which reduces to

$$\Psi_g(X) = \int_0^\infty g(\overline{F}(x))dx \quad \text{for } X \geq 0.$$

Formulae (7.39) and (7.40) extend the partial integration formulae

$$EX = -\int_{-\infty}^0 F(x)dx + \int_0^\infty \overline{F}(x)dx$$

by reweighting the distribution respectively the survival functions. Any increasing
concave function g gives more weight to the tails than the identity $\Psi_0(x) = x$. In
this limit case we obtain $\varrho_{\Psi_0}(X) = E(-X)$. Examples are the power transform
$g(x) = x^r$ or $g(x) = \Phi(\Phi^{-1}(x) + \lambda)$ (see Denneberg (1990) and Wang (1996)).
One can interpret $\varrho_g(X)$ also as a Choquet integral

$$\varrho_g(X) = \int (-X)dc_g \tag{7.41}$$

with the Choquet capacity defined via g,

$$c_g(A) := g(P(A)).$$

Any spectral risk measure ϱ_μ has a representation as distortion risk measure and conversely

$$\varrho_\mu(X) = \varrho_g(X),\qquad(7.42)$$

where $g'_+(t) = \int_{(t,1]} s^{-1} \mu(ds)$ describes a one-to-one correspondence between the probability measures μ and the concave distortions g (see Föllmer and Schied (2011, Theorem 4.64)). It is obvious that spectral risk measures and as a consequence also distortion risk measures are coherent and comonotone additive and law invariant. An interesting theorem going back to Kusuoka (2001) states:

Theorem 7.6 (Comonotone additive risk measures). *A law invariant risk measure ϱ on L^∞ is comonotone additive if and only if $\varrho = \varrho_\mu$ is a spectral risk measure.*

The second important result in the Kusuoka paper is the following representation result for law invariant coherent risk measures.

Theorem 7.7 (Kusuoka representation of law invariant coherent risk measures). *A coherent risk measure ϱ on L^∞ is law invariant if and only if for some subclass $\mathcal{Q} \subset M^1([0,1])$ it holds that*

$$\varrho(X) = \sup_{\mu \in \mathcal{Q}} \varrho_\mu(X).\qquad(7.43)$$

In the Kusuoka paper as an additional assumption the Fatou-continuity of ϱ was postulated. This condition was removed in Jouini et al. (2006). We will discuss some extensions of these results to L^p-spaces and convex risk measures in Section 7.2. Theorem 7.7 implies that the average value at risk measures are the basic building blocks of the class of law invariant risk measures and it also stresses the importance of the spectral risk measures.

Zero Utility Premiums

The classical zero utility premium introduced by Bühlmann (1970) (see also Gerber (1979) and Goovaerts et al. (1984)) is based on expected utility. Let u be a "utility function" on $\mathbb{R}$, i.e. u is increasing and concave as for example the exponential utility $u(x) = \frac{1}{a}(1 - e^{-ax})$. Then the risk premium $\Psi(X) = \Psi_u(X)$ is defined as a solution of the equation

$$Eu(z + \Psi_u(X) - X) = u(z).\qquad(7.44)$$

Here z describes the endowment of the insurer. Adding the premium $\Psi_u(X)$ minus the risk X should not change the expected utility. Like the other classical examples of premium calculation $\bigl($net premium $\Psi(X) = EX$, standard deviation principle $\Psi(X) = EX + \beta\sqrt{\mathrm{Var}(X)}$, or Escher principle $\Psi(X) = \frac{EXe^{hX}}{Ee^{hX}}\bigr)$ the zero utility principle is law invariant. It is also a convex risk measure (premium principle). Deprez and Gerber (1985) considered the zero utility principle (indifference premium) with a random endowment Z and determined the premium $\Psi_u(X)$ as a solution of

$$Eu(Z + \Psi_u(X) - X) = Eu(Z). \tag{7.45}$$

Here the solution $\Psi_u(X)$ no longer is law invariant but depends on the endowment Z. In fact the construction of $\Psi_u(X)$ includes a hedging argument as described in the introduction of Section 7. If the endowment Z is positive dependent on X then it can be used to hedge against the risk X. $\Psi_u(X)$ is the first not law invariant (version dependent) convex risk measure which has been given in the literature. For the exponential utility function $u(x) = \frac{1}{a}(1 - e^{-ax})$ the solution of (7.45) can be explicitly given depending on the parameter a,

$$\Psi_a(X) = \frac{1}{a}\ln\frac{Ee^{a(X-Z)}}{Ee^{-aZ}}. \tag{7.46}$$

For $a \to 0$ one obtains the expansion

$$\Psi_a(X) \approx EX + \frac{a}{2}\mathrm{Var}(X) - a\,\mathrm{Cov}(X, Z), \tag{7.47}$$

which shows the effect that the solution not only depends on the distribution of X but on the joint distribution of (X, Z) and positive correlation of X, Z reduces the risk.

The paper of Deprez and Gerber (1985) also discusses several general criteria for establishing convexity of risk measures and criteria for the optimal cooperation (allocation) problem. This article precedes by more than 10 years the introduction of convex risk measures in mathematical finance.

The zero utility (indifference) principle can also be applied to risk measures different from expected utility. For a distorted (insurance) risk measure (as in (7.40))

$$\Psi_g(X) = -\int_{-\infty}^{0}(1 - g(\overline{F}(x)))dx + \int_{0}^{\infty}g(\overline{F}(x))dx \tag{7.48}$$

which reduces to $\Psi_g(X) = \int_0^\infty g(\overline{F}(x))dx$ for $X \geq 0$, the resulting indifference price principle defined as a solution $\Pi_g(X)$ of the equation

$$\Psi_g(z + \Pi_g(X) - X) = \Psi_g(z), \tag{7.49}$$

then inherits the properties from Ψ_g.

Inf-Convolution Risk Measures

One class of interesting risk measures related to the risk measure in (7.1), defined as the minimal price of a hedging action which makes the risk acceptable, are inf-convolution risk measures. For two risk measures ϱ_1, ϱ_2 on $\mathcal{X}$ define the "inf-convolution" of ϱ_1, ϱ_2

$$\varrho_1 \,\square\, \varrho_2(X) = \inf\{\varrho_1(Y) + \varrho_2(X - Y);\ Y \in \mathcal{X}\} \tag{7.50}$$

where we assume that $\mathcal{X}$ is closed under differences. This class of risk measures was studied in Barrieu and El Karoui (2005) and convexity and continuity properties as well as the acceptability set of $\varrho_1 \,\square\, \varrho_2$ is described.

A particular case arises if $\mathcal{C}$ is some (convex) set of hedging actions and

$$\varrho_2(Y) = \begin{cases} 0 & \text{if } Y \in \mathcal{C}, \\ \infty & \text{if } Y \notin \mathcal{C}. \end{cases}$$

Then

$$\varrho_1 \,\square\, \varrho_2(X) = \varrho^{\mathcal{C}}(X) := \inf\{\varrho_1(X - Y);\ Y \in \mathcal{C}\} \tag{7.51}$$

is the "remaining risk" after pure hedging. Here the cost of the hedgings is zero. For example in finance the hedging actions might arise from pure trading in some basic price process S, $\mathcal{C} = \{\int_0^T \varphi_u \cdot dS_u;\ \varphi \text{ admissible}\}$ over all admissible trading strategies. Taking $\mathcal{C}_1 = \mathbb{R} + \mathcal{C} = \{x + \int_0^T \varphi_u dS_u;\ \varphi \text{ admissible}, x \in \mathbb{R}\}$ and with

$$\varrho_2(Y) = \begin{cases} x & \text{if } Y = x + \int_0^T \varphi_u \cdot dS_u \in \mathcal{C}_1 \\ \infty & \text{if } Y \notin \mathcal{C}_1 \end{cases}$$

$$=: \pi(Y) \text{ defining the price of } Y,$$

we obtain

$$\varrho_1 \,\square\, \varrho_2(X) = \inf\{\pi(Y) + \varrho_1(X - Y);\ Y \in \mathcal{C}_1\}. \tag{7.52}$$

The risk of X has two components, one measuring the price of a hedging action and one measuring the remaining risk. (7.52) is thus an extension of the idea of construction of risk measures in (7.1).

7.2 Representation and Continuity Properties of Convex Risk Measures on L^p-Spaces

A basic result in the theory of coherent and convex risk measures ϱ on L^∞ is the "robust representation" of ϱ in the form

$$\varrho(X) = \sup_{Q \in \mathcal{D}} (E_Q(-X) - \gamma(Q)), \tag{7.53}$$

where $\mathcal{D} = ba$ is the set of finitely additive measures and $\gamma : \mathcal{D} \to (-\infty, \infty]$ is a penalty term. For coherent risk measures this representation reduces to

$$\varrho(X) = \sup_{Q \in \mathcal{Q}} E_Q(-X), \tag{7.54}$$

where $\mathcal{Q} = \{Q \in \mathcal{D}; \gamma(Q) = 0\}$. Thus coherent risk measures can be described as worst case expectations w.r.t. some class $\mathcal{Q}$ of scenario measures. The penalty term in (7.53) gives different weights to the possible scenarios – some of them are more plausible than others. Under additional continuity assumptions on ϱ one can replace the finitely additive scenario measures in (7.53) and (7.54) by probability measures and can establish existence results for "worst case scenario measures" $\mathcal{Q}$. This extended representation theory for convex risk measures is due to Delbaen (2000, 2002), Kusuoka (2001), Föllmer and Schied (2011), and Frittelli and Rosazza Gianin (2002) and is represented in all details in Föllmer and Schied (2011).

Since risks in finance and insurance typically are not bounded it is of interest to extend the robust representation theory to spaces of unbounded random variables. Delbaen (2002) and Cheridito et al. (2006) show that nontrivial finite convex risk measures on L^0 do not exist. Frittelli and Rosazza Gianin (2002) give an abstract robust representation of convex risk measures on L^p spaces (without discussing continuity issues). Several aspects of this extension problem are discussed in Cherny (2006) and Rockafellar et al. (2006). Ruszczyński and Shapiro (2006) discuss $\overline{\mathbb{R}}$ valued risk measures on general vector spaces. Detailed representation results on L^p-spaces are given in Kaina and Rü (2009), Cheridito and Li (2009), and Filipović and Svindland (2012); the paper of Cheridito and Li (2009) considers more generally risk measures on Orlicz spaces. The following part is essentially based on Cheridito and Li (2009) and Kaina and Rü (2009).

7.2.1 Convex Duality and Continuity Results

The robust representation theorems as in (7.53) or (7.54) are most directly obtained as consequences of the Fenchel–Moreau Theorem in convex analysis (for reference see Ekeland and Teman (1974)).

Theorem 7.8 (Fenchel–Moreau Theorem). *Let (E, τ) be a locally convex topological vector space with topological dual E^*. Let $f : E \to \mathbb{R} \cup \{\infty\}$ be proper (i.e. $f \not\equiv \infty$) convex and lower semicontinuous. Then f is identical to the doubly conjugate f^{**} i.e.*

$$f(x) = \sup_{x^* \in E^*} \left(\langle x^*, x \rangle - f^*(x^*) \right) \quad \text{for all } x \in E, \tag{7.55}$$

where $f^*(x^*) = \sup_{x \in E} (\langle x^*, x \rangle - f(x))$, $x^* \in E^*$, is the conjugate of f.

To apply this representation theorem to convex risk functionals the following extension of a classical theorem of Namioka stating that positive linear functionals on a Fréchet lattice are continuous is useful. (For references see e.g. Biagini and Frittelli 2006.)

Theorem 7.9 (Extended Namioka Theorem). *Let (E, τ) be a Fréchet lattice and $f : E \to \mathbb{R} \cup \{\infty\}$ be a proper, convex, increasing function. Then f is continuous on $I_f := \text{int}(\text{dom } f)$, where $\text{dom } f = \{x \in E; \ f(x) < \infty\}$ denotes the domain of f.*

As a consequence this result implies the following continuity properties of convex risk measures.

Corollary 7.10. *Let $\varrho : L^p \to \mathbb{R} \cup \{\infty\}$, $1 \le p \le \infty$, be a proper convex risk measure. Then*

1. *ϱ is continuous on I_ϱ w.r.t. relative norm topology.*
2. *Any finite convex risk measure on L^p, $1 \le p \le \infty$ is continuous on L^p.*
3. *If $\varrho : L^\infty \to \mathbb{R} \cup \{\infty\}$ is a proper convex risk measure, then ϱ is finite and continuous on L^∞.*

The continuity properties in Theorem 7.9, Corollary 7.10 can be further refined (see Cheridito and Li 2009). The "algebraic interior core", core(A) of a subset A of a topological vector space is the set of all $x \in A$ that have an algebraic neighbourhood contained in A. By definition the interior of A is contained in the core(A), $\text{int}(A) \subset$ core(A).

Lemma 7.11. *If $f : E \to (-\infty, \infty]$ is an increasing function on a Banach lattice E, then*
$$\text{core}(\text{dom } f) = \text{int}(\text{dom } f).$$

Definition 7.12 (Subdifferentiability). A convex function $f : E \to (-\infty, \infty]$, E a topological vectorspace, is called "subdifferentiable" in $x \in E$ if there exists an element x^* in the topological dual E^* such that
$$x^*(y) \le f(x + y) - f(x), \quad \forall y \in E.$$

By definition of the conjugate $f^*(x^*) = \sup_{x \in E}(x^*(x) - f(x))$ we obtain that f^* is $\sigma(E^*, E)$-lower semicontinuous and further
$$f(x) \ge f^{**}(x) := \sup_{x^* \in E^*} \{x^*(x) - f^*(x^*)\}, x \in E. \tag{7.56}$$

Furthermore,
$$f(x) = \max_{x \in E^*} \{x^*(x) - f^*(x^*)\} \tag{7.57}$$

for all $x \in E$ such that f is subdifferentiable in x.

This leads to the following refinement of Theorem 7.9. It also gives a useful criterion when the dual representation of f is attained.

Theorem 7.13. *Let E be a Banach lattice and let $f : E \to (-\infty, \infty]$ be increasing and convex. Then for all $x \in \text{core}(\text{dom } f) = \text{int}(\text{dom } f)$ holds:*

(a) f is Lipschitz continuous w.r.t. the norm in a neighbourhood of x,
(b) f is subdifferentiable in x,
(c) $f(x) = \max_{x^* \in E^*}\{x^*(x) - f^*(x^*)\}$.

Proof. Continuity of f and subdifferentiability of f in x follows from Ruszczyński and Shapiro (2006, Proposition 3.1). Thus (b) holds and (c) follows from (7.57).

Since f is continuous there exists a neighbourhood U of x such that f is bounded on U. Thus (a) follows from Zălinescu (2002, Corollary 2.2).   □

For a probability measure P on $(\Omega, \mathfrak{A})$ we consider as basic lattice the L^p-space $E = L^p(\Omega, \mathfrak{A}, P)$. We identify the normed positive part of the dual space of L^p with

$$\mathcal{Q}_p := \mathcal{M}_1^q = \left\{ Q \in M_1; \frac{dQ}{dP} \in L^q \right\} \quad \text{for } 1 \le p < \infty, \tag{7.58}$$

where q is the conjugate index and

$$\mathcal{Q}_\infty := \mathcal{M}_{1,f} \tag{7.59}$$
$$= \{Q \in \text{ba}(P); \; Q \text{ is a normed finite additive } P\text{-continuous measure}\}$$

in case $p = \infty$. We call a risk measure ϱ on L^p *representable* if ϱ has a representation of the form

$$\left\{ \begin{array}{l} \varrho(X) = \sup_{Q \in \mathcal{Q}} (E_Q(-X) - \alpha(Q)) \text{ for some } \mathcal{Q} \subset \mathcal{Q}_p \\ \text{and } \alpha : \mathcal{Q} \to \mathbb{R} \cup \{\infty\}, \text{ such that } \inf_{Q \in \mathcal{Q}} \alpha(Q) \in \mathbb{R} \end{array} \right\}. \tag{7.60}$$

As a consequence of the Fenchel–Moreau theorem one obtains the following representation result of convex risk measures (see Inoue (2003), Nakano (2004), Dana (2005), and Biagini and Frittelli (2006)).

Theorem 7.14 (Representation result on L^p).

(a) Let $\varrho : L^p \to \mathbb{R} \cup \{\infty\}$ be a proper, convex, lower semicontinuous (w.r.t. $\|\cdot\|_p$) risk measure on L^p, then

(1) $\varrho(X) = \sup_{Q \in \mathcal{Q}_p} (E_Q(-X) - \varrho^*(Q)), X \in L^p \text{ for } 1 \le p \le \infty.$ (7.61)
(2) $\varrho^*(Q) = \sup_{X \in \mathcal{A}_\varrho} E_Q(-X).$ (7.62)

(b) A monetary risk measure ϱ on L^p, $1 \le p \le \infty$, is representable if and only if ϱ is convex and lower semicontinuous w.r.t $\|\cdot\|_p$.

Remark 7.15. (a) In comparison to the case $p = \infty$ where the representation is based on finitely additive measures the representation in the case $p < \infty$ is restricted to probability measures. From this point of view the case $1 \le p < \infty$ is more pleasant than the case $p = \infty$. All finite convex risk measures on L^p, $p < \infty$, are norm continuous by Corollary 7.10 and thus have a representation as in (7.61).

(b) In the case $p = \infty$, $\mathcal{Q}_\infty = \mathcal{M}_{1,f}$ is weak-*-compact and any convex risk measure is upper semi-continuous w.r.t. weak-*-topology and thus the sup in (7.61) is attained and

$$\varrho(X) = \max_{Q \in \mathcal{Q}_\infty} (E_Q(-X) - \varrho^*(Q)), \quad X \in L^\infty, \tag{7.63}$$

see Föllmer and Schied (2011, Theorem 4.15).   ◊

7.2.2   Representation of Coherent and Convex Risk Measures on L^p

Coherent risk measures on L^p have a simplified representation compared with Theorem 7.14 and allow to show that the sup is attained as in (7.63) for the L^∞-case. The proof of these properties is based on the following lemmas.

Lemma 7.16. *Let $f : E \to \mathbb{R}$ be a convex, positively homogeneous function on a normed Fréchet lattice $(E, \|\cdot\|)$. Then f is Lipschitz-continuous, i.e. there exists a $C < \infty$ such that*

$$|f(x) - f(y)| \le C\|x - y\|, \quad \forall x, y \in E. \tag{7.64}$$

Proof. By the extended Namioka Theorem (Theorem 7.9) f is $\|\cdot\|$-continuous. This implies similarly to the proof of boundedness of continuous linear functionals that f is bounded. Therefore, for some constant $C < \infty$

$$f(x) \le |f(x)| \le C\|x\|, \quad x \in E.$$

This implies using convexity and positive homogeneity

$$f(x) - f(y) \le 2f\left(\frac{1}{2}x - \frac{1}{2}y\right) = f(x - y) \le C\|x - y\|$$

and

$$f(y) - f(x) \le f(y - x) \le C\|x - y\|,$$

i.e. (7.64) follows.   □

For the case $E = L^p$ and for coherent risk measures the equivalence of continuity and the Lipschitz property was already stated in Inoue (2003, Lemma 2.1).

Corollary 7.17. *Any finite coherent risk measure* $\varrho : L^p \to \mathbb{R}$ *is Lipschitz-continuous.*

The following two results generalize the corresponding results in Föllmer and Schied (2011, Corollary 4.8) stated there for the case $E = L^\infty$.

Proposition 7.18. *Let* $f : E \to \mathbb{R} \cup \{\infty\}$ *be a proper convex, lower semicontinuous (lsc) positive homogeneous function on a locally convex topological vector space* (E, τ). *Then*

$$f^*(x^*) \in \{0, \infty\}, \quad \forall x^* \in E^*.$$

Proof. For $x^* \in E^*$ it holds that

$$\begin{aligned}
f^*(x^*) &= \sup_{x \in E}(x^*(x) - f(x)) \\
&= \sup_{\lambda x \in E}(x^*(\lambda x) - f(\lambda x)) \\
&= \lambda f^*(x^*), \quad \forall \lambda > 0.
\end{aligned}$$

Thus $f^*(x^*) \in \{0, \infty\}$. ☐

Proposition 7.19. *Let* $f : E \to \mathbb{R} \cup \{\infty\}$ *be a proper convex, lsc, positively homogeneous function on a normed vectorspace* $(E, \|\cdot\|)$. *Then*

$$f(x) = \max_{x^* \in Q} x^*(x) \tag{7.65}$$

with $Q := \{x^* \in E^*;\ f^*(x^*) = 0\}$.

Proof. By the Fenchel–Moreau Theorem (Theorem 7.8) and Proposition 7.18

$$f(x) = \sup\{x^*(x);\ x^* \in E^*,\ f^*(x^*) = 0\}. \tag{7.66}$$

We have to prove that the sup is attained. By Brezis (1999, Proposition I.9) f is $\sigma(E^*, E)$ lower semicontinuous. Thus

$$\{x^* \in E^*;\ f^*(x^*) = 0\} = \{x^* \in E^*;\ f^*(x^*) \le 0\} \tag{7.67}$$

is $\sigma(E^*, E)$-closed. Further,

$$\{x^* \in E^*;\ f^*(x^*) = 0\} \quad \text{is } \|\cdot\|_{E^*}\text{-bounded.} \tag{7.68}$$

To argue for (7.68) note that

$$f^*(x^*) = 0 \Leftrightarrow x^*(x) \le f(x), \quad \forall x \in E$$

and

$$f^*(x^*) \leq 0 \Leftrightarrow \sup_{x \in E}(x^*(x) - f(x)) = 0$$

$$\Leftrightarrow \inf_{x \in E}(-x^*(x) + f(x)) = 0$$

$$\Leftrightarrow x^*(-x) \geq -f(x), \quad \forall x \in E$$

$$\Leftrightarrow x^*(x) \geq -f(-x), \quad \forall x \in E.$$

Thus for $x \in E$, $\{x^*(x); \ x^* \in E^*, \ f^*(x^*) = 0\}$ is bounded below by $-f(-x)$ and above by $f(x)$. But pointwise boundedness of Q implies norm-boundedness of Q w.r.t. $||\cdot||_{E^*}$.

By Alaoglu's Theorem Q is $\sigma(E^*, E)$-compact. Thus for all $x \in E$ the continuous functional $x^* \to x^*(x)$ attains its supremum on Q.   □

As a consequence we now obtain a more specific version of the representation result in (7.61) on L^p in the case of coherent risk measures.

Theorem 7.20 (Representation of coherent risk measures on L^p).   *Let ϱ :* $L^p \to \mathbb{R} \cup \{\infty\}$, $1 \leq p \leq \infty$. *Then*

(a) ϱ is a proper, $||\cdot||_p$-lsc, coherent risk measure
   $\Leftrightarrow \exists Q \subset Q_p$ such that

$$\varrho(X) = \max_{Q \in Q} E_Q(-X), \quad X \in L^p. \tag{7.69}$$

   $\Leftrightarrow \varrho$ is a finite, $||\cdot||_p$-continuous, coherent risk measure.
(b) In case $p = \infty$ holds:

   ϱ is a finite coherent risk measure   (7.70)
   $\Leftrightarrow \varrho(X) = \max_{Q \in Q} E_Q(-X)$ for some $Q \subset Q_\infty$.

Proof. (a) Denote the equivalences by (1)–(3). Then

   (1) $\Rightarrow$ (2) follows from Propositions 7.19 and 7.18.
   (2) $\Rightarrow$ (3) The properties of a coherent risk measure are easy to establish. Finiteness of ϱ follows by definition and thus $I_\varrho = \{X \in L^p; \ \varrho(X) < \infty\} = L^p$. By the extended Namioka Theorem (Theorem 7.9) we obtain that ϱ is $||\cdot||_p$-continuous.
   (3) $\Rightarrow$ (1) is obvious.

(b) Since coherent risk measures on L^∞ are $||\cdot||_\infty$-continuous (see Corollary 7.10) (b) follows from (a).   □

A general extension of the attainment result as in Theorem 7.20 to convex risk measures on L^p has been given in Cheridito and Li (2009, Theorem 4.3)

Theorem 7.21 (Representation result for convex risk measures). *Let $\varrho : L^p \to$* $(-\infty, \infty]$, $1 \leq p < \infty$ *be a convex risk measure with* core(dom ϱ) $\neq \phi$. *Then ϱ is real valued and*

$$\varrho(X) = \max_{Q \in Q_p} \{E_Q(-X) - \varrho^*(Q)\}, \quad X \in L^p \qquad (7.71)$$

with penalty function $\varrho^*(Q) = \sup_{X \in A_\varrho} E_Q(-X)$.

Remark 7.22. (a) For the proof of Theorem 7.21 Cheridito and Li (2009) show that the condition core(dom ϱ) $\neq \phi$ is equivalent to a lower bound condition on the penalty function $\gamma = \varrho^*$ of the form

$$\gamma(Q) \geq a + b \left\| \frac{dQ}{dP} \right\|_p \qquad \text{for some } b > 0. \qquad (7.72)$$

This is further equivalent to real valuedness and Lipschitz continuity of ϱ_γ in a neighbourhood of any $X \in L^p$. The real valuedness of ϱ is then obtained by an abstract separation theorem. The attainment is a consequence of Theorem 7.13.
(b) Under the assumption that ϱ is real valued an attainment result under the "strong continuity" condition on ϱ stating that the representation set Q considered as subset in L^q,

$$\mathcal{D} = \left\{ \frac{dQ}{dP}; \ Q \in Q \right\} \subset L^q,$$

is weakly compact was given in Kaina and Rü (2009). Strong continuity holds for example for real valued coherent risk measures.   ◊

7.2.3 Continuity Results for Risk Measures on L^p

For applications also approximation and continuity properties of risk measures are of interest. We concentrate mostly on the case $1 \leq p < \infty$. The case $p = \infty$ is studied in detail in Föllmer and Schied (2011). For more details see Kaina and Rü (2009). We consider the following list of continuity properties.

Definition 7.23 (Continuity properties). Let ϱ be a risk functional on L^p,

$$\varrho : L^p \to \mathbb{R} \cup \{\infty\}, \quad 1 \leq p \leq \infty.$$

(a) ϱ is called "continuous from above", if $(X_n) \subset L^p$, $X_n \downarrow X$ P-a.s. to some $X \in L^p$ implies that $\lim \varrho(X_n) = \varrho(X)$.
(b) ϱ is called "continuous from below", if $(X_n) \subset L^p$, $X_n \uparrow X$ P-a.s. to some $X \in L^p$ implies that $\lim \varrho(X_n) = \varrho(X)$.
(c) ϱ is called "Fatou-continuous" (has the Fatou-property), if $(X_n) \subset L^p$, $|X_n| \leq Y$ P-a.s. for some $Y \in L^p$ and $X_n \to X$ P-a.s. for some $X \in L^p$ implies $\varrho(X) \leq \liminf \varrho(X_n)$.
(d) ϱ is called "Lebesgue-continuous" if $(X_n) \subset L^p$, $X_n \to X$ P-a.s., $X \in L^p$ and $|X_n| \leq Y$ P-a.s. for some $Y \in L^p$ implies $\lim \varrho(X_n) = \varrho(X)$.

The following theorem shows that for the class of finite convex risk measures on L^p, $p < \infty$ all pointwise continuity properties are fulfilled.

Theorem 7.24 (Continuity of finite convex risk measures). *Let $\varrho : L^p \to \mathbb{R}$, $1 \le p < \infty$, be a finite convex risk measure. Then it holds that:*

(a) ϱ is $\sigma(L^p, L^q)$-lsc.   *(b) $\mathcal{A}_\varrho$ is $\sigma(L^p, L^q)$-closed.*

(c) ϱ has the Fatou-property.   *(d) ϱ is continuous from above.*

(e) ϱ is continuous from below.   *(f) ϱ is Lebesgue-continuous.*

Proof. (a) By the extended Namioka Theorem (Theorem 7.9), ϱ is $\|\ \|_p$-continuous. From Aliprantis and Border (1994, Corollary 4.73 and Example 4.67) τ-lower semicontinuity of a function $f : E \to \mathbb{R} \cup \{\infty\}$ on a topological vector space (E, τ) is equivalent to $\sigma(E, E^*)$-lsc of f. Thus a) follows.
(b) $\mathcal{A}_\varrho = \{X \in L^p; \varrho(X) \le 0\}$ is by (a) closed.
(c) and (f) If $X_n \to X$ P-a.s. and $|X_n| \le Y$ P-a.s. for some $Y \in L^p$ then by the majorized convergence theorem of Lebesgue $X_n \to X$ in L^p. As ϱ is $\|\ \|_p$-continuous, $\lim \varrho(X_n) = \varrho(X)$, ϱ has the Fatou-property and ϱ is even Lebesgue-continuous.
(d) follows from c) observing that for $X_n \downarrow X$ P-a.s. $|X_n| \le \max\{|X_1|, |X|\} \in L^p$.
(e) similarly follows from (c).   □

For not necessarily finite risk measures the following lemma extends Föllmer and Schied (2011, Lemma 4.20), who consider the case $p = \infty$.

Lemma 7.25. *Let $\varrho : L^p \to \mathbb{R} \cup \{\infty\}$ be a proper, convex risk measure on L^p, $1 \le p \le \infty$. Then it holds that:*

$$\varrho \text{ is continuous from above} \Leftrightarrow \varrho \text{ has the Fatou-property.}$$

Proof. "$\Rightarrow$"   Is similar to the proof of the majorized convergence theorem which is based on the monotone convergence theorem.
"$\Leftarrow$"   For $(X_n) \subset L^p$, $X_n \downarrow X$ P-a.s. for some $X \in L^p$ holds by monotonicity of ϱ, $\varrho(X_n) \le \varrho(X)$, for all $n \in \mathbb{N}$. On the other hand by the Fatou-property $\liminf \varrho(X_n) \ge \varrho(X)$ since $|X_n| \le \max(|X_1|, |X|)$, $\forall n \in \mathbb{N}$. Together we obtain $\lim \varrho(X_n) = \varrho(X)$.   □

The following result shows that $\|\ \|_p$-lsc convex risk measures ϱ are continuous from above and thus have the Fatou-property also for non-finite ϱ.

Theorem 7.26. *Let $\varrho : L^p \to \mathbb{R} \cup \{\infty\}$, $1 \le p \le \infty$ be a proper, convex risk measure on L^p. Then the following are equivalent:*

(a) ϱ is $\sigma(L^p, L^q)$-lsc

(b) ϱ is $\|\cdot\|_p$-lsc

(c) $\varrho(X) = \sup_{Q \in \mathcal{M}_1^q}(E_Q(-X) - \varrho^*(Q))$, $\forall X \in L^p$

(d) ϱ is continuous from above

(e) ϱ has the Fatou-property.

Proof. (a) ⇔ (b)   This holds true using the same argument as in the proof of a) of
Theorem 7.24.

(b) ⇔ (c)   holds by Theorem 7.14.

(d) ⇔ (e)   holds by Lemma 7.25.

(b) ⇔ (d)   follows as in the proof of Lemma 7.25.   □

For completeness reasons we state the essential continuity results for the case of
$p = \infty$ from Föllmer and Schied (2011, Theorem 4.31).

Theorem 7.27 (Continuity and representation). *Let* $\varrho : L^\infty \to \mathbb{R}$ *be a finite
convex risk measure. Then the following are equivalent:*

(a) ϱ *has a representation on* $\mathcal{M}_1$*, the set of all* P*-continuous probability measures,
i.e.*

$$\varrho(X) = \sup_{Q \in \mathcal{M}_1} (E_Q(-X) - \varrho^*(Q)), \quad X \in L^\infty. \tag{7.73}$$

(b) ϱ *is continuous from above.*

(c) ϱ *is Fatou-continuous.*

(d) ϱ *is* $\sigma(L^\infty, L^1)$*-lsc.*

(e) $\mathcal{A}_\varrho$ *is* $\sigma(L^\infty, L^1)$*-closed.*

For the following interesting result on law invariant risk measures on L^∞ it is
assumed that the underlying probability space $(\Omega, \mathfrak{A}, P)$ is an atomless, separable
complete metric space with Borel σ-algebra $\mathfrak{A}$.

Theorem 7.28 (Law invariant risk measures, Jouini et al. (2006)). *Let* $\varrho : L^\infty \to
\mathbb{R}$ *be a finite convex, law invariant, risk measure. Then*

$$\varrho(X) = \sup_{Q \in \mathcal{M}_1} (E_Q(-X) - \varrho^*(Q)), \quad X \in L^\infty \tag{7.74}$$

and ϱ *is* $\sigma(L^1, L^\infty)$*-lsc.*

Remark 7.29. Theorem 7.28 is the basis to show that any law invariant, convex risk
measure on L^∞ has a Kusuoka type representation via mixtures of average value at
risk measures (see Theorem 7.7).   ◊

Finally, the following result of Föllmer and Schied (2011, Theorem 4.31) and
Jouini et al. (2006, Theorem 5.2), combines continuity properties of convex risk
measures on L^∞ with representability on $\mathcal{M}_1$ and attainment of the supremum.

Theorem 7.30 (Convex risk measures on L^∞). *Let* $\varrho : L^\infty \to \mathbb{R}$ *be a convex
risk measure on* L^∞ *with* $\sigma(L^\infty, (L^\infty)')$*-conjugate* $\varrho^* : (L^\infty)' \to \mathbb{R} \cup \{\infty\}$*. Then
it holds that:*

(a) The following conditions are equivalent:

1. ϱ *is Lebesgue-continuous.*

2. $\varrho(X) = \sup_{Q \in \mathcal{M}_1}(E_Q(-X) - \varrho^*(Q)), \quad X \in L^\infty,$ \tag{7.75}

$\{\varrho^* \leq c\} \subset L^1$ for all $c \in \mathbb{R}$ and $\{\varrho^* \leq c\}$ is uniformly integrable for all $c > \max\{0, -\varrho(0)\}$.

3. ϱ is continuous from below.

4. ϱ is Fatou-continuous and $\operatorname{dom} \varrho^* = \{\varrho^* < \infty\} \subset L^1$.

5. ϱ is Fatou-continuous and $\{\varrho^* \leq c\}$ is a $\sigma(L^1, L^\infty)$-compact subset of L^1 for all $c > -\varrho(0)$.

(b) 1.–5. imply that the sup in (7.75) is attained

(c) If L^1 is separable, then 1.–5. are equivalent with the attainment of the sup in (7.75).

Several examples of extensions and robust representation of classical convex risk measures on L^∞ to L^p are discussed in Kaina and Rü (2009).

Example 7.31. (a) Conditional tail expectation: The conditional tail expectation

$$CTE_\alpha(X) = \frac{1}{\alpha}\{EXI_{\{X \leq q_\alpha^+\}} + q_\alpha^+(\alpha - P(X \leq q_\alpha^-))\} \qquad (7.76)$$

(see Section 7.1.2) is a coherent risk measure on L^1. CTE_α is law invariant and has a representation as average value at risk, $CTE_\alpha(X) = \mathrm{AV@R}_\alpha(X)$. The robust representation of CTE_α when considered on L^∞-risks (see Föllmer and Schied (2011, Theorem 4.7) extends to the case of L^1-risks (see Kaina and Rü (2009), Theorem 4.1).

Proposition 7.32 (Representation of the CTE_α on L^1).   *The conditional tail expectation $CTE_\alpha : L^1 \to \mathbb{R}$ is a coherent risk measure on L^1 with representation*

$$CTE_\alpha(X) = \max_{Q \in \mathcal{Z}_\alpha} E_Q(-X), \qquad (7.77)$$

where $\mathcal{Z}_\alpha := \{Q \in \mathcal{M}_1; \frac{dQ}{dP} \leq \frac{1}{\alpha} P\text{-a.s.}\}$. *In particular CTE_α has all the continuity properties stated in Theorem 7.24.*

(b) **Shortfall risk:** The shortfall risk is an extension of the tail expectation. For a convex, increasing loss function $\ell : \mathbb{R} \to \mathbb{R}$, ℓ not identically constant, assume that for some $1 \leq p \leq \infty$, $X \in L^p$ implies

$$E\ell(-X) < \infty. \qquad (7.78)$$

Define for some $x_0 \in \mathbb{R}$ the acceptance set

$$\mathcal{A}_p := \{X \in L^p; \ E\ell(-X) \leq x_0\} \qquad (7.79)$$

and denote by

$$SR_p(X) := \inf\{m \in \mathbb{R}; \ X + m \in \mathcal{A}_p\}, \quad X \in L^p \qquad (7.80)$$

the generated risk measure on L^p. $SR_p(X)$ is called the "shortfall risk". For $p = \infty$ this risk measure has been investigated in detail in Föllmer and Schied

(2011, Section 4.9). We restrict to the case $p < \infty$ (see Kaina and Rü (2009, Proposition 4.2)).

Proposition 7.33 (Shortfall risk). *Under assumption* (7.78) *the shortfall risk* SR_p, $1 \leq p < \infty$, *is a finite, convex,* $\|\cdot\|_p$*-continuous risk measure on* L^p. *Thus* SR_p *has all the pointwise continuity properties in Theorem 7.24. Further* SR_p *has a representation of the form*

$$SR_p(X) = \max_{Q \in \mathcal{Q}_p} (E_Q(-X) - SR_p^*(Q)), \quad X \in L^p. \tag{7.81}$$

Proof. Since $\mathcal{A}_p$ is convex and monotone it follows from (7.78) that SR_p defines a finite convex risk measure on L^p. Thus by Theorem 7.24 SR_p is continuous from above, from below and Lebesgue continuous and the acceptance set $\mathcal{A}_p$ is $\sigma(L^p, L^q)$-closed. The representation property in (7.81) is then a consequence of Theorems 7.14 and 7.20. $\qquad\square$

Remark 7.34. Föllmer and Schied (2011, Theorem 4.16) establish in the case $p = \infty$ that the minimal penalty function (the conjugate) $(SR_p)^* = \alpha_{\min}^p$ is given by

$$(SR_p)^*(Q) = \inf_{\lambda > 0} \frac{1}{\lambda}\left(x_0 + E\ell^*\left(\lambda \frac{dQ}{dP}\right)\right), \quad Q \in \mathcal{M}_1. \tag{7.82}$$

Formula (7.82) also extends in the same form to $1 \leq p < \infty$. For the proof note that for any $\lambda > 0$ and X in the acceptance set $\mathcal{A}_\varrho^p$ with $\varphi = \frac{dQ}{dP}$ it holds that

$$-X\varphi = \frac{1}{\lambda}(-X)(\lambda\varphi) \leq \frac{1}{\lambda}(\ell(-X) + \ell^*(\lambda\varphi)).$$

In consequence

$$\alpha_{\min}^p(Q) = (SR_p)^*(Q) = \sup_{X \in \mathcal{A}_\varrho^p} E_Q(-X) \leq \frac{1}{\lambda}(x_0 + E\ell^*(\lambda\varphi)). \tag{7.83}$$

On the other hand since $\mathcal{A}_\varrho^p = \{X \in L^p; \; SR_p(X) \leq 0\} \supset \mathcal{A}_\varrho^\infty$ if follows that

$$\alpha_{\min}^p(Q) = \sup_{X \in \mathcal{A}_\varrho^p} E_Q(-X) \geq \sup_{X \in \mathcal{A}_\varrho^\infty} E_Q(-X) = \alpha_{\min}^\infty(Q).$$

This implies that (7.82) also holds for $1 \leq p < \infty$. $\qquad\Diamond$

Corollary 7.35. *The minimal penalty function of the shortfall risk* SR_p *on* L^p *is given by*

$$(SR_p)^*(Q) = \inf_{\lambda > 0} \frac{1}{\lambda}\left(x_0 + E\ell\left(\lambda \frac{dQ}{dP}\right)\right), \quad Q \in \mathcal{M}_p. \tag{7.84}$$

Example 7.36. As a particular example consider $\ell(x) = \max(\frac{1}{p}x^p, 0)$, $p > 1$. Then it holds that

$$\ell^*(z) = \begin{cases} \frac{1}{q}z^q, & z \geq 0, \\ \infty, & \text{else.} \end{cases}$$

For $Q \in \mathcal{M}_p$ it holds that $\varphi = \frac{dQ}{dP} \in L^q$ and the infimum in (7.84) is attained for $\lambda_Q = (\frac{px_0}{E\varphi^q})^{\frac{1}{q}}$. Thus one obtains the explicit result:

$$(SR_p)^*(Q) = (px_0)^{\frac{1}{p}} \left[E\left(\frac{dQ}{dP}\right)^q \right]^{\frac{1}{q}} < \infty, \quad Q \in \mathcal{M}_p \qquad (7.85)$$

(see also Föllmer and Schied (2011, Example 4.109)).   ◊

A series of further examples of convex risk measures as presented in Föllmer and Schied (2011) for risks in L^∞ can be transferred in a similar way to L^p-risks.   ◊

Chapter 8
Risk Measures for Portfolio Vectors

The main aim to study risk measures for portfolio vectors $X = (X_1, \ldots, X_d)$ is
to measure not only the risk of the components X_i separately but to measure the
joint risk of X caused by the variation of the components and by their dependence.
From this point of view an important property of risk measures is consistency w.r.t.
various classes of convex orders and of dependence orders. After the introduction of
basic properties and classes of examples of risk measures for portfolio vectors we
determine robust representation properties and consistency properties of portfolio
risk measures. Many of the basic one-dimensional concepts of risk measures have a
natural extension to the multivariate case. So one can naturally extend the concept
of convex risk measures. The hedging based definition of a risk measure for real
risks in (7.1) is naturally extended to multivariate risks. For a class $\mathcal{C}$ of available
(multivariate) hedging actions including adding capital to the components or trading
actions in each (or some) of the components supplied with a price functional π :
$\mathcal{C} \to \mathbb{R}$ we can consider hedging based risk measures defined by

$$\varrho(X) = \inf \{\pi(Y); \ Y \in \mathcal{C}, X + Y \in \mathcal{A}\}, \qquad (8.1)$$

where $\mathcal{A}$ is the class of acceptable risks. Thus as in dimension $d = 1$ $\varrho(X)$ is the
minimal price of a hedging action which added to the risk X makes it acceptable.
Some new problems arise in the multivariate case caused by the missing linear order
on $\mathbb{R}^d$. There is not a natural and simple generating class of convex functions and of
quantiles as in dimension one where these allow to introduce the expected shortfall
or the value at risk.

In the following sections we introduce basic properties of risk measures, introduce some classes of portfolio risk measures, and describe consistency properties
w.r.t. convex and dependence orders. We also establish robust representation results
for convex portfolio risk measures.

L. Rüschendorf, *Mathematical Risk Analysis*, Springer Series in Operations Research
and Financial Engineering, DOI 10.1007/978-3-642-33590-7_8,
© Springer-Verlag Berlin Heidelberg 2013

8.1   Basic Properties of Portfolio Risk Measures

For portfolio vectors $X \in L_d^\infty(P) = \prod_{i=1}^d L^\infty(P)$ coherent and convex risk measures and their representation results were introduced in Jouini et al. (2004) and in Burgert and Rü (2006a). Jouini et al. (2004) defined coherent risk measures as set valued functions $R : L_d^\infty(P) \to \mathcal{P}(\mathbb{R}^n)$, where $n \leq d$ represents n aspects of the risk describing for example risk components in different currencies. $R(X)$ represents the set of all capital vectors $x \in \mathbb{R}^n$ such that $X + \bar{x}$ is acceptable, where $\bar{x} = (x, 0)$ is the vector x imbedded in $\mathbb{R}^d$.

Their definition postulates that any entry in the i-th position $i \geq n + 1$ can be substituted by some entry in the first position, i.e. $\forall i \geq n + 1$ there exist $\alpha, \beta > 0$ such that

$$\alpha e_1 - e_i \in K \text{ and } e_i - \beta e_1 \in K, \tag{8.2}$$

where $e_i \in \mathbb{R}^d$ are the unit vectors and $K \subset \mathbb{R}^d$ is an order defining convex cone with $\mathbb{R}_+^d \subset K, K \neq \mathbb{R}^d$. The order is defined on $\mathbb{R}^d$ by

$$x \geq 0 \Leftrightarrow x \in K, \tag{8.3}$$

which is extended to $L_d^\infty(P)$ by

$$X \succeq 0 \Leftrightarrow X \in K[P]. \tag{8.4}$$

Monotonicity of the risk measure is defined via this cone order. Jouini et al. (2004) establish a representation result for the class of (d, n)-coherent risk measures, i.e. those which are monotone, homogeneous, translation invariant and subadditive.

Convex risk measures on $L_d^\infty(P)$ were introduced in Burgert and Rü (2006a).

Definition 8.1 (Convex risk measures on $L_d^\infty(P)$).   Let $\preceq$ be a cone order on $\mathbb{R}^d$. A functional $\varrho : L_d^\infty(P) \to \mathbb{R}$ is a convex risk measure if for $X, Y \in L_d^\infty(P)$ it holds that

(R1)   **Monotonicity:** $X \succeq Y \Rightarrow \varrho(X) \leq \varrho(Y)$

(R2)   **Translation invariance:** $\varrho(X + me_i) = -m + \varrho(X)$ for $m \in \mathbb{R}, 1 \leq i \leq d$

(R3)   **Convexity:** $\varrho(\alpha X + (1 - \alpha)Y) \leq \alpha\varrho(X) + (1 - \alpha)\varrho(Y)$ for all $\alpha \in (0, 1)$.

Convex risk measures on $L_d^\infty(P)$ are monotone, translation invariant convex real functions on $L_d^\infty(P)$. In this definition there is only one aspect of risk. Translation invariance (R2) implies in particular

$$\varrho(X + \varrho(X)e_i) = 0.$$

Thus the capital $\varrho(X)$ added to any component makes the risk acceptable. A weaker translation invariance property is postulated in Ekeland and Schachermayer (2011).

(R2′) $$\varrho(X + m \cdot e) = -m + \varrho(X), \quad m \in \mathbb{R}. \tag{8.5}$$

If ϱ satisfies (R1), (R2), (R3) then $\varrho' = \frac{1}{d}\varrho$ satisfies (R1), (R2′), (R3).

In the typical case the cone ordering $\preceq$ is the usual partial ordering on $\mathbb{R}^d$ which is excluded by the substitution postulate (8.2) in the definition of Jouini et al. (2006). Again as in Chapter 7 we can consider an insurance version of Definition 8.1 by using the (increasing) monotonicity condition

(R̄1) $$X \preceq Y \Rightarrow \Psi(X) \le \Psi(Y). \tag{8.6}$$

One can switch from Ψ to ϱ by the relation

$$\varrho(X) = \Psi(-X). \tag{8.7}$$

As in the real case convex risk measures can be equivalently defined in terms of acceptance sets. The risk of a portfolio X is the smallest amount which has to be added to X, such that the portfolio $X + me_i$ is acceptable.

Definition 8.2. A subset $\mathcal{A} \subset L_d^\infty(P)$ is called a (convex) acceptance set if

(A1) $\mathcal{A}$ is closed and convex
(A2) $X, Y \in L_d^\infty(P), X \succeq Y$ and $Y \in \mathcal{A}$ implies $X \in \mathcal{A}$
(A3) $X + me_i \in \mathcal{A} \Leftrightarrow X + me_j \in \mathcal{A}$ for all i, j
(A4) $\mathbb{R}^d \not\subset \mathcal{A}$.

For any acceptance set $\mathcal{A}$ we define a risk measure $\varrho_{\mathcal{A}}$ by

$$\varrho_{\mathcal{A}}(X) := \inf\{m \in \mathbb{R}; \ X + me_1 \in \mathcal{A}\}. \tag{8.8}$$

Then as in the one-dimensional case it holds that

Proposition 8.3. (a) If $\mathcal{A}$ is a convex acceptance set, then $\varrho_{\mathcal{A}}$ is a convex risk measure.
(b) If ϱ is a convex risk measure, then

$$A_\varrho := \{X \in L_d^\infty(P); \ \varrho(X) \le 0\} \tag{8.9}$$

is a convex acceptance set.

Let

$$L_d^\infty(K) = L_d^\infty(K, P) = \{X \in L_d^\infty(P); \ X \in K\} \tag{8.10}$$

and let $\mathrm{ba}_d(P)$ denote the finitely additive measures on $L_d^\infty(P)$ absolutely continuous w.r.t. P, which are the positive part of the dual space of $L_d^\infty(P)$. We use the notation

$$Q(X) = E_Q(X) = \sum_{i=1}^d E_{Q_i} X_i \tag{8.11}$$

for $Q \in \mathrm{ba}_d(P)$ and define the elements of $\mathrm{ba}_d(P)$ which are positive on K by

$$\mathrm{ba}_d(K) = \mathrm{ba}_d(K, P) = \{Q \in \mathrm{ba}_d(P);\ E_Q X \geq 0, \forall X \in L_d^\infty(K)\}. \quad (8.12)$$

Then the following dual representation of convex risk measures is stated in Burgert and Rü (2006a).

Theorem 8.4 (Dual representation of convex risk measures on $L_d^\infty(P)$). *A functional $\varrho : L_d^\infty(P) \to \mathbb{R}$ is a convex risk measure if and only if there exists a function $\alpha : \mathrm{ba}_d(K) \to (-\infty, \infty]$ such that*

$$\varrho(X) = \sup_{Q \in \mathrm{ba}_d(K)} \{E_Q(-X) - \alpha(Q)\}. \quad (8.13)$$

α can be chosen as the Legendre–Fenchel inverse of ϱ

$$\alpha(Q) = \sup_{X \in L_d^\infty(K)} (E_Q(-X) - \varrho(X))$$

$$= \sup_{X \in A_\varrho} E_Q(-X). \quad (8.14)$$

The proof in Burgert and Rü (2006a) uses the separation theorem for convex sets. It can also be given as an application of the Fenchel duality theorem in Theorem 7.8. To restrict the representing scenario set $\mathrm{ba}_d(K)$ to P-continuous σ-additive measures positive on K the following version of Fatou-continuity of risk measures ϱ is needed.

Definition 8.5. A functional $\varrho : L_d^\infty(P) \to \mathbb{R}$ has the "Fatou-property" if for any uniformly bounded sequence $(X_n) \subset L_d^\infty(P)$ with $X_n \xrightarrow{P} X$ for some $X \in L_d^\infty(P)$ it holds that

$$\varrho(X) \leq \varliminf_{n \to \infty} \varrho(X_n).$$

The class of σ-additive P-continuous measures, positive on K, can be represented by the corresponding class $L_d^1(K)$ of P-densities $f = (f_1, \ldots, f_d)$.

Theorem 8.6. *Let $\varrho : L_d^\infty \to \mathbb{R}$ be a convex risk measure. Then the following are equivalent:*

(1) The class $\mathrm{ba}_d(K)$ in the representation (8.13) of ϱ can be replaced by the class $L_d^1(K)$ of σ-additive measures.
(2) The acceptance set $A_\varrho = \{X \in L_d^\infty(P); \varrho(X) \leq 0\}$ is w^*-closed in $L_d^\infty(P)$.
(3) ϱ has the Fatou-property.

Proof. The proof is similar to the one-dimensional case. If the acceptance sets A_ϱ are closed in the w^*-topology on $L_d^\infty(P)$ then the continuous linear functionals can be identified with σ-additive P-continuous measures and thus (2) implies (1) while (1) $\Rightarrow$ (2) is obvious. Also the equivalence of (2) and (3) is as in the one-dimensional case. $\qquad \square$

Remark 8.7. Under the natural normalizing condition that $\varrho(x) = -\sum_{i=1}^{n} x_i$ we obtain a representation in Theorems 8.4 and 8.6 with scenario $Q = (Q_1, \ldots, Q_d)$ such that Q_i are normed, i.e. Q_i are normed contents respectively probability measures on $\mathbb{R}^1$. ◇

Definition 8.8 (Coherent risk measure on $L_d^\infty(P)$). A risk measure ϱ : $L_d^\infty(P) \to \mathbb{R}$ is called a "coherent risk measure" if (R1) and (R2) hold and if further

(R4) "positive homogeneity" $\varrho(\alpha X) = \alpha \varrho(X)$ for all $\alpha > 0, X \in L_\alpha^\infty(P)$, and

(R5) "subadditivity" $\varrho(X + Y) \le \varrho(X) + \varrho(Y)$ for all $X, Y \in L_d^\infty(P)$

hold.

As in dimension $d = 1$ one sees that for coherent risk measures ϱ the penalty function α in the representation (8.13) takes only the values 0 and ∞. Thus the representations in Theorem 8.4 respectively in Theorem 8.6 become more specific.

Theorem 8.9. *(a) A risk measure ϱ on $L_d^\infty(P)$ is a coherent risk measure if and only if ϱ has a representation of the form*

$$\varrho(X) = \sup_{Q \in \mathcal{Q}} E_Q(-X) \tag{8.15}$$

for some $\mathcal{Q} \subset \mathrm{ba}(K, P)$.
(b) ϱ is a Fatou-continuous, coherent risk measure on $L_d^\infty(P)$ if and only if the set of scenarios $\mathcal{Q}$ in (8.15) can be chosen in the set of finite σ-additive P-continuous measures $M(K, P)$.

The robust representation properties of convex and coherent risk functionals extend to risk functional on more general domains of risks. In Section 8.3 we will give some details for the case that the domain $\mathcal{X}$ of the portfolio risk functional is $L_d^p(P) = \prod_{i=1}^{d} L^p(P)$. Generally we can consider risk functionals ϱ defined on some domain $\mathcal{X} \subset L_d^0 = \prod_{i=1}^{d} L^0(P)$.

Definition 8.10 (Law dependence). A risk measure $\varrho : \mathcal{X} \to \mathbb{R} \cup \{\infty\}, \mathcal{X} \subset L_d^0$, is called "law invariant (law dependent)" if $X, Y \in \mathcal{X}, X \stackrel{d}{=} Y$ implies

$$\varrho(X) = \varrho(Y).$$

Law dependent risk measures only depend on the distribution of X but not on the random vectors X themselves. Thus they can equivalently be defined on the induced set of laws $\mathcal{L}(\mathcal{X}) = \{P^X; X \in \mathcal{X}\}$.

This allows to establish empirical versions respectively statistical estimators of $\varrho(X)$ from risk data $X_1, \ldots, X_n$.

Remark 8.11 (Law invariance and Fatou-property). Jouini et al. (2006) established that law invariant convex risk measures on $L^\infty(P)$ have the Fatou-property and thus the representation property w.r.t. σ-additive measures. An extension of this result to $L_d^\infty(P)$ and $L_d^p(P)$ is stated in Rü (2006, 2009). A detailed proof in the case $L_d^\infty(P)$ is given in Ekeland and Schachermayer (2011). $\Diamond$

For the class of law invariant (law dependent) risk measures one can also introduce natural consistency postulates with various stochastic and convex orders and w.r.t. positive dependence orders as for example the increasing stochastic order $\leq_{st}$, the convex order $\leq_{cx}$ and the positive dependence orderings $\leq_{sm}$, $\leq_{dcx}$ induced by the supermodular and by the directionally convex functions (see Sections 6.1 and 6.2).

The postulate that a risk measure ϱ is monotone w.r.t. the convex ordering $\leq_{cx}$ on $\mathcal{X}$ is a mathematical natural postulate and well founded since $X \leq_{cx} Y$ means that the vector Y is more diffuse than X. This interpretation is clear from the diffusion representation of this ordering (Strassen's representation result in Corollary 3.36). On the other hand the convexity postulate of a risk measure is based on an economic diversification argument. In the real case it was shown in Chapter 6 that diversification is not always a good property. In the following we will see that convexity of a risk measure is a consequence of the natural monotonicity w.r.t. convex ordering $\leq_{cx}$. This conclusion also describes the domain of the convexity postulate since convex ordering on $\mathbb{R}^d$ is naturally restricted to $L_d^1(P)$.

We assume in the following again that the underlying probability space $(\Omega, \mathfrak{A}, P)$ is non-atomic even if this property is not always needed.

Let $\mathcal{F}_i$ (respectively $\mathcal{F}_{de}$) denote the class of increasing (respectively decreasing) functions on $\mathbb{R}^d$, i.e.

$$x \leq y \text{ implies that } f(x) \leq f(y). \tag{8.16}$$

Here $\leq$ denotes the ordering on $\mathbb{R}^d$ induced by a closed cone K. The stochastic ordering $\leq_{st}$ is defined for random vectors X, Y on $(\Omega, \mathfrak{A}, P)$ by

$$X \leq_{st} Y \text{ if } Ef(X) \leq Ef(Y) \tag{8.17}$$

for all $f \in \mathcal{F}_i \cap L^1(\{P^X, P^Y\})$ or equivalently for all bounded increasing functions f. The monotonicity condition (R1) for a law invariant risk measure ϱ implies monotonicity w.r.t. the stochastic order, i.e. larger financial risk vectors have smaller risks.

Proposition 8.12. *Let ϱ be a law invariant risk measure on $\mathcal{X} \subset L_d^0(P)$ satisfying (R1), then ϱ is consistent w.r.t. stochastic ordering, i.e., for $X, Y \in \mathcal{X}$*

$$X \leq_{st} Y \text{ implies } \varrho(Y) \leq \varrho(X).$$

Proof. Strassen's Theorem which is valid for closed orderings $X \leq_{st} Y$ implies the existence of versions $\widetilde{X} \overset{d}{=} X$ and $\widetilde{Y} \overset{d}{=} Y$ such that pointwise ordering $\widetilde{X} \preceq \widetilde{Y}$ holds. Therefore, by the monotonicity condition (R1) and the law invariance

$$\varrho(Y) = \varrho(\widetilde{Y}) \leq \varrho(\widehat{X}) = \varrho(X). \qquad \square$$

The class $\mathcal{F}_{cx}$ of convex functions on $\mathbb{R}^d$ is suitable to measure diffusiveness as in $d = 1$. Let $\mathcal{F}_{icx}$ and $\mathcal{F}_{decx}$ denote the class of increasing respectively decreasing convex functions on $\mathbb{R}^d$. The induced "convex stochastic order" is defined for random vectors X, Y by

$$X \leq_{cx} Y (\text{respectively } X \leq_{icx} Y \text{ respectively } X \leq_{decx} Y) \qquad (8.18)$$

if $Ef(X) \leq Ef(Y)$ for all $f \in \mathcal{F}_{cx}$ (respectively $\mathcal{F}_{icx}$ respectively $\mathcal{F}_{decx}$) such that $f(X)$ and $f(Y)$ are in $L^1(P)$.

The ordering $-X \leq_{icx} -Y$ is equivalent to $X \leq_{decx} Y$. In the one-dimensional case $d = 1$ this ordering is also called "second order stochastic dominance". In contrast to $d = 1$ there is in $d \geq 2$ no simple and natural generating class of $\mathcal{F}_{cx}$ respectively $\mathcal{F}_{decx}$ like $f_t(x) = (t - x)_+$, which relates the one-dimensional case uniquely to the tail probabilities and quantiles. Convex ordering is well defined only on $L_d^1(P)$. Convex law invariant risk measures on $L_d^1(P)$ are consistent with the convex order.

Theorem 8.13 (Consistency w.r.t. convex order). *Let ϱ be a convex law invariant risk measure on $L_d^1(P)$. Then ϱ is consistent with the convex order $\leq_{cx}$ and $\leq_{decx}$, i.e.,*

$$X \leq_{cx} Y \text{ implies } \varrho(X) \leq \varrho(Y) \qquad (8.19)$$

and

$$X \leq_{decx} Y \text{ implies } \varrho(X) \leq \varrho(Y). \qquad (8.20)$$

Proof. By an approximation argument as in Lemma 2.2 in Schied (2004) extended to the multivariate case we obtain that for any convex law invariant risk measure ϱ on $L_d^1(P)$ and any $X, Y \in L_d^1(P)$ it holds that:

$$\varrho(X) \geq \varrho(E(X \mid Y)). \qquad (8.21)$$

(1) If $X \leq_{cx} Y$ then by Strassens's a.s. representation result (see e.g. Corollary 3.36) there exist versions $\widetilde{X} \stackrel{d}{=} X$, $\widetilde{Y} \stackrel{d}{=} Y$ on Ω such that $E(\widetilde{Y} \mid \widetilde{X}) = \widetilde{X}\,[P]$. Therefore, by (8.21)

$$\varrho(Y) = \varrho(\widetilde{Y}) \geq \varrho(E(\widetilde{Y} \mid \widetilde{X})) = \varrho(\widetilde{X}) = \varrho(X).$$

(2) If $X \leq_{decx} Y$, then there are versions $\widetilde{X} \stackrel{d}{=} X$, $\widetilde{Y} \stackrel{d}{=} Y$ on Ω such that $E(\widetilde{Y} \mid \widetilde{X}) \leq \widetilde{X}\,[P]$. Therefore, $\varrho(\widetilde{Y}) \geq \varrho(E(\widetilde{Y} \mid \widetilde{X})) \geq \varrho(\widetilde{X})$. $\qquad \square$

Remark 8.14. (a) In the one-dimensional case the convex ordering result has been proved as a consequence of the Kusuoka representation result for convex law invariant risk measures on L^∞ in Föllmer and Schied (2011, Corollary 4.59).

(b) **(Convexity of risk measures)** The consistency of convex risk measures with respect to the convex order $\leq_{cx}$ respectively $\leq_{decx}$ is an important motivation for the convexity condition (R3) posed on risk measures on $L_d^1(P)$. Note that obviously also the converse implication holds, i.e.

> If ϱ is a monetary risk measure on $L_d^1(P)$ which is consistent w.r.t. the convex order $\leq_{cx}$ respectively $\leq_{decx}$, then ϱ is a convex risk measure.   (8.22)

Convex ordering of random vectors is well defined only on $L_d^1(P)$. Thus the motivation of convexity of risk measures by consistency is restricted to $L_d^1(P)$. We have seen that the "diversification argument" as motivation for convexity of risk measures goes wrong when we leave $L_d^1(P)$ as domain of a risk measure.
$$\diamond$$

Further consistency properties of portfolio risk measures w.r.t. dependence orderings will be considered in connection with some classes of concrete examples in Section 8.2.

8.2 Classes of Examples of Portfolio Risk Measures

In this section we introduce several classes of examples of portfolio risk measures and we discuss for some of them consistency properties w.r.t. dependence orderings.

8.2.1 Aggregation Type Risk Measures

A natural and important class of examples of portfolio risk measures is obtained by applying one-dimensional risk measures to some real aggregation of the risk vector X. The most interesting examples of this construction method are risk measures of the form

$$\Psi(X) = \Psi_1\left(\sum_{i=1}^{d} X_i\right) \tag{8.23}$$

or

$$\Psi(X) = \Psi_2\left(\max_{i \leq d} X_i\right), \tag{8.24}$$

where Ψ_1, Ψ_2 are suitable one-dimensional risk measures like the expected shortfall, the value at risk, the average value at risk, or a distorted risk measure. Ψ measures the risk of the aggregated joint portfolio $\sum_{i=1}^{d} X_i$ in (8.23) respectively of the aggregated maximal risk component in (8.24). Some general classes of aggregation portfolio risk measures are introduced and investigated in Burgert and Rü (2006a).

One may consider combinations of both aggregation risk measures in (8.23) and (8.24) given by

$$\Psi(X) = \alpha \Psi_1 \left(\sum_{i=1}^{d} X_i \right) + \beta \Psi_2 \left(\max_{i \leq d} X_i \right) \tag{8.25}$$

or

$$\Psi(X) = \max \left(\alpha \Psi_1 \left(\sum_{i=1}^{d} X_i \right), \beta \Psi_2 \left(\max_{i \leq d} X_i \right) \right). \tag{8.26}$$

An extension of the idea of measuring the risk of the joint portfolio or the maximal portfolio as in (8.23) and (8.24) is to introduce some class $\mathcal{F}_0 = \{ f_\alpha; \alpha \in A \}$ or real aggregation functions on $\mathbb{R}^d$ and to measure the risk of the real "aspects" $f_\alpha(X)$ of X, $\alpha \in A$, by

$$\Psi_A := \sup_{\alpha \in A} \Psi_\alpha(f_\alpha(X)), \tag{8.27}$$

and

$$\Psi_M := \sup_{\mu \in M} \int \Psi_\alpha(f_\alpha(X)) d\mu(\alpha). \tag{8.28}$$

Here M is some class of weighting measures on A and $\{ \Psi_\alpha \}$ is a class of one-dimensional risk measures for $\alpha \in A$. Thus we are measuring the maximal risk of the real aspects $f_\alpha(X)$ or the maximal average risk over some weighting class M. If e.g. $A = \Delta := \{ \xi \in \mathbb{R}^d_+; \ \sum \xi_i = 1 \}$, then

$$\Psi_\Delta := \sup_{\alpha \in \Delta} \Psi_1(\alpha \cdot X) \tag{8.29}$$

is the maximal risk of X over all positive directions ξ and

$$\Psi_\mu(X) := \int_\Delta \Psi_1(\xi \cdot X) d\mu(\xi) \tag{8.30}$$

is the risk of X averaged over all positive directions.

Concrete examples of the procedure are

$$\Psi_\xi^1(X) = \text{ess sup} \sum_{i=1}^{d} \xi_i X_i \tag{8.31}$$

$$\Psi_\xi^2(X) = \text{VaR}_\alpha \left(\sum_{i=1}^{d} \xi_i X_i \right) \tag{8.32}$$

$$\Psi_\xi^3(X) = ES_\alpha \left(\sum_{i=1}^{d} \xi_i X_i \right) \tag{8.33}$$

or

$$\Psi_\xi^4(X) = \max_{1 \le i \le k} \{\beta_i \, \mathrm{VaR}_{\alpha_i}(\xi \cdot X)\} \tag{8.34}$$

$$\Psi_\xi^5(X) = \sum_{i=1}^k \beta_i \, \mathrm{VaR}_{\alpha_i}(\xi \cdot X) \tag{8.35}$$

where $\alpha_1 < \ldots < \alpha_k$ are k levels of quantiles and β_i are weights for these levels.

An advantage of the class Ψ_A, Ψ_M of portfolio risk measures is that the one-dimensional representation results can be used for these classes and yield scenario type representations.

A way to single out useful classes of aggregation functions F_α is to consider consistency properties of the induced risk measures Ψ_A, Ψ_M w.r.t. convex and dependence orderings.

Convex Order $\le_{cx}$

In order to obtain an ordering consistency result w.r.t. convex ordering $\le_{cx}$ for the risk measures Ψ_A, Ψ_M we need the following assumption on the one-dimensional law invariant risk measures Ψ_α.

Assumption (A_{icx}). *Let $\{\Psi_\alpha\}$ be one-dimensional law invariant risk measures such that Ψ_α is monotone and preserves the increasing convex order $\le_{\mathrm{icx}}$ for all $\alpha \in A$.*

It is easy to see that for $\mathcal{F} = \mathcal{F}_{cx}$ the class of convex functions $f : \mathbb{R}^d \to \mathbb{R}^1$, and for $f \in \mathcal{F}_{cx}$ it holds that:

$$X \le_{cx} Y \Rightarrow f(X) \le_{\mathrm{icx}} f(Y). \tag{8.36}$$

As a consequence of (8.36) we obtain the following proposition.

Proposition 8.15 (Consistency with the convex order). *We assume condition A_{icx} on $\{\Psi_\alpha\}$.*

(a) If the class of aggregation functions $\mathcal{F}_0 = \{f_\alpha, \alpha \in A\} \subset \mathcal{F}_{cx}$, then Ψ_A and Ψ_M are consistent with the convex order, i.e.

$$X \le_{cx} Y \Rightarrow \Psi_A(X) \le \Psi_A(Y) \quad and \quad \Psi_M(X) \le \Psi_M(Y). \tag{8.37}$$

(b) If $\mathcal{F}_0 \subset \mathcal{F}_{\mathrm{icx}}$, then Ψ_A, Ψ_M are consistent with $\le_{\mathrm{icx}}$.

Remark 8.16. In particular it follows from (8.37) that the portfolio risk measures $\Psi_1\left(\sum_{i=1}^d X_i\right)$, $\Psi_2(\max_i X_i)$ in (8.23) and (8.24), $\Psi_\Delta(X) = \sup_{\alpha \in \Delta} \Psi_1(\alpha \cdot X)$, and $\Psi_\mu(X) = \int_\Delta \Psi_1(\xi \cdot X) d\mu(\xi)$ in (8.29) and (8.30) are risk measures consistent with the convex order if Ψ_1, Ψ_2 satisfy A_{icx}. $\Diamond$

Supermodular and Directionally Convex Ordering

The supermodular ordering and the directionally convex ordering are particular interesting positive dependence orderings (for definition see Sections 6.1 and 6.3). Twice differentiable functions f are "supermodular" if

$$\frac{\partial^2}{\partial x_i \, \partial x_j} f(x) \geq 0 \quad \text{for all } x \text{ and } i < j; \tag{8.38}$$

f is "directionally convex" if

$$\frac{\partial^2 f}{\partial x_i \, \partial x_j}(x) \geq 0 \quad \forall \, i \leq j. \tag{8.39}$$

If random vectors X, Y are comparable with respect to the supermodular ordering $\leq_{\text{sm}}$, then necessarily the marginals are identical, i.e. $X_i \overset{d}{=} Y_i$, $1 \leq i \leq d$. The comparison w.r.t. the directionally convex order $X \leq_{\text{dcx}} Y$ is possible if the marginals increase convexly i.e. $X_i \leq_{\text{cx}} Y_i$, $1 \leq i \leq d$. Similarly, comparison w.r.t. the increasing directionally convex order $\leq_{\text{idcx}}$ is possible if $X_i \leq_{\text{icx}} Y_i$, $1 \leq i \leq d$.

Proposition 8.17 (Consistency w.r.t. supermodular and directionally convex ordering). *Let the one-dimensional risk measures $\{\Psi_\alpha\}$ fulfil assumption A_{icx}.*

(a) *If $X \leq_{\text{ism}} Y$, then $f(X) \leq_{\text{icx}} f(Y)$ for all $f \in \mathcal{F}_{\text{ism}}$ and for $\mathcal{F}_0 \subset \mathcal{F}_{\text{ism}}$ it holds that:*

$$\Psi_A(X) \leq \Psi_A(Y), \quad \Psi_M(X) \leq \Psi_M(Y), \tag{8.40}$$

i.e. Ψ_A, Ψ_M are consistent w.r.t. increasing supermodular ordering $\leq_{\text{ism}}$.

(b) *If $X \leq_{\text{idcx}} Y$, then $f(X) \leq_{\text{icx}} f(Y)$ for all $f \in \mathcal{F}_{\text{idcx}}$ and for $\mathcal{F}_0 \subset \mathcal{F}_{\text{idcx}}$ it holds that:*

$$\Psi_A(X) \leq \Psi_A(Y), \quad \Psi_M(X) \leq \Psi_M(Y). \tag{8.41}$$

Proof. By an approximation argument it is sufficient for the ordering result to consider twice differentiable functions f. Then for any $h \in \mathcal{F}_{\text{icx}} \cap C^2(\mathbb{R}^1)$ it holds that

$$\frac{\partial^2 h \circ f}{\partial x_i \, \partial x_j} = h'' \circ f \, \frac{\partial f}{\partial x_i} \frac{\partial f}{\partial x_j} + h' \circ f \, \frac{\partial^2 f}{\partial x_i \, \partial x_j}. \tag{8.42}$$

Since $h'' \circ f \geq 0$, $h' \circ f \geq 0$, we obtain that $\frac{\partial^2 h \circ f}{\partial x_i \, \partial x_j}$ is positive for $i \neq j$ and $f \in \mathcal{F}_{\text{dcx}} \cap C^2(\mathbb{R}^d)$ and, therefore, (8.40) follows. The proof of (8.41) is similar. $\qquad\qquad\square$

Remark 8.18. Sufficient conditions for $\leq_{\text{sm}}$ and $\leq_{\text{dcx}}$ were given in Sections 6.1 and 6.3. If for example for $d = 2$ we assume:

$$\overline{F}_X(u, v) - \overline{F}_{X_1}(u)\overline{F}_{X_2}(v) \leq \overline{F}_Y(u, v) - \overline{F}_{Y_1}(u)\overline{F}_{Y_2}(v), \tag{8.43}$$

then

$$X_i \leq_{cx} Y_i, \quad i = 1, 2 \Rightarrow X \leq_{dcx} Y$$
$$X_i \leq_{icx} Y_i, \quad i = 1, 2 \Rightarrow X \leq_{idcx} Y.$$

In particular (8.40) and (8.41) imply results of the type that more positive dependence of random vectors leads to higher risks. The classical result in this direction is that

$$X \leq_{sm} X^c := \left(F_1^{-1}(U), \ldots, F_d^{-1}(U) \right), \tag{8.44}$$

where X^c is the comonotonic vector to X and F_i are the distribution functions of X_i (see Sections 2.2 and 3.1). Equation (8.44) yields in particular the consequence that the comonotonic vector is most risky for the joint portfolio

$$\sum_{i=1}^{d} X_i \leq_{icx} \sum_{i=1}^{d} F_i^{-1}(U), \tag{8.45}$$

and

$$\Psi_M(X) \leq \Psi_M(X^c), \quad \Psi_A(X) \leq \Psi_A(X^c). \tag{8.46}$$

The comonotonic risk vector has the highest risk in the Fréchet class of X with respect to all risk measures of the form Ψ_M, Ψ_A under the conditions made on the aggregation functions. These ordering results imply arguments for the choice of aggregation functions as described in Proposition 8.17. $\Diamond$

Schur Convex and Symmetric Convex Ordering

Let $\prec_S$ denote the Schur order on $\mathbb{R}^d$,

$$x \prec_S y \Leftrightarrow \sum_{i=1}^{k} x_{(i)} \leq \sum_{i=1}^{k} y_{(i)}, \quad 1 \leq k \leq d - 1 \text{ and } \sum_{i=1}^{d} x_i = \sum_{i=1}^{d} y_i \tag{8.47}$$

and the increasing Schur order

$$x \preceq_S y \Leftrightarrow \sum_{i=1}^{k} x_{(i)} \leq \sum_{i=1}^{k} y_{(i)}, \quad 1 \leq k \leq d, \tag{8.48}$$

where $x_{(1)} \geq \ldots \geq x_{(d)}$ is the ordered vector. The corresponding stochastic orders are called the majorization order respectively weak majorization order (see Section 3.2) and are also denoted by $\prec_m$ and $\preceq_m$. These are relevant diffusion orders. A random vector Y is more diffuse and therefore more risky than X if $X \prec_m Y$. Note that choosing the convex ordering cone K as

$$K = \left\{ x \in \mathbb{R}^d; \ 0 \preceq_S x \right\} \tag{8.49}$$

we have that $\mathbb{R}_+^d \subset K$. Monotonicity of a risk measure Ψ w.r.t. the cone ordering induced by K means that comparability of random vectors X in the weak majorization order $\preceq_m$ implies comparability of the risks

$$X \preceq_m Y \Rightarrow \Psi(X) \le \Psi(Y). \tag{8.50}$$

Thus studying consistency w.r.t. Schur ordering is equivalent to studying monotonicity w.r.t. the cone ordering induced by K. As shown in Section 3.2 the Schur order is closely connected with ordering w.r.t. symmetric convex functions $\mathcal{F}_{\text{sym, cx}}$ respectively their increasing variants.

Proposition 8.19 (Consistency w.r.t. Schur order).   *Let $\{\Psi_\alpha\}$ fulfil assumption A_{icx}.*
(a) $$X \prec_m Y \ \text{implies} \ X \le_{\text{sym, cx}} Y$$
$$X \preceq_m Y \ \text{implies} \ X \le_{\text{isym, cx}} Y \tag{8.51}$$

(b) If $X \preceq_m Y$ and $\mathcal{F}_0 \subset \mathcal{F}_{\text{isym, cx}}$, then

$$\Psi_A(X) \le \Psi_A(Y) \ \text{and} \ \Psi_M(X) \le \Psi_M(Y). \tag{8.52}$$

Proof. If $X \prec_m Y$, then by Strassen's representation result (see Section 3.4) there exist versions $\widetilde{X} \overset{d}{=} X, \widetilde{Y} \overset{d}{=} Y$ such that $\widetilde{X} \prec_s \widetilde{Y}$ and, therefore, $\widetilde{X} = \sum_{\pi \in S_d} \alpha_\pi \widetilde{Y}_\pi$ for some random $\alpha_\pi \ge 0$ with $\sum_{\pi \in S_d} \alpha_\pi = 1$, where $\widetilde{Y}_\pi$ is the reordered vector. If $f \in \mathcal{F}_{\text{sym, cx}}$ then $f(\widetilde{X}) \le \sum \alpha_\pi f(\widetilde{Y}_\pi) = f(\widetilde{Y})$ and thus $Ef(X) \le Ef(Y)$ i.e. $X \le_{\text{sym, cx}} Y$.

If $X \preceq_m Y$ then by a variant of Strassen's representation theorem there exist $\widetilde{X} \overset{d}{=} X, \widetilde{Y} \overset{d}{=} Y$ and Z such that $\widetilde{X} \prec_s Z \le \widetilde{Y}$. If $f \in \mathcal{F}_{\text{isym, cx}}$, then

$$Ef(X) = Ef(\widetilde{X}) \le Ef(Z)$$
$$\le Ef(\widetilde{Y}) = Ef(Y), \quad \text{i.e. } X \le_{\text{isym, cx}} Y.$$

The other conclusions are similar to the corresponding ones in Propositions 8.15 and 8.17 $\square$

By the connection between the (weak) majorization order $\preceq_m$ and the (weak) Schur order $\preceq_s$ we identify in the following $\preceq_s$ with $\preceq_m$ also on the level of random vectors. Proposition 8.19 gives conditions on the choice of aggregation functions which imply consistency of the risk measures Ψ_A, Ψ_M w.r.t. Schur convex order.

Conclusion

(a) The risk measures Ψ_A, Ψ_M define meaningful and easy to interpret classes of risk measures, consistent with respect to the various classes of convex and dependence orderings under suitable choice of the aggregation functions.

In order to get monotonicity (R1) the aggregation functions have to be chosen monotone. The simple way of construction allows one easily to establish further relevant risk properties of Ψ_A, Ψ_M, which are inherited from the basic one-dimensional risk measures $\{\Psi_\alpha\}$ used for their construction. If we take e.g. $\Psi_A(X) = \sup_{\alpha \in A} \Psi_1(\alpha \cdot X)$ from (8.29) then Ψ_A is a convex risk measure if Ψ_1 is convex and Ψ_A is a coherent risk measure if Ψ_1 is coherent.

(b) To measure the risk caused by dependence of the components of X it is natural to consider the difference

$$\widehat{\Psi}(X) := \Psi(X) - \Psi(X^{\perp}) \tag{8.53}$$

where $X^{\perp}$ is the vector with independent components $X_i^{\perp} \overset{d}{=} X_i, 1 \leq i \leq d$.

8.2.2 Multivariate Distortion and Quantile-Type Risk Measures

In dimension $d = 1$ basic classes of risk measures are defined via quantiles as the value at risk VaR_α, the AVaR_α, and more generally the spectral risk measures and the distortion risk measures (see (7.38)–(7.40)). The Kusuoka representation result (Theorem 7.7) stresses the importance of these classes of risk measures. Some extensions of these risk measures to portfolio vectors have been introduced in Rü (2006). We concentrate in the following on the insurance case with non-negative risk vectors $X \geq 0$.

Multivariate Distortion Risk Measure

A direct extension of distortion risk measures for $d = 1$ to $d \geq 1$ is obtained by the following definition. Let X be a risk vector with distribution function $F = F_X$, let $g : [0, 1] \to [0, 1]$ be an increasing, concave distortion function with $g(0) = 0$, $g(1) = 1$ and let μ be a measure on $\mathbb{R}_+^d$. Then we define the "multivariate distortion risk measure" $\Psi_\mu = \Psi_{\mu,g}$ by

$$\Psi_\mu(X) = \int_{\mathbb{R}_+^d} g(\overline{F}(x)) d\mu(x), \tag{8.54}$$

where $\overline{F}(x) = P(X \geq x)$ is the multivariate survival function describing high risk in $[x, \infty)$. Alternatively one could also consider

$$\Psi_\mu(X) = \int_{\mathbb{R}_+^d} g(1 - F(x)) d\mu(x) \tag{8.55}$$

considering high risk in any of the d-components. (8.54) is more suitable to detect positive dependence effects on the risk.

More generally one could also consider some class $A \subset M(\mathbb{R}_+^d)$ of weighting measures on $\mathbb{R}_+^d$ and define

$$\Psi_A(X) := \sup_{\mu \in A} \Psi_\mu(X). \tag{8.56}$$

Useful choices of weighting measures μ are for example the Lebesgue measure $\mu = \lambda_+^d$ or discrete measures $\mu = \sum_{i=1}^m \alpha_i \varepsilon_{x_i}$, $x_i \in \mathbb{R}^d$. For the discrete measure μ the distortion risk measure $\Psi_\mu^1(X) = \sum_{i=1}^m \alpha_i g(\overline{F}(x_i))$ is a weighted mean of distorted survival probabilities.

Multivariate Quantile Functionals

For $0 < \alpha < 1$ we define the risk sets

$$A_\alpha := \left\{ x \in \mathbb{R}_+^d; \ \overline{F}(x) > 1 - \alpha \right\} \tag{8.57}$$

where $\overline{F}(x)$ again is either taken as survival function $P(X \geq x)$ or as $1 - F(x)$. The boundary set $\partial A_\alpha = \overline{A}_\alpha - \overset{\circ}{A}_\alpha$ represents multivariate quantiles.

Any measure μ on $\mathbb{R}_+^d$ induces a "multivariate quantile functional"

$$\overline{F}_\mu^{-1}(X) := \mu(A_\alpha). \tag{8.58}$$

In case $d = 1$ and $\mu = \lambda_+$ it holds that

$$\overline{F}_\mu^{-1}(\alpha) = \lambda_+\left(\left\{ x \in \mathbb{R}_+; \ \overline{F}(x) \geq 1 - \alpha \right\}\right) = F^{-1}(\alpha).$$

Thus $\overline{F}_\mu^{-1}(\alpha)$ is the α-quantile of the distribution function $F = F_X$.

The multivariate distortion risk measure $\Psi_\mu = \Psi_{\mu,g}$ in (8.54) respectively (8.55) then can be also written as a multivariate quantile functional

$$\Psi_\mu(X) = \int_0^1 g(t) d\overline{F}_\mu^{-1}(t). \tag{8.59}$$

Further from partial integration we also get a representation as weighted quantile risk functional

$$\Psi_\mu(X) = -\int_0^1 \overline{F}_\mu^{-1}(t) dg(t). \tag{8.60}$$

Remark 8.20 (Multivariate average value at risk). If g is absolutely continuous and concave, then we can write (8.60) also in the form

$$\Psi_\mu(X) = \int_0^1 \mathrm{AVaR}_\alpha^\mu(X) d\nu(\alpha), \tag{8.61}$$

where

$$\mathrm{AVaR}_\alpha^\mu(X) := \frac{1}{\alpha} \int_0^\alpha \overline{F}_\mu^{-1}(1-u) du, \tag{8.62}$$

and ν is the probability measure defined by $d\nu(s) = s d\widetilde{\nu}(s)$, where $\widetilde{\nu}((s,1]) := g_+'(s)$. Note that $\int_0^1 t d\widetilde{\nu}(t) = \int_0^1 \widetilde{\nu}((s,1]) ds = \int_0^1 g'(s) ds = g(1) - g(0) = 1$.

Thus $\mathrm{AVaR}_\alpha^\mu(X)$ plays the role of an "average value at risk measure" in $d \geq 1$. $\Diamond$

Example 8.21. We consider the special case that μ is the Lebesgue measure on the positive diagonal, $\mu = \lambda^\pi$, where $\pi : [0,\infty) \to \mathbb{R}_+^d, t \to t \cdot 1$ is a parameterization of the diagonal.

Then with $\overline{F} = 1 - F$ we obtain

$$\psi_\mu(X) = \int_0^\infty g(1 - F(t \cdot 1)) dt, \tag{8.63}$$

and

$$\overline{F}_\mu(t \cdot 1) = 1 - F(t \cdot 1) = P(\max_{i \leq d} X_i > t) = \overline{F}_{\max X_i}(t). \tag{8.64}$$

Further

$$\overline{F}_\mu^{-1}(u) = \lambda(\{0 \leq t; \ \overline{F}_\mu(t \cdot 1) \leq u\}) = \overline{F}_{\max X_i}^{-1}(u)$$

and thus we obtain

$$\Psi_\mu(X) = -\int_0^1 \overline{F}_{\max X_i}^{-1}(u) dg(u). \tag{8.65}$$

Thus in this special case $\Psi_\mu(X)$ is identical to the one-dimensional distortion risk measure applied to $\max_{i \leq d} X_i$. $\Diamond$

Generalized Distortion Risk Measures

The class of distortion type risk measures defined in (8.54) and (8.55) is weighting the risk sets

$$A_x := \{X \leq x\}^c \text{ respectively } A_x := \{X \geq x\}, \quad x \in \mathbb{R}_+^d \tag{8.66}$$

by the distortion

$$c_g := g \circ P \tag{8.67}$$

of the probability measure P. In the multivariate case the distribution function is however no longer simple to calculate and thus the calculation of Ψ_μ in (8.54) and (8.55) poses a considerable problem. It is also not the case that the risk sets of the form A_x represent the only relevant class of risk sets.

We next consider an extension of the class of distortion risk measures defined in (3.1) by allowing more general classes of relevant risk sets. We restrict to one-parametric classes of risk sets $(A_t)_{t \geq 0} \subset \mathbb{R}_+^d$ in order to induce the order structure from $\mathbb{R}_+$ and to get not too complicated forms. We assume the following conditions on the class of risk sets $(A_t)_{t \geq 0} \subset \mathbb{R}_+^d$.

Risk sets: $(A_t)_{t \geq 0} \subset \mathbb{R}_+^d$ is called a family of "risk sets", if
(R1) A_t are monotone sets, $t \geq 0$, i.e. $x \in A_t$ and $y \geq x \Rightarrow y \in A_t$,
(R2) (A_t) is decreasing in t,
(R3) $A_0 = \mathbb{R}_+^d$, $\lim_{t \to \infty} A_t = \phi$,
(R4) (A_t) is right continuous, i.e. $A_{t+\varepsilon} \uparrow A_t$ as $\varepsilon \downarrow 0$.

As a consequence of (R1)–(R4) we may introduce a generating "risk function" $U :$ $\mathbb{R}_+^d \to \mathbb{R}_+$ by

$$U(x) := \inf \{s; \ x \in A_s\}. \tag{8.68}$$

We obtain a representation of the risk sets A_t as level sets of the risk function U:

$$A_t = \{U \geq t\}. \tag{8.69}$$

Our generalized class of multivariate distortion risk measures is induced by a class of risk sets (A_t) satisfying (R1)–(R4) and by a concave distortion function g. It is given by

$$\Psi^g(X) := \int_0^\infty g \circ P(X \in A_t) d\lambda(t). \tag{8.70}$$

Since

$$\overline{F}(t) := P(X \in A_t) = P(U(X) \geq t) = \overline{F}_{U(X)}(t) \tag{8.71}$$

we obtain

$$\Psi^g(X) = \int_0^\infty g(\overline{F}_{U(X)}(t)) dt = -\int_0^1 \overline{F}_{U(X)}^{-1}(t) dg(t) = \Psi_g(U(X)). \tag{8.72}$$

Thus $\Psi^g(X)$ is identical to the one-dimensional distortion risk measures Ψ_g applied to the (real) transformation $U(X)$ of X. In dimension $d = 1$ we thus connect with the "rank dependent expected distortions" (see Denuit et al. (2005, p. 88–90)). As a consequence Ψ^g can be subsumed in the classes Ψ_A, Ψ_M of risk measures introduced in Section 8.2.1.

If U is a convex function, then Ψ^g is a convex risk measure. Let e.g. $U(x) = \sum_{i=1}^n x_i^2$, then $A_t = \{x; \sum_{i=1}^d x_i^2 \geq t\}$ and Ψ^g is based on weighting the radial part of the risk X. Further interesting choices of U are $\sum a_i x_i^2$, or $\max a_i X_i$, $a_i \geq 0$, which lead to convex risk measures Ψ^g as well.

8.3 Representation and Continuity of Convex Risk Measures on L_d^p

Convex and coherent risk measures on $L_d^p(P)$, $1 \le p \le \infty$ are motivated as in $d = 1$ by the unboundedness of typical risk models. They have in dimension $d \ge 1$ similar representation and continuity properties as in $d = 1$. Some of the basic representation and continuity properties have already been stated in Section 7.2 in a general context, like the Fenchel–Moreau Theorem 7.8, the extended Namioka Theorem 7.9, the Lipschitz continuity property in Theorem 7.13 and Remark 7.22a). In this section we state some of the basic continuity and representation results in the context of convex risk measures on $L_d^p = L_d^p(P)$ in order to have a reference available in the application of risk measures in later sections. On L_d^p we consider the product norm $\|X\|_p := \max_{1 \le i \le d} \|X_i\|_p$ for $X = (X_1, \ldots, X_d) \in L_d^p$. Then $(L_d^p, \|\ \|_p)$ is a Banach lattice and the $\|\ \|_p$-norm induces the product topology on L_d^p:

$$X_n \to X \text{ w.r.t. } \|\ \|_p \text{ if and only if } X_{i,n} \to X_i$$

w.r.t. the p-norm $\|\ \|_p$ of the components.

As a corollary of the Namioka Theorem 7.9 we obtain

Corollary 8.22. *Let* $\varrho : L_d^p \to \mathbb{R} \cup \{\infty\}$, $1 \le p \le \infty$, *be a proper convex risk measure. Then*

(a) *ϱ is continuous on $I_\varrho = \text{int dom } \varrho$*
(b) *If ϱ is a finite convex risk measure on L_d^p, $1 \le p \le \infty$ then ϱ is continuous on L_d^p.*

We denote the (positive part of the) dual spaces of $L_d^p = L_d^p(P)$ by

$$\mathcal{Q}_{d,p} := \{Q = (Q_1, \ldots, Q_d);\ Q_i \in M, \tfrac{dQ_i}{dP} \in L^q\}, 1 \le p < \infty \qquad (8.73)$$

and

$$\mathcal{Q}_{d,\infty} := \{Q = (Q_1, \ldots, Q_d);\ Q_i \in \text{ba}_d(P), 1 \le i \le d\}, p = \infty, \qquad (8.74)$$

where $\text{ba}_d(P)$ is the set of finite additive, positive measures with P-continuous components Q_i on L_d^∞. By $\mathcal{M}_{d,\infty} := \{Q = (Q_1, \ldots, Q_d);\ Q_i \in M_d(P)\}$ we denote the σ-additive P-continuous measures on L_d^∞. We also identify $\mathcal{M}_{d,\infty}$ with the positive part of L_d^1. The value of the linear functional l represented by $Q = (Q_1, \ldots, Q_d)$ at X is given by

$$\ell(X) = E_Q(X) = \sum_{i=1}^d E_{Q_i} X_i$$

and $E_{Q_i} X_i = \langle Q_i, X \rangle$ denotes the integral of X_i w.r.t. Q_i. For the representation results we additionally add the normalization condition $Q_i \in M_1$ respectively $Q_i(\mathbb{R}) = 1$ if the risk measures ϱ are assumed to be normed

$$\varrho(m) = -\sum_{i=1}^{d} m_i, \quad m \in \mathbb{R}^d.$$

The corresponding dual classes we denote $\mathcal{Q}_{d,p}^1$ respectively $\mathcal{Q}_{d,\infty}^1$. The following representation result corresponds to Theorem 7.14 and is a consequence of the Fenchel–Moreau Theorem 7.8.

Theorem 8.23 (Representation of convex risk measures on L_d^p). *If $\varrho : L_d^p \to \mathbb{R} \cup \{\infty\}$ is a proper, convex $\| \ \|_p$-lower semicontinuous risk measure on L_d^p, $1 \le p \le \infty$, then*

$$\varrho(X) = \sup_{Q \in \mathcal{Q}_{d,p}} (E_Q(-X) - \varrho^*(Q)), \quad X \in L_d^p. \tag{8.75}$$

Remark 8.24 (Proper convex risk measures on L_d^∞). If ϱ is a proper convex risk measure on L_d^∞, then

(a) ϱ is finite and Lipschitz continuous

(b)
$$\varrho(X) = \max_{Q \in \mathcal{Q}_{d,\infty}} (E_Q(-X) - \varrho^*(Q)), \tag{8.76}$$

i.e. the sup in (8.56) is attained.

The proof is similar to the case $d = 1$. (2) is a consequence of weak $*$-compactness of bounded closed sets in the dual space $\mathcal{Q}_{d,\infty}$. $\diamond$

As in Theorem 7.20 the representation simplifies for coherent risk measures on L_d^p.

Theorem 8.25 (Representation of coherent risk measures on L_d^p). *Let $\varrho : L_d^p \to \mathbb{R} \cup \{\infty\}$ be a risk functional on L_d^p, $1 \le p \le \infty$. Then the following are equivalent:*

(a) ϱ is a proper, $\| \ \|_p$-lower semicontinuous, coherent risk measure on L_d^p.
(b) ϱ has a representation of the form

$$\varrho(X) = \max_{Q \in \mathcal{Q}} E_Q(-X) \tag{8.77}$$

for some closed scenario set $\mathcal{Q} \subset \mathcal{Q}_{d,p}$.

The proof uses Propositions 7.18 and 7.19, which already are formulated in general form, and the representation then follows from Theorem 8.23. Also Theorem 7.13 on Lipschitz continuity, subdifferentiability and the attainment of the dual representation is already given for general Banach lattices. As a consequence we obtain an analogue of the representation result in Theorem 7.21 (compare the proof of Theorem 4.3 in Cheridito and Li (2009)).

Theorem 8.26 (Representation of convex risk measures on L_d^p, $1 \leq p < \infty$).
Let $\varrho : L_d^p \to (-\infty, \infty]$, $1 \leq p < \infty$ be a convex risk measure on L_d^p with $I_\varrho = \text{core}(\text{dom}\,\varrho) \neq \phi$. Then ϱ is real valued and

$$\varrho(X) = \max_{Q \in \mathcal{Q}_{d,p}} \{E_Q(-X) - \varrho^*(Q)\}, \quad X \in L_d^p \tag{8.78}$$

with penalty function ϱ^* given by $\varrho^*(Q) = \sup_{X \in A_\varrho} E_Q(-X)$.

The representation properties of convex risk measures on L_d^p are more pleasant for $p < \infty$ than for $p = \infty$. The dual spaces are already σ-additive measure spaces for $p < \infty$. For $p = \infty$ one can restrict to σ-additive scenarios under additional continuity assumptions.

In the following we state continuity results for risk measures on L_d^p. These are of relevance in applications since they allow approximation arguments and are useful in consistency results. The following definitions and results correspond to continuity results in Section 7.2.3 in case $d = 1$. To keep the presentation in this book in a balanced frame we do not give the extended proofs here which are given in detail in Kaina (2007).

Definition 8.27 (Continuity properties of convex risk measures on L_d^p). Let $\varrho : L_d^p \to \mathbb{R}$ be a convex risk measure on L_d^p. Then ϱ is called

(a) "continuous from above" if $(X_n) \subset L_d^p$, $X_n \downarrow X$ P-a.s. for some $X \in L_d^p$ implies $\lim \varrho(X_n) = \varrho(X)$
(b) "Fatou-continuous" if $X, (X_n) \subset L_d^p$, $|X_n| \leq Y$ componentwise for some $Y \in L_d^p$ and $X_n \to X$ in L_d^p implies that

$$\varrho(X) \leq \liminf \varrho(X_n) \tag{8.79}$$

(c) "continuous from below" if $X, (X_n) \subset L_d^p$ and $X_n \uparrow X$ P-a.s. implies $\lim \varrho(X_n) = \varrho(X)$
(d) "Lebesgue-continuous" if $X, (X_n) \subset L_d^p$. $X_n \to X$ P-a.s., $X \in L_d^p$ and $|X_n| \leq Y$ componentwise for some $Y \in L^p$ implies $\lim \varrho(X_n) = \varrho(Y)$.

Remark 8.28. For proper convex risk measures on L_d^p, $1 \leq p \leq \infty$ it holds that:

$$\varrho \text{ is continuous from above if and only if } \varrho \text{ is Fatou-continuous.} \tag{8.80}$$
$$\Diamond$$

In case $1 \leq p < \infty$ we have an analogue of the continuity results in Theorem 7.24 to the portfolio case.

Theorem 8.29 (Continuity of finite convex risk measures on L_d^p). *Let $\varrho : L_d^p \to \mathbb{R}$, $1 \leq p < \infty$ be a finite convex risk measure. Then the following properties hold:*

(1) ϱ is $\sigma(L_d^p, L_d^q)$-lower semicontinuous

(2) $\mathcal{A}_\varrho$ is $\sigma(L_d^p, L_d^q)$-closed

(3) ϱ has the Fatou-property

(4) ϱ is continuous from above

(5) ϱ is continuous from below

(6) ϱ is Lebesgue-continuous.

The following result concerns the nonfinite case and supplements Theorem 8.23.

Theorem 8.30 (Infinite convex risk measures on L_d^p). *For a proper convex risk measure $\varrho : L_d^p \to \mathbb{R} \cup \{\infty\}$, $1 \le p < \infty$ the following are equivalent:*

(1) ϱ is $\sigma(L_d^p, L_d^q)$-lower semicontinuous

(2) $\varrho(X) = \sup_{Q \in \mathcal{Q}_{d,p}}(E_Q(-X) - \varrho^*(Q)), \quad X \in L_d^p$

(3) ϱ is continuous from above

(4) ϱ is Fatou-continuous.

The following continuity and representation results concern the case $p = \infty$. As in the case $d = 1$ the following characterization holds. It says that under Fatou-continuity the dual representation can be restricted to σ-additive measures. For details of the proof see Kaina (2007).

Theorem 8.31 (σ-additive representation of convex risk measures on L_d^∞). *Let ϱ be a proper convex risk measure on L_d^∞. Then the following are equivalent:*

(1) ϱ is continuous from above

(2) ϱ is Fatou-continuous

(3)
$$\varrho(X) = \sup_{Q \in \mathcal{M}_{d,\infty}} (E_Q(-X) - \varrho^*(Q)), \quad X \in L_d^\infty(P) \qquad (8.81)$$

(4) ϱ is $\sigma(L_d^\infty, L_d^1)$-lsc on L_d^∞.

Finally we have the following attainment result in the class of σ-additive measures for convex risk measures on L_d^∞.

Theorem 8.32 (Attainment result for convex risk measures on $L_d^\infty(P)$). *Let $\varrho : L_d^\infty \to \mathbb{R}$ be a finite, normed, convex risk measure on L_d^∞ with conjugate $\varrho^* : (L_d^\infty)' \to \mathbb{R} \cup \{\infty\}$ w.r.t. $\sigma(L_d^\infty, (L_d^\infty)')$-topology.*

(a) The following are equivalent:

   (1) ϱ is Lebesgue-continuous

   (2) ϱ is continuous from below

   (3) ϱ is Fatou-continuous and $\operatorname{dom} \varrho^* = \{\varrho^* < \infty\} \subset L_d^1$

   (4) ϱ is Fatou-continuous and $\{\varrho^* \le c\}$ is a $\sigma(L_d^1, L_d^\infty)$-compact subset of L_d^1 for any $c > -\varrho(0)$.

Further (1)–(4) imply the attainment in

(5) $$\varrho(X) = \max_{Q \in \mathcal{M}_{d,\infty}} (E_Q(-X) - \varrho^*(Q))$$ (8.82)

(b) *If L^1 is separable then the properties (1)–(4) in (a) are equivalent to the attainment in (5).*

The proof of part (b) follows from the following variant of James Theorem:

Theorem 8.33 (James Theorem).   *Let $(E, \| \ \|)$ be a separable Banach space and let $f : E \to \mathbb{R} \cup \{\infty\}$ be proper, convex, and lsc with dom(f) bounded in E. If for some $c \in \mathbb{R}$, $\{f \le c\}$ is not $\sigma(E, E^*)$-compact, then there exists an element $x^* \in E^*$ such that the sup in the conjugate $f^*(x^*) = \sup_{x \in E}(x^*(x) - f(x))$ is not attained.*

Remark 8.34. The original version of "James Theorem" states for a bounded, $\sigma(E, E^*)$-closed subset $A \subset E$:

A is $\sigma(E, E^*)$-compact $\Leftrightarrow$ any continuous linear functional f on
A attains its supremum.   (8.83)

Note that in part (b) of Theorem 8.32 dom(ϱ^*) is bounded since for $X \in L_d^\infty$ and for all $Q \in \mathcal{M}_{d,\infty}^1$ it holds that

$$|E_Q(-X)| \le \|Q\|_{L_d^1} \|X\|_{L_d^\infty} \le d \|X\|_{L_d^\infty}.$$ $\Diamond$

Chapter 9
Law Invariant Convex Risk Measures on L_d^p and Optimal Mass Transportation

As explained before law invariant risk measures are of particular importance by the property that they allow empirical versions and statistical estimates. In dimension $d = 1$ law invariant convex risk measures on $L^\infty(P)$ have been characterized by Kusuoka (2001) for the coherent case and by Frittelli and Rosazza Gianin (2005) in the convex case by a representation of the form

$$\varrho(X) = \sup_{\mu \in M^1([0,1])} \left(\int_{[0/1]} \varrho_\lambda(X) d\mu(\lambda) - \beta(\mu) \right) \qquad (9.1)$$

where $\varrho_\lambda(X) = \text{AV@R}_\lambda(S)$ is the average value at risk. Thus the average value at risk risk measures can be seen as building blocks of the class of all law invariant convex risk measures. A natural question is: Which class of multivariate risk measures plays the same role in $d \geq 1$ as the average value at risk plays in $d = 1$?

An answer to this question was given in Rü (2006) in the case L_d^∞ and in Rü (2012a) in the case L_d^p. See also the recent more detailed study on the case L_d^∞ in Ekeland and Schachermayer (2011). The role of the average value at risk is taken in $d \geq 1$ by the "max correlation risk measures" introduced in Rü (2006). These risk measures connect law invariant convex risk measures with optimal mass transportation. This connection not only is relevant for the representation of law invariant convex risk measures but also explains why there is no general notion of comonotonicity describing worst case dependence structures in the multivariate case. The worst case dependence structure depends crucially on the risk measure taken. We identify the worst case dependence structure w.r.t. a given convex law invariant risk measure later on.

L. Rüschendorf, *Mathematical Risk Analysis*, Springer Series in Operations Research and Financial Engineering, DOI 10.1007/978-3-642-33590-7_9,
© Springer-Verlag Berlin Heidelberg 2013

9.1   Law Invariant Risk Measures and Optimal Mass Transportation

The aim of this section is to extend the Kusuoka type representation of law invariant risk measures from $d = 1$ to $d \geq 1$ and to introduce the max correlation risk measures which is the suitable generalization of the average value at risk. We state the results in the insurance form. The basic tools are the representation results for coherent and convex risk measures on L_d^p in Section 8.3. We restrict to normalized risk measures Ψ with $\Psi(m) = \sum_{i=1}^d m_i$, $m \in \mathbb{R}^d$. For convex risk measures on L_d^p, $1 \leq p < \infty$ with $I_\varrho = \mathrm{core}(\mathrm{dom}\,\varrho) \neq \phi$ we obtain by Theorem 8.26 a representation result

$$\Psi(X) = \max_{Q \in \mathcal{Q}_{d,p}} \{E_Q(X) - \alpha(Q)\}, \quad X \in L_d^p \tag{9.2}$$

with penalty function $\alpha(Q) = \varrho^*(Q) = \sup_{x \in A_\varrho} E_Q(-X)$ and for $p = \infty$ for Fatou-continuous risk measures Ψ a representation result (see Theorem 8.31)

$$\Psi(X) = \sup_{Q \in \mathcal{M}_{d,\infty}} \{E_Q(X) - \alpha(Q)\}, \quad X \in L_d^\infty, \tag{9.3}$$

as well as a corresponding attainment result in Theorem 8.32. Note that law invariant risk measures on L_d^∞ are Fatou-continuous (see Jouini et al. (2006) and Rü (2006)) and therefore Theorem 8.31 is applicable.

For normed risk measures we can restrict to $Q \in \mathcal{Q}_{d,p}$ with probability measures Q_i as components and thus we can equivalently restrict to the class of corresponding P-densities for $1 \leq p \leq \infty$

$$D_{d,p} = \{Y = (Y_1, \ldots, Y_d); \quad Y_i \geq 0 \ P\text{-a.s.},$$
$$Y_i \in L^q, E_P Y_i = 1, 1 \leq i \leq d\} \subset L_d^q. \tag{9.4}$$

Thus the representations in (9.2) and (9.3) can be equivalently written as

$$\Psi(X) = \sup_{Y \in D_{d,p}} \{EX \cdot Y - \alpha(Y)\}, \quad X \in L_d^p, \tag{9.5}$$

where $\alpha(Y) = \sup_{X \in A_\varrho} EX \cdot Y$ and $X \cdot Y$ is the dot-product.

The following simple observation in Rü (2006, 2012a) is the basis of the representation result for law invariant convex risk measures. Let for $X \in L_d^p$

$$A(X) := \{\widetilde{X} \in L_d^p; \quad \widetilde{X} \sim X\}, \tag{9.6}$$

where $\sim$ denotes equality in distribution. $A(X)$ is the class of all $\widetilde{X}$ in L_d^p, which have the same distribution as X, i.e. $A(X)$ is the class of all versions of the random vector X.

Proposition 9.1. *Let* Ψ *be a convex risk measure on* L_d^p, $1 \leq p \leq \infty$. *Then*

(a)
$$\widehat{\Psi}(X) := \sup \left\{ \Psi(\widetilde{X}); \quad \widetilde{X} \in A(X) \right\} \tag{9.7}$$

is a convex, law invariant risk measure on L_d^p.

(b)
$$\Psi \text{ is law invariant} \Leftrightarrow \Psi = \widehat{\Psi}. \tag{9.8}$$

Proof. Obviously $\widehat{\Psi}$ is law invariant. For $X, Y \in L_d^p(P), \alpha \in (0,1)$ and with $Z := \alpha X + (1-\alpha)Y$ it holds that

$$\widehat{\Psi}(\alpha X + (1-\alpha)Y) = \sup_{\widetilde{Z} \sim Z} \Psi(\widetilde{Z}).$$

Since $\widetilde{Z} \sim Z = h(X, Y)$, with $h(x, y) := \alpha x + (1-\alpha)y$, by the result on solutions of stochastic equations in Theorem 1.21 there exist random variables $(\widetilde{X}, \widetilde{Y}) \sim (X, Y)$ such that $\widetilde{Z} = h(\widetilde{X}, \widetilde{Y}) \, [P]$.

Therefore,

$$\widehat{\Psi}(\alpha X + (1-\alpha)Y) \tag{9.9}$$

$$= \sup \left\{ \Psi(\alpha \widetilde{X} + (1-\alpha)\widetilde{Y}); \ (\widetilde{X}, \widetilde{Y}) \sim (X, Y) \right.$$

$$\left. \text{and } \widetilde{Z} := \alpha \widetilde{X} + (1-\alpha)\widetilde{Y} \sim Z \right\}$$

$$\leq \sup \left\{ \alpha \Psi(\widetilde{X}) + (1-\alpha)\Psi(\widetilde{Y}); \ \widetilde{X} \sim X, \widetilde{Y} \sim Y \right.$$

$$\left. \text{and } \alpha \widetilde{X} + (1-\alpha)\widetilde{Y} \sim Z \right\}$$

$$\leq \alpha \widehat{\Psi}(X) + (1-\alpha)\widehat{\Psi}(Y),$$

i.e. $\widehat{\Psi}$ is a convex, law invariant risk measure. Equation (9.8) is obvious. □

Thus for any convex risk measure ϱ respectively Ψ we obtain by the process in (9.7) a law invariant convex risk measure and the mapping $\Psi \to \widehat{\Psi}$ from the class of convex risk measures to the class of convex, law invariant risk measures is surjective.

For a scenario density vector $Y \in D_{d,p}$ we define

$$\Psi_Y : L_d^p \to \mathbb{R}, \quad \Psi_Y(X) := EX \cdot Y. \tag{9.10}$$

$\Psi_Y(X)$ is up to normalization the correlation coefficient of X, Y. Ψ_Y is a coherent risk measure on L_d^p. The corresponding law invariant risk measure $\widehat{\Psi}_Y$ from Proposition 9.1 then is given by

$$\widehat{\Psi}_Y(X) = \sup_{\widetilde{X} \sim X} E\widetilde{X} \cdot Y = \sup_{\substack{\widetilde{X} \sim X \\ \widetilde{Y} \sim Y}} E\widetilde{X} \cdot \widetilde{Y}$$

$$= \widehat{\Psi}_{\widetilde{Y}}(X) \text{ for any } \widetilde{Y} \sim Y. \tag{9.11}$$

Thus $\widehat{\Psi}_Y$ depends only on the P-law $\mu = P^Y$ of Y. We therefore also use the notation

$$\Psi_\mu(X) = \widehat{\Psi}_Y(X). \tag{9.12}$$

Definition 9.2 (max correlation risk measure). The risk measure

$$\Psi_\mu(X) = \widehat{\Psi}_Y(X) = \sup_{\widetilde{X} \sim X} E\widetilde{X} \cdot Y \tag{9.13}$$

is called the "max correlation risk measure". $\Psi_\mu = \widehat{\Psi}_Y$ is a coherent, law invariant risk measure on L_d^p.

Any scenario density vector $Y \in \mathcal{D}_{d,p}$ corresponds to the scenario vector $Q = (Q_1, \ldots, Q_d)$ with $Y_i = \frac{dQ_i}{dP}$, $1 \leq i \leq d$. All scenario density vectors $Z \in A(Y)$ describe different scenarios $Z_i = \frac{d\widetilde{Q}_i}{dP}$ but generate the same law invariant max correlation risk measure Ψ_μ, where $\mu = P^Y$. We denote by

$$\mathcal{D}_{d,p}^l = \{\mu; \ \exists Y \in \mathcal{D}_{d,p}, \mu = P^Y\} \tag{9.14}$$

the class of all laws of scenario density measures. For any $\mu \in D_{d,p}^l$ and $X \in L_d^p$ there is a $Y \in D_{d,p}$ such that

$$\Psi_\mu(X) = EX \cdot Y = E_Q X \tag{9.15}$$

with $Q_i = Y_i P$, the scenario with density Y_i w.r.t. P.

Some properties of the max correlation risk measures are stated in the following remark.

Remark 9.3 (max correlation risk measures).

(a) Since $\Psi_\mu = \widehat{\Psi}_Y$ is a coherent, law invariant risk measure on L_d^p it is consistent w.r.t. increasing convex ordering $\leq_{\mathrm{icx}}$, i.e.

$$X_1 \leq_{\mathrm{icx}} X_2 \text{ implies } \Psi_\mu(X_1) \leq \Psi_\mu(X_2) \tag{9.16}$$

(see Theorem 8.13).

(b) In dimension $d = 1$ it follows from the Hoeffding–Fréchet bounds that

$$\Psi_\mu(X) = \widehat{\Psi}_Y(X) = \int_0^1 F_X^{-1}(u) F_Y^{-1}(u) du. \tag{9.17}$$

By partial integration $\widehat{\Psi}_Y$ has a representation as spectral risk measure

$$\widehat{\Psi}_Y(X) = \int_{[0,1]} \varrho_\lambda(X) d\tau(\lambda) \tag{9.18}$$

with $\varrho_\lambda = \mathrm{AVaR}_\lambda$ (see Föllmer and Schied (2011, Corollary 4.58) or (8.61)).

(c) "Optimal mass transportation": In $d \geq 1$ the defining optimization problem (9.11) respectively (9.13) for the max correlation risk measure, i.e. to determine $\sup_{\widetilde{X} \sim X, \widetilde{Y} \sim Y} E\widetilde{X} \cdot \widetilde{Y}$ is the basic instance of the optimal mass transportation problem. A pair of solutions $(\overline{X}, \overline{Y})$ of (9.11) is called a pair of optimal couplings. If $X \sim P_1$, $Y \sim P_2$ then problem (9.11) is equivalent to the problem

$$\ell_2^2(P_1, P_2) := \inf\{E\|\widetilde{X} - \widetilde{Y}\|^2; \ \widetilde{X} \sim P_1, \widetilde{Y} \sim P_2\}, \tag{9.19}$$

$\ell_2(P_1, P_2)$ is the "minimal ℓ_2-metric" defined on the set of probability measures on $\mathbb{R}^d$ with finite second moments. Optimal couplings are therefore those couplings which minimize the L^2-distance.

An optimal coupling X, Y of P_1, P_2 is denoted by $X \underset{\mathrm{oc}}{\sim} Y$.

(d) The maximal correlation risk measure has the following interpretation. It describes the maximal possible risk over all possible distributional versions $\widetilde{X} \sim X$ averaged over all directions y according to the scenario measure $\mu = P^Y$. This interpretation results from the presentation

$$\widehat{\Psi}_Y(X) = \sup_{\widetilde{X} \sim X} \int E(\widetilde{X}|y) \cdot y \, dP^Y(y). \tag{9.20}$$

Here $E(\widetilde{X}|y) \cdot y$ is the conditional risk of $\widetilde{X}$ in direction $y \in \mathbb{R}_+^d$. Thus $\widehat{\Psi}_Y(X)$ describes the average risk of X in random direction Y. $\Diamond$

The following result is an extension of the Kusuoka representation theorem, giving the representation of law invariant coherent risk measures on L^∞ in $d = 1$, to the case of law invariant convex risk measures on L_d^p.

Theorem 9.4 (Law invariant convex risk measures on L_d^p, Rü (2006, 2012a)). *Let $\Psi : L_d^p \to \mathbb{R} \cup \{\infty\}$ be a proper, convex risk measure on L_d^p with penalty function α. Then the following are equivalent:*

(1) Ψ is law invariant

(2) The penalty function $\alpha : \mathcal{D}_{d,p} \to \mathbb{R} \cup \{\infty\}$ is law invariant, i.e. $\alpha(Y) = \alpha(\widetilde{Y})$ if $Y \sim \widetilde{Y}$

(3) Ψ has a representation of the form

$$\Psi(X) = \sup_{\mu \in \mathcal{D}_{d,p}^\ell} (\Psi_\mu(X) - \gamma(\mu)) \tag{9.21}$$

with penalty function $\gamma : \mathcal{D}_{d,p}^\ell \rightarrow \mathbb{R} \cup \{\infty\}$ *which can be chosen as*

$$\gamma(\mu) = \sup_{X \in \mathcal{A}_\Psi} \widehat{\Psi}_Y(X) = \sup_{X \in \mathcal{A}_\Psi} \Psi_\mu(X).$$

Proof. By Remark 8.11 a proper, law invariant convex risk measure Ψ on L_d^p is Fatou-continuous. Therefore by Theorem 8.23 in case $p < \infty$ and by Theorem 8.31 in case $p = \infty$ Ψ has a presentation of the form

$$\Psi(X) = \sup_{Q \in \mathcal{Q}_{d,p}} (E_Q X - \beta(Q)), \quad \beta = \Psi^*,$$

$$= \sup_{Y \in \mathcal{D}_{d,p}} (EX \cdot Y - \alpha(Y)) \tag{9.22}$$

where in case $p = \infty$ the scenario set can be replaced by the σ-additive scenario measures $\mathcal{M}_{d,\infty}$ respectively the scenario densities $Y \in \mathcal{D}_{d,1}$. If Ψ is law invariant, then by Proposition 9.1

$$\Psi(X) = \sup_{\widetilde{X} \sim X} \sup_{Y \in \mathcal{D}_{d,p}} (E\widetilde{X} \cdot Y - \alpha(Y))$$

$$= \sup_{Y \in \mathcal{D}_{d,p}} (\widehat{\Psi}_Y(X) - \alpha(Y))$$

$$= \sup_{\mu \in \mathcal{D}_{d,p}^\ell} (\Psi_\mu(X) - \gamma(\mu)),$$

i.e. the representation in (9.21) holds and γ can be chosen as $\gamma(\mu) = \inf_{\widetilde{Y} \sim \mu} \alpha(\widetilde{Y})$. Furthermore,

$$\alpha(Y) = \sup_{X \in L_d^p} (EX \cdot Y - \Psi(X))$$

$$= \sup_{X \in L_d^p} \sup_{\widetilde{X} \sim X} (E\widetilde{X} \cdot Y - \Psi(\widetilde{X}))$$

$$= \sup_{X \in L_d^p} (\widehat{\Psi}_Y(X) - \Psi(X))$$

$$= \alpha(\widetilde{Y}) \text{ for all } \widetilde{Y} \sim Y,$$

since $\widehat{\Psi}_Y = \widehat{\Psi}_{\widetilde{Y}} = \Psi_\mu$. Thus α is law invariant.

If conversely α is law invariant, then for $\widetilde{X} \sim X$ using Proposition 9.1 it holds that

$$\Psi(X) = \sup_{Y \in \mathcal{D}_{d,p}} (EX \cdot Y - \alpha(Y))$$

$$= \sup_{Y \in \mathcal{D}_{d,p}} (\sup_{\widetilde{Y} \sim Y} EX \cdot \widetilde{Y} - \alpha(Y))$$

$$= \sup_{Y \in \mathcal{D}_{d,p}} (\widehat{\Psi}_Y(X) - \alpha(X))$$

$$= \sup_{Y \in \mathcal{D}_{d,p}} (\widehat{\Psi}_Y(\widetilde{X}) - \alpha(Y))$$

$$= \Psi(\widetilde{X}).$$

Thus Ψ is law invariant and the presentation of Ψ in (9.21) holds.   □

As corollary the law invariant coherent risk measures have a representation determined by a class of scenario distributions $A \subset \mathcal{D}_{d,p}^{\ell}$.

Corollary 9.5 (Law invariant coherent risk measures). *The class of proper, law invariant coherent risk measures is given by* $\{\Psi_A; \ A \subset \mathcal{D}_{d,p}^{\ell}\}$ *where*

$$\Psi_A(X) = \sup_{\mu \in A} \Psi_\mu(X). \tag{9.23}$$

Remark 9.6. (a) Attainment result: Under the condition of Theorem 8.25 in the coherent case respectively under the condition $I_\varrho \neq \phi$ in the convex case, $p < \infty$ as in Theorem 8.26 respectively in the convex case $p = \infty$ as in Theorem 8.32 the sup in the representation (9.23) is attained by some $Y_0 \in \mathcal{D}_{d,p}$ which implies that the sup in (9.21) is also attained in $\mu_0 = P^{Y_0}$ and we have a representation of the form

$$\Psi(X) = \max_{\mu \in \mathcal{D}_{d,p}^{\ell}} (\Psi_\mu(X) - \gamma(\mu)) \tag{9.24}$$

in the convex case respectively

$$\Psi(X) = \max_{\mu \in A} \Psi_\mu(X) \tag{9.25}$$

in the coherent case.

For finite lower semicontinuous coherent risk measures Ψ the scenario set $A \subset \mathcal{D}_{d,p}^{\ell}$ can be chosen weakly compact in L^q (identifying $\mu \in A$ with a corresponding scenario density in Y). Similarly for convex law invariant risk measures the scenario set can be chosen weakly compact if the convex conjugate Ψ^* of Ψ is bounded above on its support. By Theorem 8.32 the attainment result for risk measures on L_d^p is equivalent to continuity from below and equivalently to Fatou-continuity and weak compactness of $\{\Psi^* \leq c\}$ for any $c > \Psi(0)$. We call a risk measure "strongly continuous" if it has a weakly compact representation set.

(b) **Convex consistency:** By Theorem 8.13 convex law invariant risk measures Ψ on L_d^p are consistent w.r.t. the increasing convex order

$$X_1 \preceq_{\mathrm{icx}} X_2 \text{ implies } \Psi(X_1) \leq \Psi(X_2). \tag{9.26}$$

Obviously also the converse conclusion holds: If Ψ is a risk measure and Ψ is $\leq_{icx}$-consistent then Ψ is a convex, law invariant risk measure.   ◊

From the representation in (9.24) and (9.25) of law invariant convex respectively coherent risk measures we obtain that the magnitude of risk $\Psi(X)$ is evaluated by a worst case scenario measure $\mu_0 \in \mathcal{D}_{d,p}^{\ell}$ in (9.24) resp $\mu_0 \in A$ in (9.25). This leads to the following definition.

Definition 9.7 (Worst case scenario measure). Let Ψ be a law invariant convex risk measure with representation $\Psi(X) = \sup_{\mu \in A}(\Psi_\mu(X) - \alpha(\mu))$ as in (9.21). Then $\mu_0 \in A$ is called the "worst case scenario measure" if

$$\Psi(X) = \Psi_{\mu_0}(X) - \alpha(\mu_0).$$

By the representation theorem for law invariant convex risk measures in Theorem 9.4 a basic role is played by the max correlation risk measures which are defined via the classical version of the mass transportation problem (see Remark 9.3d). In order to determine the max correlation risk measure $\Psi_\mu(X)$ we have to determine optimal couplings between $P_1 = \mu$ and $P_2 = P^X$ the distribution of X. The basic result of optimal L^2-mass transportation describing these optimal couplings is the following theorem.

Theorem 9.8 (Optimal L^2-mass transportation).   *Let P_1, $P_2 \in M^1(\mathbb{R}^d, \mathcal{B}^d)$ be probability measures on $\mathbb{R}^d$ with finite second moments, $\int \|x\|^2 dP_i(x) < \infty$.*

(a) There exists an optimal L^2-coupling of P_1, P_2, i.e. $\exists X_i \sim P_i$, $i = 1, 2$ such that

$$EX_1 \cdot X_2 = \sup_{Y_i \sim P_i} EY_1 \cdot Y_2 \qquad (9.27)$$

or equivalently

$$E\|X_1 - X_2\|^2 = \inf_{Y_i \sim P_i} E\|Y_1 - Y_2\|^2. \qquad (9.28)$$

(b) $X_i \sim P_i$, $i = 1, 2$ are an optimal coupling of P_1, P_2 if and only if there exists a convex, lsc function $f \in L^1(P_1)$ such that X_2 is a.s. in the subgradient of f at X_1,

$$X_2 \in \partial f(X_1) \text{ a.s.} \qquad (9.29)$$

(c) If $P_1 \ll \lambda^d$ then for f as in b) it holds that $\partial f(X) = \{\nabla f(X)\}$ a.s. and the pair $(X, \nabla f(X))$ is an optimal coupling.

(d) If $P_1 \ll \lambda^d$, then there exists a P_1-a.s. unique gradient ∇f of a convex lsc function f with

$$P_1^{\nabla f} = P_2. \qquad (9.30)$$

Remark 9.9. (a) Historical note on the mass transportation problem: The existence result is a consequence of compactness of $M(P_1, P_2)$ in the topology of weak convergence and lower semicontinuity of $P \to \int |x_1 - x_2|^2 dP(x_1, x_2)$. The main part of Theorem 9.8 is part (b) on the characterization of optimal

couplings. The important necessary part is due to Rü and Rachev (1990) and
Brenier (1991). This part is a consequence of the duality theorem for Fréchet
bounds as in Chapter 2. The sufficiency in part (b) is due to earlier work in Knott
and Smith (1984, 1987).

Part (c) follows as special case from the necessary direction in part (b) by the
classical Rademacher theorem on a.s. differentiability of convex functions using
the simple to verify existence result in part (a). Part (c) implies that $(X, \nabla f(X))$
is also a solution of the related "Monge problem" to solve for $X \sim P_1$,

$$E\|X - T(X)\|^2 = \inf$$

over all mappings T which transfer mass P_1 to P_2, i.e. $T(X) \sim P_2$. By part (c)
one obtains the existence and the structure of a solution of the Monge problem.
This result has been of fundamental importance in the treatment of nonlinear
PDEs, as the Monge–Ampère equation, in modern analysis (see Villani (2003,
2009)). Part (d) is due to Brenier (1991).

(b) **General coupling functions:** An extension of the characterization of optimal
couplings to the case of general coupling functions $c(x, y)$ (here formulated as
sup-problem) is to determine

$$\sup\left\{\int c(x, y)d\mu(x, y); \quad \mu \in M(P_1, P_2)\right\}. \tag{9.31}$$

A solution to this problem was given in Rü (1991a, 1995b).
A pair $X \sim P_1$, $Y \sim P_2$ is an "optimal c-coupling", i.e. $\mu = P^{(X,Y)}$
solves (9.31) if and only if

$$Y \in \partial_c f(X) \text{ a.s.,} \tag{9.32}$$

where f is a "c-convex" function, i.e. f is a function of the form

$$f(x) = \sup_{y \in A \subset \mathbb{R}^d} (c(x, y) + a(y)).$$

$\partial_c f$ is the set of "c-subgradients" of f, i.e.

$$\partial_c f(x) = \left\{y \in \mathbb{R}^d; f(z) - f(x) \geq c(z, y) - c(x, y) \text{ for all } z\right\}.$$

For these notions we refer to the above literature or to Villani (2003, 2009).
A presentation of the development and applications in probability is given in
Rachev and Rü (1998a, 1998b) and Rü (2007).

(c) There have been several extensions of the mass transportation theory, dismissing
for example with the second moment assumption in Theorem 9.8 and concerning the generality of the cost function c (see Rachev and Rü (1998a, 1998b) and
Villani (2003, 2009), and references given there).   ◊

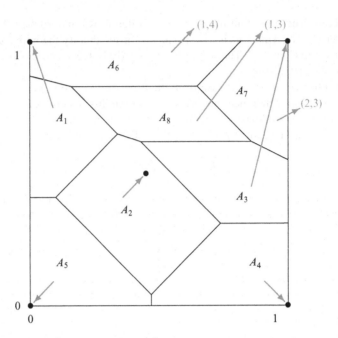

Figure 9.1 Optimal L^2-coupling between $U_{[0,1]}^2$ and a discrete distribution on $n = 8$ points

Example 9.10 (Optimal coupling of uniform and discrete distribution). The following example gives the exact optimal coupling between $P_1 = U_{[0,1]^2}$, the uniform distribution on $[0,1]^2$, and a discrete distribution $Q = \sum_{j=1}^{n} \alpha_j \varepsilon_j$. Note that for general cost c by the criterion in (9.29) we can restrict to c-convex functions of the form $f(x) = \sup_{j \leq n}(c(x, x_j) + a_j)$ and the subgradients of f in x are given by

$$\{x \in [0,1]^2; x_j \in \partial_c f(x)\} = \{x; \ f(x) = c(x, x_j) + a_j\}$$

$$=: A_j.$$

The points in the Voronoi cells A_j are optimally coupled with their subgradients x_j. The problem is to find the shifts a_j, such that $P(A_j) = \alpha_j$. The example in Figure 9.1 with $n = 8$ and $c(x, y) = \|x - y\|^2$ has been calculated in explicit form in Rü and Uckelmann (2000).

For larger n there is an efficient gradient descent algorithm for this problem. $\Diamond$

9.2   Multivariate Comonotonicity and the n-Coupling Problem

By the basic representation result Theorem 9.4 law invariant convex risk measures on L_d^p are based on the max correlation risk measures $\Psi_\mu(X)$. These are given by $\Psi_\mu(X) = E\widetilde{X} \cdot \widetilde{Y}$ for an optimal L^2-coupling $\widetilde{X} \underset{\mathrm{oc}}{\sim} \widetilde{Y}$ of $P_1 = P^X$ and $P_2 = \mu$.

The optimal coupling $\widetilde{X}$, $\widetilde{Y}$ is characterized in the optimal L^2-mass transportation problem in Theorem 9.8.

In Section 9.3 we determine worst case dependence structures for the joint portfolio $\Psi(\sum_{i=1}^{n} X_i)$ and some law invariant, convex risk measure Ψ. The L^2-risk $\Psi(X) = (E\|X\|^2)^{1/2}$ with $\| \| = \| \|_2$ the Euclidean norm is a coherent but not translation invariant and not monotone risk measure. The worst case dependence problem for the L^2-risk leads to an extension of the L^2-mass transportation problem which is called the "n-coupling problem":

$$E\left\| \sum_{i=1}^{n} X_i \right\|^2 = \sup_{X_i \sim P_i}, \tag{9.33}$$

where $P_1, \ldots, P_n \in M^1(\mathbb{R}^d, \mathfrak{B}^d)$ are given probability measures on $\mathbb{R}^d$ with finite second moments. In this section we will consider the n-coupling problem and related results on optimal couplings and multivariate comonotonicity.

In dimension $d = 1$ problem (9.33) is solved by the comonotonic vector

$$X^c = (F_1^{-1}(U), \ldots, F_d^{-1}(U)), \tag{9.34}$$

where F_i are the distribution functions of P_i and $U \sim U(0, 1)$. In this case all pairs (P_i, P_j) are optimally coupled by the components X_i^c, X_j^c of the comonotonic vector X^c.

Problem (9.33) is equivalent to

$$E \sum_{i<j} \|X_i - X_j\|^2 = \inf_{X_i \sim P_i} \tag{9.35}$$

and also to

$$E \sum_{i<j} X_i \cdot X_j = \sup_{X_i \sim P_i}. \tag{9.36}$$

In particular if for some coupling vector X of $P_1, \ldots, P_n$ all pairs X_i, X_j are optimal couplings of P_i, P_j, i.e. $X_i \underset{oc}{\sim} X_j$ for all $i \neq j$, then X obviously solves (9.33) and is an optimal solution of the n-coupling problem. X then would describe a worst case dependence structure w.r.t. Ψ. We then might call X an "L^2-comonotonic vector". In general such L^2-comonotonic vectors however do not exist.

Proposition 9.11 (Nonexistence of L^2-comonotonic vectors). *Let $P_i = N(0, \Sigma_i)$, $1 \leq i \leq n$, $n \geq 3$ be multivariate normal vectors with regular covariance matrices. Then there exists an L^2-comonotonic coupling $X = (X_1, \ldots, X_n)$ of $P_1, \ldots, P_n$ if and only if*

$$\Sigma_i \Sigma_j = \Sigma_j \Sigma_i \text{ for all } i \neq j. \tag{9.37}$$

Proof. The proof follows from some facts from mass transportation.

(1) If $X_i \sim N(0, \Sigma_i)$ and

$$T_{ij} := \Sigma_j^{1/2}(\Sigma_j^{1/2}\Sigma_i\Sigma_j^{1/2})\Sigma_j^{1/2}, \tag{9.38}$$

then $T_{i,j}X_i \sim N(0, \Sigma_j)$ and $(X_i, T_{i,j}X_i)$ is an optimal coupling of $N(0, \Sigma_i)$ and $N(0, \Sigma_j)$. This follows from the characterization of optimal L^2-couplings in Theorem 9.8(b).

(2) Since $P_j \ll \lambda^d$ by the uniqueness result in Theorem 9.8(c) $T_{i,j}$ is the P_1-a.s. unique optimal map from P_i to P_j.

As consequence we obtain that if $X = (X_1, \ldots, X_n)$ is a coupling of $P_1, \ldots, P_n$ such that $X_i \underset{\text{oc}}{\sim} X_j$ for all $i \neq j$, then $X_j = T_{i,j}X_i$ for all $i \neq j$. This implies that the covariance matrices Σ_i commute, i.e. (9.37) holds. $\square$

In fact the situation in Proposition 9.11 is typical and examples in $d \geq 2$ with L^2-comonotonic couplings are rare. From this example we can conclude that we cannot expect in $d \geq 2$ the existence of a notion of comonotonicity which serves as universal worst case dependence structure for law invariant convex risk measures as the comonotonicity notion does in dimension $d = 1$.

An equivalent problem to the n-coupling problem is

$$\sum_{i=1}^n E\|X_i - S_n\|^2 = \inf, \tag{9.39}$$

where $S_n = \sum_{i=1}^n X_i$ is the sum of the X_i. This equivalence let Knott and Smith (1994) to the interesting idea to reduce the n-coupling problem (9.33) to n two-coupling problems namely to construct a coupling $X_1, \ldots, X_n$ of $P_1, \ldots, P_n$ such that all X_i are optimally coupled to the (unknown) distribution of the sum P^{S_n}. Knott and Smith (1994) in fact considered the case $n = 3$ and $P_i = N(0, \Sigma_i)$, $1 \leq i \leq 3$. Their idea is called "coupling to the sum principle". A coupling $X_1, \ldots, X_n$ is conjectured to be an optimal n-coupling if

$$X_i \underset{\text{oc}}{\sim} S_n, \quad 1 \leq i \leq n, \tag{9.40}$$

i.e. if all X_i are optimally coupled to the sum. This property can be described by the notion of "μ-comonotonicity" coined in Ekeland et al. (2010).

Definition 9.12 (μ-comonotonicity). Let $X_1, \ldots, X_n$ be random vectors in $\mathbb{R}^d$ and let μ be a probability measure on $\mathbb{R}^d$. Then $X_1, \ldots, X_n$ are called "μ-comonotone" if there exists a random vector $Y \sim \mu$, such that all X_i are optimally coupled to Y, i.e. $X_i \underset{\text{oc}}{\sim} Y$, $1 \leq i \leq n$.

Remark 9.13. (a) Optimal coupling to the sum principle: The optimal coupling to the sum principle can be described as follows. Let $X_1, \ldots, X_n$ be a coupling to $P_1, \ldots, P_n$. The coupling to the sum principle says:

If $X_1, \ldots, X_n$ are μ-comonotone with $\mu = P^{S_n}$ then $X_1, \ldots, X_n$ are optimal n-couplings.

(b) Using this optimal coupling to the sum principle Knott and Smith (1994) constructed in the case $n = 3$, $P_i = N(0, \Sigma_i)$, $1 \le i \le 3$ an optimal 3-coupling in the regular normal case under the assumption that a positive definite solution of the matrix equation

$$\sum_{i=1}^{3} \left(\Sigma_0^{1/2} \Sigma_i \Sigma_0^{1/2} \right)^{1/2} = \Sigma_0 \tag{9.41}$$

can be found. This nonlinear matrix equation is a consequence of the "coupling to the sum" idea.

For the construction of an optimal triple (X, Y, Z) let T be a random vector, $T \stackrel{d}{=} N(0, \Sigma_0)$ and define

$$S_i = \Sigma_i^{1/2} \left(\Sigma_i^{1/2} \Sigma_0 \Sigma_i^{1/2} \right)^{-1/2} \Sigma_i^{1/2}. \tag{9.42}$$

S_i is the optimal coupling mapping between $N(0, \Sigma_0)$ and $N(0, \Sigma_i)$. Then defining

$$X := S_1 T, \qquad Y := S_2 T, \qquad Z := S_3 T \tag{9.43}$$

(9.41) implies that $X + Y + Z = T$ and Knott and Smith (1994) proved by analytic inequalities that this triple is optimal and

$$E \, \|T\|^2 = \mathrm{tr}\,(\Sigma_0). \tag{9.44}$$

To prove the existence of a solution of (9.42) turned out however to be difficult. For $d = 2$ there was some indication by an iterative algorithm that a solution of (9.41) and (9.42) exists. ◊

Even if the optimal coupling to the sum principle in general turned out to be not true it gave an important clue to the solution of the n-coupling problem. The following results are based on Rü and Uckelmann (2002). At first we note that optimal coupling to the sum is a necessary condition but it is in general not a sufficient condition.

Proposition 9.14 (Necessity of optimal coupling to the sum). $X_i \sim P_i$, $1 \le i \le n$, and let $X_1, \ldots, X_n$ be an optimal n-coupling for $P_1, \ldots, P_n$, then with $T_i := \sum_{j \ne i} X_j$, $T := \sum_{j=1}^{n} X_j$, X_i is optimally coupled to the sum T_i as well as to T, $1 \le i \le n$.

Proof. Consider w.l.g. the case $i = 1$. The n-coupling problem (9.33) is equivalent to (9.36) i.e. to

$$E\left(\langle X_1, T_1\rangle + \sum_{i=2}^n \left\langle X_i, \sum_{j>i} X_j\right\rangle\right) = \text{max!} \qquad (9.45)$$

The second term depends only on $X_2, \dots, X_n$. If X_1 was not optimally coupled to T_1, it would be possible to find a strict improvement of (9.45).

Furthermore, (9.45) is also equivalent to

$$E\left(\langle X_1, T\rangle + \sum_{i=2}^n \left\langle X_i, \sum_{j>i} X_j\right\rangle\right) = \text{max!}$$

(the difference depends only on the marginal distribution). Therefore, by the same argument, X_1 has to be optimally coupled to the sum T as well. $\qquad\square$

We next prove that Knott and Smith's idea of optimal coupling to the sum leads to a complete characterization of solutions in the normal case $P_i = N(0, \Sigma_i)$, $1 \le i \le n$. Note that by a simple compactness argument optimal n-couplings exist.

Theorem 9.15 (Optimal n-coupling of multivariate normal distributions). *Let $P_i = N(0, \Sigma_i)$ with $\Sigma_i > 0$ positive definite, $1 \le i \le n$.*

(a) *An optimal n-coupling $X = (X_1, \dots, X_n)$ of $P_1, \dots, P_n$ exists.*
(b) *A coupling X of $P_1, \dots, P_n$ is an optimal n-coupling if and only if $\Sigma_0 = \text{Cov } T$, $T = \sum_{j=1}^n X_j$ is a positive definite solution of*

$$\sum_{i=1}^n \left(\Sigma_0^{1/2} \Sigma_i \Sigma_0^{1/2}\right)^{1/2} = \Sigma_0. \qquad (9.46)$$

(c) *There exists a solution Σ_0 of (9.46). With*

$$S_i = \Sigma_i^{1/2}\left(\Sigma_i^{1/2}\Sigma_0\Sigma_i^{1/2}\right)^{-1/2}\Sigma_i^{1/2} \text{ and } X_i \overset{d}{=} N(0, \Sigma_i)$$

one obtains a solution of the n-coupling problem in functional form

$$X_i = S_i\, S_1^{-1} X_1 \text{ a.s.,} \quad 1 \le i \le n. \qquad (9.47)$$

Proof. Let $X = (X_1, \dots, X_n)$ be an optimal n-coupling; then we may assume w.l.g. that (X_i) are jointly normal distributed. Otherwise replace X by an n-tuple with joint normal distribution and identical covariance matrix. This implies that also $T_n = \sum_{j<n} X_j$ and $T = \sum_{j=1}^n X_j$ are normal.

By Proposition 9.14, T_n and X_n are optimally coupled i.e. (X_n, T_n) is an optimal pair for $N(0, \Sigma_n)$ and $Q := N(0, \Sigma_{T_n})$, where $\Sigma_{T_n} = \text{Cov}(T_n)$. Note that it is not obvious that Σ_{T_n} is positive definite. An optimal coupling between $N(0, \Sigma_n)$ and Q is given by the pair (X_n, AX_n) where

$$A = \Sigma_n^{-1/2} \left(\Sigma_n^{1/2} \Sigma_{T_n} \Sigma_n^{1/2} \right)^{1/2} \Sigma_n^{-1/2}.$$

Positive definiteness of Σ_n implies uniqueness of optimal pairs then (see Theorem 9.8(b) or Cuesta-Albertos et al. (1997) and Gangbo and McCann (1996)). Therefore, $T_n = A X_n$ a.s.

This implies that

$$T = \sum_{j=1}^{n} X_j = (A + I) X_n \text{ a.s.}$$

Since A is positive semidefinite and

$$\langle x, (A + I)x \rangle = \langle x, Ax \rangle + \langle x, x \rangle \geq \langle x, x \rangle > 0 \text{ for } x \neq 0,$$

$A + I$ is positive definite and, therefore,

$$\Sigma_0 = \text{Cov}(T_n + X_n) = (A + I)\Sigma_n (A + I)^T > 0.$$

Since $N(0, \Sigma_0)$ and $N(0, \Sigma_i)$ are optimally coupled by the mappings S_i, i.e. $(T, S_i T)$ is an optimal pair for $(N(0, \Sigma_0), N(0, \Sigma_i))$, and since optimal coupling to the sum is a necessary condition by Proposition 9.14 we obtain from the same uniqueness result that $X_i = S_i T$ a.s.

This implies that

$$T = \sum_{j=1}^{n} X_j = \left(\sum_{j=1}^{n} S_j \right) T, \qquad \text{i.e.} \quad \sum_{j=1}^{n} S_j = I.$$

By some simple algebra this shows that Σ_0 is a solution of equation (9.46).

For the converse direction of Theorem 9.15 the inequalities used in the proof of Knott and Smith (1994) for the case $n = 3$ can easily be extended to general n.

The existence of an optimal n-coupling is proved as in the case $n = 2$.   □

Remark 9.16. Theorem 9.15 in particular implies existence of a positive solution Σ_0 of (9.41) respectively (9.46). In order to find a positive definite solution Σ_0 of (9.46)

$$\sum_{i=1}^{n} \left(\Sigma_0^{1/2} \Sigma_i \Sigma_0^{1/2} \right)^{1/2} = \Sigma_0,$$

Knott and Smith (1994) suggest for $n = 3$ to consider the equivalent problem to find a positive definite solution K_0,

$$\sum_{i=1}^{n} (K_0 K_i^2 K_0)^{1/2} = K_0^2, \quad \text{where } K_i := \Sigma_i^{1/2}.$$

They suggested to use the iterative procedure

$$K_0^{(k+1)} = \left(\sum_{i=1}^n \left(K_0^{(k)} K_i^2 K_0^{(k)} \right)^{1/2} \right)^{1/2}. \tag{9.48}$$

It turns out by extensive simulations (for $n = 3$) with random initial matrices that the iteration converges in dimension $d = 2$ (typically one needs about 100 iteration steps for exactness up to 8 digits). But for dimension $d = 3$ only for favourable initial matrices convergence is observed. $\Diamond$

The following result says that optimal coupling to the sum is a sufficient condition for optimal n-coupling if the distribution P^T of the sum is Lebesgue continuous and the support is simply connected.

Theorem 9.17 (Coupling to the sum). *Let P_i be distributions on $\mathbb{R}^d$, $1 \le i \le n$ with finite second moments and convex supports and let $X_i \sim P_i$, $1 \le i \le n$, be such that X_i are optimally coupled to the sums $T_i = \sum_{j \ne i} X_j$. If P^T is Lebesgue-continuous, and the interior of the support of T is nonempty and simply connected, then $X = (X_1, \dots, X_n)$ is an optimal n-coupling of $(P_1, \dots, P_n)$.*

Proof. From the characterization of optimal couplings in Theorem 9.8 there exist convex functions g_i such that $T_i \in \partial g_i(X_i)$ a.s. Therefore, with $\overline{g_i}(x) = g_i(x) + \frac{1}{2}\|x\|^2$ we obtain $T = T_i + X_i \in \partial \overline{g_i}(X_i)$ and so we obtain $X_i \in \partial f_i(T)$ a.s. where $f_i = (\overline{g_i})^*$. f_i is convex and continuously differentiable on the interior of $rg(T)$ since $\overline{g_i}$ is strictly convex (see f.e. Gangbo and Święch (1998)). Since $P^T \ll \lambda^d$ this implies that

$$X_i = \nabla f_i(T) =: \Phi_i(T) \quad \text{a.s.} \tag{9.49}$$

From $T = \sum_{i=1}^n X_i = \sum \nabla f_i(T)$ and regularity of $\Phi_i = \nabla f_i$ we conclude that $\sum_{i=1}^n \nabla f_i(t) = \nabla \sum_{i=1}^n f_i(t) = t$ on the interior of the support of T. By assumption $A := \text{int}(\text{spt}\,T) \ne \phi$ and A is simply connected. Therefore, by Poincaré's lemma we can conclude that $\sum_{i=1}^n f_i(t) = \frac{1}{2}\|t\|^2$ on A.

By definition of the convex conjugate functions f_i^* for $x_i \in \mathbb{R}^d$, and with $t = \sum_{i=1}^n x_i$ it holds that

$$\langle x_i, t \rangle \le f_i(t) + f_i^*(x_i). \tag{9.50}$$

Therefore, for $t \in A$ we have

$$\|t\|^2 = \langle t, t \rangle = \left\langle \sum_i x_i, t \right\rangle$$

$$\le \sum_i f_i(t) + \sum_i f_i^*(x_i)$$

$$= \frac{1}{2}\|t\|^2 + \sum_i f_i^*(x_i), \quad \text{i.e.}$$

$$\frac{1}{2}\|t\|^2 \le \sum_i f_i^*(x_i). \tag{9.51}$$

The condition $X_i \in \partial f_i(T)$ a.s. implies that

$$\langle X_i, T \rangle = f_i(T) + f_i^*(X_i) \text{ a.s.} \tag{9.52}$$

i.e. equality holds in (9.50) and, therefore, $\frac{1}{2}||T||^2 = \sum_i f_i^*(X_i)$ a.s.

This implies that $X = (X_i)$ is an optimal n-coupling, since for any $Y_i \sim P_i$ it holds that

$$E\frac{1}{2}\left\|\sum_{i=1}^{n} Y_i\right\|^2 \le E \sum_i f_i^*(Y_i) = E \sum_i f_i^*(X_i) = \frac{1}{2}E||T||^2, \tag{9.53}$$

observing that $\sum_{i=1}^{n} Y_i$ is a.s. contained in A. This implies optimality of $(X_1, \ldots, X_n)$. $\qquad\square$

In the normal case $P_i = N(0, \Sigma_i)$ we have obtained in (9.47) that the optimal solution of the n-coupling problem is given by a Monge-solution

$$X_i = \Phi_i(X_1), \quad 1 \le i \le n, \tag{9.54}$$

i.e. a solution in functional form. This structure of optimal Monge solutions had been established for the multivariate Fréchet problem $M(P_1, \ldots, P_n)$ in Gangbo and Święch (1998). It also holds for the optimal n-coupling problem.

Theorem 9.18 (Monge solutions for the n-coupling problem). *Let P_i, $1 \le i \le n$, vanish on $(d-1)$-rectifiable sets and have finite second moments. Then there exists a Monge solution of the form $(X_1, \Phi_2(X_1), \ldots, \Phi_n(X_1))$, $X_1 \sim P_1$, of the n-coupling problem.*

Proof. Let $X = (X_1, \ldots, X_n)$ be a solution of the n-coupling problem (P_i), $X_i \sim P_i$. Then by Proposition 9.14 X_i are optimally coupled to the sums $T_i = \sum_{j \ne i} X_j$ and so by Theorem 9.8

$$T_i \in \partial f_i(X_i) \text{ a.s.}$$

for some convex functions f_i. From an extension of Rademacher's theorem on the structure of singular sets of convex functions (see Alberti (1994), Gangbo and Święch (1998)), the f_i are differentiable with the exception of a $(d-1)$-rectifiable set and, therefore, by the assumption

$$\partial f_i(X_i) = \{\nabla f_i(X_i)\} \text{ a.s.},$$

i.e. $T_i = \nabla f_i(X_i)$ a.s.

Therefore, defining

$$\overline{f_i}(x) := f_i(x) + \frac{1}{2}||x||^2$$

we obtain $\overline{f_i}$ is strictly convex, $\nabla \overline{f_i}$ exists λ^d a.s. and is invertible and

$$T = T_i + X_i = \nabla \overline{f_i}(X_i) \text{ a.s.}$$

This implies that $X_i = (\nabla \overline{f_i})^{-1}(T)$ a.s. for all i and, therefore,

$$X_i = \left(\nabla \overline{f_i} \right)^{-1} (\nabla f_1(X_1)) = \Phi_i(X_1) \text{ a.s.} \qquad (9.55)$$

which is the stated Monge solution. □

Note that the proof of Theorem 9.18 is not constructive. If we take convex functions f_i and define X_i by (9.55) we do generally not obtain optimal n-couplings. Improved constructive versions of Monge solutions are given in Rü and Uckelmann (2002).

In this section we have determined optimal n-couplings w.r.t. L^2-norm which can be considered as risk measure $\Psi(X) = (E\|X\|^2)^{1/2}$. We have seen that "$L^2$-comonotonic" optimal couplings $X_1, \ldots, X_n$ of $P_1, \ldots, P_n$ typically do not exist. An optimal coupling however is under regularity conditions determined by an optimal coupling to the sum principle or equivalently by μ-comonotonicity of $X_1, \ldots, X_n$ with $\mu = P^T$ the distribution of the sum. In the final remark we note that optimal n-couplings respectively worst case dependence structures also heavily depend on the convex law invariant risk measure Ψ chosen. Thus also from this point of view a universal "comonotone" notion in $d \geq 2$ does not exist anymore. We verify this statement by choosing as risk measure the p-norms

$$\Psi_p(X) = (E\|X\|_p^p)^{1/p}, \quad p \geq 1. \qquad (9.56)$$

Remark 9.19 (Optimal couplings in $d \geq 2$ depend on the risk measure chosen).
For $P_1, P_2 \in M^1(\mathbb{R}^d, \mathfrak{B}^d)$ define the "minimal L_p-metric"

$$\ell_p(P_1, P_2) := \inf \{(E\|X - Y\|_p^p)^{1/2}; \quad X \sim P_1, Y \sim P_2\}, \qquad (9.57)$$

where $\|x\|_p = \left(\sum_{i=1}^d |x_i|^p \right)^{1/p}$ is the p-norm on $\mathbb{R}^d$. Couplings X, Y of P_1, P_2 which solve (9.57) are called optimal L_p-couplings. In dimension $d = 1$ the optimal L_p-couplings are given by the comonotonic dependence structure independent of p, i.e. for $d = 1$ it holds that

$$\ell_p(P_1, P_2) = \left(\int_0^1 |F_1^{-1}(u) - F_2^{-1}(u)|^p du \right)^{1/p}, \qquad (9.58)$$

where F_i are the distribution functions of P_i.

For $d > 1$ the optimal L_p-couplings depend on p. In the case $d = 2$, $P_1 = U_{[0,1]^2}$, $P_2 = \sum_{i=1}^n \alpha_i \varepsilon_{x_i}$ the optimal coupling between the uniform distribution on $[0, 1]^2$ and a discrete distribution in $n = 15$ points is given in Figure 9.2 (compare Example 9.10). The optimal couplings depend heavily on the norm $\| \|_p$.

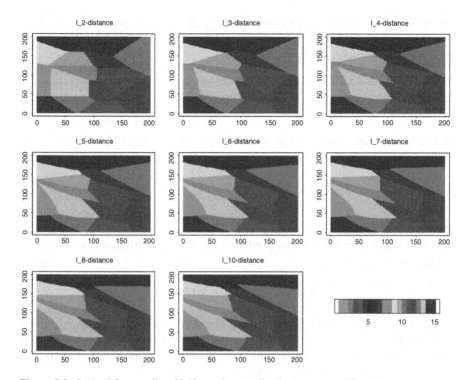

Figure 9.2 Optimal L_p-coupling: Uniform, discrete distribution on $n = 15$ points

Similar dependence on the norm arises when we consider the problem

$$\Psi_p\left(\sum_{i=1}^{n} X_i\right) = \sup_{X_i \sim P}, \tag{9.59}$$

i.e. the problem to find the worst case dependence structure of the joint portfolio w.r.t. the risk measure Ψ_p. This problem will be dealt with in Section 9.3.   ◇

9.3   Worst Case Portfolio Vectors and Diversification Effects

By the results in Section 9.2 there does not exist a worst case dependence structure (respectively a worst case portfolio vector) maximizing the risk $\Psi(\sum_{i=1}^{n} X_i)$ of the joint portfolio uniformly over all law invariant convex risk measures. Thus "comonotonicity" as universal worst case dependence does not exist in $d \geq 2$. In consequence in this section we consider the problem to determine to a given law invariant, convex risk measure Ψ a worst case dependence structure. The following results depend heavily on the representation results in Section 9.1 for law invariant convex risk measures on L_d^p. They are mostly given in Rü (2012a).

Definition 9.20 (Worst case portfolio vector). Let Ψ be a law invariant, convex risk measure on L_d^p. A portfolio vector $X = (X_1, \ldots, X_d) \in L_d^p$, $X_i \sim P_i$ is called a "worst case portfolio" w.r.t. Ψ (WCP$_\Psi$) if

$$\Psi\left(\sum_{i=1}^n X_i\right) = \sup_{\widetilde{X}_i \sim P_i} \Psi\left(\sum_{i=1}^n \widetilde{X}_i\right). \tag{9.60}$$

X is called a "worst case average portfolio" w.r.t. Ψ (WCAP$_\Psi$) if

$$\Psi\left(\frac{1}{n}\sum_{i=1}^n X_i\right) = \sup_{\widetilde{X}_i \sim P_i} \Psi\left(\frac{1}{n}\sum_{i=1}^n \widetilde{X}_i\right). \tag{9.61}$$

Remark 9.21 (WCAP$_\Psi$ and WCP$_\Psi$). In the case Ψ is a coherent risk measure, a portfolio vector X is a worst case portfolio w.r.t. Ψ if and only if it is a worst case average portfolio, i.e.

$$X \in \text{WCP}_\Psi \Leftrightarrow X \in \text{WCAP}_\Psi. \tag{9.62}$$

Since $\Psi\left(\sum_{i=1}^n X_i\right) = \Psi\left(\frac{1}{n}\sum_{i=1}^n nX_i\right)$ we obtain from a worst case average portfolio for $P_1^n, \ldots, P_n^n$ where $P_i^n \sim nX_i$ a worst case portfolio for $P_1, \ldots, P_n$ and conversely. Thus we can restrict to one type of worst case portfolios. For convex risk measures the WCAP$_\Psi$ property has some advantage. $\Diamond$

We need the following simple property of optimal couplings which has already been used in the coupling to the sum principle in Section 9.2.

Proposition 9.22. *Let* $X_i \sim P_i$, $1 \le i \le n$ *and let for some random vector* Y, $X_i \underset{oc}{\sim} Y$, *then*

$$\sum_{i=1}^n X_i \underset{oc}{\sim} Y.$$

Proof. We have to show

$$\sup\left\{E\left(\sum_{i=1}^n X_i\right) \cdot \widetilde{Y}; \widetilde{Y} \sim Y\right\} = E\left(\sum_{i=1}^n X_i\right) \cdot Y. \tag{9.63}$$

But this follows since

$$E\left(\sum_{i=1}^n X_i\right) \cdot \widetilde{Y} = \sum_{i=1}^n EX_i \cdot \widetilde{Y}$$

$$\le \sum_{i=1}^n EX_i \cdot Y = E\left(\sum_{i=1}^n X_i\right) \cdot Y$$

using the assumption $X_i \underset{oc}{\sim} Y$, $1 \le i \le n$. $\qquad\square$

In other words Proposition 9.22 says:

If $\mu \in M^1(\mathbb{R}^d, \mathfrak{B}^d)$ and if $X_1, \ldots, X_n$ are μ-comonotone, then

$$\sum_{i=1}^n X_i \underset{oc}{\sim} Y \quad \text{for some } Y \sim \mu. \tag{9.64}$$

As consequence of this observation we obtain the following characterization of the worst case dependence structure of max correlation risk measures (see Ekeland et al. (2010, Theorem 1) in case $p = q = 2$).

Proposition 9.23. $(X_1, \ldots, X_n)$ *is a worst case dependence structure for* Ψ_μ *if and only if* $X_1, \ldots, X_n$ *are* μ*-comonotone.*

Further the worst case risk of the joint portfolio is given by:

$$\sup_{\widetilde{X}_i \sim X_i} \Psi_\mu\left(\sum_{i=1}^n \widetilde{X}_i\right) = \Psi_\mu\left(\sum_{i=1}^n X_i\right) = \sum_{i=1}^n \Psi_\mu(X_i). \tag{9.65}$$

For general law invariant convex risk measures Ψ we have by Theorem 9.4 a representation of the form

$$\Psi(X) = \sup_{\mu \in A}(\Psi_\mu(x) - \alpha(\mu)), \quad X \in L_d^p, \tag{9.66}$$

where $A \subset \mathcal{D}_{d,p}^\ell$ is a weakly closed set of scenario measures. For coherent, law invariant risk measures this representation simplifies to the form in (9.66) with $\alpha(\mu) = 0$,

$$\Psi(X) = \sup_{\mu \in A} \Psi_\mu(X). \tag{9.67}$$

A set of conditions is given in Remark 9.6, which implies that the sup in (9.66) and (9.67) is attained.

Definition 9.24 (Worst case scenario).   For a risk measure Ψ as in (9.66) respectively (9.67) and a portfolio $X_i \in L_d^p$, $1 \le i \le n$ we define the "average risk functional"

$$F(\mu) := \frac{1}{n}\sum_{i=1}^n \Psi_\mu(X_i) - \alpha(\mu), \quad \mu \in A. \tag{9.68}$$

A scenario measure $\mu_0 \in A$ is called a "worst case scenario measure" if it maximizes the average risk functional, i.e.

$$F(\mu_0) = \sup_{\mu \in A} F(\mu). \tag{9.69}$$

Note that the average risk functional $F(\mu)$ depends only on the marginal distribution P_i, it does not involve their joint distribution of $(X_1, \ldots, X_n)$. Under the attainment conditions in Remark 9.6 a worst case scenario measure exists.

Theorem 9.25 (Worst case average joint portfolio).   *Let $X_i \sim P_i$, $1 \le i \le n$ be a portfolio in L_d^p and consider a finite, convex, law invariant risk measure Ψ on L_d^p as in (9.66).*

(a) The worst case risk is given by the sup of the average risk functional

$$\sup_{\widetilde{X}_i \sim X_i} \Psi\left(\frac{1}{n}\sum_{i=1}^{n}\widetilde{X}_i\right) = \sup_{\mu \in A} F(\mu). \tag{9.70}$$

(b) Let $X_i^* \sim P_i$, $1 \le i \le n$ be a coupling of P_i and let μ_0 be a worst case scenario measure. Then $X_1^*,\ldots,X_n^*$ is a worst case average portfolio if and only if $X_1^*,\ldots,X_n^*$ are μ_0-comonotone.

(c) If Ψ is strongly continuous then there exists a worst case scenario measure $\mu_0 \in A$ such that

$$F(\mu_0) = \sup_{\mu \in A} F(\mu). \tag{9.71}$$

Proof. (a) Generally for any $\widetilde{X}_i \sim X_i$ and for any law invariant convex risk measures Ψ holds the inequality

$$\Psi\left(\frac{1}{n}\sum_{i=1}^{n}\widetilde{X}_i\right) \le \frac{1}{n}\sum_{i=1}^{n}\Psi(\widetilde{X}_i) = \frac{1}{n}\sum_{i=1}^{n}\Psi(X_i). \tag{9.72}$$

Furthermore, by Proposition 9.23 and by the representation in (9.66) it holds that

$$
\begin{aligned}
\sup_{\widetilde{X}_i \sim X_i} \Psi\left(\frac{1}{n}\sum_{i=1}^{n}\widetilde{X}_i\right) &= \sup_{\widetilde{X}_i \sim X_i} \sup_{\mu \in A}\left(\Psi_\mu\left(\frac{1}{n}\sum_{i=1}^{n}\widetilde{X}_i\right) - \alpha(\mu)\right) \\
&= \sup_{\mu \in A}\left(\sup_{\widetilde{X}_i \sim X_i} \Psi_\mu\left(\frac{1}{n}\sum_{i=1}^{n}\widetilde{X}_i\right) - \alpha(\mu)\right) \\
&= \sup_{\mu \in A}\left(\frac{1}{n}\sum_{i=1}^{n}\Psi_\mu(X_i) - \alpha(\mu)\right) \\
&= \sup_{\mu \in A} F(\mu). \tag{9.73}
\end{aligned}
$$

(b) If μ_0 is a worst case scenario measure and if $X_i^* \sim X_i$ are μ_0-comonotone, i.e. $X_i^* \underset{\mathrm{oc}}{\sim} Y$, $1 \le i \le n$, for some $Y \sim \mu_0$, then by Proposition 9.22

$$\frac{1}{n}\sum_{i=1}^{n}X_i^* \underset{\mathrm{oc}}{\sim} Y$$

and

$$\Psi\left(\frac{1}{n}\sum_{i=1}^{n}X_i^*\right) = F(\mu_0) = \sup_{\mu\in A}F(\mu).$$

Therefore, (X_i^*) is a worst case portfolio. Conversely if (X_i^*) is a worst case portfolio then

$$\sup_{\mu\in A}F(\mu) = F(\mu_0) = \Psi_{\mu_0}\left(\frac{1}{n}\sum_{i=1}^{n}X_i^*\right) - \alpha(\mu_0)$$

$$\leq \frac{1}{n}\sum_{i=1}^{n}\Psi_{\mu_0}(X_i) - \alpha(\mu_0). \tag{9.74}$$

Equality in (9.74) implies that, for some $Y \sim \mu_0$,

$$EX_i^* \cdot Y = \sup_{\widetilde{X}_i \sim P_i} E\widetilde{X}_i \cdot Y, \quad 1 \leq i \leq n$$

and, therefore, all X_i are optimally coupled to the same Y, i.e. they are μ_0-comonotone. If Ψ is strongly continuous, then the scenario set A is weakly compact in $\mathcal{M}_{d,p}$. Since the function $\mu \to \Psi_\mu(X) = \sup\{EX \cdot \widetilde{Y}; \widetilde{Y} \sim \mu\}$ is usc in the weak topology on L_d^q it follows that the sup in a) is attained at some $\mu_0 \in A$. Thus a worst case scenario measure $\mu_0 \in A$ exists. $\square$

Remark 9.26 (Worst case total risk).   An analogue result to Theorem 9.25 in the case of coherent risk measure Ψ holds for the worst case joint portfolio WCP$_\psi$. Defining the "total risk functional" $F_c(\mu) := \sum_{i=1}^{n}\Psi_\mu(X_i)$, then the worst case risk is given by

$$\sup_{\widetilde{X}_i \sim P_i}\Psi\left(\sum_{i=1}^{n}\widetilde{X}_i\right) = \sup_{\mu\in A}F_c(\mu). \tag{9.75}$$

If μ_0 is a worst case scenario measure i.e.

$$F_c(\mu_0) = \sup_{\mu\in A}F_c(\mu), \tag{9.76}$$

then $X_1^*, \ldots, X_n^*$ is a worst case dependence structure if and only if $X_1^*, \ldots, X_n^*$ is comonotone w.r.t. μ_0.

For a convex, law invariant risk measure the same result holds for the worst case total cost,

$$\sup_{\widetilde{X}_i \sim P_i}\Psi\left(\sum_{i=1}^{n}\widetilde{X}_i\right) = \sup_{\mu\in A}F_n(\mu) \tag{9.77}$$

with

$$F_n(\mu) = \frac{1}{n} \sum_{i=1}^{n} \Psi_\mu(n X_i) - \alpha(\mu) = \sum_{i=1}^{n} \Psi_\mu(X_i) - \alpha(\mu). \qquad (9.78)$$
$$\diamond$$

For general convex risk measures on L_d^p

$$D_a^\Psi(X) := \frac{1}{n} \sum_{i=1}^{n} \Psi(X_i) - \Psi\left(\frac{1}{n} \sum_{i=1}^{n} X_i\right) \qquad (9.79)$$

is called the "average diversification effect" and

$$D^\Psi(X) := \sum_{i=1}^{n} \Psi(X_i) - \Psi\left(\sum_{i=1}^{n} X_i\right) \qquad (9.80)$$

the "diversification effect" of the portfolio $X = (X_i)$. By convexity the diversification effects are positive

$$D_a^\Psi(X) \geq 0, \quad D^\Psi(X) \geq 0. \qquad (9.81)$$

For "typical" convex law invariant risk measures the "worst case diversification effects"

$$D_{w,a}^\Psi(X) := \sup_{\widetilde{X}_i \sim X_i} D_a(\widetilde{X}) \quad \text{respectively} \quad D_w^\Psi(X) := \sup_{\widetilde{X}_i \sim X_i} D^\Psi(\widetilde{X}) \qquad (9.82)$$

are strictly positive. An exception are the max correlation risk measures Ψ_μ. By Proposition 9.23 the worst case diversification effect of max correlation risk measures is zero. It is interesting that these are essentially the only risk measures such that the worst case diversification effect is zero. This converse result was proved in Ekeland et al. (2010) for convex law invariant risk measures on L_d^2. A simplified proof for strongly continuous convex risk measures was given in Rü (2012a) and a detailed result in the case L_d^∞ was given in Ekeland and Schachermayer (2011). The following result is formulated for strongly continuous convex law invariant risk measures, i.e. the scenario set A can be chosen weakly compact. In particular finite, coherent, law invariant risk measures are strongly continuous.

Theorem 9.27 (Worst case diversification effect). *Let Ψ be a strongly continuous, convex law invariant risk measure on L_d^p. Then it it holds that:*
Ψ has no worst case diversification effect, i.e., for all portfolios (X_i) it holds that

$$\sup_{\widetilde{X}_i \sim X_i} \Psi\left(\frac{1}{n} \sum_{i=1}^{n} \widetilde{X}_i\right) = \frac{1}{n} \sum_{i=1}^{n} \Psi(X_i) \qquad (9.83)$$

if and only if Ψ *is a translated max correlation risk measure,*

$$\Psi = \Psi_\mu - \alpha(\mu)$$

for some scenario measure $\mu \in \mathcal{M}_{d,p}$ *and* $\alpha(\mu) \in \mathbb{R}^1$.

Proof. By Proposition 9.23 any translated max correlation risk measure has no worst case diversification effect.

For the converse direction assume that Ψ has no worst case diversification effect. Define

$$\mathcal{M}(\Psi, X) = \{\mu \in A;\ \Psi_\mu(X) - \alpha(\mu) = \Psi(X) = \sup_{\widetilde{\mu} \in A}(\Psi_{\widetilde{\mu}}(X) - \alpha(\widetilde{\mu}))\}$$

$$= \mathcal{M}(\Psi, P^X). \tag{9.84}$$

Since for any $X \in L_d^p$ the mapping $\widetilde{\mu} \to \Psi_{\widetilde{\mu}}(X) - \alpha(\widetilde{\mu})$ is usc with respect to the weak topology on $\mathcal{D}_d^q$ and since $A \subset \mathcal{M}_{d,p}^\ell$ is weakly compact it follows that $\mathcal{M}(\Psi, X) \neq \phi$ is a nonempty closed subset of A.

For $X_1, \ldots, X_n \in L_d^p$ with distributions $P_1, \ldots, P_n \in \mathcal{M}_{d,p}$ let $\mu_0 \in A$ be a worst case scenario measure for the portfolio (X_i). Thus from our assumption (9.83) we obtain

$$\sup_{\widetilde{X}_i \sim X_i} \Psi\left(\frac{1}{n}\sum_{i=1}^n \widetilde{X}_i\right) = F(\mu_0) = \frac{1}{n}\sum_{i=1}^n(\Psi_{\mu_0}(X_i) - \alpha(\mu_0))$$

$$= \frac{1}{n}\sum_{i=1}^n \Psi(X_i). \tag{9.85}$$

This implies that $\Psi(X_i) = \Psi_{\mu_0}(X_i) - \alpha(\mu_0)$ for $1 \le i \le n$ and thus

$$\mu_0 \in \bigcap_{i=1}^n \mathcal{M}(\Psi, X_i) = \bigcap_{i=1}^n \mathcal{M}(\Psi, P_i),$$

i.e. finite intersections of $\mathcal{M}(\Psi, P_i)$, $P_i \in \mathcal{M}_{d,p}$, $1 \le i \le n$ are nonempty. By weak compactness of A this implies

$$\bigcap_{P \in \mathcal{M}_{d,p}} \mathcal{M}(\Psi, P) \neq \phi.$$

Thus there exists some $\mu \in A$ such that

$$\Psi_\mu(X) - \alpha(\mu) = \sup_{\widetilde{\mu} \in A}(\Psi_{\widetilde{\mu}}(X) - \alpha(\widetilde{\mu})) = \Psi(X)$$

i.e., Ψ is a translated max correlation risk measure. $\qquad\square$

9.4 Examples of Worst Case Risk Portfolios and Worst Case Diversification Effects

By Theorem 9.25 the worst case portfolio (dependence) structure for portfolio distributions $P_1, \ldots, P_n$ and a risk measure Ψ is given by comonotone random vectors $X_1, \ldots, X_n$, $X_i \sim P_i$ with respect to a worst case scenario measure $\mu_0 \in A$. By the basic characterization of optimal couplings w.r.t. L^2-distance in Theorem 9.8 it holds that:

$X_i \sim P_i$, $Y \sim \mu_0$ is an optimal coupling, $X_i \underset{\text{oc}}{\sim} Y$ if and only if X_i lies a.s. in the subgradient

$$X_i \in \partial f_i(Y) \quad \text{a.s.} \tag{9.86}$$

of some lsc convex function f_i. Criterion (9.86) holds without any continuity assumptions on the scenario measure μ_0 and Y can be chosen independent of the index i. If μ_0 is absolutely continuous the subgradient reduces to the gradient a.s. and then (9.86) is equivalent to

$$X_i = \nabla f_i(Y) \quad \text{a.s.,} \tag{9.87}$$

X_i is given by the gradient of f_i applied to Y. Thus determination of the worst case portfolio structure is reduced by Theorem 9.8 to an optimal coupling problem w.r.t. the worst case scenario measure μ_0 as in (9.86) and (9.87).

The examples of worst case portfolios and of worst case diversification effects in this section are given in Rü (2012a).

(a) Discrete distributions and approximation. For the numerical analysis it is important that the optimal coupling problem can be approximated by optimal couplings between discrete distributions. For the case that $Q = \Sigma_i \alpha_i \varepsilon_{\{y_i\}}$ is discrete, where Q stands for some portfolio measure P_i, one can restrict in (9.86) and (9.87) to convex functions f of the form

$$f(x) = \max_i (\langle x, y_i \rangle + a_i) = f_{(a_i)}(x) \tag{9.88}$$

Then with $A_i := \{x; \ f(x) = \langle x, y_i \rangle + a_i\}$ it holds that

$$y_i \in \partial f(x) \text{ if and only if } x \in A_i, \tag{9.89}$$

see Example 9.10. The optimal shifts a_i can be determined by the condition

$$P(A_i) = \alpha_i. \tag{9.90}$$

Numerically this can be done most efficiently by a gradient approach to the minimization of the convex function of the shifts (a_i).

$$f_{(a_i)}(x) - \sum_i \alpha_i a_i = \inf_{(a_i)}, \tag{9.91}$$

as was observed in a related problem on the combinatorial Voronoi type partitioning problem in Aurenhammer et al. (1998), see also Rü and Uckelmann (2000) and Ekeland et al. (2010). The solutions (a_i) of the optimization problem (9.91) determine the optimal shifts (a_i) and thus by (9.86) respectively (9.87) they determine the optimal couplings $X \sim Q, Y \sim \mu$ by the rule

$$X = y_i \quad \text{implies} \quad Y \in A_i. \tag{9.92}$$

If μ is absolutely continuous, then (9.92) determines X uniquely as a function of Y: a.s. it holds that

$$X = y_i \quad \text{if and only if} \quad Y \in A_i. \tag{9.93}$$

If μ is not continuous, then X has to be chosen on the boundaries of A_i such that additionally (9.90) holds (which is typically an easy task).

The procedure above allows a numerical solution of the optimal coupling problem and has been applied successfully in a series of examples (see Rü and Uckelmann (2000) and Ekeland et al. (2010)). Iteratively applying this procedure to all pairs (P_i, μ) we obtain as a result approximatively a worst case portfolio $X_1^*, \ldots, X_n^*$.

(b) Location – scale families, elliptical distributions. For a random vector $X \in \mathbb{R}^d$ with distribution Q, $X \sim Q$ consider the generated location scale family

$$\mathcal{Q} := \left\{ Q_{a,B}; \ B \in \mathcal{A}, a \in \mathbb{R}^d \right\} \tag{9.94}$$

where $Q_{a,B} \sim X_{a,B} := BX + a$ and where $\mathcal{A}$ is some set of $d \times d$ scaling matrices. Consider the scenario measure $\mu = Q \equiv Q_{0,I}$, $X \sim Q$ and assume that the portfolio distributions $P_i = Q_{a_i, B_i} \in \mathcal{Q}, 1 \leq i \leq n$ are in the generated scale family $\mathcal{Q}$.

(b$_1$) If $\mathcal{A} \subset NN(d)$ i.e. $\mathcal{A}$ lies in the class of positive semidefinite matrices then by the optimal coupling criterion (9.86) it holds that

$$X_i := X_{a_i, B_i} \underset{\text{oc}}{\sim} X, \quad 1 \leq i \leq n \tag{9.95}$$

and

$$X_1, \ldots, X_n \text{ are } \mu\text{-comonotone.} \tag{9.96}$$

Further in this case the worst case risk of the portfolio $P_1, \ldots, P_n$ w.r.t. Ψ_μ is given by

$$\sup_{\widetilde{X}_i \sim X_i} \Psi_\mu \left(\sum_{i=1}^n \widetilde{X}_i \right) = \Psi_\mu \left(\sum_{i=1}^n X_i \right) = \operatorname{tr} \left(\left(\sum_{i=1}^n B_i \right) \Sigma \right) \tag{9.97}$$

with $\Sigma = \operatorname{Cov} X$ the covariance matrix of $X \sim Q$ and tr the trace operator.

(b$_2$)   Assume that the basic measure Q in (b$_1$) is invariant under orthogonal transformations like e.g. the normal distribution $N(0, I)$ or the uniform distribution on a ball around 0. Then we can extend in (b$_1$) to general affine linear transformations $Q_{a,B} \sim a + BX$, $X \sim Q$, $B \in A = M(d, \mathbb{R})$. By the polar factorization theorem it holds that

$$B = PO \tag{9.98}$$

where P is positive semidefinite and O is orthogonal. Therefore,

$$BX \sim POX \sim PY, \tag{9.99}$$

where $Y := OX$, $X \sim Y$. Thus the optimal coupling problem in this case is reduced to the optimal coupling in the positive semidefinite case.

Interesting examples of (b$_1$), (b$_2$) are multivariate normal distributions $N(\mu, \Sigma)$, uniform distributions on ellipses and general elliptical distributions. The optimal coupling results available for multivariate normal distributions (see Proposition 9.11) extend in the same form to these scale families. Thus for a scenario measure $\mu \sim Q$ and $P_i \in Q$ we explicitly obtain the worst case dependence structure. In terms of the covariances Σ_i of P_i, Σ_0 of Q and a scenario vector $T \sim Q$ the worst case portfolio for location scale families is given by

$$X_i = S_i T \tag{9.100}$$

where $S_i = \Sigma_i^{1/2}(\Sigma_i^{1/2}\Sigma_0\Sigma_i^{1/2})^{-1/2}\Sigma_i^{1/2}$.

If the class of scenario measures A is a subclass of Q, then the determination of the worst case scenario measure reduces to a standard optimization problem of the form

$$\mathrm{tr}\left[\left(\sum_{i=1}^{n} S_i^\top\right)B\Sigma_0\right] = \sup_{B \in A} \tag{9.101}$$

where $\Sigma_0 = \mathrm{Cov}(T)$ is the covariance matrix of Q.

(c) Coupling to the sum. In some cases even if the explicit representation of the convex risk measures is not known explicitly it is possible to determine the worst case dependence structure and the worst case scenario measure. We consider as an example the L^2-variation risk measure

$$\Psi(X) = (E\|X\|^2)^{1/2} \tag{9.102}$$

where $\|X\|$ is the usual Euclidean norm of X. Then it has been shown in Section 9.2 that the property of worst case dependence of a portfolio $X_i \sim P_i$, $1 \le i \le n$, is closely related to the fact that all X_i are optimally coupled to their sum $T = \sum_{i=1}^{n} X_i$. More precisely, optimal coupling to the sum is a necessary condition and together with a regularity condition on the support of T also a sufficient condition for a worst case portfolio (see Theorem 9.17). In our context this means that the worst case scenario measure μ_0 is given by the distribution of the (worst case) sum T, $\mu_0 = P^T$. The worst case dependence structure is given by μ_0-comonotone vectors X_i. Since the L^2-variation risk measure is not monotone we leave in this example formally the framework of convex risk measures and have to allow also non-positive directions as scenarios.

In the case of normal distributions $X_i \sim P_i$, $P_i = N(0, \Sigma_i)$, $1 \leq i \leq n$, Σ_i positive definite covariance matrices, the worst case scenario measure μ_0 is given by

$$\mu_0 = N(0, \Sigma_0), \tag{9.103}$$

where Σ_0 is a (unique) positive definite solution of the matrix equation

$$\sum_{i=1}^{n} (\Sigma_0^{1/2} \Sigma_i \Sigma_0^{1/2})^{1/2} = \Sigma_0. \tag{9.104}$$

The worst case dependence structure is given by

$$X_1 \sim N(0, \Sigma_1), \quad X_i = S_i S_1^{-1} X_1, \quad i = 2, \ldots, n, \tag{9.105}$$

where $S_i = \Sigma_i^{1/2} (\Sigma_i^{1/2} \Sigma_0 \Sigma_i^{1/2})^{1/2} \Sigma_i^{1/2}$ (see Proposition 9.11). Since the optimal coupling property is a property of the couplings (mappings) and not of the underlying distributions it follows that this determination of the worst case dependence structure extends in the same way to location scale families, in particular to elliptical distributions as in (b_1), (b_2).

(d) Distributions of spherical type. Let U be a random vector on the unit sphere in $\mathbb{R}^d$ w.r.t. Euclidean distance and let $X = RU$ with some scaling real random variable $R > 0$ independent of U. Then we call X of "Euclidean spherical type". Special cases of distributions of Euclidean spherical type are "spherical invariant" distributions which are invariant under orthogonal transformations. In this case U is uniformly distributed on the Euclidean unit sphere. Examples are uniform distributions on spheres or on balls and normal distributions $N(0, \sigma^2 I_d)$ which have exponential tails. But also interesting unsymmetric distributions are of Euclidean spherical type. If R has polynomial tails like for positive stable distributions then also $X = RU$ has polynomial tails and forms a class of distributions of interest in extreme value theory.

Assume that a portfolio is given by $X_i = R_i \cdot U_i$, $1 \leq i \leq n$ where $U_1 \overset{d}{=} U_2 \overset{d}{=} \ldots \overset{d}{=} U_n \overset{d}{=} U$, $R_i \geq 0$ are independent of U_i with distribution functions F_i. Denote by P_i the distribution of X_i.

Proposition 9.28 (Euclidean spherical type portfolio). *Let* $X_i = R_i \cdot U_i$, $1 \leq i \leq n$ *be a portfolio of Euclidean spherical type with*

$$U_i \overset{d}{=} U, \quad 1 \leq i \leq n.$$

Define $X_i^* = F_i^{-1}(V) \cdot U$, $1 \leq i \leq n$, for some uniformly on $(0, 1)$ distributed random variable V independent from U. Then $X_1^*, \ldots, X_n^*$ is a worst case portfolio structure with respect to the L^2-variation risk measure $\Psi_2(X) = (E \|X\|_2^2)^{1/2}$, where $\| \ \|_2$ is the Euclidean norm on $\mathbb{R}^d$, and the worst case risk is given by

$$\sup_{\widetilde{X}_i \sim X_i} \Psi_2\Big(\sum_{i=1}^n \widetilde{X}_i \Big) = \Psi_2\Big(\sum_{i=1}^n X_i^* \Big) = \Big(E \Big(\sum_{i=1}^n F_i^{-1}(V) \Big)^2 \Big)^{1/2}. \qquad (9.106)$$

Proof. By definition $\|X_i\| = R_i$, $i \leq i \leq n$, and with $R_i^* := F_i^{-1}(V)$ it it holds that $\|X_i^*\| = R_i^*$. Thus we obtain for $i \neq j$,

$$\|X_i - X_j\|_2 \geq \big| \|X_i\|_2 - \|X_j\|_2 \big| = \big| R_i - R_j \big|.$$

This implies by a well-known one-dimensional coupling result

$$E \|X_i - X_j\|_2^2 \geq E |R_i - R_j|^2 \geq E |R_i^* - R_j^*|^2 = E \|X_i^* - X_j^*\|_2^2. \qquad (9.107)$$

In consequence all pairs X_i^*, X_j^* are optimally coupled, $X_i^* \underset{oc}{\sim} X_j^*$. This implies directly that $X_1^*, \ldots, X_n^*$ is a worst case portfolio with respect to Ψ_2 and (9.106) follows from (9.107). $\square$

Remark 9.29. Obviously in the Euclidean spherical portfolio above all X_i^* are optimally coupled to the spherical part U of the distribution and also are optimally coupled to the sum $\sum_{i=1}^n X_i^*$. Thus μ the distribution of U is a worst case scenario measure in this situation. From (9.106) we see that typically a worst case diversification effect arises in this class of distributions. We remark that a similar coupling result has also been discussed in Cuesta-Albertos et al. (1993). $\lozenge$

The argument for the Euclidean spherical type portfolio also extends to distributions of spherical type with respect to other norms on $\mathbb{R}^d$. Consider for example the one-norm $\|x\|_1 = \sum_{i=1}^n |x_i|$. Let U be distributed on the unit one-sphere and let $X_i = R_i U_i$, $1 \leq i \leq n$ be a portfolio of spherical type (w.r.t. the one-sphere) with $U_i \sim U$. In case that U_i are uniformly distributed on the one-sphere $\{x \in \mathbb{R}^d; \|x\|_1 = 1\}$ we obtain a class of Archimedean distributions, i.e. distributions which have Archimedean copulas. And conversely to any Archimedean copula C one can give explicitly a radial part R such that RU has copula C (see McNeil and Nešlehová (2010)). Therefore, we call this class of spherical type distributions "distributions of Archimedean type". "Archimedean type" distributions have been used a lot in recent dependence modelling. Consider the risk measure

$$\Psi_1(X) := E \|X\|_1 \qquad (9.108)$$

defined by the L^1-norm.

Proposition 9.30 (Archimedean type portfolio). *Let $X_i = R_i U_i$, $1 \leq i \leq n$, be an Archimedean type portfolio, with R_i independent of U_i and $U_i \overset{d}{=} U$, $1 \leq i \leq n$, are distributed as U. Let (R_i) be independent of U. Then $X_i^* := R_i U$, $1 \leq i \leq n$, is a worst case portfolio w.r.t. Ψ_1 and the worst case risk is given by*

$$\sup_{\widetilde{X}_i \sim X_i} \Psi_1\left(\sum_{i=1}^n \widetilde{X}_i\right) = \Psi_1\left(\sum_{i=1}^n X_i^*\right) = E \sum_{i=1}^n R_i = \sum_{i=1}^n \Psi_1(X_i). \tag{9.109}$$

There is no worst case diversification effect.

Proof. For the proof note that $X_i^* = R_i U$ have the correct portfolio distribution, $X_i^* \sim P_i$. Furthermore, we obtain for any portfolio $X_i \sim P_i$ by Minkowski's inequality

$$\Psi_1\left(\sum_{i=1}^n X_i\right) = E\left\|\sum_{i=1}^n X_i\right\|_1 \tag{9.110}$$

$$\leq E \sum_{i=1}^n R_i = E\left\|\sum_{i=1}^n R_i U\right\|_1 = \Psi_1\left(\sum_{i=1}^n X_i^*\right).$$

Thus (X_i^*) is a worst case portfolio. Since $X_i^* = R_i U$ we obtain further

$$\Psi_1\left(\sum_{i=1}^n X_i^*\right) = \sum_{i=1}^n E R_i = \sum_{i=1}^n \Psi_1(X_i),$$

i.e. there is no worst case diversification effect. $\qquad\square$

For the L^1-norm risk we obtain by a classical result, that $\Psi_1(X)$ is identical to a max-correlation risk measure with worst case scenario given by the sign of X. This explains the disappearance of the worst case diversification effect in (9.109). Note that in this case again we have to allow negative scenarios.

In fact the arguments of Proposition 9.30 can be generalized to the following general spherical equivalent models. Let $\| \ \|$ be any norm on $\mathbb{R}^d$ and let $X = R \cdot U$ be the polar representation of X with radial part $R = \|X\|$ and spherical part $U = X/\|X\|$. We assume that the radial part R is independent of the spherical part U. A random vector Y is called "spherically equivalent" to X if the spherical part of Y is identically distributed to that of X,

$$Y/\|Y\| \overset{d}{=} X/\|X\| \tag{9.111}$$

and the radial part of Y is independent of the spherical part.

Let f be a convex nondecreasing function $f : [0, \infty) \to [0, \infty)$ and let g be a nondecreasing function $g : [0, \infty) \to [0, \infty)$ and consider a risk measure Ψ of the form

$$\Psi(X) = g(Ef(\|X\|)). \tag{9.112}$$

Examples are the p-norms, $p \geq 1$, i.e. $\Psi(X) = \|X\|_p$. Choosing norms of the form $\|x\| = x^\top A x$ with some positive semidefinite A, we get as examples models of elliptical type distributions as in b).

Theorem 9.31 (Spherically equivalent portfolios). *Let $X_i = R_i U_i$ be a portfolio of spherical type $X = RU$ and let Ψ be a risk measure as in (9.112). Let V be independent of U uniformly distributed on $(0, 1)$ and define*

$$R_i^* := F_i^{-1}(V), \quad X_i^* := R_i^* U, \quad 1 \leq i \leq n. \tag{9.113}$$

Then $X_1^*, \ldots, X_n^*$ is a worst case portfolio structure with respect to Ψ and the worst case risk is given by

$$\sup_{\widetilde{X}_i \sim X_i} \Psi\left(\sum_{i=1}^n \widetilde{X}_i\right) = \Psi\left(\sum_{i=1}^n X_i^*\right) \tag{9.114}$$

$$= g\left(Ef\left(\sum_{i=1}^n R_i^*\right)\right).$$

Proof. All elements X_i^* have the same spherical part U and by construction X_i^* have the correct marginal distributions, $X_i^* \sim P_i$. Furthermore, by Minkowski's inequality we have for any portfolio (X_i) with marginals (P_i)

$$\left\|\sum_{i=1}^n X_i\right\| \leq \sum_{i=1}^n \|X_i\| = \sum_{i=1}^n R_i,$$

while for the portfolio (X_i^*) it holds that $\|\sum_{i=1}^n X_i^*\| = \sum_{i=1}^n R_i^*$. By our construction $R_i^* \stackrel{d}{\simeq} R_i$. Thus we can apply Theorem 3.5 on the worst case character of the comonotonic dependence structure which implies optimality of the comonotonic vector (R_i^*) and thus

$$Ef\left(\left\|\sum_{i=1}^n X_i\right\|\right) \leq Ef\left(\sum_{i=1}^n R_i\right)$$

$$\leq Ef\left(\sum_{i=1}^n R_i^*\right) = Ef\left(\left\|\sum_{i=1}^n X_i^*\right\|\right).$$

As consequence we obtain that (X_i^*) is a worst case dependence structure

$$\sup_{\widetilde{X}_i \sim X_i} \Psi\left(\sum_{i=1}^n \widetilde{X}_i\right) = \Psi\left(\sum_{i=1}^n X_i^*\right). \qquad \square$$

Remark 9.32. In typical cases Theorem 9.31 implies a worst case diversification effect. An exception is the situation of an Archimedean type portfolio with one-norm as in Proposition 9.30. Theorem 9.31 gives a tool to calculate worst case portfolios in some classes of examples and to determine the corresponding worst case diversification effect.   ◊

Part III
Optimal Risk Allocation

The optimal risk allocation respectively risk sharing problem has a long history in mathematical economics and insurance and is of considerable practical and theoretical interest. The basic problem can be described as follows. We consider a "market" given by a probability space $(\Omega, \mathfrak{A}, P)$, n economic agents (traders) supplied with risks $X_1, \ldots, X_n$ and possibly different risk measures $\varrho_1, \ldots, \varrho_n$ for the evaluation of their risks. The problem is to redistribute the total risk $X = \sum_{i=1}^{n} X_i$ to the traders by a reallocation $X = \sum_{i=1}^{n} Y_i$ such that the risk vector $(\varrho_i(Y_i))$ is Pareto optimal in the class of all admissible allocations of X or such that the total risk $\sum_{i=1}^{n} \varrho_i(Y_i)$ is minimal under all admissible allocations. In some variants of the problem additional well-motivated constraints are put on the allocation problem as for example side constraints of the form $Y_i \geq A_i$ or $Y_i \geq X_i - A_i$, which limit the magnitude of exchange of agent i or upper bounds $Y_i \leq X_i + B_i$ respectively $\leq A_i + B_i$ which protect the liquidity of the individual traders. An alternative restriction is the "individual rationality condition"

$$\varrho_i(Y_i) \leq \varrho_i(X_i) \quad \text{or} \quad Y_i \leq_{\text{icx}} X_i;$$

only those changes are admissible and acceptable for trader i which are preferable compared with the actual risk X_i.

The classical examples of risk exchange are from insurance where a risk X is redistributed between two traders, an insurer and a reinsurer, both supplied with their own risk measures or equivalently with their specific utility functions. The classical insurance respectively reinsurance contracts like linear quota sharing or stop-loss contracts can be derived in this way as optimal reallocation (reinsurance) contracts.

In a series of classical papers Borch (1960a,b, 1962), Du Mouchel (1968), and Gerber (1978) showed that based on utility functions Pareto optimal risk exchanges can be characterized and in many cases lead to familiar linear quota-sharing of the total pooled losses or to stop-loss contracts and to mixtures of both. Solutions are however typically not uniquely determined which may lead to the necessity to arrange substantial side payments in order to make these solutions acceptable.

In several papers authors have added game-theoretic considerations or additional concepts (like the concept of fairness) to arrive at a specific element in the set of Pareto optimal rules (see Borch (1960b), Lemaire (1977), and Bühlmann and Jewell (1979)).

Since risk pools redistribute only actual losses and possibly the associated premiums but not the individual wealth of the company it is natural as mentioned above to include side constraints in the exchange protocol. The importance of side constraints has been suggested by Borch (1968) and has formally been introduced and applied in Gerber (1978, 1979).

Several authors have extended the framework to include the presence of background risk and have considered the allocation problem also in the context of financial risks (see Leland (1980), Chevallier and Müller (1994), and Barrieu and El Karoui (2004, 2005), Dana and Scarsini (2007), Chateauneuf et al. (2000), Denault (2001) and references therein). A main motivation comes from portfolio optimization problems which can be considered in a joint market model as generalized form of the risk exchange problem. Here more general exchange mechanisms described by trading strategies are considered. Also more general types of risk measures (distortion type, coherent, convex, comonotone risk measures) have been considered for the allocation problem. For the background literature on risk measures and their applications to finance and insurance we refer to Part II as well as to Deprez and Gerber (1985), Kaas et al. (2001), Delbaen (2000, 2002), and Föllmer and Schied (2011).

In Chapter 10 we consider the unrestricted optimal risk allocation problem. The problem is to characterize optimal allocations of a risk $X \in L^p(P)$ to the n traders, i.e. to determine solutions of the problem

$$\sum_{i=1}^{n} \varrho_i(X_i) = \inf! \tag{III.1}$$

under all allocations of X to the traders, i.e. under all decompositions $X = \sum_{i=1}^{n} X_i$, $X_i \in L^p(P)$. Solutions of the risk allocation problem are not unique but in fact are given under an equilibrium condition by the set of all Pareto optimal allocations, as follows from a general separation argument and the translation invariance of the ϱ_i (see Gerber (1979, pp. 88–96)). Thus the optimal allocation problem can be interpreted as a problem to minimize the total risk of a risk sharing contract but also as a basic tool to determine Pareto optimal allocations. The value of the optimal allocation problem is given by the infimal convolution $\widehat{\varrho} = \varrho_1 \wedge \ldots \wedge \varrho_n$ defined for $X \in L^p(P)$ by

$$\widehat{\varrho}(X) = \inf\left\{ \sum_{i=1}^{n} \varrho_i(X_i); \ X_i \in L^p(P), \sum_{i=1}^{n} X_i = X \right\}. \tag{III.2}$$

We show that the general formulation of the optimal risk allocation problem in (III.1) and (III.2) makes sense only under a Pareto equilibrium condition (E). In vague form (E) can be stated as follows:

A market is in equilibrium if in a balance of supply and demand it is not possible to lower the risk of some traders without increasing that of some other traders. This equilibrium condition (E) has been characterized for coherent risk measures $\varrho_1, \ldots, \varrho_n$ in Heath and Ku (2004) and in Burgert and Rü (2005/2008) in terms of the scenario measures of the ϱ_i. An extension of this characterization is given in Burgert and Rü (2006b). The Pareto equilibrium condition (E) implies that $\widehat{\varrho}$ is a convex risk measure. In the coherent case also the converse relation holds true.

We discuss various monotonicity results for optimal risk allocations, stating that optimal allocations $X = \sum_{i=1}^{n} X_i$ of X can be found in the class of allocations X_i such that X_i are comonotone to X respectively where $X_i = f_i(X)$ are monotonically increasing functions of X. There are several related monotonicity results on the construction and design of optimal options in mathematical finance which we discuss briefly.

We introduce a class of well-motivated restrictions on the allocations which leads to a meaningful version of the allocation problem also without an equilibrium condition. The idea of introducing this class of restrictions is connected with a similar idea in portfolio theory, where one considers (lower bounded) admissible strategies in order to exclude strategies which allow arbitrage. As a consequence we obtain a convex risk measure – the convex infimal admissible convolution risk measure – which describes the optimal total admissible risk $\sum_{i=1}^{n} \varrho_i(X_i)$ avoiding the possibility of risk arbitrage.

Chapter 11 is concerned with some generalizations of the classical characterization results of optimal allocations due to Borch (1960b) and others. These allow to derive several of the well-known reinsurance contracts as optimal solutions of the allocation problems.

In the final part of this chapter we give extensions to the risk allocation problem for portfolio vectors $X \in \mathbb{R}^d$. Optimal allocations in this context are connected with the worst case dependence structure and μ-comonotonicity. These connections are inherited from the representation results for risk measures for portfolio vectors.

Chapter 12 is concerned with applications of the dependence bounds in Part I to the construction of optimal contingent claims and portfolios and also of optimal (re-)insurance contracts. We deal in particular with the construction of efficient portfolios and of optimal robust insurance contracts.

Chapter 10
Optimal Allocations and Pareto Equilibrium

In Section 1 of this chapter we introduce the notion of Pareto equilibrium and give its characterization by Heath and Ku (2004) and Burgert and Rü (2005/2008, 2006a). This game theoretic notion is naturally associated with properties of some risk measures which describe the optimal risk sharing problem in a market with n traders as described in the introduction. The main result in this section shows that the optimal risk sharing problem without constraints is well defined if and only if the Pareto equilibrium condition holds. We also briefly indicate an extension to the case of incomplete markets with restrictions on the admissible allocations. In the second section we introduce a senseful extension of the risk allocation problem in the case that the equilibrium condition is not fulfilled. This is done by restricting to a class of suitable admissible allocations.

We consider in Sections 1 and 2 the case that the risk measures ϱ_i are coherent. In Section 3 we discuss extensions to the case of convex risk measures ϱ_i. The main result in Section 4 is the comonotone improvement result stating that any allocation can be improved by a comonotone allocation. We generally assume that the underlying probability space $(\Omega, \mathcal{A}, P)$ is non-atomic.

10.1 Pareto Equilibrium and Related Risk Measures in the Coherent Case

In this section we consider the case that the risk measures $\varrho_i : L^p(P) \to \mathbb{R}$ are coherent and lsc with representation of the form

$$\varrho_i(X) = \sup_{Q \in \mathcal{P}_i} E_Q(-X), \tag{10.1}$$

where $\mathcal{P}_i$ are convex, $\sigma(L^q, L^p)$-closed representing scenario sets in $\mathcal{M}_1^q$ respectively $\mathcal{M}_1$ in case $p = \infty$. Also in case $p = \infty$ we assume that the representation

L. Rüschendorf, *Mathematical Risk Analysis*, Springer Series in Operations Research and Financial Engineering, DOI 10.1007/978-3-642-33590-7_10,
© Springer-Verlag Berlin Heidelberg 2013

is based on σ-additive measures (see the discussion in Section 8.1). The acceptance sets are denoted by

$$\mathcal{A}_{\varrho_i} = \{X \in L^p(P); \ \varrho_i(X) \leq 0\}. \tag{10.2}$$

One can consider the risk allocation problem as a game in the sense of game theory. From this point of view Heath and Ku (2004) introduced the notion of Pareto equilibrium for the allocation problem.

Definition 10.1 (Pareto equilibrium). A market model with risk measures $\varrho_1, \ldots,$ ϱ_n is in Pareto equilibrium, if

(E) $X_i \in L^p(P)$ with $\sum\limits_{i=1}^{n} X_i = 0$ and $\varrho_i(X_i) \leq 0, \quad 1 \leq i \leq n,$

implies $\varrho_i(X_i) = 0, \quad 1 \leq i \leq n.$

In a balance of supply and demand it is in equilibrium not possible to lower the risk of some traders without increasing that of others. Vaguely one could say that there is *no arbitrage* situation concerning risk. The equilibrium condition implies that the trivial decomposition $0 = 0 + \cdots + 0$ is a Pareto optimal decomposition of zero. We define as the (unrestricted) "optimal risk allocation problem" the problem to determine the set of allocations (X_i) of X that minimize $\sum_{i=1}^{n} \varrho_i(X_i)$.

The optimal total risk of the "optimal risk allocation problem" is given by the "inf-convolution" $\widehat{\varrho}$

$$\widehat{\varrho}(X) = \inf \left\{ \sum_{i=1}^{n} \varrho_i(X_i); \ X_i \in L^p(P), \sum_{i=1}^{n} X_i = X \right\}. \tag{10.3}$$

The set of "allocations" of $X \in L^p(P)$ is defined as

$$A(X) := \left\{ (X_i) \in (L^p(P))^n; \ \sum_{i=1}^{n} X_i = X \right\}. \tag{10.4}$$

Definition 10.2 (Pareto optimal allocation). An allocation $(X_i) \in A(X)$ is called "Pareto optimal" if the risk vector $(\varrho_i(X_i))$ is a minimal element of the risk set

$$\mathcal{R}(X) = \{(\varrho_i(Y_i)); \ (Y_i) \in A(X)\}, \tag{10.5}$$

i.e. there exists no allocation $(Y_i) \in A(X)$ with $\varrho_i(Y_i) \leq \varrho_i(X_i)$ for all i and $\varrho_{i_0}(Y_{i_0}) < \varrho_{i_0}(X_{i_0})$ for some $i_0 \leq n$.

In this section we give a characterization of the Pareto equilibrium condition (E) due to Heath and Ku (2004) in the case of finite Ω and to Burgert and Rü (2005/2008) respectively Jouini et al. (2008) for $L^\infty(P)$ on general Ω. Our derivation makes use of properties of risk measures which are naturally associated to the optimal risk allocation problem like the inf-convolution $\widehat{\varrho}$ and a risk measure defined in terms of acceptance sets $\mathcal{A}_{\varrho_i}$ of ϱ_i. Under the Pareto equilibrium condition $\widehat{\varrho}(X)$

is finite and the set of optimal solutions of the total risk optimization in (10.3) coincides with the set of all Pareto optimal allocations. Thus minimizing the total risk of allocations is equivalent to determining Pareto optimal allocations under the equilibrium condition.

To derive this connection we first introduce a seemingly stronger version of the Pareto equilibrium condition (E).

(SE)   **Strong equilibrium**

$$\text{If } X_i \in L^p(P) \text{ with } \sum_{i=1}^n X_i = 0, \text{ then } \sum_{i=1}^n \varrho_i(X_i) \geq 0.$$

It is immediate to see that (SE) $\Rightarrow$ (E). Therefore, we call this condition strong equilibrium. But in fact both conditions are equivalent.

Proposition 10.3. *The equilibria conditions (E) and (SE) are equivalent.*

Proof. Assume that for some $X_i \in L^p(P)$ with $\sum_{i=1}^n X_i = 0$ it holds that

$$\sum_{i=1}^n \varrho_i(X_i) =: c < 0.$$

Then with $c_i := \varrho_i(X_i)$ we introduce a rebalancing and define $Z_i := X_i + c_i - \frac{c}{n}$, then $\sum_{i=1}^n Z_i = 0$ and

$$\varrho_i(Z_i) = \varrho_i(X_i) - c_i + \frac{c}{n} = \frac{c}{n} < 0, \quad 1 \leq i \leq n.$$

In consequence we obtain a contradiction to (E). $\square$

Thus under the Pareto equilibrium condition (E) the sum of all risks in a balance of supply and demand situation is non-negative. The equilibrium condition (E) is closely connected with the following risk measure $\varrho_{\mathcal{A}}$ induced by the acceptance set

$$\mathcal{A} := \text{cone}\left(\bigcup_{i=1}^n \mathcal{A}_{\varrho_i} \right) \tag{10.6}$$

by

$$\varrho_{\mathcal{A}}(X) := \inf\{m \in \mathbb{R}; \ X + m \in \mathcal{A}\}. \tag{10.7}$$

Then $\mathcal{A}$ is the acceptance set of $\varrho_{\mathcal{A}}$. Risk positions are acceptable w.r.t. $\varrho_{\mathcal{A}}$ if they are acceptable for any of the traders in the market. Thus it seems natural that $\varrho_{\mathcal{A}}$ is connected with an optimistic view towards risk and thus with the optimal risk sharing problem to minimize $\sum_{i=1}^n \varrho_i(X_i)$ over all allocations (X_i) of X. The equilibrium condition (E) can be described in terms of the risk measure $\varrho_{\mathcal{A}}$.

Proposition 10.4.

$$\varrho_{\mathcal{A}} \text{ is a coherent risk measure} \Leftrightarrow \varrho_{\mathcal{A}}(0) = 0$$

$$\Leftrightarrow \text{ The equilibrium condition (E) holds.}$$

Proof. The first equivalence follows directly from the definition of $\mathcal{A}$ and $\varrho_{\mathcal{A}}$. For the proof of the second equivalence assume that (E) holds. Then by Proposition 2.5 also (SE) holds. By definition

$$\varrho_{\mathcal{A}}(0) = \inf\left\{ m \in \mathbb{R}; \ \exists X_i \in \mathcal{A}_{\varrho_i}, 1 \le i \le n, m = \sum_{i=1}^{n} X_i \right\}.$$

By (SE) for any $X_i \in \mathcal{A}_{\varrho_i}$ with

$$\sum_{i=1}^{n} X_i - m = (X_1 - m) + \sum_{i=2}^{n} X_i = 0$$

it holds that

$$\varrho_1(X_1 - m) + \sum_{i=2}^{n} \varrho_i(X_i) = \sum_{i=1}^{n} \varrho_i(X_i) + m \ge 0,$$

i.e. $m \ge -\sum_{i=1}^{n} \varrho_i(X_i) \ge 0$, since $\varrho_i(X_i) \le 0$. Thus we obtain $\varrho_{\mathcal{A}}(0) = 0$.

Conversely, if $\varrho_{\mathcal{A}}(0) = 0$ and if $X_i \in \mathcal{A}_{\varrho_i}$ are in balance of supply and demand, $\sum_{i=1}^{n} X_i = 0$, then using that

$$\varrho_{\mathcal{A}} \le \varrho_i, \quad 1 \le i \le n, \tag{10.8}$$

we obtain

$$0 = \varrho_{\mathcal{A}}\left(\sum_{i=1}^{n} X_i\right) \le \sum_{i=1}^{n} \varrho_{\mathcal{A}}(X_i) \le \sum_{i=1}^{n} \varrho_i(X_i) \le 0.$$

This implies $\varrho_i(X_i) = 0$, $1 \le i \le n$, i.e. (E) holds. $\square$

As a corollary we obtain a characterization of $\varrho_{\mathcal{A}}$ under the Pareto equilibrium condition as largest coherent risk measure below $\min \varrho_i(X)$.

Corollary 10.5. *Let the market* $((\Omega, \mathfrak{A}, P), \varrho_1, \ldots, \varrho_n)$ *with coherent risk measures* ϱ_i *be in equilibrium (i.e. condition (E) holds). Then* $\varrho_{\mathcal{A}}$ *is the largest coherent risk measure* ϱ *such that*

$$\varrho(X) \le \min\{\varrho_i(X); \ 1 \le i \le n\}. \tag{10.9}$$

Proof. Let ϱ be a coherent risk measure $\varrho \le \min_{1 \le i \le n} \varrho_i$. Then $X \in \mathcal{A}_{\varrho_i}$ implies that $X \in \mathcal{A}_{\varrho}$ and thus

$$\mathcal{A}_{\varrho} \supset \mathrm{cone}\left(\bigcup_{i=1}^{n} \mathcal{A}_{\varrho_i}\right) = \mathcal{A}. \tag{10.10}$$

This implies that $\varrho \le \varrho_{\mathcal{A}}$. $\square$

The equivalence of (E) and (SE) suggests to consider the infimal convolution $\widehat{\varrho} = \varrho_1 \wedge \cdots \wedge \varrho_n$ defined on $L^p(P)$ by

$$\widehat{\varrho}(X) := \inf \left\{ \sum_{i=1}^{n} \varrho_i(X_i); \ X_i \in L^p(P), \sum_{i=1}^{n} X_i = X \right\}. \tag{10.11}$$

$\widehat{\varrho}$ is the risk measure that describes the *optimal* total risk of all risk allocations of X to the traders in the market. $\varrho_{\mathcal{A}}$ evaluates the risk from a viewpoint of the traders in the market who can decompose a present risk X as $(X - Y) + Y$ for some acceptable $Y \in \mathcal{A}$ such that $X - Y$ is acceptable and therefore $\varrho_{\mathcal{A}}(X) \leq \varrho_{\mathcal{A}}(X - Y) + \varrho_{\mathcal{A}}(Y) \leq \varrho_{\mathcal{A}}(Y)$. The choice of suitable Y such that $X - Y$ is acceptable can be seen as a risk exchange procedure. Therefore, it is natural to expect a close connection of $\varrho_{\mathcal{A}}$ and the inf convolution $\widehat{\varrho}$.

It is easy to check that $\widehat{\varrho}$ satisfies all axioms of a coherent risk measure except possibly the condition $\widehat{\varrho}(0) = 0$.

Proposition 10.6. *(a)* $\widehat{\varrho}$ *is a coherent risk measure* $\Leftrightarrow$ *The equilibrium condition* (E) *holds* $\Leftrightarrow \widehat{\varrho}(0) = 0$.
(b) *Under the equilibrium condition* (E) *it holds that:*

$$\widehat{\varrho} = \varrho_{\mathcal{A}}. \tag{10.12}$$

Proof.

(a) If $\widehat{\varrho}$ is a coherent risk measure then obviously $\widehat{\varrho}(0) = 0$. By the definition of the strong equilibrium condition (SE) this is equivalent to the (SE)-condition. Thus by Proposition 10.3 it implies the equilibrium condition (E).

   If conversely (E) holds and thus (SE), it follows that $\widehat{\varrho}(0) = 0$. This is the missing part which implies that $\widehat{\varrho}$ is a coherent risk measure on $L^p(P)$.

(b) Assume that (E) holds true. Then by (a) $\widehat{\varrho}$ is a coherent risk measure and thus by Corollary 10.5 we have $\widehat{\varrho} \leq \varrho_{\mathcal{A}}$. Conversely, for any decomposition $X = \sum_{i=1}^{n} X_i$ it holds that – using $\varrho_{\mathcal{A}} \leq \min\{\varrho_i\}$ –

$$\sum_{i=1}^{n} \varrho_i(X_i) \geq \sum_{i=1}^{n} \varrho_{\mathcal{A}}(X_i) \geq \varrho_{\mathcal{A}} \left(\sum_{i=1}^{n} X_i \right) = \varrho_{\mathcal{A}}(X),$$

i.e. $\widehat{\varrho}(X) \geq \varrho_{\mathcal{A}}(X)$. $\square$

Remark 10.7. As a consequence of Proposition 10.6 we obtain that the optimal unconstrained allocation problem makes sense only under the Pareto equilibrium condition (E). Without condition (E) the optimal risk allocation problem leads to the inconsistency that $\widehat{\varrho}(0) = -\infty$. In particular without the Pareto equilibrium condition (E) it is not possible to determine Pareto optimal allocations in the optimal allocation problem and the domain of $\widehat{\varrho}$ is empty. $\Diamond$

It is therefore of interest to understand this equilibrium condition. By the proof of the representation theorem of coherent risk measures based on the Fenchel–Moreau theorem the set of scenarios $\mathcal{P} \subset \mathcal{M}_1^q(P)$ can be chosen as the set of normed elements of the polar set $\mathcal{A}^0$ of the acceptance set $\mathcal{A} = \mathcal{A}_\varrho$ of ϱ, i.e.

$$\mathcal{P} = \left\{ Q \in \mathcal{M}_1^q(P); \ Q1 = 1, E_Q X \geq 0, \ \forall X \in \mathcal{A} \right\} = \mathcal{P}_\varrho. \tag{10.13}$$

Our aim is to describe the equilibrium condition (E) in terms of the scenario set $\mathcal{P}$ of ϱ_A. The following proposition says that $\mathcal{P}$ contains exactly those scenario measures which are common to all risk measures.

Proposition 10.8. *Consider the market model with coherent risk measures ϱ_i and polar representation sets $\mathcal{P}_i \subset \mathcal{M}_1^q(P)$, $1 \leq i \leq n$ and assume that the Pareto equilibrium condition (E) holds.*
Then $\bigcap_{i=1}^n \mathcal{P}_i \neq \phi$ and the polar representation set of $\mathcal{P}$ of $\widehat{\varrho} = \varrho_A$ is given by

$$\mathcal{P} = \bigcap_{i=1}^n \mathcal{P}_i. \tag{10.14}$$

Proof. For $\mathcal{P} = \mathcal{P}_{\varrho_A} = \mathcal{P}_{\widehat{\varrho}}$ it holds that

$$\mathcal{P} = \left\{ Q \in \mathcal{M}_1^q(P); \ E_Q X \geq 0, \ \forall X \in \mathrm{cone}\left(\bigcup_{i=1}^n \mathcal{A}_{\varrho_i} \right) \right\}$$

$$= \left\{ Q \in \mathcal{M}_1^q(P); \ \forall i \leq n \text{ it holds that: } E_Q X \geq 0, \ \forall X \in \mathcal{A}_{\varrho_i} \right\}$$

$$= \bigcap_{i=1}^n \left\{ Q \in \mathcal{M}_1^q(P); \ E_Q X \geq 0, \ \forall X \in \mathcal{A}_{\varrho_i} \right\}$$

$$= \bigcap_{i=1}^n \mathcal{P}_i. \qquad \square$$

As a consequence of Propositions 10.6 and 10.8 we obtain the characterization result for equilibrium in terms of the scenarios of the risk measures ϱ_i. This result was stated in Heath and Ku (2004) for finite models Ω and finitely generated scenario sets $\mathcal{P}_i$ and in general form in Burgert and Rü (2005/2008) and Jouini et al. (2008).

Theorem 10.9 (Characterization of equilibrium). *Consider the market with coherent risk measures $\varrho_1, \ldots, \varrho_n$. Then the equilibrium condition (E) is equivalent to the condition*

$$\bigcap_{i=1}^n \mathcal{P}_i \neq \phi, \tag{10.15}$$

i.e., there exists a scenario measure Q which is shared by all traders in the market.

Proof. If condition (E) holds, then by Proposition 10.8 we obtain that

$$\mathcal{P} = \bigcap_{i=1}^{n} \mathcal{P}_i \neq \phi.$$

Conversely, if $\bigcap_{i=1}^{n} \mathcal{P}_i \neq \phi$ and $Q \in \bigcap_{i=1}^{n} \mathcal{P}_i$, then

$$\widehat{\varrho}(0) = \inf \left\{ \sum_{i=1}^{n} \varrho_i(X_i); \sum_{i=1}^{n} X_i = 0 \right\}$$

$$\geq \inf \left\{ \sum_{i=1}^{n} E_Q(-X_i); \sum_{i=1}^{n} X_i = 0 \right\} = 0.$$

This implies that $\widehat{\varrho}(0) = 0$ and thus by Proposition 10.6 condition (E) holds.   □

Remark 10.10. (a) For the proof of Theorem 10.9 Heath and Ku (2004) reduce the characterization problem to an application of the duality theorem of linear programming. The above given proof of Burgert and Rü (2005/2008) is based on properties of the derived risk measures ϱ_A and $\widehat{\varrho}$ which are naturally associated to the equilibrium problem.

(b) The infimal convolution $\widehat{\varrho}$ and the risk measure ϱ_A have been introduced in the literature and applied to the problem of risk transfer in Barrieu and El Karoui (2004, 2005), and in Delbaen (2000). In these papers also related results on the acceptance set and representation set of $\widehat{\varrho}$ are given. In particular one finds there also an investigation of the Fatou-property of $\widehat{\varrho}$. The application of these risk measures to derive the equilibrium characterization result is from the paper of Burgert and Rü (2005/2008).

(c) An important early paper on the allocation problem is Deprez and Gerber (1985). In that paper Deprez and Gerber characterize for convex premium principles Pareto optimal allocations (generalization of Borch's Theorem). Moreover, for the class of those premium principles, which are based on a generalized principle of utility in that paper a *no trade equilibrium* premium notation is introduced. The existence of a no trade equilibrium premium is equivalent to the Pareto equilibrium notion in Definition 10.1 (see Theorems 16 and 17 in Deprez and Gerber (1985)). So their paper is an original source of the notion of convex risk measure and also of the allocation problem.   ◊

Remark 10.11 (Pareto equilibrium in incomplete models).   The results of this section extend directly to incomplete models where trading of the i-th trader is restricted to linear subspaces $M_i \subset L^p(P)$, $1 \leq i \leq n$. There are various motivations for considering restricted classes of trading sets in the literature like restricted resources, regulatory restrictions, technical restrictions. For some motivation and further references in the context of the related assignment problem we refer to Ramachandran and Rü (2002). We assume that the constants are contained in the trading sets M_i, $\mathbb{R} \subset M_i$, and define risk measures on the trading space $M := \sum_{i=1}^{n} M_i$, by defining

$$\overline{\mathcal{A}_i} := \{X_i \in M_i; \ \varrho_i(X_i) \le 0\}$$

$$\overline{\mathcal{A}_M} := \text{cone}\left(\bigcup_{i=1}^{n} \overline{\mathcal{A}_i}\right).$$

For $X \in M$ we introduce the modified version of the risk measure $\varrho_{\mathcal{A}}$ defined by

$$\varrho_M(X) := \inf\left\{m \in \mathbb{R}; \ m + X \in \overline{\mathcal{A}_M}\right\}, \qquad (10.16)$$

$$\widehat{\varrho}_M(X) := \inf\left\{\sum_{i=1}^{n} \varrho_i(X_i); \ X_i \in M_i, \sum_{i=1}^{n} X_i = X\right\}. \qquad (10.17)$$

Then we obtain as in the unrestricted case:

$$\varrho_M \text{ is a coherent risk measure on } M \Leftrightarrow \varrho_M(0) = 0 \qquad (10.18)$$

$$\widehat{\varrho}_M \text{ is a coherent risk measure on } M \Leftrightarrow \widehat{\varrho}_M(0) = 0. \qquad (10.19)$$

The "Pareto equilibrium condition" for the incomplete market case is defined by

$$(\text{E}_M) \qquad X_i \in M_i \text{ with } \sum_{i=1}^{n} X_i = 0 \text{ and } \varrho_i(X_i) \le 0 \text{ for all } i \qquad (10.20)$$

$$\text{implies } \varrho_i(X_i) = 0 \text{ for all } i.$$

The Pareto equilibrium condition (E_M) is equivalent to $\varrho_M(0) = 0$.   (10.21)

The corresponding strong equilibrium condition is defined by

$$(\text{SE}_M) \qquad X_i \in M_i \text{ and } \sum_{i=1}^{n} X_i = 0 \text{ implies } \sum_{i=1}^{n} \varrho_i(X_i) \ge 0. \qquad (10.22)$$

The strong equilibrium condition (SE_M) is equivalent to $\widehat{\varrho}_M(0) = 0$.   (10.23)

As a consequences we obtain for the incomplete case the following conclusions in a similar way as for the complete case.

Proposition 10.12. *Under the Pareto equilibrium condition (E_M) holds for the incomplete model:*

(a) ϱ_M is the largest coherent risk measure on M such that $\varrho_M/_{M_i} \le \varrho_i/_{M_i}, 1 \le i \le n$, where $\varrho_M/_{M_i}$ and $\varrho_i/_{M_i}$ denote the restrictions of ϱ_M and ϱ_i to the trading sets M_i.

(b) *The risk measures* $\widehat{\varrho}_M$ *and* ϱ_M *are identical. The corresponding scenario set is given by*

$$\mathcal{P}_{\widehat{\varrho}_M} = \mathcal{P}_{\varrho_M} = \left\{ Q \in \mathcal{M}_{1,s}^q(P); \ \mathcal{Q}/_{M_i} \in \mathcal{P}_i/_{M_i}, \ 1 \le i \le n \right\} \quad (10.24)$$

where $\mathcal{M}_{1,s}^q(P)$ *is the set of signed* P*-continuous* σ*-additive measures with finite* q*-th moments.*

The restriction to subspaces M_i does in general not imply positivity of the representing measures – as in the complete market. As consequence of Proposition 10.12 we obtain an extension of the characterization of Pareto equilibria in incomplete models which for finite models was given in Heath and Ku (2004).

Theorem 10.13. *In the incomplete market case the equilibrium condition* (E_M) *is equivalent to the following condition:*

$$\exists Q_i \in \mathcal{P}_i \ and \ \exists Q \in \mathcal{M}_{1,s}^q(P) \ such \ that \ \mathcal{Q}/_{M_i} = \mathcal{Q}_i/_{M_i}, \ 1 \le i \le n, \quad (10.25)$$

i.e., there exists a common scenario Q *on the trading spaces* M_i. Q *is a signed measure with finite* q*-th moments which is positive on* M_i. $\diamond$

10.2   Optimal Allocations Under Admissibility Restrictions

The allocation results for the infimal convolution risk measure $\widehat{\varrho}$ in Section 10.1 leave open the question how to allocate optimally risk when the equilibrium condition does not hold. It is of interest to note that in many situations of practical relevance the equilibrium condition (E) does not hold. The most simple example of this type is the case where $\varrho_i(X) = E_{Q_i}(-X)$, $1 \le i \le n$, where Q_i are P-continuous probability measures, which represent the view towards risk of the i-th trader. If not all views Q_i are identical, then the equilibrium condition does not hold. The equilibrium condition (E) does not hold by Theorem 10.9 if the views of the traders towards risk are too different.

In the other direction the equilibrium condition does hold if for some $c > 0$

$$\varrho_i(X_i) \ge cE(-X_i), \quad 1 \le i \le n, \quad (10.26)$$

the expectation w.r.t. P, because then for any decomposition (X_i) of zero $\sum_{i=1}^{n} X_i = 0$ it holds that

$$\sum_{i=1}^{n} \varrho_i(X_i) \ge cE \sum_{i=1}^{n} (-X_i) = 0$$

and thus $\widehat{\varrho}(0) = 0$.

In particular, for all law invariant convex risk measures ϱ_i, where $\varrho_i(X)$ only depends on the law of X w.r.t. P, condition (10.26) holds. Thus the equilibrium condition (E) holds, if $\varrho_1, \ldots, \varrho_n$ are all law invariant.

It is a problem of interest to modify the risk allocation problem so that it makes sense also in case the equilibrium condition does not hold. The main idea to deal with this situation is to restrict the class of allowed allocations in order to admit no "pathological" allocations. This is similar to the restriction to admissible strategies in portfolio theory in order to avoid that effects like doubling strategies may occur in risk allocation. Some types of restrictions have been introduced in the insurance literature (see the introduction to this chapter). Our aim is to introduce restrictions as weak as possible which still yield a senseful version of the optimal allocation problem and thus allow to determine w.r.t. this class Pareto optimal allocations respectively risk sharing strategies.

Definition 10.14 (Admissible allocation).   An allocation $(X_i) \in A(X)$, $X \in L^p(P)$, is called "admissible" if

$$X(\omega) \geq 0 \text{ implies that} \qquad 0 \leq X_i(\omega) \leq X(\omega)$$

$$\text{and } X(\omega) \leq 0 \text{ implies that} \quad X(\omega) \leq X_i(\omega) \leq 0. \qquad (10.27)$$

Let $A^a(X) := \{(X_i) \in A(X);\ (X_i) \text{ is admissible}\}$ denote the set of "admissible allocations" of X.

The restrictions in (10.27) are natural in reinsurance contracts where the total risk $X \geq 0$ is divided into a part X_1 taken by an insurer and some parts X_i, $i \geq 2$ taken by some reinsurers. In finance the admissibility restrictions in (10.27) prevent uncontrolled "borrowing" as e.g. by unrestricted puts and calls and as a consequence prevent risk arbitrage.

Definition 10.15 (Admissible infimal convolution).   Given a market with risk measures $\varrho_1, \ldots, \varrho_n$ on $L^p(P)$ we define the "admissible infimal convolution" ϱ_* on $L^p(P)$ by

$$\varrho_*(X) = \inf \left\{ \sum_{i=1}^n \varrho_i(X_i);\ (X_i) \in A^a(X) \right\}. \qquad (10.28)$$

In this section we consider as in Section 10.1 the case that ϱ_i are coherent risk measures on $L^p(P)$ with representation sets $\mathcal{P}_i \subset \mathcal{M}_1^q$. The admissible infimal convolution has the following simple to establish properties.

Proposition 10.16. *Let ϱ_i be coherent risk measures on $L^p(P)$, then*

(a) ϱ_* is a subadditive, homogeneous, monotone risk functional on $L^p(P)$

(b) $$\varrho_* \leq \min_{1 \leq i \leq n} \varrho_i, \quad \varrho_*(X) \geq \sum_{i=1}^n \varrho_i(|X|). \qquad (10.29)$$

In particular $\varrho_*(1) \leq -1$ and $\varrho_*(X)$ is finite.

The admissible inf-convolution ϱ_* avoids the risk arbitrage which arises for the inf-convolution $\widehat{\varrho}$ when the equilibrium condition does not hold. In general ϱ_* is not translation invariant and, therefore, is not a coherent risk measure.

The following theorem gives an essentially simplified dual representation of ϱ_* in terms of the scenario measures $\mathcal{P}_i$ of ϱ_i. To obtain this representation we make use of an alternative description of the admissible decompositions like in multiple decision problems in terms of multiple test functions φ_i, $0 \leq \varphi_i \leq 1$:

$$A^a(X) = \left\{ (\varphi_i X);\ 0 \leq \varphi_i \leq 1, \sum_{i=1}^{n} \varphi_i = 1 \right\}. \tag{10.30}$$

For $P_i \in \mathcal{P}_i$ let $P_1 \wedge \cdots \wedge P_n$ denote the lattice infimum of (P_i) in the lattice $\mathcal{M}_1^p(P)$ and let $P_1 \vee \cdots \vee P_n$ denote the lattice supremum of (P_i) in $\mathcal{M}_1^p(P)$. If P_i are probability measures with densities f_i w.r.t. P then $P_1 \wedge \cdots \wedge P_n$ respectively $P_1 \vee \cdots \vee P_n$ have densities $\min\{f_i\}$ respectively $\max\{f_i\}$ w.r.t. P. The admissible infimal convolution ϱ_* admits the following useful dual representation of ϱ_* using the lattice infima and suprema. This representation simplifies essentially the calculation of ϱ_* and is useful also in the following.

Theorem 10.17. *Let $\varrho_j = \varrho_{\mathcal{P}_j}$ be coherent risk measures.*

(a) The admissible infimal convolution ϱ_* has the dual representation

$$\varrho_*(X) = \sup \left\{ \int X_- d \bigwedge_j P_j - \int X_+ d \bigvee_j P_j;\ P_j \in \mathcal{P}_j, 1 \leq j \leq n \right\}. \tag{10.31}$$

(b)
$$\mathcal{A}_{\varrho_*} = \left\{ X \in L^p(P);\ such\ that \int X_- d \bigwedge P_j \leq \int X_+ d \bigvee P_j \right.$$

$$\left. for\ all\ P_j \in \mathcal{P}_j \right\}. \tag{10.32}$$

Proof. (a) For $P_i \in \mathcal{P}_i$, $1 \leq i \leq n$ and $Y := -X \in L^p(P)$ it holds that

$$a_{P_1,\dots,P_n}(Y) := \inf \left\{ \sum_{i=1}^{n} \int \varphi_i Y \, dP_i;\ 0 \leq \varphi_i, \sum_{i=1}^{n} \varphi_i = 1 \right\}$$

has a solution (φ_i^*) and

$$if\ Y(\omega) > 0,\ then\ \{\varphi_i^* > 0\} \subset \{P_i = \bigwedge_{j=1}^{n} P_j\} \quad and$$

$$if\ Y(\omega) < 0,\ then\ \{\varphi_i^* > 0\} \subset \{P_i = \bigvee_{j=1}^{n} P_j\}.$$

Thus $$a_{P_1,\dots,P_n}(Y) = \int_{Y \geq 0} Y\, d \bigwedge_{j=1}^{n} P_j + \int_{Y < 0} Y\, d \bigvee_{j=1}^{n} P_j$$

$$= \int Y_+ \, d \bigwedge P_j - \int Y_- \, d \bigvee P_j$$

$$= \int X_- \, d \bigwedge P_j - \int X_+ \, d \bigvee P_j.$$

Therefore, we obtain

$$\varrho_*(X) = \inf_{(\varphi_i)} \sum_i \varrho_i(\varphi_i X) = \inf_{(\varphi_i)} \sum_i \sup_{P_i \in \mathcal{P}_i} \int (-\varphi_i X)\, dP_i$$

$$= \inf_{(\varphi_i)} \left[-\sum_{i=1}^{n} \inf_{P_i \in \mathcal{P}_i} \int \varphi_i X\, dP_i \right]$$

$$= -\sup_{(\varphi_i)} \sum_i \inf_{P_i \in \mathcal{P}_i} \int \varphi_i X\, dP_i.$$

We now apply a useful and general version of the minimax theorem (see Müller (1971)):

Minimax Theorem. *Let* $f : A \times B \to \overline{\mathbb{R}}$, $A, B \neq \phi$ *be a game of concave–convex type, i.e.*

(1) $\forall\, b_1, b_2 \in B$, $\alpha \in [0, 1]$ *there exists* $a, b \in B$ *such that for all* $a \in A$ *it holds that*

$$f(a, b) \leq (1 - \alpha) f(a, b_1) + \alpha f(a, b_2).$$

(2) $\forall\, a_1, a_2 \in A$, $\alpha \in [0, 1]$ *there exists an* $a \in A$ *such that for all* $b \in B$ *it holds that*

$$f(a, b) \geq (1 - \alpha) f(a_1, b) + \alpha f(a_2, b).$$

If $f < \infty$ *and for some topology* τ *on* A *it holds that* A *is* τ-*compact and* $\forall\, b \in B$, $f(\cdot, b) : A \to \mathbb{R}$ *is upper semicontinuous, then*

$$\inf_{b \in B} \sup_{a \in A} f(a, b) = \sup_{a \in A} \inf_{b \in B} f(a, b). \tag{10.33}$$

We choose $A = \{(\varphi_i); 0 \leq \varphi_i, \sum \varphi_i = 1\}$, which is compact in weak $*$-topology, $B = \mathcal{P}_1 \times \cdots \times \mathcal{P}_n$ and $f((\varphi_i), (P_i)) = \sum_{i=1}^{n} \int \varphi_i X dP_i$. By linearity of f and convexity of $\mathcal{P}_i$ and A the conditions of the minimax theorem are fulfilled and we obtain from the first part of the proof

$$\varrho_*(X) = -\inf_{P_i \in \mathcal{P}_i} \sup_{(\varphi_i)} \sum_i \int \varphi_i X \, dP_i$$

$$= -\inf_{P_i \in \mathcal{P}_i} \left(\int X_+ d \bigvee P_i - \int X_- d \bigwedge P_i \right)$$

$$= \sup_{P_i \in \mathcal{P}_i} \left(\int X_- d \bigwedge P_i - \int X_+ d \bigvee P_i \right).$$

(b) follows from (a).   □

An interesting consequence of Theorem 10.17 is the following characterization of the Pareto equilibrium condition (E) in terms of the admissible infimal convolution ϱ_*.

Proposition 10.18. *Let ϱ_i be coherent risk measures and let ϱ_* denote the admissible infimal convolution, then it holds that:*

$$\varrho_* \text{ is a coherent measure } \Leftrightarrow \varrho_*(1) = -1$$

$$\Leftrightarrow \text{ The Pareto equilibrium condition (E) holds.}$$

Under (E) we have $\varrho_* = \widehat{\varrho} = \varrho_{\mathcal{A}}$.

Proof. The first equivalence is obvious since the condition $\varrho_*(1) = -1$ implies translation invariance of ϱ_*. By Theorem 10.17

$$\varrho_*(1) = \sup \left\{ -|\bigvee_j P_j|; \ P_j \in \mathcal{P}_j \right\}$$

$$= -\inf \left\{ |\bigvee_j P_j|; \ P_j \in \mathcal{P}_j \right\} \leq -1.$$

Thus $\varrho_*(1) = -1$ if and only if there exists a common scenario measure $Q \in \bigcap_{i=1}^n \mathcal{P}_i$. By Theorem 10.9 this is equivalent to the Pareto equilibrium condition (E).   □

In the case without equilibrium condition we modify ϱ_* to obtain a coherent risk measure which we call coherent admissible infimal convolution.

Definition 10.19 (Coherent admissible infimal convolution). The coherent admissible infimal convolution risk measure $\widehat{\varrho}_*$ is defined by

$$\widehat{\varrho}_*(X) := \inf \{ m \in \mathbb{R}; \ X + m \in \mathcal{A}_{\varrho_*} \} \tag{10.34}$$

$$= \inf \{ m \in \mathbb{R}; \ \varrho_*(X + m) \leq 0 \}$$

where $\mathcal{A}_{\varrho_*} = \{ X \in L^p(P); \ \varrho_*(X) \leq 0 \}$.

From the definition $\widehat{\varrho}_*$ is a coherent risk measure with acceptance set $A_{\widehat{\varrho}_*} \supset A_{\varrho_*}$ and

$$\widehat{\varrho} \le \widehat{\varrho}_* \le \varrho_* \le \min \varrho_i. \tag{10.35}$$

The following theorem says that $\widehat{\varrho}_*$ is the largest coherent risk measure $\varrho \le \min \varrho_i$. This is a justification for our choice of restrictions on the class of decompositions. Essentially less severe restrictions do not lead to a coherent allocation problem and thus admit pathological decompositions.

Theorem 10.20. *Let $\varrho_1, \dots, \varrho_n$ be coherent risk measures, then:*

(a) $\widehat{\varrho}_*$ *is the largest coherent risk measure ϱ such that $\varrho \le \min_i \varrho_i$.*
(b) *Under the equilibrium condition (E) it holds that*

$$\widehat{\varrho}_* = \widehat{\varrho} = \varrho_A.$$

Proof. (a) If ϱ is a coherent risk measure $\varrho \le \min \varrho_i$ and if $X \in L^p(P)$ has a decomposition $X = \sum_{i=1}^n \varphi_i X$, $0 \le \varphi_i$, $\sum \varphi_i = 1$, then

$$\varrho(X) \le \sum_i \varrho(\varphi_i X) \le \sum_i \varrho_i(\varphi_i X).$$

This implies that $\varrho(X) \le \varrho_*(X)$ and thus

$$A_\varrho = \{X \in L^p(P); \ \varrho(X) \le 0\}$$
$$\supset \{X \in L^p(P); \ \varrho_*(X) \le 0\}$$
$$= A_{\varrho_*}.$$

As a consequence we obtain

$$\varrho(X) = \inf \{m \in \mathbb{R}; \ X + m \in A_\varrho\}$$
$$\le \inf \{m \in \mathbb{R}; \ X + m \in A_{\varrho_*}\} = \widehat{\varrho}_*(X).$$

(b) By construction $\widehat{\varrho}_*$ is the largest coherent risk measure ϱ with $\varrho \le \varrho_*$. Under the equilibrium condition (E) also the infimal convolution $\widehat{\varrho}$ is a coherent risk measure $\widehat{\varrho} \le \min \varrho_i$. By a similar argument as in (a) any coherent risk measure $\varrho \le \min \varrho_i$ satisfies that $\varrho \le \widehat{\varrho}$. Thus by (a) $\widehat{\varrho} = \widehat{\varrho}_* = \varrho_*$. □

Remark 10.21 (Allowance of restricted borrowing). Consider the following enlarged class $A_*(X)$ of admissible allocations allowing restricted borrowing such that the components X_i satisfy $|X_i| \le |X|$. $A_*(X)$ is defined by

$$A_*(X) = \{(X_i) \in A(X); \ |X_i| \le |X|, 1 \le i \le n\}. \tag{10.36}$$

Then $\mathcal{A}_*(X)$ has the equivalent representation

$$\mathcal{A}_*(X) = \left\{ (\varphi_i X); \ |\varphi_i| \le 1, \sum_{i=1}^{n} \varphi_i = 1 \right\}. \tag{10.37}$$

The corresponding infimal convolution $\widetilde{\varrho}_*$ is given by

$$\widetilde{\varrho}_*(X) = \inf \left\{ \sum_{i=1}^{n} \varrho_i(X_i); \ (X_i) \in \mathcal{A}_*(X) \right\}. \tag{10.38}$$

Then $\widetilde{\varrho}_*(X) \le \varrho_*(X)$ and $\widetilde{\varrho}_*$ is a finite, subadditive, homogeneous, monotone risk functional on $L^p(P)$. $\widetilde{\varrho}_*$ is a coherent risk measure if and only if

$$\widetilde{\varrho}_*(1) = -1. \tag{10.39}$$

A dual representation as in Theorem 10.17 takes in this case the following form. Let $P_i = f_i \mu$ have densities $f_1, \ldots, f_n$ w.r.t. μ and let $f_{(1)} \le \cdots \le f_{(n)}$ denote the densities ordered in increasing order. Then define

$$g_n^{(P_j)} := \begin{cases} \displaystyle\sum_{i=1}^{m+1} f_{(i)} - \sum_{i=m+2}^{n} f_{(i)} & \text{if } n = 2m+1, \\ \displaystyle\sum_{i=1}^{m} f_{(i)} - \sum_{i=m+2}^{n} f_{(i)} & \text{if } n = 2m. \end{cases} \tag{10.40}$$

Then we have

Proposition 10.22. *Let $\varrho_i = \varrho_{P_i}$ be coherent risk measures. Then the extended infimal convolution $\widetilde{\varrho}_*$ has the following dual representation*

$$\widetilde{\varrho}_*(X) = \sup_{P_i \in \mathcal{P}_i} \int X_- g_n^{(P_j)} d\mu - \inf_{P_i \in \mathcal{P}_i} \int X_+ d \bigvee P_i. \tag{10.41}$$

Proof. Similarly as in the proof of Theorem 10.17 we obtain

$$\widetilde{\varrho}_*(X) = - \inf_{P_i \in \mathcal{P}_i} \sup_{\substack{-1 \le \varphi_i \le 1 \\ \sum \varphi_i = 1}} \left(\sum_i \int \varphi_i X_+ dP_i - \sum_i \int \varphi_i X_- dP_i \right)$$

$$= - \inf_{P_i \in \mathcal{P}_i} \int X_+ d \bigvee_{i=1}^{n} P_i + \sup_{P_i \in \mathcal{P}_i} \inf_{\substack{-1 \le \varphi_i \le 1 \\ \sum \varphi_i = 1}} \sum_i \int \varphi_i X_- dP_i.$$

It is easy to see that the last inf is attained in case $n = 2m + 1$ by choosing the φ_i corresponding to the $m + 1$ smallest densities $f_{(1)}, \ldots, f_{(m+1)}$ equal to 1 and those corresponding to $f_{(m+2)}, \ldots, f_{(n)}$ equal to -1. The case $n = 2m$ is obtained similarly.   □

The dual representation in (10.41) implies that in general $\widetilde{\varrho}_*(X)$ is strictly smaller than $\varrho_*(X)$. For non-negative $X \geq 0$ it holds that

$$\widetilde{\varrho}_*(X) = \varrho_*(X) = - \inf_{P_j \in \mathcal{P}_j} \int X d \bigvee P_j,$$

but for $X = -X_-$ we have

$$\varrho_*(X) = \sup_{P_j \in \mathcal{P}_j} \int X_- d \bigwedge P_j \geq \widetilde{\varrho}_*(X) = \sup_{P_j \in \mathcal{P}_j} \int X_- g_n^{(P_j)} d\mu, \qquad (10.42)$$

since $f_{(1)} \geq g_n^{(P_j)}$ and the inequality will be typically strict. $\Diamond$

The (admissible) infimal convolution is relevant from the point of view of traders. From a regulatory point of view a regulator would like to take care for the worst case of allocation of risk and the corresponding necessary capital to meet this situation.

A first choice in this direction is

$$\varrho_{\max}(X) := \max \varrho_i(X) \qquad (10.43)$$

(see Delbaen (2000) and Föllmer and Schied (2011)). It is easy to check that $\varrho_{\max}$ is a coherent risk measure. The acceptance set is given by

$$\begin{aligned}
\mathcal{A}_{\varrho_{\max}} &= \{X \in L^p(P); \ \varrho_{\max}(X) \leq 0\} \\
&= \{X \in L^p(P); \ \varrho_i(X) \leq 0, \ \forall i\} \\
&= \bigcap_{i=1}^{n} \mathcal{A}_{\varrho_i}.
\end{aligned} \qquad (10.44)$$

The representing scenario set is given by

$$\mathcal{P}_{\varrho_{\max}} = \text{conv} \left(\bigcup_{i=1}^{n} \mathcal{P}_i \right) \qquad (10.45)$$

since

$$\begin{aligned}
X \in \mathcal{A}_{\varrho_{\max}} &\Leftrightarrow X \in \mathcal{A}_{\varrho_i}, 1 \leq i \leq n \\
&\Leftrightarrow \forall \, Q_i \in \mathcal{P}_i : E_{Q_i} X \geq 0, 1 \leq i \leq n \\
&\Leftrightarrow \forall \, Q \in \text{conv} \left(\bigcup_{i=1}^{n} \mathcal{P}_i \right) : E_Q X \geq 0.
\end{aligned}$$

$\varrho_{\max}$ is obviously the smallest coherent risk measure majorizing ϱ_i, $1 \leq i \leq n$. No equilibrium condition is necessary for the application of $\varrho_{\max}$.

When the risk is possibly allocated to the traders in the most unfavourable way the regulator would apply the supremal convolution.

Definition 10.23 (Supremal convolution). For coherent risk measures $\varrho_1, \ldots, \varrho_n$ define the supremal convolution $\widehat{\tau} := \varrho_1 \vee \cdots \vee \varrho_n$ by

$$\widehat{\tau}(X) := \sup \left\{ \sum_{i=1}^{n} \varrho_i(X_i); \ \sum_{i=1}^{n} X_i = X \right\}. \tag{10.46}$$

$\widehat{\tau}$ satisfies all conditions of a coherent risk measure except possibly the condition $\widehat{\tau}(0) = 0$. $\widehat{\tau}(0)$ can be interpreted as a possible hidden risk position in the market when decomposing the zero position in an unfavourable way to the traders. To investigate this we introduce as in Section 10.1 Pareto equilibrium conditions:

$$(E_\tau) \qquad \sum_{i=1}^{n} X_i = 0 \text{ and } \varrho_i(X_i) \geq 0, \ \forall\, i \text{ implies } \varrho_i(X_i) = 0, \ \forall\, i. \tag{10.47}$$

The corresponding strengthened version of (E_τ) is

$$(SE_\tau) \qquad \sum_{i=1}^{n} X_i = 0 \text{ implies } \sum_{i=1}^{n} \varrho_i(X_i) \leq 0. \tag{10.48}$$

As in Proposition 10.3 we find:

Proposition 10.24. *(a) The equilibrium conditions (E_τ) and (SE_τ) are equivalent.*

(b) $\widehat{\tau}$ is a coherent risk measure $\Leftrightarrow \widehat{\tau}(0) = 0$

$\Leftrightarrow$ *The equilibrium condition (E_τ) holds.*

In the following proposition we establish that – in spite of the formal similarity of condition (E_τ) to (E) – only in an exceptional case the equilibrium condition (E_τ) holds.

Proposition 10.25. *Consider coherent risk measures $\varrho_i = \varrho_{\mathcal{P}_i}$. Then the equilibrium condition (E_τ) holds if and only if $\mathcal{P}_1 = \mathcal{P}_2 = \cdots = \mathcal{P}_n$ and $|\mathcal{P}_i| = 1$.*

Proof. We consider at first the case $n = 2$. Then (E_τ) is equivalent to:

$$\forall X \in L^p(P): \ \sup_{Q \in \mathcal{P}_1} E_Q X - \inf_{Q \in \mathcal{P}_2} E_Q X \leq 0.$$

This is further equivalent to:

$$\forall X \in L^p(P) \text{ holds}: \quad \sup_{Q \in \mathcal{P}_1} E_Q X \leq \inf_{Q \in \mathcal{P}_2} E_Q X$$

$$\text{and} \quad \sup_{Q \in \mathcal{P}_2} E_Q X \leq \inf_{Q \in \mathcal{P}_1} E_Q X$$

considering X as well as $-X$.

Thus we obtain equivalence to

$$\sup_{Q \in \mathcal{P}_1} E_Q X \leq \inf_{Q \in \mathcal{P}_2} E_Q X$$

$$\leq \sup_{Q \in \mathcal{P}_2} E_Q X \leq \inf_{Q \in \mathcal{P}_1} E_Q X, \quad \forall X \in L^p(P)$$

or equivalently

$$\sup_{Q \in \mathcal{P}_1} E_Q X = \inf_{Q \in \mathcal{P}_1} E_Q X \qquad (10.49)$$

$$= \inf_{Q \in \mathcal{P}_2} E_Q X = \sup_{Q \in \mathcal{P}_2} E_Q X, \quad \forall X \in L^p(P).$$

Equation (10.49) again is equivalent to

$$\mathcal{P}_1 = \mathcal{P}_2 \text{ and } |\mathcal{P}_1| = |\mathcal{P}_2| = 1.$$

In the case $n \geq 2$ we obtain from the recursive structure of the supremal convolution

$$\widehat{\tau} = (\varrho_1 \vee \cdots \vee \varrho_{n-1}) \vee \varrho_n. \qquad (10.50)$$

Thus the result for the case $n = 2$ implies:

$$\mathcal{P}_{\varrho_n} = \mathcal{P}_{\varrho_1 \vee \cdots \vee \varrho_{n-1}} = \mathcal{P}_{\varrho_{n-1}} = \cdots = \mathcal{P}_{\varrho_1} \text{ and } |\mathcal{P}_{\varrho_i}| = 1 \qquad (10.51)$$

is equivalent to the equilibrium condition (E$_\tau$).   □

Remark 10.26 (Risk arbitrage). As a consequence of Proposition 10.25 we find that the equilibrium condition (E$_\tau$) is only satisfied in the exceptional case that $\mathcal{P}_1 = \mathcal{P}_2 = \cdots = \mathcal{P}_n = \{P_1\}$. In all other situations it follows from the homogeneity of $\widehat{\tau}$ that

$$\widehat{\tau}(0) = \infty. \qquad (10.52)$$

Thus for any $K > 0$ there exist pathological risk decompositions (X_i) of 0

$$0 = \sum_{i=1}^{n} X_i \text{ with } \sum_{i=1}^{n} \varrho_i(X_i) \geq K. \qquad (10.53)$$

Thus in the market there are hidden allocations of the risk 0 to the traders with arbitrary large sum of risks $\sum \varrho_i(X_i)$ if any decomposition of X is taken into consideration.   ◊

As a consequence of Proposition 10.25 it is necessary to prevent the regulatory agent to be too strict since this would lead to "artificial" worst case scenarios with too high risk, i.e. the supremal risk measure is no reasonable risk measure. It seems natural to restrict the class of admissible decompositions as in the inf-convolution case. Let

$$A^a(X) := \left\{ (X_i) = (\varphi_i X); \ 0 \le \varphi_i \le 1, \sum_{i=1}^{n} \varphi_i = 1 \right\} \tag{10.54}$$

denote the class of admissible decompositions and define the "admissible supremal convolution"

$$\varrho^*(X) := \sup \left\{ \sum_{i=1}^{n} \varrho_i(X_i); \ (X_i) \in A^a(X) \right\}. \tag{10.55}$$

Then ϱ^* is a subadditive homogeneous monotone risk functional (with $\varrho^*(0) = 0$) but ϱ^* is not translation invariant in general. Obviously

$$\varrho_{\max} \le \varrho^* \le \widehat{\tau}. \tag{10.56}$$

As in Theorem 10.17, ϱ^* can be calculated explicitly in terms of the representation scenarios $\mathcal{P}_i$ of ϱ_i.

Theorem 10.27. *Let $\varrho_i = \varrho_{\mathcal{P}_i}$ be coherent risk measures and let ϱ^* be the corresponding admissible supremal convolution. Then*

(a) $\quad \varrho^*(X) = \sup \left\{ \int X_- d \bigvee_i P_i - \int X_+ d \bigwedge_i P_i; \ P_i \in \mathcal{P}_i, 1 \le i \le n \right\},$

(b) $\quad A_{\varrho^*} = \left\{ X \in L^p(P); \ such\ that \int X_- d \bigvee_i P_i \le \int X_+ d \bigwedge_i P_i, \ \forall P_i \in \mathcal{P}_i \right\}.$

Proof. (a) As in the proof of Theorem 10.17 we obtain

$$\varrho^*(X) = \sup_{(\varphi_i)} \sum_i \varrho_i(\varphi_i X) = \sup_{(\varphi_i)} \sum_i \sup_{P_i \in \mathcal{P}_i} E_{P_i} \varphi_i Y, \quad \text{with } Y := -X$$

$$= \sup_{(P_i) \in (\mathcal{P}_i)} \sup_{(\varphi_i)} \sum_i \int \varphi_i Y \, dP_i$$

$$= \sup_{(P_i)} \left(\int Y_+ d \bigvee_i P_i - \int Y_- d \bigwedge_i P_i \right)$$

$$= \sup_{(P_i)} \left(\int X_- d \bigvee_i P_i - \int X_+ d \bigwedge_i P_i \right).$$

(b) follows from (a).

$\square$

Remark 10.28. (a) As a consequence of Theorem 10.27 we see that

$$\varrho^*(1) = - \inf \left\{ \left| \bigwedge_i P_i \right|; \ P_i \in \mathcal{P}_i \right\} \tag{10.57}$$

and thus

$$\varrho^*(1) = -1 \Leftrightarrow |\bigwedge P_i| = 1, \quad \forall P_i \in \mathcal{P}_i \tag{10.58}$$
$$\Leftrightarrow P_1 = P_2 = \cdots = P_n \text{ and } |\mathcal{P}_i| = 1$$
$$\Leftrightarrow \text{The Pareto equilibrium condition (E}_\tau) \text{ holds.}$$

(b) **Enlarged class of admissible strategies.** As in the infimal convolution case we can enlarge the class of admissible decompositions to

$$\widetilde{A}(X) = \left\{ (X_i); \ \sum_{i=1}^n X_i = X \text{ and } |X_i| \le |X| \right\}.$$

We then get a corresponding risk functional

$$\widetilde{\varrho}^*(X) := \sup \left\{ \sum \varrho_i(X_i); \ (X_i) \in \widetilde{A}(X) \right\}. \tag{10.59}$$

Similarly to Theorem 10.20 we obtain an explicit representation of $\widetilde{\varrho}^*$ by

$$\widetilde{\varrho}^*(X) = \sup_{P_i \in \mathcal{P}_i} \int X_- d\bigvee P_i - \inf_{P_i \in \mathcal{P}_i} \int X_+ g_n^{(P_j)} d\mu \tag{10.60}$$

with $g_n^{(P_j)}$ defined as in (10.40).

As a consequence we obtain that similar to the infimum case in Remark 10.11 the enlargement of admissible strategies leads to a different risk measure

$$\varrho^*(X) \le \widetilde{\varrho}^*(X) \tag{10.61}$$

and equality holds only under the Pareto equilibrium condition (E$_\tau$).   ◊

Definition 10.29 (Coherent admissible supremal convolution).
The coherent admissible supremal convolution risk measure $\widehat{\varrho}^*$ is defined by

$$\widehat{\varrho}^*(X) = \inf\{m \in \mathbb{R}; \ X + m \in A_{\varrho*}\} = \inf\{m \in \mathbb{R}; \ \varrho^*(X + m) \le 0\}, \tag{10.62}$$

where $A_{\varrho*} = \{X \in L^p(P); \ \varrho^*(X) \le 0\}$.

In the following proposition we characterize $\widehat{\varrho}^*$ by a maximality property.

Proposition 10.30. *(a) The coherent admissible supremal convolution risk measure $\widehat{\varrho}^*$ is the largest coherent risk measure ϱ such that*

$$\varrho^* \ge \varrho \ge \varrho_{max}.$$

(b) $A_{\widehat{\varrho}*} = A_{\varrho*}$ and the scenarios representation set $\mathcal{P}^*$ of $\widehat{\varrho}^*$ is given by

$$\mathcal{P}^* = \overline{\text{conv } Q}^{L^q(P)},$$

where $\mathcal{Q} = \left\{ \frac{Q}{|Q|} \in M^q(P); \exists P_i \in \mathcal{P}_i, 1 \leq i \leq n, \text{ with } \bigwedge P_i \leq Q \leq \bigvee P_i \right\}$ *and* $M^q(P)$ *denotes the P-continuous probability measures with finite q-th moments of* $\frac{dQ}{dP}$.

Proof. (a) $\widehat{\varrho}^*$ is the largest coherent risk measure ϱ with $\varrho \leq \varrho^*$.

(b) By the bipolar theorem and the representation of ϱ^* in Theorem 10.27 it is to show that

$$\int X_- d \bigvee P_i \leq \int X_+ d \bigwedge P_i, \quad \forall P_i \in \mathcal{P}_i$$

is equivalent to $\int X_- dQ \leq \int X_+ dQ, \quad \forall Q \in \mathcal{P}^*$.

If $Q \in \mathcal{P}^*$ and $\bigwedge P_i \leq \alpha Q \leq \bigvee P_i$ with some positive constant α then

$$\alpha \int X_- dQ \leq \int X_- d \bigvee P_i \leq \int X_+ d \bigwedge P_i \leq \alpha \int X_+ dQ.$$

For the other direction let $P_i \in \mathcal{P}_i, 1 \leq i \leq n$ and define

$$Q := \mathbb{1}_{\{X \geq 0\}} \bigwedge P_i + \mathbb{1}_{\{X < 0\}} \bigvee P_i.$$

Then $\frac{Q}{|Q|} \in \mathcal{P}^*$ and

$$\int X_- d \bigvee P_i = \int X_- dQ \leq \int X_+ dQ = \int X_+ d \bigwedge P_i.$$

Thus $\mathcal{P}^*$ is the representation set of $\widehat{\varrho}^*$ since $\mathcal{A}_{\widehat{\varrho}^*} = \mathcal{A}_{\varrho^*} = (\mathcal{P}^*)^0$ and thus

$$\text{conv}\left(\{0\} \cup \mathcal{P}^*\right) = (\mathcal{A}_{\widehat{\varrho}^*})^0. \qquad \square$$

Resumée of Section 10.2: The classical risk allocation (risk sharing) problem is only well defined if the view towards risks of the traders is not too different, in mathematical terms if the Pareto equilibrium condition holds. Thus it is of interest to extend the allocation problem in a senseful way to the general case. An effective way for the formulation of a general allocation problem is to introduce restrictions on the class of admissible allocations in a suitable way. This is done for the risk sharing problem from the viewpoint of the traders and also from the viewpoint of a regulatory agent. As a consequence we obtain new relevant coherent risk measures for allocation problems. A simplified dual representation for these intrinsic risk measures is derived. The choice of the restriction conditions is justified by the fact that essentially less restrictive conditions would allow "pathological" allocations ("risk arbitrage") and in particular would not allow to obtain Pareto optimal allocations.

10.3   Pareto Equilibrium for Convex Risk Measures

For convex risk measures ϱ_i on $L^p(P)$ the equilibrium condition and optimal allocations take a somewhat different form compared with Sections 10.1 and 10.2. We
consider lower semicontinuous convex risk measures on $L^p(P)$ with representation
of the form

$$\varrho_i(X) = \sup_{Q \in \mathcal{P}_i} (E_Q(-X) - \alpha_i(Q)), \qquad (10.63)$$

where $\mathcal{P}_i$ are convex $\sigma(L^q, L^p)$-closed representing scenario sets in $\mathcal{M}_1^q$ respectively $\mathcal{M}_1$ in case $p = \infty$ and $\alpha_i(Q) : L^q(P) \to (-\infty, \infty]$ are convex penalty
functions which can be chosen as Fenchel–Legendre conjugates of ϱ_i

$$\alpha_i(Q) = \inf_{X \in L^p(P)} (E_Q(-X) - \varrho_i(X)). \qquad (10.64)$$

The acceptability sets $\mathcal{A}_{\varrho_i} = \{X \in L^p(P); \varrho_i(X) \leq 0\}$ are $\sigma(L^p(P), L^q(P))$-
closed under the above assumptions. In general $\varrho_i(0)$ are not necessarily equal to
zero.

The optimal risk allocation problem as in (10.3) is not well defined in general but
is connected with a Pareto equilibrium condition.

Definition 10.31 (Pareto equilibrium).   A market model with risk measures
$\varrho_1, \ldots, \varrho_n$ is in Pareto equilibrium, if

(E) $X_i \in L^p(P)$ with $\sum_{i=1}^{n} X_i = 0$ and $\varrho_i(X_i) \leq \varrho_i(0), \quad 1 \leq i \leq n,$

   implies $\varrho_i(X_i) = \varrho_i(0), \quad 1 \leq i \leq n.$

In a balance of supply and demand in the market it is in Pareto equilibrium not
possible to lower the risk of some traders without increasing that of others. Vaguely
Definition 10.31 says that there is a *no arbitrage* situation concerning risk. There is
a seemingly stronger version of the equilibrium condition:

(SE)   **Strong equilibrium**

   If $X_i \in L^p(P)$ with $\sum_{i=1}^{n} X_i = 0$, then $\sum_{i=1}^{n} \varrho_i(X_i) \geq \sum_{i=1}^{n} \varrho_i(0).$

It is immediate to see that (SE) $\Rightarrow$ (E), but in fact both conditions are equivalent.

Proposition 10.32. *The equilibria conditions* (E) *and* (SE) *are equivalent.*

Proof. Assume that for some $X_i \in L^p(P)$ with $\sum_{i=1}^{n} X_i = 0$ it holds that

$$\sum_{i=1}^{n} (\varrho_i(X_i) - \varrho_i(0)) =: c < 0.$$

Then with $c_i := \varrho_i(X_i) - \varrho_i(0)$ and $Z_i := X_i + c_i - \frac{c}{n}$ it holds that:

$$\sum_{i=1}^{n} Z_i = 0 \quad \text{and}$$

$$\varrho_i(Z_i) = \varrho_i(X_i) - c_i + \frac{c}{n} = \varrho_i(0) + \frac{c}{n} < \varrho_i(0), \quad 1 \le i \le n.$$

In consequence we obtain a contradiction to (E). □

Thus under the Pareto equilibrium condition (E) the sum of all risks in a balance situation is greater than the cumulative risk of zero. The infimal convolution $\widehat{\varrho} = \varrho_1 \wedge \cdots \wedge \varrho_n$ as defined in (10.3) describes the optimal allocation of risk. $\widehat{\varrho}$ itself is, as a functional of X, a risk measure.

If $\widehat{\varrho}$ is a convex risk measure, then by Theorem 3.6 in Barrieu and El Karoui (2005) the penalty function $\alpha_{\widehat{\varrho}}$ of $\widehat{\varrho}$ is given by

$$\alpha_{\widehat{\varrho}} = \sum_{i=1}^{n} \alpha_i.$$

A characterization of Pareto equilibrium is given in the following proposition in terms of the penalty functions α_i.

Proposition 10.33. *Let* $\mathcal{P}_i := \{Q \in \mathcal{M}_1^q(P); \alpha_i(Q) < \infty\}$ *denote the "scenario sets" of* ϱ_i, $1 \le i \le n$. *Then the following conditions (1)–(3) are equivalent:*

(1) $\widehat{\varrho}$ *is a convex risk measure.*

(2) $\widehat{\varrho}(0) > -\infty$.

(3) $\bigcap_{i=1}^{n} \mathcal{P}_i \ne \phi$.

Further conditions (4) and (5) are equivalent and imply (1)–(3):

(4) $\widehat{\varrho}(0) = \sum_{i=1}^{n} \varrho_i(0)$.

(5) *The equilibrium condition (E) holds.*

Proof. The first group of equivalences is similar to those in Proposition 10.4. The equivalence of (4) and (5) follows from the definition of $\widehat{\varrho}$ since

$$\widehat{\varrho}(0) = \inf\left\{\sum_{i=1}^{n} \varrho_i(X_i); \sum_{i=1}^{n} X_i = 0\right\} = \sum_{i=1}^{n} \varrho_i(0)$$

is equivalent to

$$\sum_{i=1}^{n} X_i = 0 \Rightarrow \sum_{i=1}^{n} \varrho_i(X_i) \ge \sum_{i=1}^{n} \varrho_i(0)$$

i.e. to condition (SE). and thus by Proposition 10.32 to the equilibrium condition (E). □

Remark 10.34. In the case of coherent risk measures ϱ_i, $\alpha_i(Q) \in \{0, \infty\}$ for all i and the scenario sets $\mathcal{P}_i = \{Q \in \mathcal{M}_1^q(P); \alpha_i(Q) = 0\}$ defined in Proposition 10.33 are identical with the sets of scenario measures of ϱ_i. In this case (1) is also equivalent to (4). This follows from the fact that under (1) the penalty function $\alpha_{\widehat{\varrho}}$ of $\widehat{\varrho}$ is given by $\alpha_{\widehat{\varrho}} = \sum_{i=1}^{n} \alpha_i$ and using the representation (10.63) of convex risk measures. Thus Proposition 10.33 implies the characterization in Theorem 10.9. The equilibrium condition (E) is equivalent to the condition $\bigcap_{i=1}^{n} \mathcal{P}_i \neq \phi$, i.e., there exists a scenario measure Q which is shared by all traders in the market. $\Diamond$

In the case that the Pareto equilibrium condition does not hold we introduce the following formulation of the optimal allocation problem which makes sense also without assuming condition (E).

Definition 10.35 (Infimal admissible convolution).

(a) For $X \in L^p(P)$ define a decomposition $X = \sum_{i=1}^{n} X_i$ to be "admissible" if $X(\omega) \geq 0$ implies that $0 \leq X_i(\omega)$ and $X(\omega) \leq 0$ implies that $X_i(\omega) \leq 0$. Let $A(X)$ denote the set of admissible decompositions (X_i) of X.

(b) The "infimal admissible convolution" ϱ_* is defined as the value of the allocation problem restricted to admissible allocations

$$\varrho_*(X) := \inf\left\{\sum_{i=1}^{n} \varrho_i(X_i); \ (X_i) \in A(X)\right\}. \tag{10.65}$$

Obviously, we have

$$\varrho_*(X) \leq \min(\varrho_i(X) - \varrho_i(0)) + \sum_{i=1}^{n} \varrho_i(0).$$

The restriction to admissible decompositions of risks excludes, similarly as the restriction to admissible strategies in portfolio theory, that effects like doubling strategies paradoxes may occur in risk allocation.

In the following theorem we derive a dual representation of the infimal admissible convolution ϱ_* in terms of the representation scenarios $\mathcal{P}_i := \{Q \in \mathcal{M}_1^q(P); \alpha_i(Q) < \infty\}$ of ϱ_i. For $P_i \in \mathcal{P}_i$ define by $P_1 \wedge \cdots \wedge P_n$ the lattice infimum of (P_i) in the lattice $\mathcal{M}_1^q(P)$ and by $P_1 \vee \cdots \vee P_n$ the lattice supremum of (P_i) in $\mathcal{M}_1^q(P)$. If P_i are probability measures with densities f_i w.r.t. μ then $P_1 \wedge \cdots \wedge P_n$ respectively $P_1 \vee \cdots \vee P_n$ have densities $\min\{f_i\}$ respectively $\max\{f_i\}$ w.r.t. μ.

Theorem 10.36. *Let ϱ_j be convex risk measures with penalty functions α_j. Then the infimal admissible convolution ϱ_* has the dual representation*

$$\varrho_*(X) = \sup\left\{\int X_- d \bigwedge_j P_j - \int X_+ d \bigvee_j P_j - \sum_j \alpha_j(P_j); \right. \tag{10.66}$$

$$\left. P_j \in \mathcal{P}_j, 1 \leq j \leq n\right\}.$$

Proof. The proof is similar to that for coherent risk measures in Theorem 10.17. The class of admissible allocations has an alternative description like in multiple decision problems:

$$A(X) = \left\{ (\varphi_i X); \ 0 \le \varphi_i \le 1, \sum_{i=1}^{n} \varphi_i = 1 \right\};$$ (10.67)

thus we consider allocations of the form $(\varphi_i X)$.

For $P_i \in \mathcal{P}_i$, $1 \le i \le n$ and $Y := -X \in L^p(P)$ it holds that

$$a_{P_1,\ldots,P_n}(Y) := \inf \left\{ \sum_{i=1}^{n} \int \varphi_i Y \, dP_i; \ 0 \le \varphi_i, \sum_{i=1}^{n} \varphi_i = 1 \right\}$$

has a solution (φ_i^*).

If $\quad Y(\omega) > 0$, then $\{\varphi_i^* > 0\} \subset \{P_i = \bigwedge_{j=1}^{n} P_j\}$
and if $\quad Y(\omega) < 0$, then $\{\varphi_i^* > 0\} \subset \{P_i = \bigvee_{j=1}^{n} P_j\}$.

Thus

$$a_{P_1,\ldots,P_n}(Y) = \int_{Y \ge 0} Y \, d \bigwedge_{j=1}^{n} P_j + \int_{Y < 0} Y \, d \bigvee_{j=1}^{n} P_j$$

$$= \int Y_+ \, d \bigwedge P_j - \int Y_- \, d \bigvee P_j$$

$$= \int X_- \, d \bigwedge P_j - \int X_+ \, d \bigvee P_j.$$

Therefore, we obtain

$$\varrho_*(X) = \inf_{(\varphi_i)} \sum_i \varrho_i(\varphi_i X) = \inf_{(\varphi_i)} \sum_i \sup_{P_i \in \mathcal{P}_i} \left(\int (-\varphi_i X) \, dP_i - \alpha_i(P_i) \right)$$

$$= \inf_{(\varphi_i)} \left[-\sum_{i=1}^{n} \inf_{P_i \in \mathcal{P}_i} \left(\int \varphi_i X \, dP_i + \alpha_i(P_i) \right) \right]$$

$$= - \sup_{(\varphi_i)} \sum_i \inf_{P_i \in \mathcal{P}_i} \left(\int \varphi_i X \, dP_i + \alpha_i(P_i) \right).$$

We now apply the "minimax theorem" for games (A, B, f) of concave–convex type, in (10.33) with $A = \{(\varphi_i); 0 \le \varphi_i, \sum \varphi_i = 1\}$, which is compact in weak *-topology, $B = \mathcal{P}_1 \times \cdots \times \mathcal{P}_n$ and

$$f((\varphi_i), (P_i)) = \sum_{i=1}^{n} \left(\int \varphi_i X \, dP_i + \alpha_i(P_i) \right).$$

By linearity of f and convexity of $\mathcal{P}_i$ and A the conditions of the minimax theorem are fulfilled and we obtain from the first part of the proof

$$\varrho_*(X) = - \inf_{P_i \in \mathcal{P}_i} \sup_{(\varphi_i)} \sum_i \left(\int \varphi_i X \, dP_i + \alpha_i(P_i) \right)$$

$$= - \inf_{P_i \in \mathcal{P}_i} \left(\int X_+ \, d\bigvee P_i - \int X_- \, d\bigwedge P_i + \sum_i \alpha_i(P_i) \right)$$

$$= \sup_{P_i \in \mathcal{P}_i} \left(\int X_- \, d\bigwedge P_i - \int X_+ \, d\bigvee P_i - \sum_i \alpha_i(P_i) \right). \qquad \square$$

As a consequence of Theorem 10.36 we obtain a further characterization of the Pareto equilibrium condition (E) and also obtain under condition (E) the identity of $\widehat{\varrho}$ and ϱ_*. Therefore, the representation in (10.66) holds true also for the infimal convolution $\widehat{\varrho}$ under the equilibrium condition (E).

Proposition 10.37. *Let ϱ_i be convex risk measures and let ϱ_* denote the admissible infimal convolution, then it holds that:*

$$\varrho_* \text{ is a convex risk measure} \Leftrightarrow \varrho_*(1) = \varrho_*(0) - 1$$

$$\Leftrightarrow \bigcap_{i=1}^{n} \mathcal{P}_i \neq \phi.$$

Under the equilibrium condition (E) we have $\varrho_* = \widehat{\varrho}$.

Proof. The first equivalence is obvious since the condition $\varrho_*(1) = \varrho_*(0) - 1$ implies translation invariance of ϱ_*. By the dual representation in Theorem 10.36

$$\varrho_*(1) = \sup \left\{ - \left| \bigvee_j P_j \right| - \sum_j \alpha_j(P_j); \ P_j \in \mathcal{P}_j \right\}$$

$$= - \inf \left\{ \left| \bigvee_j P_j \right| + \sum_j \alpha_j(P_j); \ P_j \in \mathcal{P}_j \right\}.$$

Thus $\varrho_*(1) = \varrho_*(0) - 1$ if and only if there exists a common scenario measure $Q \in \bigcap_{i=1}^{n} \mathcal{P}_i$. By Proposition 10.33 the equilibrium condition (E) implies that $\bigcap \mathcal{P}_i \neq \phi$. $\qquad \square$

ϱ_* is a convex monotone risk functional – in particular $\varrho_*(0) = \sum_{i=1}^{n} \varrho_i(0)$ – but ϱ_* is not translation invariant in general. From the definition of ϱ_* we obtain

$$\varrho_*(1) = \inf \left\{ \sum_{i=1}^{n} \varrho_i(X_i); \ 0 \le X_i, \sum_{i=1}^{n} X_i = 1 \right\} \qquad (10.68)$$

$$\le \varrho_1(1) + \sum_{i=2}^{n} \varrho_i(0) = \varrho_*(0) - 1.$$

As we have seen the modified allocation problem given by ϱ_* leads to a senseful version of the allocation problem also when (E) does not hold. We finally get by a modification of ϱ_* a convex risk measure $\widehat{\varrho}_*$ which we call convex infimal admissible convolution.

Definition 10.38 (Convex infimal admissible convolution). We define the convex infimal admissible convolution risk measure $\widehat{\varrho}_*$ by

$$\widehat{\varrho}_*(X) := \inf\{m \in \mathbb{R};\ X + m \in \mathcal{A}\} \qquad (10.69)$$

$$= \inf\{m \in \mathbb{R};\ \varrho_*(X + m) \le 0\}$$

where $\mathcal{A} = \mathcal{A}_{\varrho_*} = \{X;\ \varrho_*(X) \le 0\}$.

From the definition $\widehat{\varrho}_*$ is a convex risk measure with $\mathcal{A}_{\widehat{\varrho}_*} \supset \mathcal{A}_{\varrho_*} = \mathcal{A}$ and

$$\widehat{\varrho}_* \le \varrho_* \qquad (10.70)$$

Theorem 10.39. Let $\varrho_1, \ldots, \varrho_n$ be convex risk measures as in (10.63), then:

(a) $\widehat{\varrho}_*$ is the largest convex risk measure ϱ with $\varrho \le \varrho_*$.
(b) Under the equilibrium condition (E) it holds that

$$\widehat{\varrho}_* = \widehat{\varrho}.$$

Proof. (a) This follows from the definition of $\widehat{\varrho}_*$.
(b) Under the equilibrium condition (E) ϱ_* is a convex risk measure by Proposition 10.37. Therefore $\widehat{\varrho}_* = \varrho_* = \widehat{\varrho}$ by (1). $\qquad \square$

From the regulatory point of view on risk allocations a cautious risk attitude is relevant. The "natural allocation problem" then is given by the supremal convolution

$$\widehat{\tau}(X) := \sup\left\{ \sum_{i=1}^n \varrho_i(X_i);\ \sum_{i=1}^n X_i = X \right\}, \qquad (10.71)$$

where ϱ_i are convex risk measures. It turns out however, that except in the trivial situation where $\mathcal{P}_1 = \mathcal{P}_2 = \cdots = \mathcal{P}_n = \{P_1\}$ we do not get by (10.71) a convex risk measure (see (10.58) and Proposition 10.25).

As in the coherent case it seems natural to restrict the class of decompositions. Let

$$A^a(X) := \left\{ (X_i) = (\varphi_i X);\ 0 \le \varphi_i \le 1, \sum_{i=1}^n \varphi_i = 1 \right\} \qquad (10.72)$$

denote the class of admissible decompositions and define the value of the allocation problem with cautious risk attitude from a regulatory point of view which we call the "admissible supremal convolution" by

$$\sigma^*(X) := \sup\left\{ \sum_{i=1}^{n} \varrho_i(X_i);\ (X_i) \in A^\alpha(X)\right\}. \tag{10.73}$$

σ^* is a convex monotone risk functional (with $\sigma^*(0) = \sum_{i=1}^{n} \varrho_i(0)$) but σ^* is not translation invariant in general. Obviously

$$\sigma^* \le \widehat{\tau}. \tag{10.74}$$

As in Theorem 10.36, σ^* can be *calculated* explicitly in terms of the representation scenario set $\mathcal{P}_i$ of ϱ_i.

Theorem 10.40. *Let ϱ_i be convex risk measures with penalty functions α_i and σ^* the corresponding admissible supremal convolution. Let $\mathcal{P}_i := \{Q \in M_1^q(P);\ \alpha_i(Q) < \infty\}$ denote the scenario set. Then*

$$\sigma^*(X) = \sup\left\{ \int X_- d \bigvee_i P_i - \int X_+ d \bigwedge_i P_i - \sum_i \alpha_i(P_i);\right.$$
$$\left. P_i \in \mathcal{P}_i, 1 \le i \le n\right\}.$$

Proof. As in the proof of Theorem 10.36 we obtain

$$\sigma^*(X) = \sup_{(\varphi_i)} \sum_i \varrho_i(\varphi_i X)$$

$$= \sup_{(\varphi_i)} \sum_i \sup_{P_i \in \mathcal{P}_i} (E_{P_i} \varphi_i Y - \alpha_i(P_i)), \quad \text{with } Y := -X$$

$$= \sup_{(P_i) \in (\mathcal{P}_i)} \sup_{(\varphi_i)} \sum_i \left(\int \varphi_i Y\, dP_i - \alpha_i(P_i)\right)$$

$$= \sup_{(P_i)} \left(\int Y_+ d \bigvee_i P_i - \int Y_- d \bigwedge_i P_i - \sum_i \alpha_i(P_i)\right)$$

$$= \sup_{(P_i)} \left(\int X_- d \bigvee_i P_i - \int X_+ d \bigwedge_i P_i - \sum_i \alpha_i(P_i)\right). \qquad \square$$

Remark 10.41. As a consequence of Theorem 10.40 we see that

$$\sigma^*(1) = - \inf\left\{ \left| \bigwedge_i P_i \right| + \sum \alpha_i(P_i);\ P_i \in \mathcal{P}_i\right\} \tag{10.75}$$

and thus

$$\sigma^*(1) = \sum \varrho_i(0) - 1 \Leftrightarrow \left| \bigwedge P_i \right| = 1, \quad \forall P_i \in \mathcal{P}_i \qquad (10.76)$$

$$\Leftrightarrow \mathcal{P}_1 = \mathcal{P}_2 = \cdots = \mathcal{P}_n \text{ and } |\mathcal{P}_i| = 1. \qquad \Diamond$$

As in the coherent case we obtain from the restricted allocation problem a meaningful convex risk measure.

Definition 10.42 (Convex admissible supremal convolution risk measure). We define the convex admissible supremal convolution risk measure $\widehat{\sigma}^*$ by

$$\widehat{\sigma}^*(X) = \inf\{m \in \mathbb{R}; \ X + m \in \mathcal{A}_{\sigma^*}\} = \inf\{m \in \mathbb{R}; \ \sigma^*(X+m) \leq 0\}, \quad (10.77)$$

where $\mathcal{A}_{\sigma^*} = \{X \in L^p(P); \ \sigma^*(X) \leq 0\}$.

As result we obtain

Proposition 10.43. *The convex admissible supremal convolution risk measure $\widehat{\sigma}^*$ is the largest convex risk measure ϱ such that*

$$\varrho_i \leq \varrho \leq \sigma^* \quad for \ 1 \leq i \leq n.$$

The admissibility of a decomposition seems to be a natural postulate for regulatory purposes. Thus the admissible supremal convolution risk measure offers a solution to the risk measure problem for markets under a regulatory attitude towards risk.

Remark 10.44 (Admissible risk transfers with finitely many instruments). Filipović and Kupper (2008) introduced an interesting and relevant class of admissible risk transfers. For a given risk allocation $X = \sum_{i=1}^n C_i$ and a given vector $Z = (Z_1, \ldots, Z_d)$ of d available random market instruments only allocations of the form

$$X = \sum_{i=1}^n X_i \quad \text{with } X_i = C_i + x_i \cdot Z \qquad (10.78)$$

with $x_i \in \mathbb{R}^d$ such that $\sum_{i=1}^n x_i \cdot Z \leq 0$ are considered as admissible transfers. Thus the optimal restricted risk allocation problem

$$\sum_{i=1}^n \varrho_i(C_i + x_i \cdot Z) = \inf_{x_i \text{ admissible}} \qquad (10.79)$$

leads to an optimization problem with vector valued variables $x_1, \ldots, x_n \in \mathbb{R}^d$ and methods from game theory can be applied to characterize optimal solutions. Problem (10.79) can be seen as a *variant* of the classical portfolio optimization problem which aims to minimize the risk $\varrho(x \cdot Z)$ over all portfolio vectors $x = (x_1, \ldots, x_d)$, $x_i \geq 0$, $\sum_{i=1}^n x_i = 1$. $\qquad \Diamond$

10.4   Pareto Optimality, Comonotonicity, and Existence of Optimal Allocations

The main result of this section is the comonotone improvement result. It states that any allocation (X_i) of X in L^1 can be improved by a comonotone allocation (X_i^*) of X, i.e. $\sum_{i=1}^{n} X_i^* = X$, $X_1^*, \dots, X_n^*$ are comonotone and $X_i^* \leq_{\mathrm{cx}} X_i$, $1 \leq i \leq n$. This result was established by Landsberger and Meilijson (1994) in the case of finite Ω. It was extended in Dana and Meilijson (2003) to the case that $X \in L^\infty(P)$ and finally in Ludkovski and Rü (2008) to the case that $X \in L^1(P)$ on a non-atomic probability space. Since convex, law invariant risk measures ϱ_i are $\leq_{\mathrm{cx}}$-consistent, it follows that $\varrho_i(X_i^*) \leq \varrho_i(X_i)$, $1 \leq i \leq n$, i.e. all components X_i^* are less risky than X_i. This section is based on the presentation in Ludkovski and Rü (2008).

We recall an equivalent definition of comonotonicity used in the following proposition.

Proposition 10.45 (Comonotonicity). *Let $X_1, \dots, X_n$ be real random variables on $(\Omega, \mathcal{A}, P)$ and $X := \sum_{i=1}^{n} X_i$, then $X_1, \dots, X_n$ are comonotone if and only if there exist nondecreasing functions $h_i : \mathbb{R} \to \mathbb{R}$ such that*

$$X_i = h_i(X), 1 \leq i \leq n \quad and \quad \sum_{i=1}^{n} h_i = \mathrm{id}_{\mathbb{R}}. \tag{10.80}$$

In particular (10.80) implies that the functions h_i are Lipschitz-continuous with constant 1. In the first step we consider the case of discrete allocations bounded below.

Proposition 10.46. *Let $Y = (Y_i) \in \mathcal{A}(X)$ be a discrete allocation of X taking a countable number of values and bounded below. Then there exists a comonotone allocation $\overline{Y} = (\overline{Y}_i)$ of X such that $\overline{Y}_i \leq_{\mathrm{cx}} Y_i$, $1 \leq i \leq n$.*

Proof. We give the proof in detail in the case $n = 2$, the general case is similar.

Thus let $Y = (Y_1, Y_2)$ and let $y = \bigcup_{i=1}^{2} y^i$ be the union of the supports of Y_i. Then y is an ordered countable set in $\mathbb{R}$, bounded from below by some $\underline{y} \in \mathbb{R}$. Consider the partition $\Omega = \bigcup_k C_k$ such that (Y_1, Y_2) are constant on each partition element C_k, and the C_ks are ordered according to the values of $Y_1 + Y_2$. Our goal is to have each of Y_1, Y_2 also ordered with respect to their values $y_1^1, y_2^1, \dots, y_1^2, y_2^2, \dots$ on C_k. Let $p_k = P(C_k) > 0$ be the weights of different pairs. If Y_1 and Y_2 are not comonotone, there must be a minimal index k such that without loss of generality $y_1^1 \leq y_2^1 \leq \cdots \leq y_k^1 \leq y_{k+1}^1$ and $y_1^2 \leq y_2^2 \leq \cdots \leq y_k^2$, but $y_{k+1}^2 < y_k^2$. Let $1 \leq j \leq k$ be the minimal index such that $y_j^2 > y_{k+1}^2$, so that the $(k+1)$-st pair violates the comonotonicity with all pairs between j and k.

Construct the update $(\widetilde{Y}_1, \widetilde{Y}_2)$ which takes on the same values $\widetilde{y}_i^1 \leftarrow y_i^1, \widetilde{y}_i^2 \leftarrow y_i^2$ as (Y_1, Y_2) except for $j \leq i \leq k+1$. Here we define

$$
\begin{cases}
\widetilde{y}^1_{k+1} \leftarrow y^1_{k+1} - (y^2_j - y^2_{k+1}) \cdot (\textstyle\sum_{i=j}^k p_i)/(\textstyle\sum_{i=j}^{k+1} p_i), \\[4pt]
\widetilde{y}^2_{k+1} \leftarrow y^2_{k+1} + (y^2_j - y^2_{k+1}) \cdot (\textstyle\sum_{i=j}^k p_i)/(\textstyle\sum_{i=j}^{k+1} p_i), \\[4pt]
\widetilde{y}^1_i \leftarrow y^1_i + (y^2_j - y^2_{k+1}) \cdot p_{k+1}/(\textstyle\sum_{i=j}^{k+1} p_i), \qquad j \leq i \leq k, \\[4pt]
\widetilde{y}^2_i \leftarrow y^2_i - (y^2_j - y^2_{k+1}) \cdot p_{k+1}/(\textstyle\sum_{i=j}^{k+1} p_i), \qquad j \leq i \leq k.
\end{cases}
\tag{10.81}
$$

This change preserves $\widetilde{Y}_1 + \widetilde{Y}_2 = Y_1 + Y_2$, maintains $E\widetilde{Y}_1 = EY_1$ and $E\widetilde{Y}_2 = EY_2$ and is an improvement of the single-crossing type, also known as mean-preserving spread (mps). It follows that $\widetilde{Y}$ is a component-wise $\leq_{cx}$ and $\leq_{mps}$ improvement of Y. Moreover, after the change, $\widetilde{y}^2_j = \widetilde{y}^2_{k+1}$ and $\widetilde{y}^1_j \leq \widetilde{y}^1_{k+1}$ so that the j-th pair is now comonotone with the $k+1$-st pair. Therefore, the next violation index pair (k, j) will be larger (in the lexicographic order) than the current one.

Iterate this argument to obtain an improvement sequence $(\widetilde{Y}^{(m)})_{m=1,2,\dots}$ of allocations. We claim that this sequence converges almost surely and in L^1 to some limit $\overline{Y}$. Indeed, let (k_m, j_m) be the violation index of the m-th update. Then $k_m \to \infty$ and therefore $p_{k_m} \to 0$. On the other hand, on a fixed subset C_k, once $k_m > k$, the value $y^{i,(m)}_k$ can change by at most

$$
|y^{i,(m+1)}_k - y^{i,(m)}_k| \leq \max_{i=1,2} |y^{i,(m)}_k - y^{i,(m)}_{k_m+1}| \cdot p_{k_m+1}/(\textstyle\sum_{j=k}^{k_m+1} p_j)
$$

$$
\leq \max_i (y^{i,(m)}_k - \underline{y}) \cdot p_{k_m+1}/p_k
$$

$$
\leq ((y^1_k + y^2_k - 2\underline{y})/p_k) \cdot p_{k_m+1},
$$

since $y^{1,(m)}_k + y^{2,(m)}_k = y^1_k + y^2_k$ for all m.

Since $p_{k_m} \to 0$, for any $\varepsilon > 0$ there is M large enough such that $\sum_{j > k_M} p_j < \varepsilon$ and the respective tail sum is then bounded by $\sum_{n=M}^{\infty} |y^{i,(n)}_k - y^{i,(n+1)}_k| \leq ((y^1_k + y^2_k - 2\underline{y})/p_k)\varepsilon$. Thus, $(Y^{i,(m)})$ converges almost surely.

Moreover, $(Y^{i,(m)})$ also converges in L^1. This is obvious if Y^i is bounded since $\sup Y^{i,(m+1)} \leq Y^{i,(m)}$. Otherwise note that for a fixed threshold index k', because of the mean-preserving spread the *tail mass* is non-increasing,

$$
\sum_{k \geq k'} p_k \cdot y^{i,(m)}_k \leq \sum_{k \geq k'} p_k \cdot y^i_k.
\tag{10.82}
$$

For a fixed level K, the tail expectation $E[Y^{i,(m)} \mathbb{1}_{\{Y^{i,(m)} > K\}}]$ will increase only if a point (y^i_k, p_k) is moved to the right of K as a result of e.g. the third line of (10.81). The algorithm of (10.81) operates by sliding points towards their average with changes in distance proportional to the weights. Thus, to slide an initial point y^i_k with mass p_k to level K requires an "energy" of $(K - y^i_k) \cdot p_k$. To do so, at least as much energy should be removed from the right of K. However, the total available energy beyond K is $\max_i E[Y^i \mathbb{1}_{\{Y^i > K\}}]$. Thus, if $(K - y_k^i)p_k > E[Y^i \mathbb{1}_{\{Y^i > K\}}]$ then the k-th point will never contribute to $E[Y^{i,(m)} \mathbb{1}_{\{Y^{i,(m)} > K\}}]$. Let $\underline{k}_K = \min\{k; (K - y_k^i)p_k < E[Y^i \mathbb{1}_{\{Y^i > K\}}]\}$, be the first index that can affect the above tail expectation. Combining with (10.82) we obtain the uniform bound

$$E[Y^{i,(m)} \mathbb{1}_{\{Y^{i,(m)} > K\}}] \leq \sum_{k \geq \underline{k}_K} p_k y_k^{i,(m)} \leq \sum_{k \geq \underline{k}_K} p_k y_k^i.$$

Finally, as $K \to \infty$, $\underline{k}_K \to \infty$ and $\sum_{k \geq \underline{k}_K} p_k y_k^i = E[Y^i \mathbb{1}_{\{Y^i > K\}}] \to 0$, establishing the uniform integrability of $(Y^{i,(m)})_m$.

The limiting $\overline{Y}$ is comonotone, since there are no comonotonicity violation pairs left in the limit and by closedness of convex ordering, $\overline{Y}_i \leq_{cx} Y_i$, $i = 1, 2, \ldots, n$. □

In the next step the comonotone improvement in Proposition 10.46 is extended to $L^\infty(P)$ (see Dana and Meilijson (2003, Proposition 5)).

Theorem 10.47. *Let for $X \in L^\infty(P)$, $Y = (Y_i) \in \mathcal{A}(X)$ be an allocation of X in $L^\infty(P)$. Then there exists a comonotone allocation $(X_i^*) \in \mathcal{A}(X)$ such that $X_i^* \leq_{cx} X_i$, $1 \leq i \leq n$.*

Proof. Let $\mathfrak{A}_m$ be the σ-algebra generated by the partition into events of the form $k2^{-m} \leq X < (k+1)2^{-m}$ and let

$$X^{(m)} := E(X \mid \mathfrak{A}_m) \text{ and } X_i^{(m)} := E(X_i \mid \mathfrak{A}_m), 1 \leq i \leq n.$$

By the martingale convergence theorem it holds that

$$X^{(m)} \to X \text{ and } X_i^{(m)} \to X_i \text{ a.s. and in } L^1(P).$$

Also by Jensen's inequality

$$X_i^{(m)} \leq_{cx} X_i.$$

Applying the comonotone improvement result in Proposition 10.46 to the discrete variables $(X_i^{(m)})$ we obtain by Proposition 10.45 1-Lipschitz functions g_{im} with $\sum_{i=1}^n g_{im} = \text{id}$ and such that $X_i^{(m)*} := g_{i,m}(X^{(m)}) \leq_{cx} X_i^{(m)}$, $1 \leq i \leq n$.

By the Ascoli theorem combined with a diagonal sequence argument there is a subsequence of g_{im}, $1 \leq i \leq n$, that converges uniformly on $[-\|X\|_\infty, \|X\|_\infty]$ to nondecreasing, 1-Lipschitz functions g_i, $1 \leq i$ for all i. Thus g_i are nondecreasing on $[-\|X\|_\infty, \|X\|_\infty]$, $\sum_{i=1}^n g_i = \text{id}$ and g_i are 1-Lipschitz.

Defining $X_i^* := g_i(X_i)$ we obtain $X_i^* \leq_{cx} X_i$ and $\sum_{i=1}^n X_i^* = X$. Thus $(X_i^*) \in \mathcal{A}(X)$ is a comonotone improvement of the allocation (X_i). □

Remark 10.48 (Pareto optimal allocations are comonotone). Theorem 10.47 also has a converse for $\varrho_i : L^p \to \mathbb{R}$ which are strictly $\leq_{cx}$-consistent, i.e. $X \preceq_{cx} Y$ and $X \not\sim_{cx} Y$ implies $\varrho_i(X) < \varrho_i(Y)$. Then any Pareto optimal allocation

$(X_i) \in L^p$ w.r.t. (ϱ_i) is comonotone (see Carlier and Dana (2003)). This follows from a simple improvement argument which allows to improve non-comonotone Pareto optimal solutions (X_i) to improved allocations $(\widetilde{X}_i)$ by an exchange step as in the proof of Proposition 10.46.   ◊

The optimality of comonotone allocations from Remark 10.48 can also be formulated in the following way. Consider on $(L^p)^n$ the partial ordering

$$(X_i) \leq_{\mathrm{cx}} (Y_i) \quad \text{if} \quad X_i \leq_{\mathrm{cx}} Y_i, \quad 1 \leq i \leq n.$$

Then we can define Pareto optimal allocations with respect to this partial ordering $\leq_{\mathrm{cx}}$ on $\mathcal{A}(X) = \mathcal{A}^p(X)$ with components in L^p and the following result holds (see Dana and Meilijson (2003)).

Theorem 10.49 (Comonotonicity of Pareto optimal allocations).
Let $(X_i) \in \mathcal{A}^p(X)$ be a $\leq_{\mathrm{cx}}$-Pareto optimal allocation of $X \in L^p$. Then (X_i) is comonotone.

Proof. Assume that (X_i) is $\leq_{\mathrm{cx}}$-Pareto optimal but not comonotone. Then by the comonotone improvement Theorem 10.47 there exists a comonotone allocation $(Y_i) \in \mathcal{A}^p(X)$ such that $(Y_i) \leq_{\mathrm{cx}} (X_i)$ and $Y_{i_0} \neq X_{i_0}$ for some $i_0 \leq n$. Then the mean $(X_i^*) := (\frac{1}{2}(X_i + Y_i)) \in \mathcal{A}^p(X)$ is an allocation of X and $(X_i^*) \leq_{\mathrm{cx}} (X_i)$ but $X_{i_0}^*$ is strictly smaller than X_{i_0} w.r.t. $\leq_{\mathrm{cx}}$ since $X_{i_0} \neq Y_{i_0}$. One can define a strictly convex risk functional ϱ_{i_0} such that $\varrho_{i_0}(X_i^*) < \varrho_{i_0}(X_i)$. This yields a contradiction.   □

For the final version of the comonotone improvement theorem we assume that the underlying probability space $(\Omega, \mathfrak{A}, P)$ is non-atomic.

Theorem 10.50 (Comonotone improvement theorem). *Let $X \in L^1(P)$ and let $Y = (Y_i) \in \mathcal{A}(X)$ be an allocation of X in $L^1(P)$. Then there exists a comonotone allocation $\overline{Y} = (\overline{Y}_i) \in \mathcal{A}(X)$ such that*

$$\overline{Y}_i \leq_{\mathrm{cx}} Y_i, \quad 1 \leq i \leq n.$$

Proof. The proof is given by reduction to Theorem 10.47. Defining

$$Y_i^{(m)} := Y_i \cdot \mathbb{1}_{\{|Y_i| \leq m\}}, \quad Y^{(m)} := \sum_{i=1}^{n} Y_i^{(m)}, \tag{10.83}$$

then $Y^{(m)} \to X$, $Y_i^{(m)} \to Y_i$, almost surely and in $L^1(P)$. By Theorem 10.47 there exists $Z^{(m)}$ comonotone, such that

$$Z_i^{(m)} \leq_{\mathrm{cx}} Y_i^{(m)} \quad \text{and} \quad \sum_{i=1}^{n} Z_i^{(m)} = Y^{(m)}. \tag{10.84}$$

Let $F_{i,m}$ be the distribution function of $Z_i^{(m)}$. Since $Z^{(m)}$ is comonotone and $(\Omega, \mathfrak{A}, P)$ is non-atomic, it follows that there exists a random variable $U \sim U(0, 1)$ such that $Z_i^{(m)} \overset{d}{=} F_{i,m}^{-1}(U)$. For each $1 \leq i \leq n$, $(Z_i^{(m)})_m$ are tight since by the convex ordering

$$E|Z_i^{(m)}| \leq E|Y_i^{(m)}| \leq E|Y_i| < \infty.$$

Therefore there exists a subsequence, again labelled $(m) \subset \mathbb{N}$ along which the distribution functions converge, $F_{i,m} \to F_i$. This implies that

$$F_{i,m}^{-1}(U) \overset{\text{a.s.}}{\longrightarrow} F_i^{-1}(U) =: Z_i,$$

and moreover

$$\sum_{i=1}^{n} Z_i^{(m)} \overset{d}{=} \sum_{i=1}^{n} F_{i,m}^{-1}(U) \to \sum_{i=1}^{n} Z_i \quad \text{a.s. and in } L^1(P).$$

On the other hand, we already had

$$\sum_{i=1}^{n} Z_i^{(m)} = \sum_{i=1}^{n} Y_i^{(m)} \to \sum_{i=1}^{n} Y_i \quad \text{a.s. and in } L^1(P).$$

Thus we obtain

$$\sum_{i=1}^{n} Z_i \overset{d}{=} \sum_{i=1}^{n} Y_i, \quad Z_i \leq_{\text{cx}} Y_i \quad \text{and in fact} \quad (Z_i^{(m)}) \overset{d}{=} (F_{i,m}^{-1}(U)) \to (Z_i) \quad \text{a.s.}$$

In particular, the limit allocation Z is comonotone. Therefore, for some measure preserving random variable U' on Ω it holds that

$$\sum_{i=1}^{n} Z_i \circ U' = \sum_{i=1}^{n} Y_i = X \quad \text{a.s.}$$

Obviously $\overline{Y}_i := Z_i \circ U'$ satisfy

$$(\overline{Y}_i) \overset{d}{=} (Z_i), \quad \sum_{i=1}^{n} \overline{Y}_i = X, \quad \text{and} \quad \overline{Y}_i \leq_{\text{cx}} Y_i. \qquad \square$$

As a consequence of the comonotone improvement result the inf-convolution $\widehat{\varrho}(X) = \varrho_1 \wedge \cdots \wedge \varrho_n(X)$ can be restricted to allocations $(X_i) \in \mathcal{A}^{\uparrow}(X)$ such that $X_i = h_i(X)$ are increasing functions of X and $\sum_{i=1}^{n} h_i = \text{id}$. This connection implies in the case that $X \in L^{\infty}(P)$ the existence of optimal allocations (see Jouini et al. (2008) and Acciaio (2007)).

Theorem 10.52 (Existence of optimal allocations in $L^\infty(P)$). *Let $X \in L^\infty(P)$ and ϱ_i be law invariant, convex risk measures on $L^\infty(P)$. Then there exists an allocation $(X_i) \in \mathcal{A}^\uparrow(X)$ such that*

$$\widehat{\varrho}(X) = \sum_{i=1}^{n} \varrho_i(X_i),$$

i.e. (X_i) is an optimal allocation of X.

Proof. We give the proof in the case $n = 2$, the case $n \geq 2$ is similar. Let (X_1^n, X_2^n) be a minimizing sequence of $\widehat{\varrho} = \varrho_1 \wedge \varrho_2$. By the comonotone improvement theorem a minimizing sequence can be taken in $\mathcal{A}^\uparrow(X)$, i.e. $X_1^n = f_1^n(X)$, $X_2^n = f_2^n(X)$ for some increasing functions $f_i^n : [a, b] \to \mathbb{R}$ with $a = \text{ess inf } X$, $b = \text{ess sup } X$. W.l.g. we can assume that ess inf $X_1^n = 0$ and thus using $X_1^n + X_2^n = X$, $0 \leq f_1^n \leq b - a$ and $0 \leq f_2^n \leq b$.

From the Ascoli theorem applied to the bounded equicontinuous family $\{f_1^n; n \in \mathbb{N}\}$ and similarly to $\{f_2^n; n \in \mathbb{N}\}$ it follows that for some subsequence $f_1^m \to f_1$ and $f_2^m \to f_2$ uniformly on $[a, b]$, where f_1, f_2 are 1-Lipschitz, increasing. By the L^∞-continuity of ϱ_1, ϱ_2 we obtain

$$\varrho_1 \wedge \varrho_2(X) = \lim_{n \to \infty} \varrho_1(X_0^n) + \varrho_2(X_1^n) = \varrho_1(f_1(X)) + \varrho_2(f_2(X)).$$

Thus the allocation $(f_1(X), f_2(X)) \in \mathcal{A}^\uparrow(X)$ is an optimal allocation of X.   □

Existence results for optimal allocations in $L^p(P)$ can be given for more general risk functionals which are not necessarily law invariant based on continuity and compactness arguments when the class of admissible allocations satisfies some restriction conditions. Let for $1 \leq p \leq \infty$

$$L_+^p := \{X \in L^p; \ X \geq 0 \text{ a.s.}\}, \quad L_-^p := \{X \in L^p; \ X \leq 0 \text{ a.s.}\}$$

and define for $X \in L^p$

$$\mathcal{A}_+^p(X) := \left\{(X_i) \in (L_+^p)^n; \ \sum_{i=1}^{n} X_i = X\right\} \tag{10.85}$$

$$\mathcal{A}_-^p(X) := \left\{(X_i) \in (L_-^p)^n; \ \sum_{i=1}^{n} X_i = X\right\} \tag{10.86}$$

and $$\mathcal{B}^p(X) := \left\{(X_i) \in (L^p)^n; \ \sum_{i=1}^{n} X_i = X, |X_i| \leq |X|, 1 \leq i \leq n\right\} \tag{10.87}$$

subclasses of the class of all allocations $\mathcal{A}^p(X)$ of X on L^p. The corresponding infimal convolutions are denoted by $\widehat{\varrho}^p_+$, $\widehat{\varrho}^p_-$, and $\widehat{\varrho}^p_{||}$,

The following extend some existence results of Carlier and Dana (2005) in the case $p = \infty$. $\mathcal{A}^p_+(X)$ and $\mathcal{A}^p_-(X)$ could also be more generally replaced by lower respectively upper bounded allocations. Detailed proofs of these results are given in Weber (2008).

In the case $p = \infty$ we make use of the following well-known closedness result (see Föllmer and Schied (2011, Satz A 64)).

Proposition 10.53 (Weak ∗-closedness of convex sets). *If $C \subset L^\infty$ is convex and for any $r > 0$, $C_r := C \cap \{X \in L^\infty; \|X\|_\infty \leq r\}$ is closed in L^1, then C is weak ∗-closed.*

The following existence result in $L^\infty(P)$ is stated for general risk functionals.

Theorem 10.54 (Existence of optimal allocations in $\mathcal{A}^\infty_+$, $\mathcal{A}^\infty_-$, and $\mathcal{B}^\infty$). *Let $\varrho_i : L^\infty \to \mathbb{R}$ be $\sigma(L^\infty, L^1)$-lsc risk functionals, $1 \leq i \leq n$. Then*

(a) For $X \in L^\infty_+$ there exists an optimal allocation (X_i) of X in $\mathcal{A}^\infty_+(X)$, i.e. $\widehat{\varrho}_+(X) = \sum_{i=1}^n \varrho_i(X_i)$.

(b) For $X \in L^\infty$ there exists an optimal allocation (X_i) of X in $\mathcal{A}^\infty_-(X)$.

(c) For $X \in L^\infty$ there exists an optimal allocation (X_i) of X in $\mathcal{B}^\infty(X)$.

Proof.   (a) The class of positive allocations $\mathcal{A}^\infty_+(X)$ in $(L^\infty)^n$ is a convex subset of the ball $B_{(L^\infty)^n}(a)$ with radius $a = n\|X\|_\infty$ in $(L^\infty)^n$. It is easy to see that for $r > 0$

$$\mathcal{A}^\infty_+ \cap \left\{ (X_i) \in (L^\infty)^n; \ \|(X_i)\|_{(L^\infty)^n} = \sum_{i=1}^n \|X_i\|_\infty \leq r \right\}$$

is closed in $(L^\infty)^n$ w.r.t. strong topology (for details see Weber (2008)). Therefore, by Proposition 10.53 $\mathcal{A}^\infty_+$ is a weak ∗-closed subset of the $\sigma((L^\infty)^n, (L^1)^n)$ compact ball $B_{(L^\infty)^n}(a)$ and thus $\mathcal{A}^\infty_+$ itself is weak ∗-compact. By assumption $\sum_{i=1}^n \varrho_i(X_i)$ is lsc on $\mathcal{A}^\infty_+$ and thus attains its infimum.

(b), (c) The proof of (b) and (c) is similar.   □

In the next step we consider the case $1 < p < \infty$.

Theorem 10.55 (Existence of optimal allocations in $\mathcal{A}^p_+$, $\mathcal{A}^p_-$, $\mathcal{B}^p$). *Let $X \in L^p$, $1 < p < \infty$ with conjugate index q, $\frac{1}{p} + \frac{1}{q} = 1$, and let ϱ_i be lsc risk functionals w.r.t. $\sigma(L^p, L^q)$ on L^p.*

Then for X in L^p_+ respectively L^p_- respectively L^p there exists an optimal allocation (X_i) in $\mathcal{A}^p_+$ respectively $\mathcal{A}^p_-$ respectively $\mathcal{B}^p$.

Proof. (a) $1 < p < \infty$: The set $\mathcal{A}_+^p$ of positive allocations is a convex subset of the ball $B_{(L^p)^n}(a)$ with radius $a = n\|X\|_+$ in $(L^p)^n$. Since for reflexive spaces as $(L^p)^n$ weak-topology is identical to weak $*$-topology and for convex sets weak and strong closedness are identical we obtain that $\mathcal{A}_+^p$ is $\sigma((L^p)^n, (L^q)^n)$-closed and therefore compact. This yields the existence of optimal allocations in $\mathcal{A}_+^p$.

The cases $\mathcal{A}_-^p$, $\mathcal{B}^p$ are similar.

(b) $p = 1$: As in the proof in (a) $\mathcal{A}_+^1(X)$ is convex and norm bounded and weakly closed in $(L^1)^n$. Furthermore, for $Y = (Y_i) \in \mathcal{A}_+^1(X)$ it holds that $\int_A |Y_i|\, dP \le \int_A X\, dP < \infty$ for $A \in \mathcal{A}$. Thus the set of all components Y_i of allocations in $\mathcal{A}_+^1(X)$ is uniformly integrable and therefore by the Eberlein–Smulian theorem weakly compact in L^1. This implies that $\mathcal{A}_+^1(X)$ is weakly compact in $(L^1)^n$ and therefore optimal allocations exist. The cases $\mathcal{A}_-^1(X)$, $\mathcal{B}^1(X)$ are similar. $\qquad\square$

Remark 10.56. Alternatively we can also consider admissible allocations $(X_i) \in \mathcal{A}^a(X)$, $X \in L^p$ with $(X_i)_+ \le X_+$ and $(X_i)_- \le X_-$ as in Theorem 10.17 and obtain in a similar way as in Theorems 10.54 and 10.55 existence of optimal admissible allocations in $\mathcal{A}^a(X)$. $\qquad\Diamond$

Chapter 11
Characterization and Examples of Optimal Risk Allocations for Convex Risk Functionals

The classical result on optimal risk sharing respectively optimal risk allocation is Borch's theorem which gives a characterization of optimal risk exchange in the context of expected utility (see Borch (1962), Bühlmann (1970), and Gerber (1979)) and as a consequence results in the derivation of optimal (reinsurance) contracts between n companies. A general differential characterization of optimal risk allocations for convex premium principles was given in Deprez and Gerber (1985). In particular in the case of dilated risk functionals $\varrho_i(X) = \lambda_i \varrho(\frac{1}{\lambda_i} X)$, $\lambda_i > 0$, $\sum \lambda_i = 1$, they derive linear quota sharing $X_i^* = \lambda_i X$ as optimal risk allocation. These results have been extended in the context of convex analysis in Barrieu and El Karoui (2005), Jouini et al. (2008), Acciaio (2007), Kiesel and Rü (2008), and Acciaio and Svindland (2009).

In the classical insurance literature the optimal allocation problem is formulated in terms of "convex premium principles" Ψ. In some of the more recent papers it is formulated in terms of "monetary utility functions" U which are concave, monotone, and translation invariant risk functionals and since they represent utility the basic problem is a maximization problem

$$\sum_{i=1}^{n} U_i(\xi_i) = \sup_{(\eta_i) \in \mathcal{A}(X)} \sum_{i=1}^{n} U_i(\eta_i). \tag{11.1}$$

With the substitution $\varrho_i = -U_i$ the optimal allocation problem in (11.1) is equivalent to the inf-formulation w.r.t. convex risk measures ϱ_i to minimize the total risk

$$\sum_{i=1}^{n} \varrho_i(\xi_i) = \bigwedge \varrho_i(X) := \inf \left\{ \sum_{i=1}^{n} \varrho_i(X_i); \ (X_i) \in \mathcal{A}(X) \right\}. \tag{11.2}$$

$\bigwedge \varrho_i$ is called the "inf-convolution" of (ϱ_i).

Based on the characterization results several of the common risk sharing contracts such as linear quota sharing, European options, stop-loss contracts, and

L. Rüschendorf, *Mathematical Risk Analysis*, Springer Series in Operations Research and Financial Engineering, DOI 10.1007/978-3-642-33590-7_11,
© Springer-Verlag Berlin Heidelberg 2013

deductibles can be derived as optimal contracts. These results are of interest in insurance in particular for reinsurance contracts, for general risk management and for the design of optimal and efficient financial products.

In Section 11.1 we introduce some basic notions of convex analysis and properties of inf-convolutions and in particular give as application an existence result for optimal allocations for general convex risk functionals ϱ_i. In this context we give in Section 11.2 the basic characterization result of optimal allocations and also consider the restricted problem of Pareto optimal allocation under the "individual rationality constraint". This principle says that for a given allocation $X = \sum_{i=1}^{n} X_i$ only those allocations $(\xi_i) \in \mathcal{A}(X)$ are considered acceptable which decrease all individual risks, i.e. such that

$$\varrho_i(\xi_i) \leq \varrho_i(X_i), \quad 1 \leq i \leq n. \tag{11.3}$$

In Section 11.3 we discuss a series of examples. An extension of the characterization results to the optimal allocation of risk vectors is given in Section 11.4. Optimal risk allocations exhibit a worst case dependence structure w.r.t. some specific max-correlation risk measure and are comonotone w.r.t. a common worst case scenario measure.

11.1 Inf-Convolution and Convex Conjugates

In this section we recall some of the basic notions of convex duality and connections with the inf convolution $\bigwedge \varrho_i$. We consider convex, lower semicontinuous (lsc) proper risk functions $\varrho_i : L^p(P) \to (-\infty, \infty]$, $1 \leq p \leq \infty$ and assume as in Chapter 10 generally that a representation of ϱ_i based on scenario measures in $\mathcal{M}_1^q$ (= $\mathcal{M}_1$ in case $p = \infty$) holds true (see (10.63)). In case $p = \infty$ the results could also be extended in a similar form to the general case with finitely additive scenario sets $\mathcal{P}_i$. This section is based on the presentation in Kiesel and Rü (2008).

For a convex proper function on a locally convex space E we denote by

$$f^* : E^* \to \overline{\mathbb{R}}, \ f^*(x^*) = \sup_{x \in E}(\langle x^*, x \rangle - f(x)), \quad x^* \in E^* \tag{11.4}$$

the convex conjugate, by

$$f^{**}(x) = \sup_{x^* \in E^*} (\langle x, x^* \rangle - f^*(x^*)), \quad x \in E \tag{11.5}$$

the bi-conjugate of f. The obvious inequality

$$f(x) + f^*(x^*) \geq (x, x^*)$$

is called "Fenchel's inequality". Further let

$$\partial f(x) = \{x^* \in E^*; f(y) - f(x) \geq \langle x^*, y - x \rangle, \ \forall y \in E\} \tag{11.6}$$

denote the set of subgradients of f in x. Then

$$\partial f(x) = \{x^* \in E^*; \ \forall y \in E, \langle x^*, y \rangle \leq D(f, x)(y)\}, \tag{11.7}$$

where $D(f, x)(y)$ is the right directional derivative of f in x in direction y. This connection is useful in the applications in order to calculate the subgradient. It also gives the connection to the differential characterization of optimal couplings as in Deprez and Gerber (1985).

For proper convex functions and $\partial f(x) \neq \phi$ it holds that

$$x^* \in \partial f(x) \Leftrightarrow \langle x^*, x \rangle = f(x) + f(x^*)$$
$$\Leftrightarrow \langle x, x^* \rangle - f(x) \geq \sup_{y \in E}(\langle y, x^* \rangle - f(y)). \tag{11.8}$$

By the Fenchel–Moreau theorem, Theorem 7.8, this is for lsc functions f further equivalent to

$$x \in \partial f^*(x^*). \tag{11.9}$$

For minimization problems of proper, convex, lsc functions f, as in our risk allocation problem (11.2) the following extension of the classical Fermat rule is of interest:

$$f(x) = \inf_{y \in E} f(y) \Leftrightarrow 0 \in \partial f(x). \tag{11.10}$$

For all results on convex duality we refer to Rockafellar (1974) and Barbu and Precupanu (1986).

The following proposition gives conditions which are needed to ensure properness of the infimal convolution $\bigwedge \varrho_i$.

Proposition 11.1. *For ϱ_i convex, lsc and proper it holds that*

$$\bigcap_{i=1}^{n} \operatorname{dom} \varrho_i^* \neq \phi \ \Rightarrow \ \operatorname{dom}(\bigwedge \varrho_i) \neq \phi. \tag{11.11}$$

Proof. For the proof we make use of the following results from convex analysis

(1) As ϱ_i are proper, convex, and lsc it follows that ϱ_i^* are proper (see Barbu and Precupanu (1986, Corollary 1.4, Chapter 2)). For proper convex ϱ_i it holds that

$$(\bigwedge \varrho_i)^* = \sum \varrho_i^*. \tag{11.12}$$

This implies that $(\bigwedge \varrho_i)^*$ is nowhere $-\infty$.

(2) Properness of f^* induces properness of f (see Barbu and Precupanu (1986, Corollary 1.4, Chapter 2)).

As a consequence of (1) and (2) we obtain

$$\text{dom}(\textstyle\bigwedge \varrho_i)^* \neq \phi \overset{(1)}{\implies} (\textstyle\bigwedge \varrho_i)^* \text{ is proper}$$

$$\overset{(2)}{\implies} (\textstyle\bigwedge \varrho_i) \text{ is proper}$$

$$\implies \text{dom}(\textstyle\bigwedge \varrho_i) \neq \phi. \qquad \square$$

Remark 11.2. For proper risk functionals ϱ_i on L^∞ respectively finite risk functionals on L^p the scenario sets satisfy

$$\mathcal{P}_i = \text{dom}\, \varrho_i^* = L^p$$

(see Theorem 7.9) and the infimal convolution is continuous. As a consequence of Corollary 1.2 in Barbu and Precupanu (1986, Chapter 2) equivalence holds in (11.11). Thus the risk allocation problem (in unrestricted form) does not make sense without the intersection conditions

(IS)
$$\bigcap_{i=1}^{n} \text{dom}\, \varrho_i^* \neq \phi, \qquad (11.13)$$

which is generally assumed in the following to hold true. $\Diamond$

An existence result for optimal allocations in $L^\infty(P)$ minimizing the total risk $\sum_{i=1}^{n} \varrho_i(\xi_i)$ over $(\xi_i) \in \mathcal{A}(X)$ for law invariant convex risk measures ϱ_i, i.e. exactness of the inf-convolution $\bigwedge \varrho_i$ has been given in Theorem 10.52. The proof there was based on the comonotone improvement theorem. The exactness of $\bigwedge \varrho_i$ for general convex risk functionals can be obtained from convex analysis.

For the existence of optimal allocations $(\xi_i) \in A(X)$ minimizing the total risk, i.e.

$$\bigwedge \varrho_i(X) = \sum \varrho_i(\xi_i) \qquad (11.14)$$

we need a strengthened version of the intersection condition (IS),

(SIS)
$$\text{dom}\, \varrho_1^* \cap \bigcap_{i=2}^{n} \text{int}(\text{dom}\, \varrho_i^*) \neq \phi. \qquad (11.15)$$

Further we assume that at least one risk functional ϱ_k is monotone (i.e. $X \leq Y$ implies $\varrho_k(X) \geq \varrho_k(Y)$). This assumption implies monotonicity of the infimal convolution $(\bigwedge \varrho_i)$ (see Acciaio (2007, Lemma 3.3, i)).

Theorem 11.3 (Existence of optimal allocations). *Let ϱ_i be proper, convex, and lsc and assume that at least one of the ϱ_i is monotone and that the strong intersection property (SIS) holds. Then for all $X \in \text{int}(\text{dom} \bigwedge \varrho_i)$ there exists an allocation $(\xi_i) \in \mathcal{A}(X)$ which minimizes the total risk i.e.*

$$\bigwedge \varrho_i(X) = \sum \varrho_i(\xi_i). \qquad (11.16)$$

Proof. For proper, convex, and monotone functions f on a Banach lattice E it holds that $\partial f(x) \neq \phi$ for all $x \in \text{int}(\text{dom } f)$ (see Theorem 7.13). Thus we obtain for all $X \in \text{int}(\text{dom} \bigwedge \varrho_i)$ that $\partial(\bigwedge \varrho_i)(X) \neq \phi$. Let $\mu \in \partial(\bigwedge \varrho_i)(X)$ and thus

$$X \in \partial(\bigwedge \varrho_i)^*(\mu) = \partial\left(\sum \varrho_i^*\right)(\mu).$$

The strong intersection property (SIS) implies additivity of the subgradient mapping

$$\partial\left(\sum \varrho_i^*\right) = \sum \partial \varrho_i^* \tag{11.17}$$

(see Barbu and Precupanu (1986, Chapter 2, Corollary 2.5 and Remark 2.8)). As a consequence

$$X \in \sum \partial \varrho_i^*(\mu). \tag{11.18}$$

Thus there exists an allocation $(\xi_i) \in A(X)$ with $\xi_i \in \partial \varrho_i^*(\mu)$. This implies that (ξ_i) minimizes the total risk as is proved in the basic characterization result of optimal allocations in Section 11.2, Theorem 11.5. $\qquad \square$

As mentioned above in the case that ϱ_i are translation invariant the intersection property (IS) is sufficient for exactness of the inf convolution $\bigwedge \varrho_i$ on L^∞ and the stronger SIS condition is not needed. Acciaio and Svindland (2009) establish exactness on L^∞ for convex, translation invariant risk functionals under the Pareto equilibrium condition (E) introduced in Sections 10.1 and 10.4.

11.2 Characterization of Optimal Allocations

The main result in this section is an extension of the characterization results of optimal allocations of Borch (1962), Deprez and Gerber (1985), and Jouini et al. (2008) and others. This result allows to characterize several of the basic contracts like linear quota sharing, stop-loss contracts, and deductibles as optimal allocations (contracts) (see Section 11.3).

Before starting the characterization result we describe in the following proposition the subgradient of the infimal convolution.

Proposition 11.4. *Let $(\zeta_i) \in A(X)$ minimize the total risk, $\bigwedge \varrho_i(X) = \sum \varrho_i(\zeta_i)$. Then*

(a) $$\forall (\eta_i) \in A(X) \text{ it holds that } \bigcap \partial \varrho_i(\eta_i) \subset \partial \bigwedge \varrho_i(X), \tag{11.19}$$

(b) $$\partial \bigwedge \varrho_i(X) = \bigcap \partial \varrho_i(\zeta_i). \tag{11.20}$$

Proof. (a) For $\mu \in \bigcap \partial \varrho_i (\eta_i)$ we obtain by the Fenchel inequality using (11.12)

$$\bigwedge \varrho_i(X) \geq \langle \mu, X \rangle - \sum \varrho_i^*(\mu) = \sum (\langle \mu, \eta_i \rangle - \varrho_i^*(\mu)) = \sum \varrho_i(\eta_i).$$

By definition of the infimal convolution therefore equality holds and thus $\mu \in \partial \bigwedge \varrho_i(X)$.

(b) If $\mu \in \partial \bigwedge \varrho_i(X)$, then

$$\sum (\langle \mu, \zeta_i \rangle - \varrho_i^*(\mu)) = \langle \mu, X \rangle - \sum \varrho_i^*(\mu)$$

$$= \langle \mu, X \rangle - \left(\bigwedge \varrho_i \right)^*(\mu) = \bigwedge \varrho_i(X) = \sum \varrho_i(\zeta_i).$$

Therefore, again from the Fenchel inequality we conclude that

$$\langle \mu, \zeta_i \rangle - \varrho_i(\mu) = \varrho_i(\zeta_i)$$

for all i and thus $\mu \in \bigcap \partial \varrho_i (\zeta_i)$. This implies that (11.20) is a consequence of (a). □

As a consequence we next obtain the basic characterization result.

Theorem 11.5 (Characterization of minimal total risk allocations). *Let ϱ_i be proper convex and lsc risk functionals on L^p at least one of them being monotone, let $X \in \text{int}(\text{dom} \bigwedge \varrho_i)$ and let $(\zeta_i) \in \mathcal{A}(X)$, then the following are equivalent:*

(a) (ζ_i) *has minimal total risk, i.e.* $\bigwedge \varrho_i(X) = \sum \varrho_i(\zeta_i)$

(b) $\exists \mu \in \mathcal{M}_1^q$ *such that* $\zeta_i \in \partial \varrho_i^*(\mu), \quad 1 \leq i \leq n$ (11.21)

(c) $\exists \mu \in \mathcal{M}_1^q$ *such that* $\mu \in \partial \varrho_i (\zeta_i), \forall i$, *i.e.* $\bigcap \partial \varrho_i (\zeta_i) \neq 0$.   (11.22)

Proof.

(a) $\Rightarrow$ (c) Since $X \in \text{int}(\text{dom} \bigwedge \varrho_i)$ and $\bigwedge \varrho_i$ is proper, convex, and monotone, it follows from Theorem 7.13 that there exists a subgradient $\mu \in \partial \bigwedge \varrho_i(X)$. Thus by Proposition 11.4 it holds that $\mu \in \bigcap_{i=1}^n \partial \varrho_i (\zeta_i)$.

(c) $\Leftrightarrow$ (b) holds by general convex duality. If $\mu \in \partial \bigwedge \varrho_i(X)$ then using (11.12) we obtain

$$\left(\bigwedge \varrho_i \right)^*(\mu) = \sum_{i=1}^n \varrho_i^*(\mu) = \sum_{i=1}^n (\langle \mu, \zeta_i \rangle - \varrho_i(\zeta_i))$$

$$\leq \langle \mu, X \rangle - \sum_{i=1}^n \varrho_i(\zeta_i).$$

Furthermore, by definition of the conjugate

$$(\bigwedge \varrho_i)^*(\mu) = \sup_{Y \in L^p} (\langle \mu, Y \rangle - \bigwedge \varrho_i(Y))$$

$$\geq \langle \mu, Y \rangle - \bigwedge \varrho_i(Y) \quad \text{for any } Y \in L^p.$$

As a consequence these two inequalities imply

$$\langle \mu, Y \rangle - \bigwedge \varrho_i(Y) \leq \langle \mu, X \rangle - \sum_{i=1}^{n} \varrho_i(\zeta_i), \quad \text{for any } Y \in L^p.$$

In particular for $Y = X$ this yields $\bigwedge \varrho_i(X) \geq \sum \varrho_i(\zeta_i)$ which implies equality i.e. (ζ_i) is an optimal allocation.   □

Remark 11.6. (a) Equation (11.21) gives an important and very useful criterion for the calculation of optimal allocations. In the classical case of differentiable risk functionals this amounts to the condition that the Gateaux gradients $D\varrho_i(\zeta_i)$ are independent of i. By Proposition 11.4 this gradient is identified with the subgradient of the infimal convolution.
(b) The monotonicity assumption on one of the risk functionals can be replaced by the assumption that

$$\partial \bigwedge \varrho_i(X) \neq \phi. \tag{11.23}$$
◇

Pareto optimal allocations are defined as those allocations $(X_i) \in \mathcal{A}(X)$ such that the corresponding risk vector $(\varrho_i(X_i))$ is a minimal element of the risk set

$$\mathcal{R} = \{(\varrho_i(Y_i)); \ (Y_i) \in \mathcal{A}(X)\} = \mathcal{R}(X).$$

To obtain a connection to Pareto optimality we need a property which we call the non-saturation property (NS). We say that ϱ has the "non-saturation property" if

(NS) $\displaystyle\inf_{X \in L^p} \varrho(X)$ is not attained.   (11.24)

(NS) is a weak property on risk measures. It is implied in particular by the cash invariance property.

Under the non-saturation condition (NS) Pareto optimality is related to the problem of minimizing the total weighted risk. This is described by the "weighted minimal convolution" $(\bigwedge \varrho_i)_\gamma(X)$, defined for weight vectors $\gamma = (\gamma_i) \in \mathbb{R}^n$ by

$$\left(\bigwedge \varrho_i\right)_\gamma(X) := \inf\left\{\sum \gamma_i \varrho_i(X_i); \ (X_i) \in A(X)\right\}. \tag{11.25}$$

This connection between Pareto optimality and minimizing total weighted risk goes back in more special situations to the early papers in insurance (see Gerber (1979)).

Theorem 11.7 (Characterization of Pareto optimal allocations). *Let ϱ_i be convex, lsc, proper risk functionals on L^p satisfying the non-saturation condition (NS). Then for $X \in L^p$ such that $\partial \bigwedge \varrho_i(X) \neq \phi$ and $(\zeta_i) \in A(X)$ the following are equivalent:*

(1) (ζ_i) is Pareto optimal

(2) $\exists \gamma = (\gamma_i) \in \mathbb{R}^n_{>0}$ such that $\sum \gamma_i \varrho_i(\zeta_i) = (\bigwedge \varrho_i)_\gamma(X)$,

(3) $\exists \gamma \in \mathbb{R}^n_{>0}$ and $\exists \mu \in \mathcal{M}_1^q(P)$ such that $\mu \in \gamma_i \partial \varrho_i(\zeta_i), \forall i$ (11.26)
*   or, equivalently, $\bigcap \gamma_i \partial \varrho_i(\zeta_i) \neq \phi$,*

(4) $\exists \mu \in \mathcal{M}_1^q(P)$ and $\exists \gamma \in \mathbb{R}^n_{>0}$ such that $\zeta_i \in \partial(\gamma_i \varrho_i)^*(\mu), \quad \forall i$.   (11.27)

Proof. The equivalences of (2)–(4) follow from Theorem 11.5 applied to the convex risk functionals $\gamma_i \varrho_i$ using lsc and the property $\partial(\gamma_i \varrho_i) = \gamma_i \partial \varrho_i$.

(1) $\Leftrightarrow$ (2) For the proof of this equivalence we use a Hahn–Banach separation argument separating $B := \mathcal{R} + \mathbb{R}^n_+$ from $C := (\varrho_i(\zeta_i)) - (\mathbb{R}^n_+ \setminus \{0\})$. By construction the vector $\lambda \in \mathbb{R}^n$ describing the separating hyperplane i.e. $\lambda^\top \cdot x \leq \lambda^\top \cdot y$ for all $x \in C, y \in B$, has non-negative coordinates, $\lambda_j \geq 0$. Now the (NS) condition is enough to imply that all components of λ are positive.

For the proof assume that $\lambda_j = 0$ for some j. We assume w.l.g. $j = n$. Then (ζ_i) minimizes $\sum_{i=1}^{n-1} \lambda_i \varrho_i(\eta_i)$ over $(\eta_i) \in A(X)$. Thus since $\lambda_n = 0$ (ζ_i) minimizes $\sum_{i=1}^{n-1} \lambda_i \varrho_i(\eta_i)$ over all $\eta_i \in L^p$. By the (NS) condition this leads to a contradiction. The direction (2) $\Rightarrow$ (1) is obvious. $\square$

Remark 11.8. (a) Using Proposition 11.4 Condition (3) in Theorem 11.7 can be reformulated as

(3') $$\exists \gamma \in \mathbb{R}^n_{>0} : \partial(\bigwedge \varrho_i)_\gamma(X) \neq \phi.$$

(b) The value of the optimal total weighted risk $(\bigwedge \varrho_i)_\gamma(X)$ is given by

$$\left(\bigwedge \varrho_i\right)_\gamma(X) = \langle \mu, X \rangle - \Sigma \gamma_i \varrho_i^*(\mu),   (11.28)$$

where μ is as in (11.26) respectively in (11.27) a solution of the dual problem.
$\Diamond$

For cash invariant convex risk functionals this characterization result implies by a simple rebalancing argument (see Jouini et al. (2008), Burgert and Rü (2006b, 2005/2008), and Acciaio (2007)) that it is enough to consider the optimal risk allocation problem for the weight vector $\gamma = (1, \dots, 1)$.

Corollary 11.9. *If ϱ_i are convex, lsc, proper cash invariant risk functionals on L^∞, then for $X \in L^p$, $(\zeta_i) \in \mathcal{A}(X)$ the following are equivalent:*

(1) (ζ_i) is Pareto optimal.

(2) (ζ_i) minimizes the total risk, i.e. $\sum \varrho_i(\zeta_i) = \bigwedge \varrho_i(X)$.   (11.29)

(3) $\bigcap \partial \varrho_i(\zeta_i) \neq \phi.$

(4) $\exists \mu \in \mathcal{M}_1^q$ *such that* $\zeta_i \in \partial \varrho_i^*(\mu), \quad \forall i.$

(5) $\exists \mu \in \mathcal{M}_1^q$ *such that* $\mu \in \partial \varrho_i(\zeta_i), \quad \forall i.$

Further for any μ as above it holds that

$$\bigwedge \varrho_i(X) = \langle \mu, X \rangle - \Sigma \varrho_i^*(\mu). \tag{11.30}$$

Our aim in the following is to prove that Pareto optimal risk allocations in the case of "strictly convex" cash invariant risk measures are unique up to rebalancing. A function is called strictly convex if the inequality in the definition of convexity is strict for any $X, Y \in E, X \neq Y$, i.e. it holds that

$$f(\alpha X + (1-\alpha)Y) < \alpha f(x) + (1-\alpha)f(Y), \quad \forall \alpha \in (0,1).$$

For the preparation of the uniqueness result we need some properties of subdifferentials.

Lemma 11.10. *Let ϱ be a finite strictly convex risk functional on L^p.*

(a) $\forall \mu \in \mathcal{M}_1^q$ *it holds that* $|\partial \varrho^*(\mu)| \leq 1$

(b) *For* $X, Y \in L^p, X \neq Y$ *it holds that* $|\partial \varrho(X) \cap \partial \varrho(Y)| = 0.$

Proof. (a) The conjugate of ϱ is defined by

$$\varrho^*(\mu) = \sup_{X \in L^p} (\langle \mu, X \rangle - \varrho(X)), \quad \mu \in \mathcal{M}_1^q.$$

Strict convexity of ϱ implies that for any $\mu \in \mathcal{M}_1^q$ there is at most one maximizer X_μ of $\langle \mu, X \rangle - \varrho(X)$, i.e. $|\partial \varrho^*(\mu)| \leq 1$.

(b) If $\mu \in \partial \varrho(X) \cap \partial \varrho(Y)$, then by (11.9) $X, Y \in \partial \varrho^*(\mu)$ in contradiction to (a). $\square$

The following property of subdifferentials is crucial for the uniqueness result.

Proposition 11.11. *If ϱ is convex, then for all $\mu \in \partial \varrho(X)$ and $v \in \partial \varrho(Y)$ by definition of the subgradient holds*

$$\langle v, X - Y \rangle \leq \varrho(X) - \varrho(Y) \leq \langle \mu, X - Y \rangle. \tag{11.31}$$

The inequalities in (11.31) are strict if ϱ is strictly convex and $X - Y \neq$ const.

Proof. Based on the Fenchel inequality and (11.8) we have

$$\varrho(X) + \varrho^*(\mu) = \langle \mu, X \rangle \quad \text{and} \quad \varrho(Y) + \varrho^*(\mu) \geq \langle \mu, Y \rangle$$

as well as $\hspace{10cm}$ (11.32)

$$\varrho(Y) + \varrho^*(v) = \langle v, Y \rangle \quad \text{and} \quad \varrho(X) + \varrho^*(v) \geq \langle v, X \rangle.$$

As a consequence we obtain

$$\varrho(X) - \varrho(Y) \le \langle \mu, X - Y \rangle \quad \text{and} \quad \varrho(X) - \varrho(Y) \ge \langle v, X - Y \rangle. \tag{11.33}$$

In the case that $X - Y = c \ne 0$ the expressions in (11.33) depend on $\langle \mu, 1 \rangle$ respectively $\langle v, 1 \rangle$ which may be different. If $X - Y \ne c$ for all $c \in \mathbb{R}$ by the assumption of strict convexity the inequalities in (11.33) are strict.

To prove this assume that one of the inequalities in (11.32) would be an equality. Then this assumption would imply that $\partial \varrho(X) \cap \partial \varrho(Y) \ne \phi$ in contradiction to Lemma 11.10. $\square$

As a consequence we get for strictly convex risk functionals uniqueness of γ-infimal convolution allocations as defined in (11.25) up to rebalancing. We call (ζ_i) a "γ-optimal allocation" if it minimizes the weighted total risk $\sum \gamma_i \varrho_i(X_i)$ over $(X_i) \in A(X)$.

Theorem 11.12. *Let ϱ_n be strictly convex risk functionals on L^p as in Theorem 11.5 and for some $\gamma \in \mathbb{R}_{>0}^n$ let $(\zeta_i) \in \mathcal{A}(X)$ be a γ-optimal allocation, i.e.*

$$\left(\bigwedge \varrho_i \right)_\gamma (X) = \sum_i \gamma_i \varrho_i(\zeta_i). \tag{11.34}$$

If (η_i) is a further γ-optimal allocation, then $\eta_n = \zeta_n + c_n$ for some constant c_n.

Proof. Assume that $\eta_n - \zeta_n \ne c$ for any $c \in \mathbb{R}$. By Theorem 11.5 applied to the risk functionals $\gamma_i \varrho_i$ there exists some $\mu \in \bigcap \partial(\gamma_i \varrho_i)(\zeta_i)$. As a consequence of Proposition 11.11 we thus obtain

$$\lambda_n \varrho_n(\zeta_n) < \langle \mu, \zeta_n - \eta_n \rangle + \lambda_n \varrho_n(\eta_n)$$

$$\lambda_i \varrho_i(\zeta_i) \le \langle \mu, \zeta_i - \eta_i \rangle + \lambda_i \varrho_i(\eta_i), \quad i \le n - 1.$$

This implies

$$\left(\bigwedge \varrho_i \right)_\gamma (X) = \sum \gamma_i \varrho_i(\zeta_i) < \sum \lambda_i \varrho_i(\eta_i)$$

and thus (η_i) is not a γ-optimal allocation. $\square$

Corollary 11.13. *If $\varrho_1, \ldots, \varrho_n$ are strictly convex lsc risk functionals on L^p as in Theorem 11.5 satisfying the (NS) condition as well as the strong intersection condition $\bigcap \operatorname{int}(\operatorname{dom}(\gamma_i \varrho_i^*)) \ne \phi$ for some $\gamma \in \mathbb{R}_{>0}^n$, then there exists up to additive constants c_i a unique γ-optimal allocation rule.*

In fact by Theorem 11.5 it is enough to postulate strict convexity only for $\varrho_1, \ldots, \varrho_{n-1}$. Theorem 11.5 does not imply a uniqueness result for cash invariant risk functionals ϱ_i since for any $X \in L^p(P)$, ϱ_i are not strictly convex on the affine subspace $X + \mathbb{R}$. To include this interesting case we define

$$L_0^p := \{ X \in L^p; \ EX = 0 \} \tag{11.35}$$

to be the class of equivalence classes of L^p modulo addition of constants. Cash invariance of ϱ implies for all $X \in L^p$, $c \in \mathbb{R}^1$

$$\partial \varrho(X) = \partial \varrho(X + c). \tag{11.36}$$

Therefore, we can extend Lemma 11.10 to the case where ϱ_i are cash invariant and strictly convex on L_0^p. We obtain for cash invariant risk measures strictly convex on L_0^p the following strong uniqueness property. As noted before for cash invariant risk functionals the intersection property (IS) is necessary and sufficient for the existence of Pareto optimal allocations.

Corollary 11.14 (Uniqueness of PO-allocations). *Let $\varrho_1, \ldots, \varrho_n$ be finite risk functionals strictly convex on L_0^p, cash invariant, lsc, finite risk functionals and at least one of them being monotone. Also assume the intersection condition $\bigcap \operatorname{dom} \varrho_i^* \neq \phi$. Then for any $X \in L^p$ there exists an up to additive constants unique Pareto optimal allocation (ζ_i) of X.*

Again it is enough that $\varrho_1, \ldots, \varrho_{n-1}$ are strictly convex on L_0^p and the monotonicity assumption can be replaced by the subgradient condition $\partial \bigwedge \varrho_i(X) \neq \phi$.

Total cost minimizing allocations respectively Pareto optimal allocations are jointly optimal but in case there is given an initial allocation $X = \sum_{i=1} X_i$ they do not necessarily improve the risk of the individual shares. To meet this condition a restriction to allocations called "individual rationality" is introduced.

Definition 11.15 (Optimal risk sharing rule (ORS)). Let $(X_i) \in \mathcal{A}(X)$ be a given initial allocation of X. Then $(\xi_i) \in \mathcal{A}(X)$ is called an "optimal risk sharing rule (ORS)" if

(1) (ξ_i) is Pareto optimal

(2) (ξ_i) satisfied the "individual rationality constraint"

(IR) $\varrho_i(\xi_i) \leq \varrho_i(X_i), \quad 1 \leq i \leq n. \tag{11.37}$

Each trader i is assumed to accept a new allocation only when it improves his part X_i of the total risk. For translation invariant risk functionals there is an easy way to improve PO-allocations $(X_i) \in \mathcal{A}(X)$ to ORS allocations $(\xi_i) \in \mathcal{A}(X)$.

Theorem 11.16. *Let $(X_i) \in \mathcal{A}(X)$ be an initial allocation of X w.r.t. translation invariant, convex risk functionals ϱ_i, $1 \leq i \leq n$. Let $(\xi_i) \in \mathcal{A}(X)$ be a PO-allocation of X. Then with $p_i := \varrho_i(X_i) - \varrho_i(\xi_i)$ the following holds:*

(a) $\sum_{i=1}^n p_i \geq 0$.
(b) *For any $\pi_i \in \mathbb{R}$ with $\sum_{i=1}^n \pi_i = 0$ the allocation $(\xi_i - \pi_i)$ is an ORS rule if and only if $\pi_i \leq p_i$, $1 \leq i \leq n$.*

Proof. (a) Pareto optimality of (ξ_i) and translation invariance of ϱ_i implies that (ξ_i) minimizes the total risk and thus

$$\sum_{i=1}^n p_i \geq 0.$$

(b) By translation invariance $(\xi_i - \pi_i)$ is Pareto optimal as well. Therefore, $(\xi_i - \pi_i)$ is an optimal risk sharing rule (ORS) if and only if

$$\varrho_i(\xi_i - \pi_i) \le \varrho_i(X_i), \quad 1 \le i \le n. \tag{11.38}$$

Equation (11.38) however is by translation invariance of ϱ_i equivalent to

$$\pi_i \le \varrho_i(X_i) - \varrho_i(\xi_i) = p_i, \quad 1 \le i \le n. \qquad \square$$

Remark 11.17. (a) Existence of ORS rules: Theorem 11.16 implies the existence of optimal risk sharing rules (ORS) under the condition that PO rules exist.

(b) **Indifference prices:** If π_i are prices to be paid for the transaction from X_i to ξ_i then $p_1, \ldots, p_n$ are called "indifference prices" since under these prices traders are indifferent to either carry out these transactions under these prices or to keep the initial risk decomposition (X_i).

If the initial allocation (X_i) is already Pareto optimal, then for any other Pareto optimal allocation $(\xi_i) \in \mathcal{A}(X)$ there is a unique vector of prices making it an ORS rule – the vector of indifference prices (p_i). Otherwise the set of possible transaction prices for a switch to an ORS rule is given by the cone

$$\prod := \left\{ (\pi_i) \in \mathbb{R}^n; \ \sum_{i=1}^{n} \pi_i = 0, \pi_i \le p_i, 1 \le i \le n \right\} \tag{11.39}$$

and further criteria have to be implemented to derive a unique price system $(\pi_i) \in \prod$ for this transaction. $\qquad \lozenge$

11.3 Examples of Optimal Risk Allocations

In this section we derive for several risk situations optimal risk allocations. In particular we get optimality results for linear quota sharing rules, for stop-loss contracts and deductibles which are classical contracts respectively options in insurance and finance.

Constant quota contracts have been obtained as optimal allocations in Borch (1962) and Gerber (1978, 1979) and others. To derive these optimality results for general lsc, convex risk measures we need the following simple to verify rules.

Lemma 11.18. *Let ϱ be convex, lsc risk functionals on L^p then for $\alpha \in \mathbb{R}_{>0}$*

(a) $(\alpha\varrho)^*(\mu) = \alpha\varrho^*\left(\frac{\mu}{\alpha}\right)$

(b) $\mathrm{dom}(\alpha\varrho)^* = \alpha \, \mathrm{dom}\, \varrho^*$

(c) $\partial(\alpha\varrho)^*(\mu) = \partial\varrho^*\left(\frac{\mu}{\alpha}\right)$ $\qquad\qquad$ (11.40)

(d) $\partial(\alpha\varrho)(X) = \alpha\partial\varrho(X).$

Remark 11.19. From Proposition 11.4 and Lemma 11.18 we obtain existence and characterization of Pareto optimal allocations under the intersection property

$$(\text{IS})_\gamma \qquad\qquad \bigcap \text{dom}(\gamma_i \varrho_i)^* \neq \phi. \tag{11.41}$$

The intersection property $(\text{IS})_\gamma$ is fulfilled under either of the following two conditions:

(a) $\exists X \in L^p$ such that $\partial(\bigwedge \varrho_i)_\gamma (X) \neq \phi$ \hfill (11.42)

(b) $\exists (\eta_i) \in A(X)$ such that $\bigcap \gamma_i \partial \varrho_i (\eta_i) \neq \phi$. \hfill (11.43)

For the proof note that by Proposition 11.4 and Lemma 11.18 it holds that for any $(\eta_i) \in A(X)$

$$\bigcap \gamma_i \partial \varrho_i (\eta_i) \subset \partial\left(\bigwedge \varrho_i\right)_\gamma (X).$$

Thus (b) is a consequence of (a).
For proper convex functions f it holds that (see Barbu and Precupanu (1986, p. 101)):

$$\begin{aligned}&\text{If } X \in \text{int dom } f, \text{ then } \partial f(X) \neq \phi;\\ &\text{if } \partial f(X) \neq \phi, \text{ then } X \in \text{dom } f.\end{aligned} \tag{11.44}$$

This implies that

$$\text{range } \partial f := \{X;\ \partial f(X) \neq \phi\} \subset \text{dom } f^*. \tag{11.45}$$

As a consequence we obtain for any $(\eta_i) \in A(X)$ the relation

$$\text{range}(\partial(\bigwedge \varrho_i)_\gamma) \subset \text{dom}(\bigwedge \varrho_i)^*_\gamma = \bigcap \gamma_i \text{ dom } \varrho_i^*. \tag{11.46}$$

Thus (a) follows. \hfill $\Diamond$

We next discuss several concrete classes of examples of risk functionals.

11.3.1 Expected Risk Functionals

Let ϱ be an expected risk functional

$$\varrho(X) = Er(X), \tag{11.47}$$

where r is a convex, strictly decreasing, differentiable function $r : \mathbb{R} \to \mathbb{R}$. This is the typical case considered in the calculation of convex principles in premium calculation (see Deprez and Gerber (1985)). Then ϱ is Gateaux differentiable and

$$\partial \varrho(X) = \{r' \circ X\} \tag{11.48}$$

is given by the Gateaux-gradient of ϱ. As a consequence we obtain from Theorem 11.28 a classical result of Borch (1962), see also Deprez and Gerber (1985) and Barrieu and Scandolo (2008).

Proposition 11.20. (ξ_i) *is a PO-allocation for the expected risk functionals* $\varrho_i(X) = Er_i(X), 1 \leq i \leq n$

$$\Leftrightarrow \exists \gamma \in \mathbb{R}_{>}^n \text{ such that}$$
$$\gamma_i r_i' \circ \xi_i = \gamma_j r_j' \circ \xi_j, \quad \forall i, j. \tag{11.49}$$

Optimal allocations (ξ_i) solving (11.49) are increasing functions of the total risk X. In the case of exponential risk functions $r_i(x) = \alpha_i \left(e^{-\frac{1}{\alpha_i} x} - 1 \right), \alpha_i > 0$ (corresponding to exponential utility functions $u_i(x) = -r_i(x)$) Eq. (11.49) and the allocation condition $\sum_{i=1}^n \xi_i = X$ lead to the solutions

$$\xi_i = \beta_i X + p_i, \quad 1 \leq i \leq n,$$

with $\beta_i := \frac{\alpha_i}{\sum_{j=1}^n \alpha_j}$ and $p_i = \alpha_i (\log \gamma_i - \sum_{j=1}^n \beta_j \log \gamma_j)$.

The optimal allocation is given by quota of the total risk plus some premium p_i depending on the risk parameter γ_i.

11.3.2 Dilated Risk Functionals

Let ϱ be a convex risk functional and define the class of dilated risk functionals ϱ_λ for $\lambda > 0$ by

$$\varrho_\lambda(X) = \lambda \varrho \left(\frac{X}{\lambda} \right). \tag{11.50}$$

Then as in Lemma 11.18 one obtains the following rules for subgradients:

(a) $\operatorname{dom} \varrho_\lambda = \lambda \operatorname{dom} \varrho$,

(b) $\varrho_\lambda^* = \lambda \varrho^*$,

(c) $\partial \varrho_\lambda(X) = \partial \varrho \left(\frac{X}{\lambda} \right)$, \hfill (11.51)

(d) $\partial \varrho_\lambda^*(\mu) = \lambda \partial \varrho^*(\mu)$.

As a consequence we obtain from the characterization in Theorem 11.7 (see Deprez and Gerber (1985, Theorem 12) and Barrieu and El Karoui (2005, Proposition 3.5) for the case of convex risk measures) the classical characterization of constant quota rules as optimal allocations.

Theorem 11.21 (Dilated risk functionals).   *Let ϱ be a convex, lsc, proper risk functional on L^p satisfying condition NS and let $\varrho_i = \varrho_{\lambda_i}, 1 \leq i \leq n$, be the corresponding dilation risk functionals, $\lambda_i > 0$. Let $\lambda := \sum \lambda_i$ and let $X \in L^p$ with $\frac{X}{\lambda} \in \operatorname{int}(\operatorname{dom} \varrho)$. Then the constant quota rules $\xi_i := \frac{\lambda_i}{\lambda} X$ define a PO-allocation of X.*

Proof. For the proof it is enough to check condition (c) of Theorem 11.5. Using (11.51) and (11.44) we obtain

$$\bigcap \partial \varrho_i (\xi_i) = \bigcap \partial \varrho \left(\frac{\xi_i}{\lambda_i} \right) = \bigcap \partial \varrho \left(\frac{X}{\lambda} \right) = \partial \varrho \left(\frac{X}{\lambda} \right) \neq \phi.$$

Thus by Theorem 11.5 and (11.46) (ξ_i) is a PO-allocation of X. □

11.3.3 Average Value at Risk and Stop-Loss Contracts

We consider the optimal allocation where $\varrho_1(X) = \text{AV@R}_\lambda(X)$, $0 < \lambda \leq 1$ and ϱ_2 is a law invariant convex translation invariant, continuous risk functional on L^p with $\varrho_2(0) = 0$. We also assume that ϱ_2 is "strictly risk averse", i.e.

$$\varrho_2(X) > \varrho_2(X 1_{A^c} + E(X \mid Y) 1_A) \tag{11.52}$$

for any A with $P(A) > 0$ and $\text{ess inf}_A X < \text{ess sup}_A X$. In this case the representation set of ϱ_1 is given by $\mathcal{P}_\lambda = \{Q \in M_1; \ 0 \leq Z = \frac{dQ}{dP} \leq \frac{1}{\lambda}\}$ and thus $\varrho_1(X) = \sup_{Q \in \mathcal{P}_\lambda} E_Q(-X)$.

The following characterization of stop-loss contracts as optimal allocations based on the characterization results in Section 11.2 is given in Jouini et al. (2008) and Acciaio (2007). An alternative derivation by stochastic ordering arguments will be given later in Section 12.2.

Theorem 11.22 (Average value at risk, stop-loss contracts). *Under the assumptions above there exists to $X \in L^p$ a unique (up to a constant) Pareto optimal allocation (X_1, X_2) in $\mathcal{A}^\uparrow(X)$ for $\varrho_1 = \text{AV@R}_\lambda$ and ϱ_2 of the form*

$$(X_1, X_2) = ((X - u)_+ - (X - l)_-, (l \vee X) \wedge u) \tag{11.53}$$

for some $l \leq u$ in $\mathbb{R}$.

Proof. By the existence and characterization results for a Pareto optimal allocation (X_1, X_2) of X there exists an element $\mu \in \mathcal{M}_1$ with density $Z = \frac{d\mu}{dP}$ such that $Z \in \partial \varrho_i(X_i)$, $i = 1, 2$. Since the average value at risk ϱ_1 is law invariant it follows from Kusuoka's representation result (respectively its version for L^p) that Z and X_1 respectively Z and X_2 are anticomonotone and thus by the description of the representation set $\mathcal{P}_\lambda$ it holds that X_1 takes it largest values at $\{Z = 0\}$ and its smallest values at $\{Z = \frac{1}{\lambda}\}$.

Furthermore, by the strict risk adverse assumption it follows that X_2 is constant on the sets $\{Z = 0\}$ respectively $\{Z = \frac{1}{\lambda}\}$ (for details see Jouini et al. (2008, Lemma 5.1)). In consequence it follows that X_1, X_2 are of the form of limited stop-loss contracts respectively its deductible in (11.53).

Any Pareto optimal allocation is by the above argument of the limited stop-loss type as in (11.53). If (Y_1, Y_2) is such an allocation with constants u_1, ℓ_1, then also the convex combination $\alpha(X_1, X_2) + (1 - \alpha)(Y_1, Y_2)$ is Pareto optimal but it is only of the form in (11.53) if $u_1 = u$, $\ell_1 = \ell$. This implies uniqueness.   □

The optimal allocation is of a typical form of an insurance contract where the insurers share X_2 has a floor ℓ and an upper limit u (a deductible with floor ℓ) and the reinsurer takes by the combination of a call and a put option the large gains and also the large losses. Thus a trader using for the evaluation of risk an average value at risk evaluation takes the extreme events (losses and gains). This nonstable behaviour is a peculiar feature of the AV@R risk measures which raises some doubt whether the choice of the AV@R is a suitable choice of risk measure for an insurance company respectively a financial institution.

11.3.4 Mean Variance Versus Standard Deviation Risk Functionals

The "mean variance principle" (MVP) ϱ_δ^{mv} is defined for $\delta > 0$ by

$$\varrho_\delta^{mv}(X) := E(-X) + \delta \operatorname{Var}(X). \tag{11.54}$$

The parameter δ reflects the degree of risk aversion. ϱ_δ^{mv} is a convex, lsc, translation invariant risk measure but ϱ_δ^{mv} is not monotone. The following lemma describes the conjugate $(\varrho_\delta^{mv})^* =: v_\delta^{mv}$ and the subgradient. We identify P continuous measures with their densities. The subgradients were given in Acciaio (2007) for the essentially equivalent utility formulation. Some related calculations of subdifferentials are also given in Ruszczyński and Shapiro (2006). We give a sketch of the proof since it shows some typical arguments for the calculation of subgradients useful also for related applications.

Lemma 11.23.
(a) $v_\delta^{mv}(\mu) = \frac{\operatorname{Var}\mu}{4\delta}$ for all $\mu \in \operatorname{dom}(v_\delta^{mv}) = \{\mu \in L^1;\ E(-\mu) = 1\} \cap L^2$
(b) $\partial\varrho_\delta^{mv}(X) = \{2\delta(X - EX) - 1\}, \quad \forall X \in L^p \cap L^2$
(c) $\partial v_\delta^{mv}(\mu) = \{\frac{\mu}{2\delta} + c;\ c \in \mathbb{R}\}, \quad \forall \mu \in \operatorname{dom}(v_\delta^{mv}).$

Proof. (b) ϱ_δ^{mv} is easily seen to be Gateaux differentiable with Gateaux gradient

$$\nabla\varrho_\delta^{mv}(X) = 2\delta(X - EX) - 1.$$

This implies (b).

(c) By lsc of ϱ_δ^{mv} we have that $\mu \in \partial\varrho_\delta^{mv}(X)$ is equivalent to $X \in \partial v_\delta^{mv}(\mu)$. Since, for $X \in L^p$ with $EX = 0$,

$$\partial\varrho_\delta^{mv}(X + c) = \partial\varrho_\delta^{mv}(X) = \{2\delta X - 1\}$$

by Corollary 4.1 in Aubin (1993) we obtain

$$\partial v_\delta^{mv}(\mu) = (\partial \varrho_\delta^{mv})^{-1}(\mu) = \left\{\frac{\mu}{2\delta} + c; \ c \in \mathbb{R}\right\}$$

for all $\mu \in \mathrm{dom}(v_\delta^{mv})$.

(a) Using that $\partial v_\delta^{mv}(\mu) = \arg\max(\langle \mu, X \rangle - \varrho_\delta^{mv}(X); \ X \in L^p)$ we obtain for $X_0 \in \partial v_\delta^{mv}(\mu)$ using (c)

$$\begin{aligned}
v_\delta^{mv}(\mu) &= \langle \mu, X_0 \rangle - \varrho_\delta^{mv}(X_0) \\
&= E\mu X_0 - E(-X_0) - \delta \,\mathrm{Var}(X_0) \\
&= EX_0(1 + \mu) - \delta \,\mathrm{Var}(X_0) \\
&= E\frac{\mu}{2\delta}(1 + \mu) + cE(1 + \mu) - \delta \,\mathrm{Var}\left(\frac{\mu}{2\delta}\right) \\
&= -\frac{1}{2\delta} + \frac{1}{2\delta}E\mu^2 - \frac{1}{4\delta}E\mu^2 + \frac{1}{4\delta} \\
&= \frac{\mathrm{Var}\,\mu}{4\delta}.
\end{aligned}$$

Since ϱ_δ^{mv} is law invariant and thus Fatou continuous we can restrict above to P-continuous measures and any $\mu \in \mathrm{dom}\, v_\delta^{mv}$ must satisfy $\mathrm{Var}\,\mu < \infty$ and thus $\mu \in L^2$.   □

The "standard deviation principle" (SDP) ϱ_δ^{sd} is defined for $\delta > 0$ by

$$\varrho_\delta^{sd}(X) = E(-X) + \delta\sqrt{\mathrm{Var}\,X}. \tag{11.55}$$

Compared with the mean variance principle it is less sensitive concerning the variance. Similarly to Lemma 11.23 one obtains the subgradients (see also Acciaio (2007)).

Lemma 11.24. *For the standard deviation principle ϱ_δ^{sd} and the corresponding conjugate v_δ^{sd} it holds that (again we identify $\mu \in M^q(P)$ with its density)*

(a) $\partial \varrho_\delta^{sd}(X) = \begin{cases} \mathrm{dom}\, v_\delta^{sd} & \text{if } X = \mathrm{const}, \\ \delta\frac{X-EX}{\|X-EX\|_2} - 1 & \text{else}, \end{cases}$ $\tag{11.56}$

(b) $\partial v_\delta^{sd}(\mu) = \mathbb{R} \cup \left\{ X \in L^p; \ \mu = \delta\frac{X - EX}{\|X - EX\|_2} - 1 \right\}.$ $\tag{11.57}$

In the following proposition we determine all Pareto optimal allocations between an SDP ϱ_1 (with $\delta_1 > 0$) and an MVP ϱ_2 (with $\delta_2 > 0$) and conjugates v_1, v_2.

Proposition 11.25 (SDP vs. MVP). *Let ϱ_1 be an SDP and ϱ_2 an MVP with $\delta_i > 0$. Then for any $X \in \mathrm{dom}(\varrho_1 \wedge \varrho_2)$, $X \not\equiv \mathrm{const}$ there exists an up to constants unique Pareto optimal allocation (ξ_1, ξ_2) which is given by:*

(a) If $0 < \frac{\delta_1}{2\delta_2\sigma(X)} < 1$, then

$$\xi_1 = \left(1 - \frac{\delta_1}{2\delta_2\sigma(X)}\right)X + c, \quad \xi_2 = \frac{\delta_1}{2\delta_2\sigma(X)}X - c \qquad (11.58)$$

and the optimal dual measure μ is given by $\mu = \delta_1 \frac{X-EX}{\sigma(X)} - 1$.

(b) If $\frac{\delta_1}{2\delta_2\sigma(X)} \geq 1$, then the allocation $(0, X)$ is up to constants the unique PO-allocation.

Proof. Both risk functionals ϱ_1, ϱ_2 are convex, lsc, translation invariant and law invariant. Thus the intersection property is fulfilled and for all $X \in \mathrm{dom}(\varrho_1 \wedge \varrho_2)$ and $\delta_1, \delta_2 > 0$ there exists a Pareto optimal allocation.

To construct a PO-allocation, we use the characterization result finding $(\xi_1, \xi_2) \in A(X)$ and a measure μ (identified with its density) such that $\mu \in \partial\varrho_1(\xi_1) \cap \partial\varrho_2(\xi_2)$, i.e. by Lemmas 11.23 and 11.24

$$\left\{\delta_1 \frac{\xi_1 - E\xi_1}{\|\xi_1 - E\xi_1\|_2} - 1\right\} = \partial\varrho_1(\xi_1) = \partial\varrho_2(\xi_2) = \{2\delta_2(\xi_2 - E\xi_2) - 1\}. \qquad (11.59)$$

In case (a) one finds a nontrivial solution as given in (11.58). In case (b) the following argument leads to the conclusion. Let $\eta_1, \eta_2 \in A(X)$ be an allocation of X, then

$$\begin{aligned}
\varrho_1(\eta_1) + \varrho_2(\eta_2) &= E(-X) + \delta_1\sigma(\eta_1) + \delta_2\,\mathrm{Var}(\eta_2) \\
&\geq E(-X) + \delta_2\,\mathrm{Var}(\eta_2) + 2\delta_2\sigma(X)\sigma(\eta_1) \\
&\hspace{4cm} (\text{as } \delta_1 \geq 2\delta_2\sigma(X)) \\
&\geq E(-X) + \delta_2\,\mathrm{Var}(\eta_2) + 2\delta_2\,\mathrm{Cov}(X, \eta_1) \\
&= E(-X) + \delta_2\,\mathrm{Var}(X - \eta_1) + 2\delta_2\,\mathrm{Cov}(X, \eta_1) \\
&= E(-X) + \delta_2(\mathrm{Var}(X) + \mathrm{Var}(\eta_1)) \\
&\geq E(-X) + \delta_2\,\mathrm{Var}(X) = \varrho_2(X).
\end{aligned}$$

Thus $(0, X)$ is an optimal allocation. Uniqueness follows from our uniqueness result in Corollary 11.14.   □

The characterization and uniqueness result allows to deal in a similar way with further convex risk functionals without assuming the monotonicity or the cash invariance property. One can e.g. consider risk functionals of the form

$$\varrho_1(X) = E(-X) + \delta E(X - q_\alpha(X))_-^p, \quad p \geq 1, \qquad (11.60)$$

where $q_\alpha(X)$ is the α-quantile of X. ϱ is convex and monotone but not cash invariant. Similarly,

$$\varrho_2(X) = E(-X) + \delta E|X - q_\alpha(X)|^p, \quad p \geq 1, \tag{11.61}$$

which is neither monotone nor cash invariant. For some explicit further examples in the case of nonmonotone risk measures see Acciaio (2007).

11.4   Optimal Allocation of Risk Vectors

In this section we give an extension of the optimal allocation results to the case of risk vectors $X = (X_1, \dots, X_d) \in L_d^p$ with $X_i \in L^p$ where L_d^p is the d-fold product of L^p, $1 \leq p \leq \infty$. Generally we consider normed convex lsc proper risk functionals $\varrho_i : L_d^p \to (-\infty, \infty]$, $1 \leq i \leq n$ that describe the risk evaluation in the market with normed means $\varrho_i(0) = 0$. We assume generally that a representation of ϱ_i based on scenario measures in $\mathcal{M}_d^q$ $(= (\mathcal{M}_1)^d$ in case $p = \infty)$ holds. Consideration of risk vectors allows to describe the risk coming from dependence between the components. As in $d = 1$ for $X \in L_d^p$

$$\mathcal{A}(X) = \left\{ (\xi_1, \dots, \xi_n); \ \xi_i \in L_d^p, \sum_{i=1}^{n} \xi_i = X \right\} = \mathcal{A}^n(X)$$

describes the (n)-allocations of the portfolio vector X.

For an allocation $(\xi_1, \dots, \xi_n) \in \mathcal{A}(X)$ trader i is exposed to the risk vector ξ_i which is evaluated by the risk functional ϱ_i. ξ_i may contain some zero components and thus trader i may only be exposed to some of the d components of risk in our formulation. Let

$$\mathcal{R} := \{(\varrho_i(\xi_i)); \ (\xi_i) \in \mathcal{A}(X)\}$$

denote the corresponding risk set. Our aim is to characterize Pareto optimal (PO) allocations $(\xi_i) \in \mathcal{A}(X)$, i.e. allocations such that the corresponding risk vectors are minimal elements of the risk set $\mathcal{R}$ in the pointwise ordering. A related optimization problem is to characterize allocations $(\eta_i) \in \mathcal{A}(X)$ which minimize the total risk, i.e.

$$\sum_{i=1}^{n} \varrho_i(\eta_i) = \inf \left\{ \sum_{i=1}^{n} \varrho_i(\xi_i); \ (\xi_i) \in \mathcal{A}(X) \right\}$$

$$=: \bigwedge \varrho_i(X).$$

As in Section 11.3 for proper, convex lsc risk functionals ϱ_i it follows that ϱ_i^* are proper and

$$(\bigwedge \varrho_i)^* = \sum_{i=1}^{n} \varrho_i^* \tag{11.62}$$

(see Barbu and Precupanu (1986, Section 2)). In consequence we obtain

$$\bigcap_{i=1}^{n} \operatorname{dom} \varrho_i^* \neq \phi \Rightarrow \operatorname{dom}(\bigwedge \varrho_i) \neq \phi, \tag{11.63}$$

and thus the inf-convolution problem makes sense. For all results on convex duality we refer to Rockafellar (1974) and Barbu and Precupanu (1986).

In this section we consider the dual pair $(L_d^p, L_d^q, \langle \cdot, \cdot \rangle_d)$ where $\langle \cdot, \cdot \rangle_d$ denotes the canonical scalar product on the product spaces

$$\langle Z, X \rangle_d := \sum_{j=1}^{d} E Z^j X^j$$

for $X = (X^1, \dots, X^d) \in L_d^p$, $Y = (Y^1, \dots, Y^d) \in L_d^q$, where q is the conjugate index to p, $\frac{1}{p} + \frac{1}{q} = 1$. We use the dual set $\mathcal{M}_d^q$ of d-tuples of normed P-continuous measures integrating $|x|^q$. In the case $p = 1, q = \infty$, $\mathcal{M}_d^\infty$ is the set of d-tuples of normed P-continuous measures. To avoid cumbersome notation we still use L_d^q in this case.

The portfolio vectors $(\xi_i)_{1 \leq i \leq n}$ are contained in the corresponding product spaces defining the dual pair $((L_d^p)^n, (L_d^q)^n, \langle \cdot, \cdot \rangle_d^n)$ where for $X = (X_1, \dots, X_n) \in (L_d^p)^n$, $Z = (Z_1, \dots, Z_n) \in (L_d^q)^n$ the scalar product is given by

$$\langle Z, X \rangle_d^n := \sum_{i=1}^{n} \langle Z_i, X_i \rangle_d.$$

We will use the notation $\langle Z, X \rangle = \langle Z, X \rangle_d^n$ when omitting the indices does not lead to confusion. The results in this section are based on Kiesel and Rü (2010).

11.4.1   Characterization of Optimal Allocations

To characterize Pareto-optimal allocations we describe at first allocations which minimize the total risk, i.e. solutions of the infimal convolution

$$\bigwedge \varrho_i(X) = \inf \left\{ \sum_{i=1}^{n} \varrho_i(\xi_i); \ (\xi_i) \in \mathcal{A}(X) \right\}. \tag{11.64}$$

The inf-convolution problem is a restricted optimization problem. It can be transformed into an unrestricted global minimization problem

$$\bigwedge \varrho_i(X) = \inf \{ \overline{\varrho}(\xi) + \mathbb{1}_{\mathcal{A}(X)}(\xi); \ \xi \in (L_d^p)^n \} \tag{11.65}$$

where $\overline{\varrho}(\xi) := \sum_{i=1}^{n} \varrho_i(\xi_i)$ and for a convex set A, $\mathbb{1}_A$ denotes the convex indicator

$$\mathbb{1}_A(x) = \begin{cases} 0, & x \in A, \\ \infty, & x \notin A. \end{cases} \tag{11.66}$$

We generally assume that there exists at least one n-allocation $\xi \in \mathcal{A}^n(X)$, where $\overline{\varrho}$ is continuous and finite. For $f : E \to \mathbb{R} \cup \{\infty\}$ the "domain of continuity" is denoted by

$$\mathrm{domc}(f) := \{x \in E; \ f \text{ is finite and continuous in } x\}.$$

The inf-convolution problem respectively the minimal total risk problem is called "well posed" for a given portfolio X if

$$\mathrm{domc}(\overline{\varrho}) \cap \mathcal{A}(X) \neq \phi. \tag{11.67}$$

The following is the basic characterization of minimal total risk allocations which extends the developments for real risks to the portfolio case.

Theorem 11.26 (Characterization of minimal total risk). *Let ϱ_i be risk functionals on L_d^p, $1 \le p \le \infty$ as described above. Let $X \in L_d^p$ be a risk portfolio such that the minimal total risk problem is well posed and let $(\eta_i) \in \mathcal{A}^n(X)$ be a risk allocation. Then the following statements are equivalent:*

(a) (η_i) has minimal total risk (w.r.t. $\varrho_1, \dots, \varrho_n$ and X)

(b) $\exists V \in L_d^q$ such that $V \in \partial \varrho_i(\eta_i)$, $1 \le i \le n$ $\tag{11.68}$

(c) $\exists V \in L_d^q$ such that $\eta_i \in \partial \varrho_i^*(V)$, $1 \le i \le n$. $\tag{11.69}$

Proof. The equivalence of (b) and (c) is a well-known result in convex analysis (see e.g. Barbu and Precupanu (1986) and Aubin (1993)). The proof of the equivalence of (a) and (b) needs in the multivariate case some additional arguments compared with the corresponding proof in the one-dimensional case in Theorem 11.5.

Let $\overline{\xi} = (\xi_1, \dots, \xi_n) \in \mathcal{A}^n(X)$ be an allocation with total minimal risk, i.e.

$$\bigwedge \varrho_i(X) = \sum_{i=1}^n \varrho_i(\xi_i).$$

Due to Fermat's rule the representation in (11.65) implies

$$0 \in \partial(\overline{\varrho} + \mathbb{1}_{\mathcal{A}^n(X)})(\overline{\xi}). \tag{11.70}$$

The infimal convolution is well posed for X. In consequence the subdifferential sum formula (see Barbu and Precupanu (1986, Section 3, Theorem 2.6)) is applicable to the right-hand side of (11.70) and yields

$$0 \in \partial\overline{\varrho}(\overline{\xi}) + \partial\mathbb{1}_{\mathcal{A}^n(X)}(\overline{\xi}). \tag{11.71}$$

Thus there exists an element $\Lambda \in (L_d^p)^n$ with

$$\Lambda \in \partial\overline{\varrho}(\overline{\xi}) \quad \text{and} \quad -\Lambda \in \partial\mathbb{1}_{\mathcal{A}^n(X)}(\overline{\xi}). \tag{11.72}$$

This leads to the equation

$$\overline{\varrho}(\overline{\xi}) + \overline{\varrho}^*(\Lambda) = \langle \Lambda, \overline{\xi} \rangle. \tag{11.73}$$

In the next step we show that (11.73) implies the existence of some $V \in L_d^q$ such that

$$\varrho_i(\xi_i) + \varrho_i^*(V) = \langle V, \xi_i \rangle, \quad \forall i, \tag{11.74}$$

i.e. all components Λ_i of Λ are identical to $V \in L_d^q$. This results from the following proposition.

Proposition 11.27. *For all $X \in L_d^p$ and $\overline{\xi} \in \mathcal{A}^n(X)$ it holds that*

$$\partial\mathbb{1}_{\mathcal{A}^n(X)}(\overline{\xi}) = \left\{ \overline{Z} \in (L_d^q)^n; \ \overline{Z} = \sum_{i=1}^n Ze_i, Z \in L_d^q \right\}, \tag{11.75}$$

where e_i is the i-th unit vector of the n-fold product space $(L_d^q)^n$. Thus the product Ze_i is understood as the element of $(L_d^q)^n$ which has Z as its i-th component and $0 \in L_d^q$ otherwise.

Proof. By definition of the subdifferential we have

$$\partial\mathbb{1}_{\mathcal{A}^n(X)}(\overline{\xi}) = \{ \overline{Z} \in (L_d^q)^n; \ \langle \overline{\eta}, \overline{Z} \rangle \le \langle \overline{\xi}, \overline{Z} \rangle, \quad \forall \overline{\eta} \in \mathcal{A}^n(X) \}.$$

If $\overline{Z} = \sum_{i=1}^n Ze_i$, with $Z \in L_d^q$ and $\overline{\eta} \in \mathcal{A}^n(X)$, then

$$\langle \overline{\eta}, \overline{Z} \rangle = \sum_{i=1}^n \langle \eta_i, Z \rangle$$

$$= \left\langle \sum_{i=1}^n \eta_i, Z \right\rangle = \left\langle \sum_{i=1}^n \xi_i, Z \right\rangle = \langle \overline{\xi}, \overline{Z} \rangle.$$

Thus $\overline{Z} \in \partial\mathbb{1}_{\mathcal{A}^n(X)}$.

Conversely for $\overline{Z} \in \partial\mathbb{1}_{\mathcal{A}^n(X)}(\xi)$ and $\overline{\eta} \in \mathcal{A}^n(X)$ it holds that

$$\langle \overline{\eta}, \overline{Z} \rangle \le \langle \overline{\xi}, \overline{Z} \rangle. \tag{11.76}$$

Choosing $\bar{\eta}$ of the form $\bar{\eta} = \bar{\xi} + \eta e_k - \eta e_\ell$ with $k, \ell \in \{1, \ldots, n\}$ and $\eta \in L_d^p$ we obtain from (11.76)

$$\langle \eta, \overline{Z}_k \rangle \leq \langle \eta, \overline{Z}_\ell \rangle.$$

Reversing the roles of k, ℓ we obtain the opposite inequality and consequently $Z_k = Z$, $1 \leq k \leq n$ for some $Z \in L_d^q$. Thus $\overline{Z} = \sum_{i=1}^n Z e_i$ is of the form as stated in (11.74). $\qquad\square$

Continuation of the proof of Theorem 11.26. From the definition of the convex conjugate it follows that

$$\bar{\varrho}^*(\overline{Z}) = \sum_{i=1}^n \varrho_i^*(\overline{Z}_i), \quad \overline{Z} = (\overline{Z}_1, \ldots, \overline{Z}_n) \in (L_d^q)^n.$$

From Proposition 11.27 we obtain for Λ as in (11.72) $\Lambda = \sum_{i=1}^n V e_i$ with $V \in L_d^q$. Equation (11.73) then implies

$$\sum_{i=1}^n (\varrho_i(\xi_i) + \varrho_i^*(V)) = \sum_{i=1}^n \langle \xi_i, V \rangle. \tag{11.77}$$

Since by the Fenchel inequalities

$$\varrho_i(\xi_i) + \varrho_i^*(V) \geq \langle \xi_i, V \rangle, \quad \forall i \tag{11.78}$$

Equation (11.77) implies equality in (11.78) and in consequence

$$V \in \partial \varrho_i(\xi_i), \quad \forall i.$$

Thus (1) implies (2). The above given proof can be reversed to yield also the opposite direction. $\qquad\square$

To obtain a connection of minimizing the total risk to Pareto-optimality we introduce as in Section 11.2 a condition called the non-saturation property. We say that ϱ has the "non-saturation property" if

(NS) $\qquad\qquad\qquad \inf_{X \in L_d^p} \varrho(X)$ is not attained. $\tag{11.79}$

The non-saturation property is a weak property of risk measures. It is implied in particular by the cash invariance property. Under the (NS) condition Pareto-optimality is related to the problem of minimizing the total weighted risk. This is described by the "weighted minimal convolution" $(\wedge \varrho_i)_\gamma(X)$ defined for weight vectors $\gamma = (\gamma_1, \ldots, \gamma_n) \in \mathbb{R}^n$ by

$$(\wedge \varrho_i)_\gamma(X) := \inf \left\{ \sum_{i=1}^n \gamma_i \varrho_i(\xi_i); \ (\xi_1, \ldots, \xi_n) \in \mathcal{A}^n(X) \right\}. \qquad (11.80)$$

The connection between Pareto-optimality and minimizing total weighted risk goes back to early papers in insurance (see e.g. Gerber (1979)).

Theorem 11.28 (Characterization of Pareto optimal allocations). *Let ϱ_i, $1 \le i \le n$, be risk functionals on L_d^p satisfying the non-saturation conditions (NS). Then for $X \in L_d^p$ and $(\xi_1, \ldots, \xi_n) \in \mathcal{A}^n(X)$ the following statements are equivalent*

(1) $(\xi_1, \ldots, \xi_n)$ is a Pareto-optimal allocation of X w.r.t. $\varrho_1, \ldots, \varrho_n$

(2) $\exists \gamma = (\gamma_1, \ldots, \gamma_n) \in \mathbb{R}^n_{>0}$ such that

$$(\wedge \gamma_i)_\gamma(X) = \sum_{i=1}^n \gamma_i \varrho_i(\xi_i) \qquad (11.81)$$

(3) $\exists \gamma = (\gamma_i) \in \mathbb{R}^n_{>0}$ and $\exists V \in L_d^q$ such that

$$V \in \gamma_i \partial \varrho_i(\xi_i), \quad \forall i \qquad (11.82)$$

(4) $\exists \gamma = (\gamma_i) \in \mathbb{R}^n_{>0}$ and $\exists V \in L_d^q$ such that

$$\xi_i \in \partial(\gamma_i \varrho_i)^*(V), \quad \forall i. \qquad (11.83)$$

Proof. The proof of Theorem 11.28 follows by a similar line of arguments as in Theorem 11.7 in the one-dimensional case. $\square$

The intersection condition (11.82) can also be described by saying that

$$V \in \partial(\wedge \varrho_i)_\gamma(X). \qquad (11.84)$$

This is a consequence of the following proposition.

Proposition 11.29. *If $(\xi_i) \in \mathcal{A}^n(X)$ minimizes the total risk w.r.t. $\varrho_1, \ldots, \varrho_n$, then*

$$\partial(\wedge \varrho_i)(X) = \bigcap_{i=1}^n \partial \varrho_i(\xi_i). \qquad (11.85)$$

Proof. Equation (11.85) is a consequence of the definition of subgradients of ϱ_i and $\wedge \varrho_i$ using the Fenchel inequality. The details are as in the one-dimensional case (see Proposition 11.4). $\square$

11.4.2   Law Invariant Risk Measures and Comonotonicity

For law invariant risk measures ϱ_i on L_d^p optimal allocations take a more specific form and are connected with multivariate comonotonicity. By the comonotone improvement theorem (Theorem 10.50) any allocation $(\xi_i) \in \mathcal{A}^n(X)$ in $d = 1$ can be improved by a comonotone allocation uniformly w.r.t. all convex law invariant risk measures ϱ_i. A uniform improvement result cannot be expected in the multivariate case. There is no notion of comonotonicity in $d \geq 1$ which is applicable to all law invariant risk measures. Several extensions of the notion of comonotonicity to the multivariate case $d \geq 1$ have been introduced in the literature (see e.g. the discussion in Puccetti and Scarsini (2010)). It turns out however that no uniform improvement result is possible in $d \geq 1$ and, therefore there is no notion of comonotonicity available which is as useful as in dimension 1. For various applications of comonotonicity in $d = 1$ see e.g. Dhaene et al. (2002).

The notion of "μ-comonotonicity" introduced in Definition 9.12 allows to construct a μ-comonotone improvement of an allocation. This allocation however is optimal only w.r.t. a specific risk measure (see the recent paper of Carlier et al. (2011)). In this section we establish that Pareto-optimal allocations w.r.t. law invariant risk measures are μ-comonotone for certain scenario measures μ.

For $1 \leq p \leq \infty$ we consider finite, law invariant convex risk measures ϱ on L_d^p. In fact we take the insurance version $\Psi(X) = \varrho(-X)$ which has a simpler monotonicity property, $\Psi : L_d^p \to \mathbb{R}^1$. Let $Y \in D_d^q$, where

$$D_d^q = \{(Y^1, \ldots, Y^d); \ Y^i \geq 0, Y^i \in L^q, E_P Y^i = 1, 1 \leq i \leq d\} \subset L_d^q \quad (11.86)$$

is the set of d-tuples of P densities and let $\mu = P^Y$ denote the distribution of Y. The maximal correlation risk measure in direction Y respectively μ is defined by

$$\widehat{\Psi}_Y(X) = \sup_{\widetilde{X} \sim X} E\widetilde{X} \cdot Y$$

$$= \sup_{\widetilde{Y} \sim \mu} EX \cdot \widetilde{Y} = \Psi_\mu(X) \quad (11.87)$$

(see Chapter 8). $\Psi_\mu = \widehat{\Psi}_Y$ is a law invariant coherent risk measure on L_d^p. For the representation in (11.87) μ is a measure in M_d^1 with marginals $(\mu_1, \ldots, \mu_d)$. With this description we assume based on the representation theorems (Theorems 8.26 and 8.32) that the law invariant, convex risk measure Ψ on L_d^p has a robust representation of the form

$$\Psi(X) = \max_{\mu \in Q}(\Psi_\mu(X) - \alpha(\mu)), \quad (11.88)$$

where Q is a weakly closed subset of $Q_{d,p} = \{Q \in M_d^1; \frac{dQ^i}{dP} \in L^q(P)\}$, $1 \leq p \leq \infty$, $\frac{1}{p} + \frac{1}{q} = 1$ and $\alpha(\cdot)$ is some law invariant penalty function. We choose in the following α as the minimal penalty function corresponding to the Fenchel conjugate ϱ^* of ϱ. Equivalently we can write (11.88) in the form

$$\Psi(X) = \max_{\mu \in A}(\Psi_\mu(X) - \alpha(\mu)), \tag{11.89}$$

where $A \subset \{\mu \in M^q(\mathbb{R}_+^d, \mathcal{B}_d^d); \exists Y \in D_d^q \text{ such that } Y \sim \mu\}$ is a weakly closed subset of distributions of density vectors with q-integrable components.

For $X \in L_d^p$ and $Y \in L_d^q$ define that X, Y are "optimally coupled", $X \sim_{oc} Y$ if

$$\widehat{\Psi}_Y(X) = \sup_{\widetilde{Y} \sim Y} EX \cdot \widetilde{Y} = EX \cdot Y, \tag{11.90}$$

where $\widetilde{Y} \sim Y$ means equality in distribution. For any $Y \in L_d^q$, $Y \sim \mu$ a symmetry relation is valid

$$\Psi_\mu(X) = \sup_{\widetilde{X} \sim X} E\widetilde{X} \cdot Y. \tag{11.91}$$

We next give a basic characterization of subgradients of convex law invariant risk measures Ψ on L_d^p with representation as in (11.88) with $\alpha = \Psi^*$. Define the risk functional

$$F(\mu) := F_X(\mu) := \Psi_\mu(X) - \Psi^*(\mu). \tag{11.92}$$

$\mu_0 \in A$ is called a "worst case scenario measure" of Ψ for X if

$$F_X(\mu_0) = \max_{\mu \in A} F_X(\mu) = \Psi(X). \tag{11.93}$$

Theorem 11.30. *For a finite, convex, law invariant risk measure Ψ on L_d^q with representation as in (11.90) and for $X \in L_d^p$, $Y_0 \in L_d^q$ with $\mu_0 = P^{Y_0} \in A$ the following statements are equivalent:*

(1) $Y_0 \in \partial \Psi(X)$

(2) (a) μ_0 is a worst case scenario measure of Ψ for X
    (b) $X \underset{oc}{\sim} Y_0$.

Proof. $(1) \Rightarrow (2)$ For $Y_0 \in \partial \Psi(X)$ we have for all $Z \in L_d^p$:

$$\Psi(X) - \Psi(Z) \le EY_0 \cdot (X - Z). \tag{11.94}$$

Thus we obtain from law invariance of Ψ

$$0 = \Psi(X) - \Psi(\widetilde{X}) \le EY_0 \cdot (X - \widetilde{X}), \quad \forall \widetilde{X} \sim X$$

or equivalently

$$EY_0 \cdot \widetilde{X} \le EY_0 \cdot X, \quad \forall \widetilde{X} \sim X.$$

This however is equivalent to

$$\Psi_{\mu_0}(X) = \sup_{\widetilde{X} \sim X} EY_0 \cdot \widetilde{X} = EY_0 \cdot X$$

and thus to $X \underset{\mathrm{oc}}{\sim} Y_0$, i.e. condition (b) holds.

For the proof of (a) note that $Y_0 \in \partial\Psi(X)$ implies also

$$\Psi(X) = EX \cdot Y_0 - \Psi^*(Y_0). \tag{11.95}$$

In consequence we obtain

$$
\begin{aligned}
F_X(\mu_0) &= \sup_{\widetilde{Y} \sim Y_0} E\widetilde{Y} \cdot X - \Psi^*(Y_0) \\
&= \max_{\mu \in A} \sup_{\widetilde{Y} \sim \mu} (E\widetilde{Y} \cdot X - \Psi^*(\mu)) = \max_{\mu \in A} F_X(\mu),
\end{aligned}
$$

i.e. μ_0 is a worst case scenario measure.

$(2) \Rightarrow (1)$ Let now $Y_0 \in L_d^q$ with $\mu_0 = P^{Y_0} \in A$ fulfil that $Y_0 \underset{\mathrm{oc}}{\sim} X$ and that μ_0 is a worst case scenario measure. Then we obtain

$$
\begin{aligned}
\Psi(X) &= \max_{\mu \in A} (\Psi_\mu(X) - \Psi^*(\mu)) \\
&= \max_{\mu \in A} F_X(\mu) = F_X(\mu_0) \\
&= \Psi_{\mu_0}(X) - \Psi^*(\mu_0) \\
&= EY_0 \cdot X - \Psi^*(Y_0)
\end{aligned}
$$

using that Ψ^* is also law invariant. Thus $Y_0 \in \partial\Psi(X)$ is a subgradient of Ψ in X. $\qquad\square$

Theorem 11.30 combined with the subgradient intersection condition in the characterization theorem of minimal risk allocations in Theorem 11.26 implies that optimal risk allocations have a specific μ-comonotonicity property where μ satisfies the intersection condition. There is also a close connection to the notion of worst case portfolios which concerns a worst case dependence structure for fixed marginal distributions. For a risk measure Ψ on L_d^p a portfolio of risk vectors $X = (X_1, \dots, X_n) \in (L_d^p)^n$ is called a "worst case portfolio" with respect to Ψ if

$$\Psi\left(\frac{1}{n} \sum_{i=1}^{n} X_i\right) = \sup_{\widetilde{X}_i \sim X_i} \Psi\left(\frac{1}{n} \sum_{i=1}^{n} \widetilde{X}_i\right) \tag{11.96}$$

(see Definition 9.20). Combining Theorems 11.26 and 11.30 with the characterization of worst case portfolios in Theorem 9.25 we obtain the following result connecting the notion of optimal allocations with μ-comonotonicity and with the notion of worst case portfolios.

Theorem 11.31 (Optimal allocations for law invariant risk measures). *Let* $\Psi_1, \ldots, \Psi_n$ *be a finite, lsc convex law invariant risk measures on* L_d^p *with representation as in* (11.89) *and scenario sets* A_i. *Let* $X \in L_d^p$ *be a risk vector such that the minimal risk allocation problem for* X *is well posed. Let* $F_i(\mu) = F_{i,\xi_i}(\mu) = \Psi_\mu(\xi_i) - \Psi_i^*(\mu)$ *denote the risk functional of* ξ_i *w.r.t.* Ψ_i.

For an allocation $(\xi_i) \in \mathcal{A}^n(X)$ *the following statements are equivalent:*

(1) (ξ_i) *has the minimal the total risk (w.r.t.* $\Psi_1, \ldots, \Psi_n$ *and* X*).*

(2) $$\exists V \in L_d^q : V \in \partial \Psi_i(\xi_i), \quad 1 \le i \le n. \tag{11.97}$$

(3) (a) $\exists$ *joint worst case scenario measure* $\mu_0 \in \mathcal{M}_d^q$ *for all* ξ_i *w.r.t.* Ψ_i, *i.e.*

$$F_i(\mu_0) = \sup_{\mu \in A_i} F_i(\mu) = \Psi_i(\xi_i), \quad 1 \le i \le n. \tag{11.98}$$

  (b) $\xi_1, \ldots, \xi_n$ *are* μ_0 *comonotone.*

(4) (a) $\exists$ *joint worst case scenario measure* μ_0 *for all* ξ_i *w.r.t.* Ψ_i.

  (b) $\xi_1, \ldots, \xi_n$ *is a worst case dependence structure for the* (11.99)
  max correlation risk measure Ψ_{μ_0}.

Remark 11.32. From the characterization of risk minimizing allocations in (11.98) we obtain for an optimal allocation $(\xi_i) \in \mathcal{A}(X)$

$$\sum_{i=1}^n \Psi_i(\xi_i) = \sum_{i=1}^n F_i(\mu_0) = \sum_{i=1}^n (\Psi_{\mu_0}(\xi_i) - \Psi_i^*(\mu_0))$$

$$= \Psi_{\mu_0}\left(\sum_{i=1}^n \xi_i\right) - \sum_{i=1}^n \Psi_i^*(\mu_0)$$

$$= \Psi_{\mu_0}(X) - \sum_{i=1}^n \Psi_i^*(\mu_0). \tag{11.100}$$

In case all Ψ_i are coherent risk measures (11.100) implies that

$$\bigwedge \Psi_i(X) = \sum_{i=1}^n \Psi_i(\xi_i) = \Psi_{\mu_0}(X). \tag{11.101}$$

$\Diamond$

11.4.3   Existence of Minimal Risk Allocations

In this section we derive a characterization of the existence of risk minimizing allocations as well as give several sufficient conditions. In the one-dimensional case existence results for optimal allocations have been based in Sections 11.1 and 11.2 on the comonotone improvement theorem which allows to restrict to allocations $\xi_i = f_i(X)$ with some monotone functions f_i. This reduction allows to apply Dini's theorem. Alternatively a strong intersection condition (SIS) from convex analysis has been used in Section 11.2.

We shall make use of the "subdifferential sum formula" for functions f, g:

$$(\text{SD}(x)) \qquad\qquad \partial(f + g)(x) = \partial f(x) + \partial g(x) \qquad\qquad (11.102)$$

which is used in convex analysis for dealing with existence results for the inf-convolution (see Barbu and Precupanu (1986)). There is a close link with the "epigraph condition" for the conjugates f^*, g^*:

$$(\text{EG}) \qquad\qquad \text{epi}(f + g)^* = \text{epi}(f^*) + \text{epi}(g^*). \qquad\qquad (11.103)$$

The following theorem of Burachik and Jeyakumar (2005, Theorem 3.1) extends previously given results and states that the EG-condition implies the subdifferential sum formula.

Theorem 11.33. *Let f, $g : X \to (-\infty, \infty]$ be proper, lsc convex functions on a Banach space X such that $\text{dom } f \cap \text{dom } g \neq \phi$. If f, g fulfil the epigraph condition (EG), then they satisfy the subdifferential sum formula $SD(x)$ for all $x \in \text{dom } f \cap \text{dom } g$.*

Subdifferentiability of $f \wedge g$ at a point x and the subdifferential sum formula $SD(x^*)$ for the conjugates f^*, g^* implies existence of a minimizer of $f \wedge g$ at x. For f, g as in Theorem 11.33 the following theorem is essentially a reformulation of Theorem 11.3.

Theorem 11.34 (Local existence). *Assume that $f \wedge g$ is subdifferentiable at x and assume that the subdifferential sum formula w.r.t. f^* and g^* holds for some $x^* \in \partial(f \wedge g)(x)$.*

$$\partial(f^* + g^*)(x^*) = \partial f^*(x^*) + \partial g^*(x^*). \qquad\qquad (11.104)$$

Then there exists an allocation $(\xi_1, \xi_2) \in \mathcal{A}^2(x)$ which minimizes the total risk,

$$f \wedge g(x) = f(\xi_1) + g(\xi_2). \qquad\qquad (11.105)$$

Next we establish that the conditions in Theorem 11.34 are in fact equivalent to the existence of a minimizer. The infimal convolution $f \wedge g$ is called "exact in x" if the inf is attained at x as in (11.105); it is called "exact" if this holds for all $x \in X$. Let f, g be functions as in Theorem 11.33.

Proposition 11.35. *The following statements are equivalent.*

(a) $f \wedge g$ is exact

(b) $f \wedge g$ is subdifferentiable at all $x \in X$ and $\forall x^* \in \partial(f \wedge g)(x)$ the subdifferential sum formula holds for f^*, g^*, i.e.

$$\partial(f^* + g^*)(x^*) = \partial f^*(x^*) + \partial g^*(x^*). \tag{11.106}$$

Proof. The direction (b) $\Rightarrow$ (a) follows from Theorem 11.33.

For the converse direction there exists to $x \in X$ an allocation $(\xi_1, \xi_2) \in \mathcal{A}^2(x)$ with minimal total risk. Then by the characterization result in Theorem 11.26 and using Proposition 11.29 there exists an element

$$x^* \in \partial f(\xi_1) \cap \partial g(\xi_2) = \partial(f \wedge g)(x). \tag{11.107}$$

Thus $f \wedge g$ is subdifferentiable in x.

To establish the subdifferential sum formula (11.106) let $x^* \in \partial(f \wedge g)(x)$. Then we obtain

$$x \in \partial(f \wedge g)^*(x^*) = \partial(f^* + g^*)(x^*) \tag{11.108}$$

(see Barbu and Precupanu (1986, Corollary 1.4, Section 2)). On the other hand

$$x^* \in \partial f(\xi_1) \cap \partial f(\xi_2) \tag{11.109}$$

for any solution (ξ_1, ξ_2) of $(f \wedge g)(x)$ by Theorem 11.26. This implies

$$\xi_1 \in \partial f^*(x^*) \text{ and } \xi_2 \in \partial g^*(x^*).$$

Thus we obtain

$$x = \xi_1 + \xi_2 \in \partial f^*(x^*) + \partial g^*(x^*). \tag{11.110}$$

Equation (11.110) implies the inclusion

$$\partial(f^* + g^*)(x^*) \subset \partial f^*(x^*) + \partial g^*(x^*).$$

Therefore, equality holds since the opposite inclusion is generally true. $\square$

By Theorem 11.33 and Proposition 11.35 the epigraph condition (EG) holding true for f^*, g^* together with subdifferentiability of $f \wedge g$ implies existence of optimal allocations. In the following we improve this statement and establish that the subdifferentiability condition can be skipped. Our proof is based essentially on the following proposition which is a restatement of Proposition 2.2 of Boţ and Wanka (2006) for the conjugates f^*, g^* of f, g.

Proposition 11.36. *Assume that* $\mathrm{dom}(f^*) \cap \mathrm{dom}(g^*) \neq \phi$. *Then the following statements are equivalent:*

(a) The epigraph condition (EG) holds for f^*, g^*, *i.e.*

$$\mathrm{epi}(f^* + g^*)^* = \mathrm{epi}(f) + \mathrm{epi}(g). \tag{11.111}$$

(b) $(f^* + g^*)^* = f \wedge g$ *and* $f \wedge g$ *is exact.*   (11.112)

Based on the equivalence in Proposition 11.36 it follows that the first condition in (2) of Proposition 5.4 of Boţ and Wanka (2006) can be omitted.

Proposition 11.37. *If the equivalent conditions of Proposition 11.35 hold true, then*

$$(f^* + g^*)^* = f \wedge g. \tag{11.113}$$

Proof. For $x \in X$ the subdifferentiability of $f \wedge g$ and the subdifferential sum formula (11.106) imply the existence of some $x^* \in \partial(f \wedge g)(x)$ such that

$$x \in \partial(f \wedge g)^*(x^*) = \partial(f^* + g^*)(x^*) = \partial f^*(x^*) + \partial g^*(x^*). \tag{11.114}$$

The exactness of $f \wedge g$ and the characterization of minimal allocations in Theorem 11.26 imply the existence of $(\xi_1, \xi_2) \in \mathcal{A}^2(x)$ such that

$$\xi_1 \in \partial f^*(x^*) \text{ and } \xi_2 \in \partial g^*(x^*). \tag{11.115}$$

In consequence we obtain from (11.114)

$$\langle x^*, x \rangle = (f^* + g^*)(x^*) + (f^* + g^*)^*(x).$$

From (11.115) we conclude

$$\langle x^*, \xi_1 \rangle = f^*(x^*) + f(\xi_1), \tag{11.116}$$

as well as

$$\langle x^*, \xi_2 \rangle = g^*(x^*) + g(\xi_2). \tag{11.117}$$

Summing up (11.116) and (11.117) and comparing this to (11.115) we conclude

$$(f^* + g^*)^*(x) = f(\xi_1) + g(\xi_2) = (f \wedge g)(x),$$

and thus (11.113) holds true.   □

As a consequence of Theorem 11.33 and of Propositions 11.35–11.37 we now obtain equivalence of exactness of $f \wedge g$ to the epigraph condition for f^*, g^*. This existence result for optimal allocations improves Theorem 11.34.

Theorem 11.38 (Existence of optimal allocations).   *Let f, g be proper lsc convex functions from a Banach space X to $(-\infty, \infty]$ such that $\operatorname{dom}(f^*) \cap \operatorname{dom}(g^*) \neq \phi$.*
Then the following statements are equivalent:

(a) $f \wedge g$ *is exact.*

(b) *The epigraph condition (EG) holds for f^*, g^*, i.e.*

$$\operatorname{epi}((f^* + g^*)^*) = \operatorname{epi} f + \operatorname{epi} g. \tag{11.118}$$

(c) $f \wedge g$ *is subdifferentiable at all $x \in X$ and for all $x^* \in \partial(f \wedge g)(x)$ the subdifferential sum formula holds*

$$\partial(f^* + g^*)(x^*) = \partial f^*(x^*) + \partial g^*(x^*).$$

By Corollary 3.1 of Burachik and Jeyakumar (2005) it is for sublinear functions even possible to omit the subdifferentiability property of $f \wedge g$ in (c):

Corollary 11.39. *Let f, g be functions as in Theorem 11.38 with the additional condition of positive homogeneity. Then the following conditions are equivalent:*

(a) *The epigraph condition (EG) holds for f^*, g^*, i.e.*

$$\operatorname{epi}((f^* + g^*)^*) = \operatorname{epi} f + \operatorname{epi} g.$$

(b) *For all $x^* \in \operatorname{dom} f^* \cap \operatorname{dom} g^*$ the subdifferential sum formula holds*

$$\partial(f^* + g^*)(x^*) = \partial f^*(x^*) + \partial g^*(x^*).$$

Thus Theorem 11.38 can be stated for positive homogeneous functions f, g following stronger version:

Proposition 11.40. *Let f, g be functions like in Theorem 11.38 with the additional condition of positive homogeneity. Then the following conditions are equivalent:*

(a) $f \wedge g$ *is exact.*

(b) *The epigraph condition (EG) holds for f^*, g^*, i.e.*

$$\operatorname{epi}((f^* + g^*)^*) = \operatorname{epi} f + \operatorname{epi} g.$$

(c) *For all $x^* \in \operatorname{dom} f^* \cap \operatorname{dom} g^*$ the subdifferential formula holds*

$$\partial(f^* + g^*)(x^*) = \partial f^*(x^*) + \partial g^*(x^*).$$

In general the epigraph condition (EG) for f^*, g^* in (11.118) is not easy to check. In the following proposition we restate some sufficient conditions for the epigraph condition (EG) in (11.118) which by Theorem 11.38 implies the existence of optimal allocations. All these sufficient conditions can be found in Boţ and Wanka

(2006). One of these conditions (the strong intersection condition (SIS)) was already used in Section 11.1.

We need some notation. For a subset $D \subset X$ denote by

$$\text{core}(D) := \{d \in D; \ \forall x \in X, \exists \varepsilon > 0, \forall \lambda \in (-\varepsilon, \varepsilon), d + \lambda x \in D\} \quad (11.119)$$

the "core" of D. Further let $\text{icr}(D)$ denote the "intrinsic core" of D relative to the "affine hull" $\text{aff}(D)$ of D. Further for D convex define the "strong quasi-relative interior" of D by

$$\text{sqri}(D) := \{x \in D; \ \text{cone}(D - x) \text{ is a closed subspace}\}. \quad (11.120)$$

Proposition 11.41 (Interior point conditions). *Any of the following interior point conditions implies the epigraph condition* (11.118):

$$\text{dom} \, f^* \cap \text{int dom} \, g^* \neq \phi \quad (11.121)$$

$$0 \in \text{core}(\text{dom} \, g^* - \text{dom} \, f^*) \quad (11.122)$$

$$0 \in \text{sqri}(\text{dom} \, g^* - \text{dom} \, f^*) \quad (11.123)$$

$$0 \in \text{icr}(\text{dom} \, g^* - \text{dom} \, f^*) \ and$$
$$\text{aff}(\text{dom} \, g^* - \text{dom} \, f^*) \ is \ a \ closed \ subspace. \quad (11.124)$$

The statements in Proposition 11.41 are given in Boţ and Wanka (2006) where also further relations between these conditions are discussed. Further sufficient conditions for (11.118) can be found in the references therein.

Next we extend the existence criteria to more than two functions. It is clear how the infimal convolution, the subdifferential sum formula and the epigraph condition are formulated for n functions. All preceding statements can be carried straight forward to this setup except the statement in Proposition 11.41. Each interior point condition has to be stated as a system of $n - 1$ conditions, to imply the epigraph condition

$$\text{epi}\left(\left(\sum_{i=1}^{n} g_i^*\right)^*\right) = \sum_{i=1}^{n} \text{epi} \, g_i. \quad (11.125)$$

Proposition 11.42. *For lower semicontinuous functions* $g_1, \ldots, g_n$ *any of the following interior point conditions implies the epigraph condition* (11.125).

(SIS) the "strong intersection condition"

$$\bigcap_{i=1}^{n-1} \text{int dom} \, g_i^* \cap \text{dom} \, g_n^* \neq \phi \tag{11.126}$$

$$0 \in \text{core} \left(\bigcap_{i=1}^{k-1} \text{dom} \, g_i^* - \text{dom} \, g_k^* \right), \quad k \in \{2, \ldots, n\} \tag{11.127}$$

$$0 \in \text{sqri} \left(\bigcap_{i=1}^{k-1} \text{dom} \, g_i^* - \text{dom} \, g_k^* \right), \quad k \in \{2, \ldots, n\} \tag{11.128}$$

$$0 \in \text{icr} \left(\bigcap_{i=1}^{k-1} \text{dom} \, g_i^* - \text{dom} \, g_k^* \right), \quad k \in \{2, \ldots, n\} \, and$$

$$\text{aff} \left(\bigcap_{i=1}^{k-1} \text{dom} \, g_i^* - \text{dom} \, g_k^* \right) \text{ is a closed subspace.} \tag{11.129}$$

Proof. First we observe that the strong intersection condition

(SIS) $$\bigcap_{i=1}^{n-1} \text{int dom} \, g_i^* \cap \text{dom} \, g_n^* \neq \phi$$

is equivalent to the following system of interior point conditions

$$\bigcap_{i=1}^{k-1} \text{int dom} \, g_i^* \cap \text{dom} \, g_k^* \neq \phi, \quad k = 2, \ldots, n. \tag{11.130}$$

The proof of Proposition 11.42 is based on the fact that the infimal convolution of the functions $g_1, \ldots, g_n$ can be solved iteratively, i.e.

$$\bigwedge_{i=1}^{n} g_i = \left(\bigwedge_{i=1}^{n-1} g_i \right) \wedge g_n. \tag{11.131}$$

Assume that any of the interior point conditions of Proposition 11.41 holds for the functions $f_{n-1}^* := \left(\bigwedge_{i=1}^{n-1} g_i \right)^* = \sum_{i=1}^{n-1} g_i^*$ and g_n^*. This means with the equivalence of (SIS) and (11.130) that in Proposition 11.42 one of the conditions holds for $k = n$, where dom $\left(\sum_{i=1}^{n-1} g_i^* \right) = \bigcap_{i=1}^{n-1} \text{dom} \, g_i^*$ respectively int dom $\left(\sum_{i=1}^{n-1} g_i^* \right) = \bigcap_{i=1}^{n-1} \text{int dom} \, g_i$. Then we get as a consequence of Proposition 11.40

$$\text{epi}((f_{n-1}^* + g_n^*)^*) = \text{epi}(g_n) + \text{epi} \left(\left(\sum_{i=1}^{n-1} g_i^* \right)^* \right). \tag{11.132}$$

Note that f_{n-1} is not necessarily lower semicontinuous. If we assume that any of the interior point conditions of Proposition 11.40 holds additionally for the functions $f_{n-2}^* := \left(\bigwedge_{i=1}^{n-2} g_i \right)^* = \sum_{i=1}^{n-2} g_i^*$ and g_{n-1}^* (which corresponds to $k = n-1$ in Proposition 11.42 we get again from Proposition 11.40

$$\text{epi}((f_{n-2}^* + g_{n-1}^* + g_n^*)^*) = \text{epi}(g_n) + \text{epi}(g_{n-1}) + \text{epi}\left(\left(\sum_{i=1}^{n-2} g_i^*\right)^*\right). \quad (11.133)$$

Proceeding further the same way and using the facts that $\text{dom}\left(\sum_{i=1}^{k-1} g_i^*\right) = \bigcap_{i=1}^{k-1} \text{dom } g_i^*$ respectively $\text{int dom}\left(\sum_{i=1}^{n-1} g_i^*\right) = \bigcap_{i=1}^{k-1} \text{int dom } g_i^*$ holds for any $k \in \{2, \ldots, n\}$ we see that any system of conditions of Proposition 11.42 implies (11.125).   □

Obviously it is sufficient for the application of Proposition 11.42 that for $k \in \{2, \ldots, n\}$ any of the interior point conditions holds.

11.4.4 Uniqueness of Optimal Allocations

Uniqueness of optimal allocations is a consequence of strict convexity.

Proposition 11.43. Let Ψ_i, $i \in \{1, \ldots, n-1\}$ be strictly convex risk functionals on L_d^p in the following sense

$$\Psi_i(\lambda X + (1 + \lambda)Y) < \lambda \Psi_i(X) + (1 - \lambda)\Psi_i(Y) \text{ for all } \lambda \in (0, 1)$$

for all $X, Y \in \text{dom } \Psi_i$. Then any optimal allocation of $X \in L_d^p$ with $\bigwedge \Psi_i(X) < \infty$ is unique.

Proof. We suppose that for $X \in L_d^p$ with $\bigwedge_{i=1}^n \Psi_i(X) < \infty$ there exist two minimizers $(X_1, \ldots, X_n) \in (L_d^p)^n$ and $(Y_1, \ldots, Y_n) \in (L_d^p)^n$ of the total risk. Then the allocation $Z_i := \lambda X_i + (1 - \lambda)Y_i$ with $\lambda \in (0, 1)$ defines an allocation of X with

$$\sum \Psi_i(Z_i) < \lambda \sum \Psi_i(X_i) + (1 - \lambda)\sum \Psi_i(Y_i) = \bigwedge_{i=1}^n \Psi_i(X).$$

This contradicts the optimality of $(X_1, \ldots, X_n)$.   □

Remark 11.44. It is obvious that in Proposition 11.43 it is necessary to postulate the strict convexity for at least $n-1$ risk functionals. If there were less than $n-1$ strict convex risk functionals then one could not exclude the existence of a rearrangement $(\overline{X}_1, \ldots, \overline{X}_n) \in (L_d^p)^n$ of an optimal allocation $(X_1, \ldots, X_n) \in (L_d^p)^n$ which is optimal, too.

If for example Ψ_{n-1} and Ψ_n are constant on L_d^p, then any rearrangement $(\overline{X}_1, \ldots, \overline{X}_n) \in (L_d^p)^n$ of $(X_1, \ldots, X_n) \in (L_d^p)^n$ defined by $\overline{X}_i = X_i, i \in \{1, \ldots, n-2\}$, $\overline{X}_{n-1} = X_{n-1} + Y, \overline{X}_n = Y_n - Y$ for any $Y \in L_d^p$ is optimal, too.   ◊

Some further uniqueness results can be transferred from Section 11.1. There in Theorem 11.12 it is shown that strict convexity of the i-th risk functional Ψ_i implies the uniqueness of the i-th risk contribution $X_i \in L_d^p$ of an optimal allocation $(X_1, \ldots, X_n) \in (L^p)^n$. Further a uniqueness result is proved for weighted versions of the allocation problem which implies uniqueness of Pareto-optimal allocations for cash invariant risk measures, which are strictly convex on L_0^∞, the class of risks $X \in L^\infty$ with $EX = 0$.

These results can be easily adapted to the multivariate case as considered in this section.

11.4.5 Examples of Optimal Allocations

Dilated Risk Functionals

Let ϱ be a convex risk functional on L_d^p. The class of dilated risk functionals ϱ_λ is defined for $\lambda > 0$ by

$$\varrho_\lambda(X) = \lambda \varrho\left(\tfrac{1}{\lambda}X\right), \tag{11.134}$$

where multiplication is componentwise. Then the following rules hold true:

$$\varrho_\lambda^* = \lambda \varrho^*, \quad \partial \varrho_\lambda(X) = \partial \varrho\left(\tfrac{1}{\lambda}X\right)$$

$$\text{and} \quad \partial \varrho_\lambda^*(V) = \lambda \partial \varrho^*(V). \tag{11.135}$$

As a consequence we obtain from the characterization of minimal risk allocations in Theorem 11.26 a simple optimal allocation rule.

Proposition 11.45 (Dilated risk functional). *Let ϱ be a convex risk functional on L_d^p, $1 \le p \le \infty$ with representation as in (11.88). Let $X \in L_d^p$ be a risk portfolio such that the total risk problem is well-posed for the dilated risk measures $\varrho_i = \varrho_{\lambda_i}$, $1 \le i \le n$, $\lambda_i > 0$, let $\lambda := \sum_{i=1}^n \lambda_i$ and assume that $\tfrac{1}{\lambda}X \in \text{int}(\text{dom}\,\varrho)$. Then the proportional allocation $\xi_i := \tfrac{\lambda_i}{\lambda}X$, $1 \le i \le n$, defines a total risk minimizing allocation of X.*

Proof. For the proof we check the intersection condition (11.69) of Theorem 11.26. This holds true by definition of the proportional allocation:

$$\bigcap_{i=1}^n \partial \varrho_i(\xi_i) = \bigcap_{i=1}^n \partial \varrho\left(\tfrac{1}{\lambda_i}\xi_i\right) = \bigcap_{i=1}^n \partial \varrho\left(\tfrac{1}{\lambda}X\right) \tag{11.136}$$

$$= \partial \varrho\left(\tfrac{1}{\lambda}X\right) \ne \phi. \qquad \square$$

In the particular case when $\varrho_i = \varrho$, $1 \le i \le n$, the allocation $\xi_i = \tfrac{1}{n}X$, $1 \le i \le n$, is risk minimizing.

Multivariate Expected Risk Function

Let $r : \mathbb{R}^d \to \mathbb{R}$ be strictly convex, continuously differentiable and satisfy the growth condition

$$|r(x)| \leq C(1 + \|x\|^p) \text{ for some } C \in \mathbb{R}, \quad p > 1. \tag{11.137}$$

Then r induces the corresponding risk functional

$$\varrho_r : L_d^p \to \mathbb{R}, \quad \varrho_r(X) := Er(X), \quad X \in L_d^p. \tag{11.138}$$

By the growth condition ϱ_r is a finite risk functional on L_d^p. To determine the subdifferential of ϱ_r on L_d^p we next prove that $\nabla r(X) \in L_d^q$.

Lemma 11.46. *Let $\frac{1}{p} + \frac{1}{q} = 1$ and let r satisfy the growth condition* (11.137). *Then for $X \in L_d^p$ it holds that*

$$\nabla r(X) \in L_d^q. \tag{11.139}$$

Proof. Define the linear operator $F : L_d^p \to \overline{\mathbb{R}}$ by

$$F(Y) := E\langle \nabla r(X), Y \rangle \tag{11.140}$$

where $\langle \cdot, \cdot \rangle$ is the Euclidean scalar product. We establish that F is well defined and norm bounded. Note that convexity of r implies

$$\|F\| = \sup_{\|Y\|_{L_d^p} \leq 1} E\langle \nabla r(X), Y \rangle$$

$$= \sup_{\|Y\|_{L_d^p} \leq 1} (E(r(X + Y) - r(X)))$$

$$\leq C_1 + \sup_{\|Y\|_{L_d^p} \leq 1} Er(X + Y)$$

$$\leq C_1 + \sup_{\|Y\|_{L_d^p} \leq 1} E\left(\frac{1}{2}r(2X) + \frac{1}{2}r(2Y)\right)$$

$$\leq C_2 + \frac{1}{2} \sup_{\|Y\|_{L_d^p} \leq 1} Er(2Y)$$

$$\leq C_2 + \frac{1}{2}CE(1 + 2\|Y\|^p) \leq C_3.$$

By the Riesz representation theorem there exists a unique $Z \in L_d^q$ such that

$$\int \langle \nabla r(X), Y \rangle dP = F(Y) = \int \langle Z, Y \rangle dP \quad \forall Y \in L_d^p. \qquad (11.141)$$

In consequence we obtain $\nabla r(X) = Z \in L_d^q$.   □

Since $\alpha \to \frac{1}{\alpha}(r(X + \alpha Y) - r(X))$ is monotone increasing for $X, Y \in L_d^p$, we obtain by the monotone convergence theorem for the directional derivative of ϱ_r in X in direction Y

$$
\begin{aligned}
D_{\varrho_r}(X, Y) &= \lim_{\alpha \downarrow 0} \frac{\varrho_r(X + \alpha Y) - \varrho_r(X)}{\alpha} \\
&= \lim_{\alpha \downarrow 0} E \frac{r(X + \alpha Y) - r(X)}{\alpha} \\
&= E \nabla r(X) \cdot Y.
\end{aligned}
\qquad (11.142)
$$

Thus ϱ_r is Gateaux differentiable with subgradient

$$\partial \varrho_r(X) = \{\nabla r(X)\} \in L_d^q. \qquad (11.143)$$

For a family $r_1, \ldots, r_n$ of convex functions as above with corresponding expected risk functionals $\varrho_{r_1}, \ldots, \varrho_{r_n}$ on L_d^p we consider the optimal risk allocation problem for $X \in L_d^p$. By Theorem 11.26 an optimal allocation $(\xi_i) \in \mathcal{A}^n(X)$ with minimal total risk is characterized by the optimality equation

$$\nabla r_i(\xi_i) = \nabla r_j(\xi_j), \quad 1 \le i \le n. \qquad (11.144)$$

This is a multivariate extension of the classical Borch theorem to $d \ge 1$. For strictly convex r_i, ∇r_i is one-to-one and as a consequence we obtain

$$\xi_i = (\nabla r_i)^{-1} \cdot \nabla r_1(\xi_1), \quad 2 \le i \le n. \qquad (11.145)$$

The critical allocation condition then becomes

$$\xi_1 + \sum_{i=2}^{n} (\nabla r_i)^{-1} \cdot \nabla r_1(\xi_1) = X. \qquad (11.146)$$

11.5   The Capital Allocation Problem

A complement to the risk allocation problem is the capital allocation problem. For a firm with N trading units there are expected future wealths $X_1, \ldots, X_N \in L^p(P)$. If risk is measured by a risk measure ϱ, then $k = \varrho(\sum_{i=1}^{N} X_i)$ is the necessary capital the firm needs to cover the total risk. The problem is to find a *fair* allocation of the risk capital $k = k_1 + \ldots + k_N$ to the N trading units. For a subadditive risk measure ϱ one can see this as the problem to distribute the gain of diversification $\sum_{i=1}^n \varrho(X_i) - \varrho(\sum_{i=1}^n X_i) \geq 0$ over the different business units of a financial institution.

An allocation $k_1, \ldots, k_N$ of the diversification gain is called "fair" if

$$\sum_{i=1}^N k_i = \varrho\left(\sum_{i=1}^N X_i\right) \tag{11.147}$$

and for all $J \subset \{1, \ldots, N\}$ it holds that

$$\sum_{j \in J} k_j \leq \varrho\left(\sum_{j \in J} X_j\right). \tag{11.148}$$

The existence of fair allocations (the Bondarava–Shapley theorem for risk measures) is proved in Delbaen (2000, Theorem 22) for coherent risk measures. If $\varrho : L_1^p \to \mathbb{R} \cup \{\infty\}$ is an lsc coherent risk measure on L_1^p, then by Theorem 8.25 ϱ has a representation $\varrho(X) = \max_{Q \in \mathcal{Q}} E_Q(-X)$ for some closed scenario set $\mathcal{Q} \subset \mathcal{Q}_{1,p}$. Under this assumption one gets in case $p = \infty$ a simple proof of this existence result and furthermore some information on the fair allocation.

Theorem 11.47. *Let ϱ be an lsc coherent risk measure on L_1^p and let $X_1, \ldots, X_N$ be N wealth variables with risk capital $k = \varrho(\sum_{i=1}^N X_i)$. Then there exists some least favourable scenario measure $Q^* \in \mathcal{Q}$ such that $E_{Q^*}(\sum_{i=1}^n X_i) = \varrho(\sum_{i=1}^n X_i)$ and $k_1^*, \ldots, k_N^*$ with $k_i^* := E_{Q^*}(-X_i)$ is a fair allocation of the risk capital k.*

Proof. By the representation of ϱ we have

$$k = \varrho\left(\sum_{i=1}^N X_i\right) = \sup_{Q \in \mathcal{Q}} E_Q\left(-\sum_{i=1}^N X_i\right) = E_{Q^*}\left(-\sum_{i=1}^N X_i\right). \tag{11.149}$$

The supremum in (11.149) is attained in Q^* and with $k_i^* = E_{Q^*}(-X_i)$ it holds that

$$k = E_{Q^*}\left(-\sum_{i=1}^N X_i\right) = \sum_{i=1}^N k_i^*. \tag{11.150}$$

Further for any $J \subset \{1, \ldots, N\}$ it holds that

$$\varrho\left(\sum_{j \in J} X_j\right) \geq E_{Q^*}\left(-\sum_{j \in J} X_j\right) = \sum_{j \in J} k_j^*.$$

Thus $k_1^*, \ldots, k_N^*$ is a fair allocation of the risk capital. $\qquad\square$

Chapter 12
Optimal Contingent Claims and (Re)insurance Contracts

We consider in this chapter various problems on the construction of optimal contingent claims (respectively options) and of optimal contracts, in particular (re)insurance contracts. Some results of this type connected with optimal risk allocation have been given in Chapters 10 and 11.

We discuss the problem of optimal contingent claims in a financial market going back to the classical Markowitz theory in the mean variance optimization context. An extension to more general utility optimization was given in He and Pearson (1991a). It was used to derive extensions of the classical results of Merton (1971) on the construction of optimal portfolios. Another classical result in insurance is the optimality of stop-loss contracts as reinsurance contracts in the mean variance frame. We reconsider the optimal reinsurance problem and derive optimality of stop-loss contracts in reinsurance based on stochastic ordering results in the general frame of law invariant convex risk measures. We also describe optimal worst case reinsurance contracts which are protections against possible dependence in the risk components of a portfolio. There are some close connections of the above-mentioned problems to the optimal investment problem which is a classical problem in portfolio theory (see Dana (2005)) as well as to minimal liability problems as described in Schied (2004).

12.1   Optimal Contingent Claims

The design and construction of optimal contingent claims equivalently of optimal investments is a basic problem in portfolio theory. There are several related variants of this optimization problem and applications concern optimal investment, minimal liability risk, optimal insurance contracts, and others. All these problems are of the form to choose a product (contingent claim) with minimal price given upper bounds on the risk or the related problem to choose contingent claims with minimal risk

L. Rüschendorf, *Mathematical Risk Analysis*, Springer Series in Operations Research and Financial Engineering, DOI 10.1007/978-3-642-33590-7_12,
© Springer-Verlag Berlin Heidelberg 2013

given upper bounds on the price (with a given budget). Instead of minimizing the risk we can consider the equivalent problem of maximizing the utility.

The classical version of these optimal investment problems is formulated in the Markowitz theory going back to the work of Markowitz (1952). The results of this theory are the basis of investment theory in mathematical economics. They describe the solutions of the optimal investment problem respectively the minimal liability problem in the case where the risk is given by the variance of the risk variable respectively by its multivariate extensions in the case of risk vectors.

12.1.1   Optimal Investment Problems

In the optimal investment problem the aim is to find to a given investment (contingent claim) X an improvement concerning price and risk. We assume that $(\Omega, \mathfrak{A}, P)$ is a non-atomic probability space and $X \in L^p(P) = L^p$, $1 \le p \le \infty$. We also assume that the pricing measure Q is given by a density $\varphi \in L^q$ w.r.t. P, i.e. $Q = \varphi P$, where q is conjugate to p.

The "optimal investment problem" is the following problem: Determine an investment $C^* \in L^p$ such that

$$E_Q C^* = \int \varphi \cdot C^* dP = \inf\{E_Q C; \ C \in L^p, C \le_{cx} X\}. \tag{12.1}$$

The price of C^* should be minimal under all contingent claims C which are less risky than X w.r.t. convex order. This implies that C^* is less risky w.r.t. all convex, law invariant risk measures ϱ (respectively ψ).

This problem was considered in Dybvig (1988) in the finite case and extended to the general case L^1, L^∞ in Jouini and Kallal (2001), Dana and Meilijson (2003), Föllmer and Schied (2011), Dana (2005), Burgert and Rü (2006b), and Rü (2012b). The solution

$$e(X, \varphi) := E_Q C^* \tag{12.2}$$

is called the "reservation price" of X. Let F_φ, F denote the distribution functions of φ respectively X and let $\tau_\varphi = \tau_\varphi(\varphi, V)$ be the distributional transform of φ (see Section 1.1). The distributional transform τ_φ turns out to be a suitable tool to describe optimal contingent claims.

Theorem 12.1 (Optimal contingent claim). *The solution of the investment problem given by the price density $\varphi \in L^q$ and the investment $X \in L^p$ is given by*

$$e(X, \varphi) = \int_0^1 F_\varphi^{-1}(1-t) F^{-1}(t) dt. \tag{12.3}$$

An optimal contingent claim C^* exists and is given by

$$C^* = F^{-1}(1 - \tau_\varphi(\varphi, V)). \tag{12.4}$$

Proof. For any claim $C \in L^p$ it holds by the lower Hoeffding–Fréchet bounds

$$E_Q C = E\varphi C \geq E F_\varphi^{-1}(1-U) F_C^{-1}(U), \qquad (12.5)$$

where $U \sim U(0,1)$ is uniformly distributed on $(0,1)$.

The function $g(u) = F_\varphi^{-1}(1-u)$ is decreasing in u. The convex order $\leq_{cx}$ is characterized by $C_1 \leq_{cx} C_2$ if and only if

$$\int_0^a F_{C_1}^{-1}(t)\,dt = \int_0^1 1_{(0,a]}(t) F_{C_1}^{-1}(t)\,dt$$

$$\geq \int_0^a F_{C_2}^{-1}(t)\,dt = \int_0^1 1_{[0,a]}(t) F_{C_2}^{-1}(t)\,dt.$$

This implies by approximation with linear combinations

$$\int g(t) F_{C_1}^{-1}(t)\,dt \geq \int g(t) F_{C_2}^{-1}(t)\,dt \qquad (12.6)$$

for all decreasing integrable functions g on $[0,1]$. In consequence we get

$$e(C_1, \varphi) = \int_0^1 F_\varphi^{-1}(1-t) F_{C_1}^{-1}(t)\,dt$$

$$\geq \int_0^1 F_\varphi^{-1}(1-t) F_{C_2}^{-1}(t)\,dt = e(C_2, \varphi).$$

This implies that the right-hand side in (12.5) is antitone w.r.t. the convex order $\leq_{cx}$ and, therefore, we get

$$E_Q C \geq E F_\varphi^{-1}(1-U) F^{-1}(U)$$

$$= \int_0^1 F_\varphi^{-1}(1-t) F^{-1}(t)\,dt = e(X, \varphi).$$

With $C^* = F^{-1}(1 - \tau_\varphi(\varphi, V))$ it holds that $C^* \sim X$, C^* has the distribution function F, the pair (φ, C^*) is antithetic, and

$$E_Q C^* = E\varphi C^* = \int_0^1 F_\varphi^{-1}(1-t) F^{-1}(t)\,dt = e(X, \varphi).$$

C^* is a random variable such that (φ, C^*) attains the lower Hoeffding–Fréchet bound. This implies the representation of the optimal claim in (12.4) by the distributional transform. $\qquad \square$

Remark 12.2. (a) (C^*, φ) is a pair of antithetic variables. The distribution of the optimal pair is unique and is given by the anticomonotone distribution. Defining

$$\widetilde{C} := E(C^* \mid \varphi) = \int_0^1 F^{-1}(1 - \tau_\varphi(\varphi, v))dv, \tag{12.7}$$

then $\widetilde{C} = g(\varphi)$ where $g \downarrow$ is a decreasing function of the price density φ alone. Further, $\widetilde{C} \leq_{cx} C$ and $E_Q \widetilde{C} = E_Q C^*$. Thus there exists an optimal investment $C^* = g^*(\varphi)$, $g^* \downarrow$ which is a decreasing function of the price density φ.

(b) **Transformed measure.** Defining the transformed measure

$$Q^* := \varphi^* P \text{ with } \varphi^* := F^{-1}(1 - \tau_F(X, V)), \tag{12.8}$$

then φ^* is decreasing in X and

$$e(X, \varphi) = E_Q C^* = E_{Q^*} X. \tag{12.9}$$

Thus the reservation price is identical to the expectation of X w.r.t. the transformed price measure Q^*. Q^* describes a worst case price density for the claim X.

(c) **Path dependent options.** Let $S = (S_t)_{0 \leq t \leq T}$ be a price process and assume that the price density φ is a function of S_T, $\varphi = \varphi(S_T)$, then

$$C^* = g(S_T). \tag{12.10}$$

Thus any path dependent option $C = f(S)$ can be improved by a European option

$$C^* = g(S_T).$$

If φ is increasing (decreasing), then g can be chosen decreasing (increasing). For this observation see Bernard et al. (2011).

(d) **Cost efficient options.** Given an option X with distribution function F we consider the class $\mathcal{C} = \mathcal{C}(F)$ of all options which have the same payoff distribution as X,

$$\mathcal{C} = \{C; F_C = F\} = \mathcal{C}(X). \tag{12.11}$$

$\diamondsuit$

As a corollary Theorem 12.1 implies.

Theorem 12.3 (Cost efficient claims). *For a given claim $X \in L^p$ and price density $\varphi \in L^q$ the claim*

$$C^* := F^{-1}(1 - \tau_\varphi(\varphi, V)) \in \mathcal{C}(X) \tag{12.12}$$

is a cost efficient claim, i.e.

$$E_Q C^* = \inf_{C \in \mathcal{C}(X)} E_Q C.$$

Proof. For the proof note that any $C \in \mathcal{C}(X)$ satisfies that $C \leq_{cx} X$. Thus Theorem 12.3 follows from Theorem 12.1. $\square$

Remark 12.4. Several examples of strict improvements for path dependent options as for lookback options or for Asian options are calculated in Bernard et al. (2012) in (single and multiasset) Black–Scholes type models. The cost optimal portfolios with the same distribution of payoffs are called (following Dybvig 1988) "cost efficient strategies". $\Diamond$

12.1.2   Minimal Demand Problem

Closely related to the optimal investment problem in Section 12.1.1 is the minimal demand problem. Here the problem is to minimize the risk of a demand (claim) C under the condition that the claim is in the budget set

$$B = B_0 \cap \{C \in L^p; \ E_Q C \leq c\} = B_c$$

where B_0 describes some additional restrictions like boundedness or distributional restrictions. We consider again the case that the price measure $Q = \varphi P$ has a price density φ and the risk of a claim is measured by a convex risk measure Ψ (the insurance version). The problem can equivalently be formulated as a maximization problem of a concave utility functional U on L^p. Considering $-X$ as a liability this problem has been discussed in Schied (2004). In Dana (2005) the equivalent utility maximization problem is considered.

The budget set is in several applications besides the cost restrictions also further restricted and assumed to be convex and closed in L^p. In connection with law invariant risk measures it is also of interest to assume that B is "law invariant", i.e., if $X \in B$ and if X, Y have the same law, $X \sim Y$, then $Y \in B$.

The "minimal demand problem" then is formulated as follows: Describe demands $C^* \in B$ such that

$$\Psi(C^*) = \inf\{\Psi(C); \ C \in B\}. \tag{12.13}$$

Proposition 12.5 (Existence of claims with minimal risk). *For an lsc convex risk measure Ψ and a convex bounded closed budget set $B \subset L^p$ there exists a solution C^* of the minimal demand problem.*

Proof. In case $p < \infty$ the budget set $B \subset L^p$ is convex and closed and thus is weakly closed. Since by assumption B is bounded it is also weakly compact and thus the lsc risk functional Ψ attains a minimal point $C^* \in B$.

In case $p = \infty$ a modification of the Komlos theorem is used.

Komlos Theorem: *For a sequence (X_n) of real random variables on $(\Omega, \mathfrak{A}, P)$ with $\sup |X_n| < K$ P-a.s. there exists a sequence $\overline{X}_n \in \text{conv}\{X_n, X_{n+1}, \dots\}$ of convex combinations and a random variable $\overline{X}$ such that*

$$\overline{X}_n \to \overline{X} \quad P\text{-a.s.} \tag{12.14}$$

Let $(X_n) \subset B$ be a sequence such that $\Psi(X_n) \to \inf\{\Psi(X); X \in B\}$. Then let $\overline{X}_n \in \text{conv}\{X_n, X_{n+1}, \dots\}$ and $\overline{X}$ be random variables such that $\overline{X}_n \to \overline{X}$ P-a.s. Since $|X_n| \leq K$, $\forall n \in \mathbb{N}$ it follows that $E_Q \overline{X}_n \to E_Q \overline{X}$ and thus $\overline{X} \in B$. Lower semicontinuity of convex risk measures on L^∞ is equivalent to Fatou-continuity. As a consequence we obtain

$$\Psi(\overline{X}) \leq \liminf \Psi(\overline{X}_n).$$

Since $\Psi(\overline{X}_n) \leq \sum_{i=0}^\infty \alpha_i \Psi(X_{n+i})$ for the α_i in the convex representation of $\overline{X}_n$ it follows that $\Psi(\overline{X}) \leq \inf\{\Psi(X); X \in B\}$ and thus $\overline{X}$ is a claim with minimal risk in B. □

Remark 12.6 (Improvement procedure). For a demand $C \in B$ such that $\overline{C} := E(C \mid \varphi) \in B_0$ it holds that

$$\overline{C} \leq_{\text{cx}} C \quad \text{and} \quad \overline{C} \in B. \tag{12.15}$$

Thus for any law invariant risk measure Ψ it holds that:

$$\Psi(\overline{C}) \leq \Psi(C), \tag{12.16}$$

i.e. $\overline{C}$ is an improvement over C concerning risk and $\overline{C}$ has the same cost as C

$$E_Q \overline{C} = E\varphi \overline{C} = E\varphi C = E_Q C.$$
◊

For the minimal demand problem similarly to the optimal contingent claim problem the search for optimal solutions C^* can be restricted to solutions of the form $C^* = h(\varphi)$ with a decreasing function h of the price density.

Theorem 12.7 (Monotone solutions of the minimal demand problem). *Consider a minimal demand problem in L^p with law invariant risk measure Ψ and with law invariant, convex, bounded set B_0 determining the budget set B. Then there exists an optimal demand $X^* \in B$ such that (X^*, φ) are antithetic.*

If $C^* := E(X^* \mid \varphi) \in B_o$ then C^* is a minimal demand and C^* is of the form $C^* = h(\varphi)$ with h decreasing.

Proof. By Proposition 12.5 there exists a solution $\widetilde{X} \in B$ of the optimal demand problem. Let $V \sim U(0, 1)$ be independent of φ, let $U := \tau_\varphi(\varphi, V) \sim U(0, 1)$ denote the distributional transform of φ and define

$$X^* := F_{\widetilde{X}}^{-1}(1 - U). \tag{12.17}$$

Then $\widetilde{X} \sim X^*$ and thus $\Psi(\widetilde{X}) = \Psi(X^*)$.

Further using that $F_\varphi^{-1}(U) = \varphi$ P-a.s. we obtain from the Hoeffding–Fréchet inequality

$$E_Q X^* = E\varphi X^* = E F_\varphi^{-1}(U) F_{\widetilde{X}}^{-1}(1 - U)$$
$$\leq E\varphi\widetilde{X} = E_Q\widetilde{X} \leq c.$$

Since $X^* \sim \widetilde{X}$ it follows by assumption that $X^* \in B_0$ and thus $X^* \in B$. Furthermore $X^* = h(\varphi, V)$, where h is decreasing in φ, V and thus X^*, φ are antithetic.

Defining $C^* := E(X^* \,|\, \varphi)$ we obtain

$$C^* = \int h(\varphi, v) dP^V(v) =: g(\varphi),$$

where g is a decreasing function of the price density φ. By assumption $C^* \in B$ and thus C^* is a minimal demand. $\qquad\square$

For some general class of risk measures Ψ the minimal demand (optimal claim in the sense of (12.13)) can be calculated in explicit form. We consider for historical reason the equivalent formulation w.r.t. a concave monetary utility functional U as a utility maximization problem:

$$U(C^*) = \sup\{U(C); \; C \in B\}, \tag{12.18}$$

where B is as usual the budget set. The transition from (12.18) to (12.13) is given by choosing $U = -\varrho$ for a convex risk measure ϱ.

More precisely we choose

$$U(X) = Eu(X), \tag{12.19}$$

as expected utility functional, where $u : \mathbb{R} \to \mathbb{R} \cup \{-\infty\}$ is a "utility function" which is assumed to be strictly increasing, strictly concave, continuously differentiable in $\operatorname{dom} u := \{x \in \mathbb{R}; \; u(x) > -\infty\}$ and to satisfy

$$u'(\infty) = \lim_{x \to \infty} u'(x) = 0, \quad u'(\overline{x}) = \lim_{x \downarrow \overline{x}} u'(x) = \infty \tag{12.20}$$

for $\overline{x} := \inf\{x \in \mathbb{R}; \; u(x) > -\infty\}$. This implies that either $\operatorname{dom}(u) = (\overline{x}, \infty)$ or $\operatorname{dom}(u) = [\overline{x}, \infty)$. Let

$$I(x) := (u')^{-1}(x)$$

denote the inverse of u'. Then by assumption (12.18) $I(\bar{x}) = \infty$ and I is a continuous, strictly decreasing function on $[\bar{x}, \infty)$ such that $I(x) \xrightarrow[x\to\infty]{} 0$.

The convex conjugate u^* of u is defined by

$$u^* : \mathbb{R}_+ \to \mathbb{R}, \quad u^*(y) = \sup_{x\in\mathbb{R}}(u(x) - xy).$$

Then the sup is attained in $I(y)$ and it holds that

$$u^*(y) = u(I(y)) - yI(y). \tag{12.21}$$

Let Q denote the pricing measure with density φ w.r.t. P and define for $x \geq \bar{x}$

$$U_Q(x) = \sup\{Eu(Y); \ Y \in L^1(Q), E_Q Y \leq x, Eu(Y)^- < \infty\}. \tag{12.22}$$

$U_Q(x)$ is the max-expected utility given the capital x and the budget constraint ensuring the existence of the expectation $Eu(Y)$.

The following result is a classical theorem in optimal portfolio theory (see for example He and Pearson (1991b), Goll and Rü (2001, Lemma 4.1), Föllmer and Schied (2011, Theorem 3.34)).

Theorem 12.8 (Solution of the minimal demand problem). *Assume that* $E_Q I(\lambda\varphi) < \infty$, $\forall \lambda > 0$, *then the following holds true:*

(a)

$$U_Q(x) = \inf_{\lambda>0} \left\{ Eu^*(\lambda\varphi) + \lambda x \right\} \text{ for all } x > \bar{x}. \tag{12.23}$$

(b) There exists a unique solution $\lambda = \lambda_Q(x)$ *of the equation* $E_Q I(\lambda\varphi) = x$. *Furthermore,*

$$U_Q(x) = Eu(I(\lambda_Q(x)\varphi)). \tag{12.24}$$

Thus $C^* = I(\lambda_Q(x)\varphi)$ *is an optimal solution of the minimal demand problem with capital* x.

Proof. Let $Y \in L^1(Q)$ be a demand (claim) with $E_Q Y \leq x$ and $Eu(Y)^- < \infty$. Then we have for $\lambda > 0$:

$$Eu(Y) \leq Eu(Y) + \lambda(x - E_Q Y)$$
$$\leq Eu^*(\lambda\varphi) + \lambda x$$
$$= Eu(I(\lambda\varphi)) + \lambda(x - E_Q I(\lambda\varphi)).$$

The inequalities hold as equalities if and only if $Y = I(\lambda_Q(x)\varphi)$.

As a consequence we obtain the representation (12.23) in (a) and the equality in (12.24). Since we have $E_Q(I(\lambda\varphi)) < \infty$ for all $\lambda > 0$, one can conclude that $E_Q(I(\lambda\varphi))$ is a continuous, monotonically decreasing function of λ with values in $(\bar{x}, \infty)$. This guarantees the existence of $\lambda_Q(x)$.

Finally we have to check that $E[u(I(\lambda_Q(x)\varphi))]^- < \infty$. From the inequality

$$u(x) - xy \leq u(I(y)) - yI(y)$$

we get that

$$E[u(I(\lambda\varphi)) - \lambda\varphi I(\lambda\varphi)]^- < \infty.$$

The inequality

$$[u(I(\lambda\varphi))]^- \leq [u(I(\lambda\varphi)) - \lambda\varphi I(\lambda\varphi)]^- + [\lambda\varphi I(\lambda\varphi)]^-$$

implies that the condition $E[u(I(\lambda_Q(x)\varphi))]^- < \infty$ is fulfilled. □

In particular the optimal claim $C^* = I(\lambda_Q(x)\varphi)$ is a decreasing function of the density in concordance with the result in Theorem 12.7. Also under the assumption of Theorem 12.7 we obtain from the representation in (12.23) that $Eu^*(\lambda\varphi) = Eu^*(\lambda\frac{dQ}{dP}) < \infty$. The representation in (12.23) connects the minimal demand problem with the notion of "f-divergence distances" which is a prominent class of distances on the class of probability measures.

Definition 12.9 (f-divergence distance). Let $Q \ll P$ and let $f : (0, \infty) \to \mathbb{R}$ be a convex function. Then the f-divergence distance between Q and P is defined as

$$f(Q\|P) := \begin{cases} \int f(\frac{dQ}{dP})dP, & \text{if the integral exists,} \\ \infty, & \text{else,} \end{cases}$$

where $f(0) = \lim_{x \downarrow 0} f(x)$.

Remark 12.10. (a) Examples of f-divergence distances are the Kullback–Leibler or entropy distance for $f(x) = x \log x$, the total variation distance for $f(x) = |x - 1|$, the Hellinger distance for $f(x) = -\sqrt{x}$, the reverse relative entropy for $f(x) = -\log(x)$ and many others (see Liese and Vajda (1987)).

(b) **Minimal demand as divergence distance:** The convex function $u_\lambda^*(y) = u^*(y)$ appearing in the representation in (12.23) is a convex function and thus generates the u_λ^*-divergence distance

$$u_\lambda^*(Q\|P) = \begin{cases} \int u_\lambda^*(\frac{dQ}{dP})dP, & \text{if the integral exists,} \\ \infty, & \text{else.} \end{cases}$$

Theorem 12.8 implies the identity

$$U_Q(x) = u_\lambda^*(Q\|P) + \lambda x \text{ with } \lambda = \lambda_Q(x). \tag{12.25}$$

The minimal demand utility is identical to the u_λ^*-divergence distance between P and Q plus λx. This allows to use known results for the f-divergence distances in order to determine the minimal demands. This connection to

f-divergence distances for existence and characterization results for optimal claims was introduced in Goll and Rü (2001).

(c) **Classical utility functions** For the classical utility functions the logarithmic utility $u(x) = \log x$, the power utility $\frac{x^p}{p}$, $p \in (-\infty, 1) \backslash \{0\}$ and the exponential utility $u(x) = 1 - e^{-px}$, $p \in (0, \infty)$ the corresponding conjugate functions u^* are given by $-\log x - 1$, $-\frac{p-1}{p} x^{\frac{p}{p-1}}$ and $1 - x + x \log x$. Hence for the exponential utility $u(x) = 1 - e^{-x}$ the u^*-divergence distance is the relative entropy distance. For the logarithmic utility $u(x) = \log x$ we obtain the reverse relative entropy. For $u(x) = -x^{-1}$ we obtain the Hellinger distance.

For these classical utility functions the optimal claim (minimal demand) $I(\lambda_Q(x)\varphi)$ can easily be determined in explicit form.

If e.g. $u(x) = \log x$, then $u'(x) = \frac{1}{x}$ and $C^* = \frac{1}{\lambda \varphi}$. From the equation $E_Q I(\lambda \varphi) = x$ we obtain $\lambda = \lambda_Q(x) = \frac{1}{x}$ and thus

$$C^* = I(\lambda_Q(x)\varphi) = \frac{1}{\varphi} x$$

is the optimal claim. As a result we obtain directly the optimal claims in the classical portfolio optimization results of Merton (1971).

In the case of a geometric Brownian motion $S = (S_t)_{0 \le t \le T}$ as underlying market process and using the arbitrage free martingale measure Q as pricing measure one obtains a pricing density $\varphi_t \sim S_t^{-b}$ for some $b > 0$. Thus the optimal claim at time T is proportional to $x S_T^b$ for logarithmic utility. A series of detailed examples is given in Duffie (1994, 2001), Miyahara (1999), Rheinländer (1999), Kallsen (2000, 2002), Goll and Rü (2001), Björk (2004) and references given there. $\lozenge$

12.2 Optimal (Re)insurance Contracts

As described in Chapter 11 optimal (re)insurance contracts can be seen as particular instances of the optimal risk-allocation (respectively risk sharing) problem. In this section we discuss some variations of the optimality of the classical stop-loss contracts which are obtained from stochastic ordering results combined with properties of convex risk measures as expanded in previous sections.

12.2.1 Optimality of Stop-Loss Contracts

In this section we describe some extensions of classical results on the optimality of stop-loss contracts.

A "stop-loss contract" for risk $X \geq 0$ is a contract of the form $(X - d)_+$ with premium $\pi_X(d) := E(X - d)_+$. The insurer retains the risk $X - (X - d)_+ = \min(X, d)$, which is bounded above by the retention limit d. The reinsurer takes the risk $(X - d)_+$ above the retention limit d.

Generally a "reinsurance contract" $I(X)$ is given by a function $I : \mathbb{R}_+ \to \mathbb{R}_+$ such that $0 \leq I(x) \leq x$ and $I(0) = 0$. Let $\mathcal{I}$ denote the set of all reinsurance contracts. $I \in \mathcal{I}$ is called an "increasing insurance contract" if

$$x - I(x) \text{ is increasing in } x. \tag{12.26}$$

If I is an increasing insurance contract, then I is 1-Lipschitz. Using the variance as risk measure and a premium to be paid by the insurer of the form

$$\pi = (1 + \delta)EI(X) = \pi(X), \tag{12.27}$$

then the optimal reinsurance contract is given by a stop-loss contract. For this classical result from insurance see Kaas et al. (2001, Theorem 1.4.1).

Theorem 12.11 (Variance optimality of stop-loss insurance). *Let $I(X)$ be a reinsurance contract and let $d \geq 0$ be a solution to*

$$E(X - d)_+ = EI(X),$$

i.e. the stop-loss contract $(X - d)_+$ and $I(X)$ have the same premium, then

$$\text{Var}(X - I(X)) \geq \text{Var}(X - (X - d)_+). \tag{12.28}$$

Proof. Let $W := X - (X - d)_+$ and $R := X - I(X)$ be the retained risks, then $EW = ER$. Furthermore, it holds that

$$|W - d| \leq |R - d|. \tag{12.29}$$

Equation (12.29) is trivial in case $X \geq d$ since then $W = d$ holds. For $X < d$ we have $W = X$ and hence

$$R - d = X - d - I(X) \leq X - d = W - d < 0$$

and thus (12.29) holds. Equation (12.29) implies

$$E(W - d)^2 \leq E(R - d)^2$$

which is equivalent to (12.28). □

As a result the stop-loss contract minimizes the variance of the retained risk $X - I(X)$ given the premium $\pi = P$.

Theorem 12.11 also implies a solution to the dual problem (see the remarks on the Markowitz theory in Section 12.1). This problem is to minimize the premium, i.e. to minimize $EI(X)$ under all contracts $I(X)$ such that the risk $\mathrm{Var}(X - I(X)) = v$ is fixed.

Note that $\mu(d) := E(X - (X - d)_+)$ and $\sigma^2(d) = \mathrm{Var}(X - (X - d)_+)$ are continuous increasing functions in d with $\mu(0) = \sigma^2(0) = 0$ and $\mu(\infty) = EX$, $\sigma^2(\infty) = \mathrm{Var}(X)$.

Theorem 12.12 (Optimal premium contracts). *Let $d \geq 0$ be a solution to $\sigma^2(d) = v$. Then for all contracts $I(X)$ with $\mathrm{Var}(X - I(X)) \leq v$ it holds that*

$$E(X - d)_+ \leq EI(X).$$

Proof. By Theorem 12.11 for any contract $I(X)$ with $\mu_I = E(X - I(X))$ it holds that the variance $\sigma_I^2 = \mathrm{Var}(X - I(X)) \geq \sigma^2(d)$ for d such that $\mu(d) = \mu_I$. Thus the increasing curve $\{(\mu(d), \sigma^2(d)); \ d \geq 0\}$ is the lower boundary of the risk set

$$\mathcal{R} = \{(\mu_I, \sigma_I^2); \ I \in \mathcal{I} \text{ a reinsurance contract}\}.$$

This implies that if $\sigma_I^2 \leq v = \sigma^2(d)$, then necessarily $\mu_I \leq \mu(d)$ – the point (μ_I, σ_I^2) has to be situated on the left-hand side of $(\mu(d), \sigma^2(d))$. In consequence we have that $E(X - d)_+ \leq EI(X)$. $\square$

Remark 12.13. The argument in the proof of Theorem 12.12 implies that the increasing curve $\{(\mu(d), \sigma^2(d)); \ d \geq 0\}$ is the lower boundary of the risk set $\mathcal{R}$. Equivalently the points $\{(\mu(d), \sigma^2(d)); \ d \geq 0\}$ form the set of Pareto optimal points of the risk set corresponding to the "efficient boundary" in the Markowitz theory. $\diamond$

The optimality of stop-loss insurance contracts holds true in a much broader sense (see Kaas et al. (2001, Example 10.4.4)).

Theorem 12.14 ($\leq_{cx}$-optimality of stop-loss contracts). *Let $I(X)$ be a reinsurance contract and let $d \geq 0$ be a solution to $E(X - d)_+ = EI(X)$. Then for the retained risks $R_d = X - (X - d)_+$, $R_I = X - I(X) = R_I(X)$ it holds that*

$$R_d \leq_{cx} R_I. \tag{12.30}$$

Proof. Since $0 \leq I(X) \leq X$ it follows that $R_I \leq X$ and thus $F_{R_I}(x) \geq F_X(x)$ for all $x > 0$. Further, using that $R_d = \min\{X, d\}$ it follows that

$$F_{R_d}(x) = F_X(x) \text{ for all } x < d \text{ and } F_{R_d}(x) = 1 \text{ for all } x \geq d.$$

In consequence the distribution functions F_{R_I}, F_{R_d} cross exactly once.

Thus the Karlin–Novikov crossing criterion (see Theorem 3.3) implies that $R_d \leq_{cx} R_I$. $\square$

As a corollary we get

Corollary 12.15 (Optimal reinsurance problem). *Let $X \in L^p$, $0 \le X$ and let Ψ be a law invariant, convex risk measure on L^p. Then the "optimal reinsurance problem"*

$$\Psi(R_I(X)) + (1 + \vartheta)EI(X)) = \inf_{I \in \mathcal{I}}, \qquad (12.31)$$

has a stop-loss solution $I_{d^*}(X) = (X - d^*)_+$ with $d^* \ge 0$ chosen as

$$d^* = \arg\min_d \Psi(\min(X, d)) + (1 + \vartheta)\pi_X(d). \qquad (12.32)$$

Remark 12.16. (a) If one fixes the premium $\pi = (1 - \delta)EI(X)$ then choosing d^* as a solution of $E(X - d)_+ = EI(X)$, we get $d^* = \pi_X^{-1}(\frac{\pi}{1+\vartheta})$. We obtain from Theorem 12.14 that the stop-loss contract $(X - d^*)_+ =: I_{d^*}$ minimizes the risk $\Psi(R_I(X))$,

$$\Psi(R_{I_{d^*}}(X))$$

$$= \inf\{\Psi(R_I(X)); \ I \text{ is an insurance contract with premium } \pi\}.$$

(b) For several risk measures Ψ the arg min in (12.32) is easy to determine.
  Let for example $\Psi(X) = E(X - q_\alpha(x) \mid X \ge q_\alpha(X))$ be the conditional tail expectation (assuming that F_X is continuous) where $q_\alpha(X) = F_X^{-1}(1 - \alpha)$ is the upper α-quantile. Then (12.32) is equivalent to minimizing

$$h(d) := d - q_\alpha(X) + (1 + \vartheta)(1 - \alpha)\pi_X(d) \text{ on } d \ge q_\alpha(X).$$

Since $\frac{\partial}{\partial t}\pi_X(t) = -\overline{F}_X(t)$ we get

$$h'(d) \ge 0 \Leftrightarrow \overline{F}_X(d) \ge \frac{1}{(1 + \vartheta)(1 - \alpha)} =: z \Leftrightarrow d \ge q_z(X)$$

$$h'(d) \le 0 \Leftrightarrow \overline{F}_X(d) \le \frac{1}{(1 + \vartheta)(1 - \alpha)} = z \Leftrightarrow d \le q_z(X). \qquad (12.33)$$

Thus in case $\alpha \le z$ we get $d^* = q_z(X)$ while in case $\alpha \ge z$ we get $d^* = q_\alpha(X)$.
(c) The results of Theorem 12.14 and Corollary 12.15 are also applicable to the case of joint portfolios $X = \sum_{i=1}^n X_i$. Given the premium π, the optimal contract is

$$I_{d^*} = \left(\sum X_i - d^*\right)_+ \qquad (12.34)$$

where $d^* = \pi_X^{-1}(\frac{\pi}{1+\vartheta})$. The excess function $\pi_X(t) = E(\sum_{i=1}^n X_i - t)_+$ can in general however be only approximated by numerical procedures. $\Diamond$

If we define as in the variance case $\mu(d)$, $\sigma_\Psi^2(d)$ as the mean and Ψ-risk of the retained risk, i.e.

$$\mu(d) = E(X - (X - d)_+) = E\min(X, d) \text{ and } \sigma_\Psi^2(d) = \Psi(X - (X - d)_+),$$

then we obtain as in the variance case that $\mu(d)$ is increasing and concave in d and $\sigma_\Psi^2(d)$ is increasing and concave in d. As a consequence the increasing curve

$$T_\Psi := \{(\mu(d), \sigma_\Psi^2(d)); \ d \geq 0\}$$

is the lower boundary of the risk set

$$\mathcal{R}_\Psi = \{(\mu_I, \sigma_\Psi^2(I)); \ I \in \mathcal{I}\}.$$

and thus T_Ψ is identical to the lower boundary of $\mathcal{R}_\Psi$ and forms a complete class of Pareto optimal contracts. As a consequence we get

Theorem 12.17 (Contracts with optimal premium). *Let $d \geq 0$ be a solution to $\sigma_\Psi^2(d) = v$. Then for all reinsurance contracts $I \in \mathcal{I}$ with $\Psi(X - I(X)) \leq v$ it holds that*

$$E(X - d)_+ \leq EI(X).$$

Proof. The proof is identical to that of Theorem 12.14.   □

12.2.2 Optimal Worst Case (Re)insurance Contracts

The results in Section 12.2.1 concern the case of an insurance contract on a single risk X. If $X = \sum_{i=1}^n X_i$ is the total risk of a portfolio, where $X_i \sim F_i$, then the optimal (re)insurance contract will depend on the dependence between the components X_i of the portfolio.

Under the assumption that the joint distribution of the portfolio vector $(X_1, \ldots, X_n)$ is not known but only the marginal distributions $X_i \sim F_i$ are known the following worst case formulation of the reinsurance problem makes sense. This robust version of the optimal reinsurance problem has been introduced in Cheung et al. (2010).

Let $F_1, \ldots, F_n$ be distribution functions on $\mathbb{R}_+$ with finite p-th moments and let Ψ be a law invariant convex risk measure on L^p. Let $\mathcal{I}_n$ denote the set of all n-tuples $(I_1, \ldots, I_n)$ with n increasing reinsurance contracts I_j. For $(X_1, \ldots, X_n)$ the Fréchet class $\mathcal{F}(F_1, \ldots, F_n)$ and $I = (I_1, \ldots, I_n) \in \mathcal{I}_n$ denote by $\pi(I) = \pi_X(I) = (1 + \vartheta) \sum_{k=1}^n EI_k(X_k)$ the premium of I.

Definition 12.18 (Optimal worst case reinsurance problem). The "worst case reinsurance problem" is defined as

$$R(\pi_0) := \inf_{\substack{I \in \mathcal{I}_n \\ \pi(I) = \pi_0}} \sup_{\substack{(X_i, \ldots, X_n) \in \\ \mathcal{F}(F_1, \ldots, F_n)}} \Psi\left(\sum_{k=1}^n (X_k - I_k(X_k))\right). \tag{12.35}$$

A solution I of (12.35) is called the "optimal worst case reinsurance contract".

The premium $\pi(I)$ depends only on the marginal distribution functions and on the reinsurance contract I. In fact Cheung et al. (2010) consider the problem to determine

$$\inf\{R(\pi_o);\ \pi_0 \le \pi^*\}$$

for some upper bound π^* on the premiums.

The first observation to solve this worst case reinsurance problem is that the inner sup-problem in (12.35) is solved by the comonotonic vector $X^c = (X_1^c, \ldots, X_n^c) \in \mathcal{F}(F_1, \ldots, F_n)$ independently of the contract I.

Proposition 12.19 (Comonotonic vector as worst case reinsurance structure).
Let Ψ be a convex law invariant risk measure on L^p and let F_i have finite p-th moments. Then for any $I = (I_1, \ldots, I_n) \in \mathcal{I}_n$ the comonotonic vector $X^c = (X_1^c, \ldots, X_n^c)$ is the worst case dependence structure for the reinsurance based on I. More precisely

$$\Psi\left(\sum_{k=1}^n (X_k^c - I_k(X_k^c))\right) = \sup_{\substack{(X_1,\ldots,X_n)\in \\ \mathcal{F}(F_1,\ldots,F_n)}} \Psi\left(\sum_{k=1}^n (X_k - I_k(X_k))\right). \tag{12.36}$$

Proof. For $I \in \mathcal{I}$ it holds that $X_k - I_k(X_k) = (\mathrm{id} - I_k)(X_k)$ is an increasing function in X_k. This implies that a comonotonic vector of $(X_k - I_k(X_k))$ is given by

$$((\mathrm{id} - I_k)(X_k^c)) = (X_k^c - I_k(X_k^c)).$$

Thus (12.36) is implied by the basic comonotonicity theorem (Theorem 2.17) on the worst case property of comonotonic vectors for portfolios.   □

With this structure the solution of the worst case insurance problem (12.35) can be reduced by means of Corollary 12.15 to the optimality of stop-loss reinsurance contracts. Let for given $d^* \ge 0$ and distribution functions $F_1, \ldots, F_n$ on $\mathbb{R}_+$ $u_0 \in [0, 1]$ be chosen such that

$$\sum_{i=1}^n F_i^{-1}(u_0-) \le d^* \le \sum_{i=1}^n F_i^{-1}(u_0)$$

and let $d_i^* \in [F_i^{-1}(u_0-), F_i^{-1}(u_0)]$ be such that

$$\sum_{i=1}^n d_i^* = d^*$$

(cf. the construction in Theorem 3.5 in the basic theorem on comonotonicity and convex order of portfolios). Let $X_k^c = F_k^{-1}(U)$, $1 \le i \le n$ denote a comonotonic vector. Then we get

Theorem 12.20 (Optimal worst case reinsurance contract). *Let $F_1, \ldots, F_n$ be distribution functions on $\mathbb{R}_+$ with finite p-th moments and let Ψ be a law invariant convex risk measure on L^p. For a given premium π_0 let d^* be a solution of $\pi_0 = E\left(\sum_{k=1}^{n} X_k^c - d^*\right)_+$. Then the optimal worst case reinsurance contracts at premium π_0 are given by the stop-loss contracts $I_k^*(x) = I_{d_k^*}(x) = (x - d_k^*)_+$, $1 \le k \le n$.*

Proof. By Proposition 12.19 we can restrict to the problem

$$R(\pi) = \inf_{\substack{I \in \mathcal{I}_n \\ \pi(I) = \pi_0}} \Psi\left(\sum_{k=1}^{n} (X_k^c - I_k(X_k^c))\right).$$

For $I \in \mathcal{I}_n$, $I = (I_1, \ldots, I_n)$ and any x of the form $x = \sum_{k=1}^{n} F_k^{-1}(u) = x(u)$ we define an insurance contract

$$\overline{I}(x) := \sum_{k=1}^{n} I_k(F_k^{-1}(u))$$

and define $\overline{I}$ by increasing continuation on $\mathbb{R}_+$.

Then we get with $X^c := \sum_{k=1}^{n} X_k^c$

$$\Psi\left(\sum_{k=1}^{n} (X_k^c - I_k(X_k^c))\right) = \Psi\left(\sum_{k=1}^{n} F_k^{-1}(U) - \sum_{k=1}^{n} I_k(F_k^{-1}(U))\right)$$

$$= \Psi\left(\sum_{k=1}^{n} F_k^{-1}(U) - \overline{I}\left(\sum_{k=1}^{n} F_k^{-1}(U)\right)\right)$$

$$\ge \inf\left\{\Psi\left(\sum_{k=1}^{n} F_k^{-1}(U) - J\left(\sum_{k=1}^{n} F_k^{-1}(U)\right)\right);\right.$$

$$\left. J \in \mathcal{I}, \pi(J) = \pi_0\right\}$$

$$= \Psi(X^c - (X^c - d^*)_+). \tag{12.37}$$

The last equality follows from Corollary 12.15. With (d_i^*) as chosen above, $\sum_{i=1}^{n} d_i^* = d^*$ we obtain from Theorem 3.5 that

$$\left(\sum_{i=1}^{n} X_i^c - d^*\right)_+ = \sum_{i=1}^{n} (X_i^c - d_i^*)_+ \quad P\text{-a.s.} \tag{12.38}$$

and therefore

$$X^c - (X^c - d^*) = \sum_{k=1}^{n} (X_k^c - (X_k^c - d_k^*)_+).$$

This implies optimality of the increasing stop-loss contracts $I_k^*(x) = (x - d_k^*)_+$. $\square$

Remark 12.21. (a) Theorem 12.20 has been established in the case of continuous strictly increasing F_i in Cheung et al. (2010). Optimal d_i, d^* are determined there explicitly in the case of the AVaR, the VaR, and the conditional tail expectation CTE risk measures. The proof in Cheung et al. (2010) is based on the Kusuoka theorem. It uses the minimax theorem to obtain a reduction to the case of the AVaR risk measures. The simplified proof above is given in Rü (2012b). It uses the classical optimality result of the stop-loss contract (Corollary 12.15) and the classical result on the worst case behaviour of the comonotonicity structure.

(b) The proof of Theorem 12.20 implies as a corollary the equality

$$R(\pi_0) = \inf_{\substack{I \in \mathcal{I} \\ \pi(I) = \pi_0}} \Psi(X^c - I(X^c)), \tag{12.39}$$

where the inf is taken over all (not necessarily increasing) reinsurance contracts of $X^c = \sum_{i=1}^{n} X_i^c$.

(c) In the case of continuous strictly increasing distribution functions F_i the determination of the retention limits d^*, d_i^* and u_0 simplifies essentially. We obtain $u_0 = F_i(d_i)$, i.e. $d_i^* = F_i^{-1}(u_0)$ and $d^* = F_{\sum_{i=1}^{n} X_i^c}^{-1}(u_0)$. As a consequence

$$d_i^* = F_i^{-1} \circ F_{\sum_{i=1}^{n} X_i^c}(d^*), \quad 1 \le i \le n. \tag{12.40}$$

We see directly that the retention limits d_i^* are increasing functions in d^* and thus in the premium π_0.

(d) The problem to determine $\inf_{\pi \le \pi_0}(R(\pi) + \pi)$ i.e. the case where an upper bound is given on the premium π leads to a (typically simple) optimization problem over the compact class of admissible d_i, d^* as in (12.38). Lower semicontinuity of the cost functional

$$\pi + R(\pi) = \Psi\left(\sum_{i=1}^{n}(X_i^c - (X_i^c - d_i)_+)\right) + (1 + \vartheta)\sum_{i=1}^{n} E(X_i^c - d_i)_+$$

in d_i^*, d^* implies existence of a solution. $\pi + R(\pi)$ is the sup of two terms:

$$\pi + R(\pi) = \Psi(X^c - (X^c - d^*)_+) + (1 + \vartheta)\pi_{X^c}(d^*). \tag{12.41}$$

The first term $\Psi(X^c - (X^c - d^*)_+)$ is concave and increasing in d^*. The second term is convex and increasing in d^*. It will typically be easier to determine d_i^*

from (12.41) i.e. to minimize $\pi + R(\pi)$ over (d_i) such that $\sum_{i=1}^{n} d_i = d^*$ than to use formula (12.40).

(e) The nonnegativity assumption on $X \geq 0$ is made throughout this section only by the motivation coming from insurance of positive risks. The optimization results only use convexity, monotonicity and comonotonicity, and therefore also hold true for general risk variables $X \in \mathbb{R}$. Also the assumption that $X \in L^p$ and Ψ is a convex risk measure on L^p can be replaced by $X \in D \subset L^0$ and Ψ is a law invariant convex risk measure on D. This is possible since we do not make use of the representation theorem of Kusuoka in our approach.   ◊

Part IV
Optimal Portfolios and Extreme Risks

Portfolio diversification is a basic approach to the reduction of non-systemic risks. The mean-variance theory of Markowitz (1952) was introduced in order to explain the effects of portfolio diversification concerning reduction of risk given the reward or in order to maximize the reward given the risk (see also Markowitz (1991)). The reward of a portfolio $\xi \cdot X$, where ξ is a portfolio vector and X is the vector of risks in the portfolio is measured by the mean $E\xi \cdot X$ while the risk is measured in this theory by the variance $\mathrm{Var}(\xi \cdot X) = \xi^{\mathsf{T}} \Sigma \, \xi$, where $\Sigma = \mathrm{Cov}(X)$ is the covariance matrix of X.

The portfolio optimization problem is then the problem to maximize the reward given bounds on the risk over all admissible portfolio vectors ξ

$$E\xi^* \cdot X = \sup_{\xi} \left\{ E\xi \cdot X; \ \ \mathrm{Var}(\xi \cdot X) \le v_0 \right\}$$

or the related problem of minimizing the risk given the reward, i.e.

$$\mathrm{Var}(\xi^* \cdot X) = \inf_{\xi} \left\{ \mathrm{Var}(\xi \cdot X); \ \ E\xi \cdot X \ge r_0 \right\}$$

or to optimize a related risk reward functional like

$$\frac{E\xi \cdot X}{(\mathrm{Var}(\xi \cdot X))^{\frac{1}{2}}} = \sup_{\xi} .$$

As in the chapter on optimal risk allocation it is also for the portfolio diversification problem well motivated to consider more relevant risk measures Ψ replacing the variance and on the other hand also more stable versions of measuring the reward $\xi \cdot X$ of a portfolio.

In this chapter we consider an approach to this diversification problem for portfolios with heavy-tailed components. Heavy-tailed portfolios even with infinite mean are common in several branches of applications in insurance or financial risks (see e.g. Moscadelli (2004) or Nešlehová et al. (2006b) for empirical evidence). So mean-variance measures or convex risk measures will not be applicable. In the framework of extreme value theory, in particular the theory of multivariate regular variation, we consider the portfolio diversification problem. Portfolio losses are compared by their sensitivity w.r.t. extremal risk events. The aim is to determine portfolios in an optimal way such that they avoid extremal risk events as much as possible.

To this aim a functional $\gamma_\xi = \gamma_\xi(\alpha, \Psi)$ is introduced which depends on the vector of portfolio weights ξ and on the distributional parameters α, Ψ where α is the tail index and Ψ is the spectral measure arising from the multivariate regular variation assumption. It is argued that the "extremal risk index" γ_ξ describes the sensitivity of the portfolio $\xi \cdot X$ concerning extremal risk events. The optimal diversification problem thereby is reduced to the optimization of the extremal risk index γ_ξ w.r.t. ξ.

An interesting effect is obtained by observing that for models with $\alpha < 1$, i.e. models with infinite mean, diversification does not improve the portfolio but it makes the portfolio worse. For $\alpha > 1$ we obtain the expected positive diversification effects while the case $\alpha = 1$ is indifferent. We also introduce empirical versions (estimators) of the optimal portfolio and the extremal risk index and establish consistency and asymptotic normality.

The second part of this chapter is concerned with a comparison of different stochastic models w.r.t. the asymptotic portfolio losses. The corresponding notion of "asymptotic portfolio loss order" is introduced and several sufficient conditions are given in order to verify this order in various classes of examples. Also connections to several further stochastic orders are elaborated.

For $\alpha < 1$ stronger positive dependence typically decreases extremal risk while for $\alpha \geq 1$ stronger positive dependence increases risk. This phenomenon for $\alpha \geq 1$ is concordant with the behaviour of convex risk measures and the related convex order $\leq_{cx}$ (see Section 3.1) for integrable risks. The examples include elliptical distributions and multivariate regularly varying models with Gumbel, Archimedean, and Galambos copulas.

Chapter 13
Optimal Portfolio Diversification w.r.t. Extreme Risks

After an introduction to some basic notions of multivariate regular variation and extreme value theory we introduce the extremal risk index γ_ξ which measures the extreme risk of the portfolio $\xi^T X$ and describe its dependence on the vector ξ of portfolio weights. In particular we describe the positive diversification effects in models with $\alpha > 1$ and negative diversification effects in models with $\alpha < 1$. Empirical estimators of the optimal portfolio weights ξ and the extremal risk index γ_ξ are introduced.

Based on empirical process theory consistency and asymptotic normality are derived. Some examples and simulations are given in the final section. This chapter is based on Mainik and Rü (2011). Some more detailed and extended results concerning also the case of negative portfolio weights is given in Mainik (2010, 2012).

13.1 Heavy-Tailed Portfolios and Multivariate Regular Variation

In this section we give a brief introduction to some basic notions and results from multivariate regular variation and the related extreme value theory. Let $X^{(1)}, \ldots, X^{(d)} \in \mathbb{R}_+$ be the losses of some risky assets and let $\xi \in \mathbb{R}_+^d$ represent the weights of the assets in the portfolio, so that the portfolio loss is given by $\xi^T X$ with $X := (X^{(1)}, \ldots, X^{(d)})^T$. It is obvious that multiplying the portfolio vector ξ by a constant factor $c > 0$ results in multiplication of the portfolio loss by c. Hence the influence of the portfolio composition on the portfolio loss can be studied by considering standardized portfolios. Following the intuition of dividing the whole capital in parts and investing them in different assets, we standardize portfolio vectors by the sums of their components. As a result, the set of portfolio vectors ξ that we need to consider is the unit simplex in $\mathbb{R}_+^d$:

$$\xi \in \Sigma^d := \left\{ x \in \mathbb{R}_+^d; \ \|x\|_1 = 1 \right\}.$$

L. Rüschendorf, *Mathematical Risk Analysis*, Springer Series in Operations Research and Financial Engineering, DOI 10.1007/978-3-642-33590-7_13,
© Springer-Verlag Berlin Heidelberg 2013

The assets $X^{(i)}$ are assumed to be (univariate) "regularly varying" with "tail index" $\alpha > 0$, i.e. $\forall x > 0$ it holds that

$$P\left(X^{(i)} > tx \mid X^{(i)} > t\right) \to x^{-\alpha}, \quad t \to \infty. \tag{13.1}$$

The tail index α characterizes the existence of absolute moments $E|X^{(i)}|^{\beta}$: for $\beta < \alpha$ they exist, whereas for $\beta > \alpha$ they explode.

It is well known that heavier tails dominate the influence of the lighter ones on the extremes, making asymptotic analysis of extreme losses trivial if the tail indices are different. Therefore only the case of equal tail indices is considered. The basic assumption made in the following is that the risk vector X is "multivariate regularly varying", i.e. there exists a sequence $a_n \to \infty$ and a Borel measure ν on $\mathcal{B}([0,\infty]^d \setminus \{0\})$ such that $\nu(\{\infty\}) = 0$ and

$$n P^{a_n^{-1} X} \xrightarrow{v} \nu \tag{13.2}$$

with $\xrightarrow{v}$ denoting the "vague convergence" of measures. Furthermore we assume that ν is non-degenerate in the following sense:

$$\nu\left(\{x \in \mathbb{R}_+^d; \ x^{(i)} > \varepsilon\}\right) > 0$$

for all $\varepsilon > 0$ and $i = 1,\ldots,d$. This assumption ensures that all components $X^{(i)}$ are relevant for the extremes of X.

The measure ν exhibits the scaling property

$$\nu(tA) = t^{-\alpha}\nu(A) \tag{13.3}$$

for all sets $A \in \mathcal{B}\left([0,\infty]^d \setminus \{0\}\right)$ that are bounded away from 0. Furthermore, for any random vector X satisfying (13.2) the limit measure ν is unique up to a constant factor. The measure ν also characterizes the asymptotic distribution of componentwise maxima

$$M_n := \left(M^{(1)},\ldots,M^{(d)}\right), \quad M^{(i)} := \max\left\{X_1^{(i)},\ldots,X_n^{(i)}\right\}$$

by the limit relation

$$P\left(\{a_n^{-1} M_n \in [0,x]\}\right) \xrightarrow{w} \exp\left(-\nu\left(\mathbb{R}_+^d \setminus [0,x]\right)\right), \quad x \in \mathbb{R}_+^d \setminus \{0\}.$$

Therefore ν is called the "exponent measure". For more details and other standardizations of the measure ν see Resnick (1987).

Another consequence of the scaling property (13.3) is the product representation of ν in polar coordinates $\tau(x) := (r,s) := (\|x\|, \|x\|^{-1}x)$:

$$\nu^{\tau}(dr \times ds) = C \cdot \varrho_\alpha(dr) \otimes \Psi(ds) \tag{13.4}$$

with some constant $C > 0$, $\varrho_\alpha(x, \infty) = x^{-\alpha}$ and a probability measure Ψ on the positive part of the sphere $\{s \in \mathbb{R}_+^d; \|s\| = 1\}$. The measure Ψ is called the "spectral measure" of ν or X.

As shown in Basrak et al. (2002), multivariate regular variation with tail index α of the loss vector X implies univariate regular variation of any portfolio loss $\xi^\top X$ with the same tail index α. This property is also inherited by the norm $\|X\|$.

Although the domain of the spectral measure Ψ depends on the norm $\|\cdot\|$ used for constructing the polar coordinates, the representation (13.4) is norm-independent in the following sense:

If (13.4) holds for some norm $\|\cdot\|$, then it also holds for any other norm $\|\cdot\|_*$ that is equivalent to $\|\cdot\|$. The tail index α is the same and the spectral measure Ψ_* on the positive part $\{s \in \mathbb{R}_+^d; \|s\|_* = 1\}$ of the unit sphere corresponding to $\|\cdot\|_*$ is obtained from Ψ by the following transformation:

$$\Psi_* = T(\Psi), \quad T(s) := \|s\|_*^{-1} s.$$

In the following we consider polar coordinates based on the sum norm $\|\cdot\|_1$ and set the constant C in (13.4) to 1, which does not lead to any loss of generality. Multivariate regular variation of X can also be written as in the equivalent form

$$\mathrm{P}^{t^{-1} X \mid \|X\|_1 > t} \xrightarrow{\mathrm{w}} \nu, \quad t \to \infty, \tag{13.5}$$

on $\{x \in \mathbb{R}_+^d; \|x\|_1 > 1\}$, where $\nu^\tau = \varrho_\alpha \otimes \Psi$.

Further details on regular variation of functions or random variables and related applications in extreme value theory can be found in Bingham et al. (1987), Resnick (1987), Basrak et al. (2002), Hult and Lindskog (2006), de Haan and Ferreira (2006), and Resnick (2007).

To compare the tail probabilities of different portfolio vectors we consider the normalized ratio

$$\frac{P\left(\{\xi^\top X > t\}\right)}{P\left(\{\|X\|_1 > t\}\right)} = P\left(\xi^\top X > t \mid \|X\|_1 > t\right);$$

the identity follows since $\xi^\top X \leq \|X\|$ and therefore

$$\{\|X\|_1 > t\} \supset \{\xi^T X > t\}.$$

With

$$A_{\xi,t} := \{x \in \mathbb{R}_+^d; \ \xi^\top x > t\} \tag{13.6}$$

and

$$A_t := \{x \in \mathbb{R}_+^d; \ \|x\|_1 > t\} \tag{13.7}$$

we thus consider $P(X \in A_{\xi,t})/P(X \in A_t)$ for $t \to \infty$ and obtain from (13.5)

$$P\left(X \in A_{\xi,t} \mid X \in A_t\right) = P\left(t^{-1} X \in A_{\xi,1} \mid \|X\|_1 > t\right)$$
$$\xrightarrow{\mathrm{w}} \nu(A_{\xi,1}). \qquad (13.8)$$

This means that under the assumption of multivariate regular variation the asymptotic behaviour of portfolio losses can be compared in terms of the functional

$$\gamma_\xi := \nu(A_{\xi,1}),$$

which characterizes the asymptotic sensitivity of the portfolio ξ to extremal events. For any pair of portfolio vectors $\xi_1, \xi_2 \in \Sigma^d$ relation (13.8) implies

$$\frac{P\left(\{\xi_1^\top X > t\}\right)}{P\left(\{\xi_2^\top X > t\}\right)} \to \frac{\gamma_{\xi_1}}{\gamma_{\xi_2}}, \quad t \to \infty.$$

Consequently, higher values of γ_ξ correspond to higher sensitivity of the portfolio to extremal events, i.e. higher conditional probability that the portfolio loss exceeds a high bound t when the sum of asset losses exceeds this bound.

Moreover, multivariate regular variation of X yields

$$P\left(\xi^\top X > rt \mid \|X\|_1 > t\right) \to \gamma_\xi \cdot r^{-\alpha}, \quad t \to \infty \qquad (13.9)$$

for all $r > 1$ and the asymptotic quantile relation

$$\frac{F_{\xi^\top X}^{-1}(1 - ut)}{F_{\|X\|_1}^{-1}(1 - t)} \to \gamma_\xi^{1/\alpha} \cdot u^{-1/\alpha}, \quad t \downarrow 0 \qquad (13.10)$$

for all $u \in (0, 1)$. Thus γ_ξ allows to order both the probabilities of extremal losses and high loss quantiles for all portfolios $\xi \in \Sigma^d$. This means that γ_ξ provides all information that is needed for comparing the influence of the portfolio vector ξ on the severity of extreme losses.

The scaling relations (13.9) and (13.10) also allow to estimate probabilities of extremal losses and high loss quantiles and to extrapolate these estimates beyond the observable area. The estimated values can be used in portfolio optimization. An empirical study based on these scaling relations is given in Hauksson et al. (2000).

13.2 Extreme Risk Index and Portfolio Diversification

The results of Section 13.1 justify the following definition.

Definition 13.1. For any portfolio vector $\xi \in \Sigma^d$ the functional

$$\gamma_\xi := \nu\left(A_{\xi,1}\right)$$

is called the "extreme risk index" of ξ.

The product structure of the measure ν in polar coordinates yields the representation

$$
\begin{aligned}
\gamma_\xi &= \int_{\Sigma^d} \int_{\mathbb{R}_+} 1\left\{\xi^\top \cdot rs > 1\right\} \varrho_\alpha(dr)\Psi(ds) \\
&= \int_{\Sigma^d} \varrho_\alpha\left(r \in \mathbb{R}_+;\ r > 1/\left(\xi^\top s\right)\right)\Psi(ds) \\
&= \int_{\Sigma^d} \left(\xi^\top s\right)^\alpha \Psi(ds).
\end{aligned}
\tag{13.11}
$$

The representation in (13.11) does not depend on the norm $\|\cdot\|$ used for the polar coordinates and the resulting spectral measure $\Psi_{\|\cdot\|}$. However, since the set $A_1 = \{x \in \mathbb{R}_+^d;\ \|x\| > 1\}$ depends on the norm, setting $\nu(A_{\xi,1}) := 1$ results in rescaling of γ_ξ by a constant factor that depends on the norm and the spectral measure Ψ. A remarkable property of the 1-norm is the fact that the extreme risk index of the equally weighted portfolio does not depend on the spectral measure:

$$
\gamma_{d^{-1}(1,\ldots,1)} = \int_{\Sigma^d} \left(d^{-1}\left(s^{(1)} + \ldots + s^{(d)}\right)\right)^\alpha \Psi(ds) = d^{-\alpha}.
$$

For the problem of finding the portfolio with lowest sensitivity we need to minimize the function $\xi \mapsto \gamma_\xi$. The resulting optimization problem is analysed in the following lemma.

Lemma 13.2. (a) *For $\alpha > 1$ the mapping $\xi \mapsto \gamma_\xi$ is convex. The convexity is strict if Ψ does not concentrate the entire mass on a linear subspace of Σ^d.*
(b) *For $\alpha = 1$ the mapping $\xi \mapsto \gamma_\xi$ is linear.*
(c) *For $\alpha \in (0,1)$ the mapping $\xi \mapsto \gamma_\xi$ is concave. The concavity is strict if Ψ does not concentrate the entire mass on a linear subspace of Σ^d.*

Proof. (a) The convexity of $\xi \mapsto \gamma_\xi$ follows from the convexity of $t \mapsto t^\alpha$ for $t > 0$ and $\alpha \geq 1$. Given $\lambda \in (0,1)$ and $\xi_1, \xi_2 \in \Sigma^d$, we obtain

$$
\begin{aligned}
\lambda\gamma_{\xi_1} + (1-\lambda)\gamma_{\xi_2} &= \int \left(\lambda\left(\xi_1^\top s\right)^\alpha + (1-\lambda)\left(\xi_2^\top s\right)^\alpha\right)\Psi(ds) \\
&\leq \int \left(\lambda\xi_1^\top s + (1-\lambda)\xi_2^\top s\right)^\alpha \Psi(ds) \\
&= \gamma_{\lambda\xi_1+(1-\lambda)\xi_2}.
\end{aligned}
$$

Strict convexity holds if the upper inequality is strict, i.e. if

$$
\int \left(\lambda\left(\xi_1^\top s\right)^\alpha + (1-\lambda)\left(\xi_2^\top s\right)^\alpha\right)\Psi(ds) < \int \left(\lambda\xi_1^\top s + (1-\lambda)\xi_2^\top s\right)^\alpha \Psi(ds)
$$

330 13 Optimal Portfolio Diversification w.r.t. Extreme Risks

for all $\xi_1, \xi_2 \in \Sigma^d$ such that $\xi_1 \neq \xi_2$. Since the mapping $t \mapsto t^\alpha$ is strictly convex for $\alpha > 1$, equality holds only if $\xi_1^\top s = \xi_2^\top s$ almost surely with respect to Ψ. This can also be written as

$$\Psi\left(\{(\xi_1 - \xi_2)^\top s = 0\}\right) = 1,$$

which means that the entire probability mass of Ψ is concentrated on $\Sigma^d \cap (\xi_1 - \xi_2)^\perp$.

(b) Is trivial since for $\alpha = 1$ the mapping $t \mapsto \alpha$ is linear and the mapping $\xi \mapsto \gamma_\xi$ is therefore a composition of linear mappings.

(c) Is analogous to part (a) due to the strict concavity of $t \mapsto t^\alpha$ for $\alpha \in (0, 1)$. □

Remark 13.3. (a) Optimal portfolio

As a consequence of Lemma 13.2 the location of the asymptotically optimal portfolio

$$\xi^{\mathrm{opt}} := \operatorname*{argmin}_{\xi \in \Sigma^d} \gamma_\xi \tag{13.12}$$

can be described as follows:

- For $\alpha > 1$ the typical location of ξ^{opt} would be in the interior of Σ^d. The optimal portfolio is unique if there is no mass concentration on linear subspaces under Ψ.
- For $\alpha \leq 1$ the minimum of γ_ξ is achieved in a corner of Σ^d, i.e. we have

$$\min_{\xi \in \Sigma^d} \gamma_\xi = \min_{i=1,\dots,d} \gamma_{e_i} \tag{13.13}$$

with e_i denoting the i-th unit vector.

Examples for the above-mentioned diversification effects are given in Figures 13.1 and 13.2.

(b) **Portfolio diversification**

The results of Lemma 13.2 and the conclusions in (13.2) have an interesting consequence: if only the losses are accounted, then portfolio diversification does not reduce the danger of extreme losses in the case $\alpha \in (0, 1]$. Moreover, for $\alpha < 1$ portfolio diversification typically increases extreme risks. The representation $\gamma_\xi = \int (\xi^\top s)^\alpha \Psi(ds)$ suggests that these negative effects are stronger in the case of low positive dependence, i.e. when Ψ concentrates the probability mass around the corners of the unit simplex Σ^d. Analogously, for $\alpha > 1$ low positive dependence makes positive diversification effects stronger.

For $d = 2$ and $\alpha > 1$ the best diversification effects are achieved if $X^{(1)}$ and $X^{(2)}$ are asymptotically independent, i.e. if $\Psi = \frac{1}{2}(\delta_{(1,0)} + \delta_{(0,1)})$, whereas the worst case is $\Psi = \delta_{(\frac{1}{2}, \frac{1}{2})}$, which corresponds to the comonotonic distribution of asset losses. While this behaviour accords with the usual intuition of diversification effects, in the case $\alpha < 1$ the situation is just the opposite. For $\alpha < 1$ diversification effects are negative or zero and the asymptotic independence of

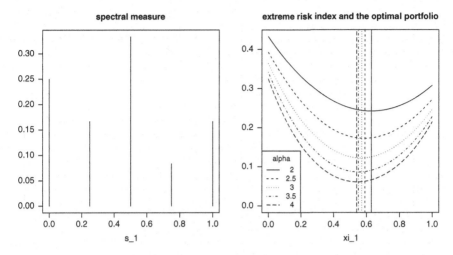

Figure 13.1 *Left*: spectral density. *Right*: the resulting extreme risk index and the optimal portfolios (*vertical lines*) for α between 2 and 4 (Published as Figure 2 in Mainik and Rüschendorf (2011))

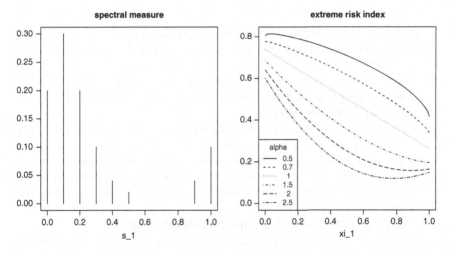

Figure 13.2 *Left*: spectral density. *Right*: the resulting extreme risk index for α between 0.5 and 2.5 (Published as Figure 3 in Mainik and Rüschendorf (2011))

asset losses is the worst case for the uniformly diversified portfolio $\xi = (\frac{1}{2}, \frac{1}{2})$, whereas the comonotonic distribution is the best case. Figure 13.3 shows plots of γ_ξ for a symmetric three-point spectral measure with mass in $(0, 1)$, $(1, 0)$ and $(\frac{1}{2}, \frac{1}{2})$ when the mass in the middle point varies between 0 and 1.

The comonotonic distribution removes all diversification effects: the positive ones for $\alpha > 1$ and the negative ones for $\alpha < 1$. This implies that in the case

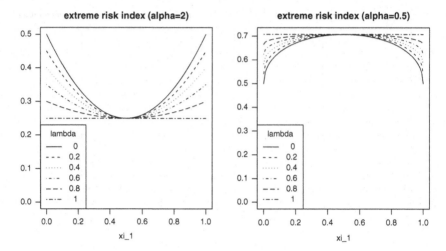

Figure 13.3 Influence of dependence on the extreme risk index for $\alpha = 2$ (*left*) and $\alpha = 0.5$ (*right*) with spectral measures $\Psi_\lambda = \lambda \delta_{(\frac{1}{2}, \frac{1}{2})} + \frac{1}{2}(1 - \lambda)(\delta_{(1,0)} + \delta_{(0,1)})$ (Published as Figure 4 in Mainik and Rüschendorf (2011))

of infinite means sensitivity to extremal events can only be optimized by minimizing the number of uncertainty sources and not by diversification. This remarkable diversification property has been repeatedly observed in settings similar to (13.57) and vividly discussed in the recent literature (Embrechts et al. (2009a,b), Barbe et al. (2006), Daníelsson et al. (2005), Wüthrich (2003)).

(c) **Accounting for gains and losses**
It should be noted that negative diversification effects for $\alpha < 1$ are restricted to models that account only the asset losses. If the profits are also incorporated, i.e. if $X^{(i)}$ can take positive and negative values as well, then a countermonotonic distribution of $(X^{(1)}, X^{(2)})$ leads to the compensation of losses from one component by the profits from the other one, so that diversification effects may become positive again.

(d) **α-stable distributions and negative diversification effects**
Negative diversification effects in infinite mean models were already noticed in the beginnings of probability theory. If, for example, $X_1, \ldots, X_n$ are iid symmetric α-stable random variables, then

$$n^{-1}(X_1 + \ldots + X_n) \overset{d}{=} n^{(1/\alpha)-1} X_1, \qquad (13.14)$$

which implies negative diversification effects for $\alpha < 1$. A general result on negative diversification effects as in (13.14) can be obtained from the Marcinkievicz–Zygmund Strong Law of Large Numbers for iid random variables, cf. Nešlehová et al. (2006a).   ◊

13.3   Estimation of the Extreme Risk Index and the Optimal Portfolio

In the following let X be a multivariate regularly varying random variable and let $X_1, \ldots, X_n$ be an iid sample of X. Our aim is the estimation of the extreme risk index γ_ξ and the optimal portfolio ξ^{opt}. The representation of γ_ξ in (13.11) suggests the following plug-in approach:

1. Estimate the tail index α by an estimator $\widehat{\alpha}$.
2. Estimate the spectral measure Ψ by an estimator $\widehat{\Psi}$.
3. Estimate γ_ξ by

$$\widehat{\gamma}_\xi := \int_{\Sigma^d} \left(\xi^{\mathsf{T}} s\right)^{\widehat{\alpha}} \widehat{\Psi}(ds). \tag{13.15}$$

4. Obtain an estimate for the optimal portfolio by minimizing $\widehat{\gamma}_\xi$:

$$\widehat{\xi}^{\text{opt}} := \underset{\xi \in \Sigma^d}{\operatorname{argmin}} \, \widehat{\gamma}_\xi. \tag{13.16}$$

Since $\widehat{\gamma}_\xi$ is obtained by plugging $\widehat{\Psi}$ and $\widehat{\alpha}$ into the representation (13.11), the minimization problem for $\widehat{\gamma}_\xi$ has the same properties as for γ_ξ and is characterized by Lemma 13.2. For $\widehat{\alpha} \leq 1$ the minimization is simplified by Lemma 13.2.

To obtain strong consistency of the estimation $\widehat{\xi}^{\text{opt}}$ we establish strong consistency of $\widehat{\gamma}_\xi$ uniformly in $\xi \in \Sigma$. The following section then will be concerned with asymptotic normality of the estimators. Let (R, S) and (R_i, S_i) denote the polar coordinates of X and X_i with respect to the 1-norm:

$$(R, S) := \left(\|X\|_1, \|X\|_1^{-1} X \right), \quad (R_i, S_i) := \left(\|X_i\|_1, \|X_i\|_1^{-1} X_i \right).$$

In order to avoid technical difficulties we assume that the distribution function of the radial parts $F_R(t) := \mathrm{P}\,(R \leq t)$ is continuous. This assumption will be fulfilled in typical applications.

As usual in extreme value theory, the estimates of tail related parameters are based on the $k = k(n)$ observations with highest absolute values, i.e. on the observations associated with the k upper order statistics $R_{n:1}, \ldots, R_{n:k}$ of the radial parts $R_1, \ldots, R_n$. We also make the usual assumptions concerning the growth of $k(n)$:

$$k(n) \to \infty, \quad \frac{k(n)}{n} \to 0.$$

Let $i(n, 1), \ldots, i(n, k)$ denote the indices corresponding to the k observations with greatest radial parts of R_i, ordered as they appear in the sample. Then we have

$$1 \leq i(n, 1) < \ldots < i(n, k) \leq n$$

and there exists a permutation π such that

$$\left(R_{i(n,1)}, \ldots, R_{i(n,k)}\right) = \left(R_{n:\pi(1)}, \ldots, R_{n:\pi(k)}\right). \tag{13.17}$$

The subsample $X_{i(n,1)}, \ldots, X_{i(n,k)}$ contains all information that is needed for estimating γ_ξ. By (13.11) γ_ξ can be written as

$$\Psi f := \int_{\Sigma^d} f(s)\Psi(ds), \quad \text{where} \quad f(s) := f_{\xi,\alpha}(s) := \left(\xi^\top s\right)^\alpha. \tag{13.18}$$

The function f is estimated by

$$\widehat{f} := f_{\xi,\widehat{\alpha}} = \left(\xi^\top s\right)^{\widehat{\alpha}}$$

with an estimator $\widehat{\alpha}$ for α obtained from the upper order statistics of the radial parts,

$$\widehat{\alpha} = \widehat{\alpha}\left(R_{n:1}, \ldots, R_{n:k}\right), \tag{13.19}$$

which can be based on various approaches (cf. Hill (1975), Pickands (1975), Smith (1987), Dekkers et al. (1989)).

The spectral measure Ψ is estimated by the empirical measure of the angular parts $S_{i(n,1)}, \ldots, S_{i(n,k)}$:

$$\widehat{\Psi} := \mathbb{P}_n := \frac{1}{k} \sum_{j=1}^{k} \delta_{S_{i(n,j)}}. \tag{13.20}$$

The following theorem states strong consistency of $\widehat{\gamma}_\xi$ uniformly in $\xi \in \Sigma^d$ and, under weaker conditions, pointwise in ξ.

Theorem 13.4. *Let $X_1, \ldots, X_n$ be iid multivariate regularly varying random variables with tail index $\alpha \in (0, \infty)$ and spectral measure Ψ and assume that the distribution function F_R of the radial parts is continuous.*

(a) If the estimator $\widehat{\alpha}$ is consistent almost surely,

$$\widehat{\alpha} \to \alpha \text{ P-a.s.}, \tag{13.21}$$

and

$$\sup_{\xi \in \Sigma^d} \left| \mathbb{E}\widehat{\Psi} f_{\xi,\alpha} - \Psi f_{\xi,\alpha} \right| \to 0, \tag{13.22}$$

then the estimator $\widehat{\gamma}_\xi$ is consistent uniformly in $\xi \in \Sigma^d$ almost surely:

$$\sup_{\xi \in \Sigma^d} \left| \widehat{\gamma}_\xi - \gamma_\xi \right| \to 0 \quad \text{P-a.s.}$$

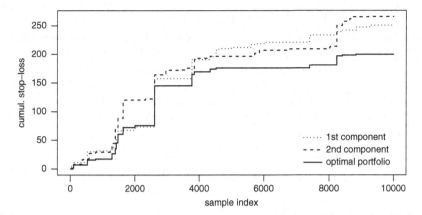

Figure 13.4 Cumulative stop loss for components and optimal portfolio

(b) If only (13.21) is satisfied, then the almost sure consistency of $\widehat{\gamma}_\xi$ holds pointwise:

$$\left| \widehat{\gamma}_\xi - \gamma_\xi \right| \to 0 \quad P\text{-a.s.} \quad \forall \xi \in \Sigma^d.$$

Remark 13.5. Since the functions $f_{\xi,\alpha}$ are bounded by 1 and for any fixed $\alpha \geq 1$ the function class $\{ f_{\xi,\alpha}; \; \xi \in \Sigma^d \}$ is uniformly Lipschitz, condition (13.22) is satisfied for any $\alpha \geq 1$. See Remark 13.11 for more details. Since for $\alpha \leq 1$ the optimization problem can be reduced to the minimization of $\widehat{\gamma}_\xi$ in the corners of Σ^d, condition (13.22) is crucial only for applications where there is no clear evidence for $\alpha > 1$ or $\alpha \leq 1$.   ◊

Uniform convergence of functions implies convergence of their minima to the minimum of the limit function in the case when the limit function has a unique minimum. Hence, as a consequence of Theorem 13.4, we obtain the following result.

Corollary 13.6. *Suppose that the conditions of Theorem 13.4(a) are satisfied and that the optimal portfolio ξ^{opt} is unique. Then the estimator $\widehat{\xi}^{\mathrm{opt}}$ and the estimated optimal risk extremal risk index $\widehat{\gamma}_{\xi^{\mathrm{opt}}}$ are consistent almost surely:*

$$\widehat{\xi}^{\mathrm{opt}} \to \xi^{\mathrm{opt}} \; P\text{-a.s.}, \quad \widehat{\gamma}_{\xi^{\mathrm{opt}}} \to \gamma_\xi^{\mathrm{opt}} \; P\text{-a.s.} \tag{13.23}$$

The simulation in Figure 13.4 by G. Mainik for a simulated data set and $d = 2$ shows the statistical effect of estimation of the optimal portfolio ξ^{opt} and the gain of the optimal estimated portfolio compared with the pure strategies.

Notions from empirical process theory For the proof of consistency of the estimators in Theorem 13.4 we recall some properties from empirical process theory. The estimator $\widehat{\gamma}_\xi$ can be written as

$$\widehat{\gamma}_\xi = \mathbb{P}_n \widehat{f} := \int_{\Sigma^d} \widehat{f}(s) \mathbb{P}_n(ds),$$

where $\widehat{f} = f_{\xi,\widehat{\alpha}}$ and $\mathbb{P}_n$ is the empirical measure of the subsample $S_{i(n,1)}, \ldots, S_{i(n,k)}$. Therefore it is natural to study $\widehat{\gamma}_\xi$ in the framework of empirical measures indexed by functions where $k = k(n)$ denotes the number of observations with large radial parts which are used for estimation. The strong consistency and the asymptotic normality of $\widehat{\gamma}_\xi$ can be viewed as special versions of the "Glivenko–Cantelli" and the "Donsker" theorems (cf. van der Vaart and Wellner (2000)).

Let $\mathbb{P}_{k,\Psi}$ denote the empirical measure corresponding to k iid random variables with probability distribution Ψ:

$$\mathbb{P}_{k,\Psi} := \frac{1}{k}\sum_{i=1}^{k}\delta_{Y_i}, \quad Y_1, \ldots, Y_k \text{ iid } \sim \Psi. \tag{13.24}$$

A function class $\mathcal{F}$ is called a "Glivenko–Cantelli class" if the Glivenko–Cantelli theorem holds for $\mathbb{P}_{k,\Psi}$ uniformly in $f \in \mathcal{F}$:

$$\mathbb{P}_{k,\Psi} \to \Psi \text{ P-a.s. in } \ell^\infty(\mathcal{F}), \quad k \to \infty. \tag{13.25}$$

Let $\mathbb{G}_{k,\Psi}$ denote the empirical process corresponding to $\mathbb{P}_{k,\Psi}$:

$$\mathbb{G}_{k,\Psi} := \sqrt{k}\left(\mathbb{P}_{k,\Psi} - \Psi\right). \tag{13.26}$$

A function class $\mathcal{F}$ is called a "Donsker class" if the Donsker theorem holds for $\mathbb{G}_{k,\Psi}$ uniformly in $f \in \mathcal{F}$,

$$\mathbb{G}_{k,\Psi} \overset{w}{\to} \mathbb{G}_\Psi \text{ in } \ell^\infty(\mathcal{F}), \quad k \to \infty, \tag{13.27}$$

where $\mathbb{G}_\Psi$ is the Brownian bridge "with time" Ψ, i.e.

$$(\mathbb{G}_\Psi f_1, \ldots, \mathbb{G}_\Psi f_m) \sim \mathcal{N}(0, C)$$

and $C = (C_{i,j})$ is given by

$$C_{i,j} := \Psi\left[(f_i - \Psi f_i)(f_j - \Psi f_j)\right] = \Psi f_i f_j - \Psi f_i \Psi f_j.$$

There are two problems that do not allow us to apply the standard Glivenko–Cantelli and Donsker theorems to the empirical measure of the subsample $S_{i(n,1)}, \ldots, S_{i(n,k)}$ and the resulting empirical process: the lack of independence between $S_{i(n,1)}, \ldots, S_{i(n,k)}$ and the fact that the underlying probability measure varies with n. Therefore a version is needed which is suitable for $\mathcal{L}(S_{i(n,1)}, \ldots, S_{i(n,k)})$. The following lemma gives conditional independence and provides a basis for the following results.

Lemma 13.7. *Suppose that the distribution function F_R of the radial part $R = \|X\|_1$ is continuous and consider the $(k+1)$-th upper order statistic of $R_1, \ldots, R_n$, transformed by F_R:*

$$U_n := F_R(R_{n:k+1}). \tag{13.28}$$

Then

$$\mathcal{L}\left(\left(S_{i(n,1)}, \ldots, S_{i(n,k)}\right) \mid U_n = u\right) = \otimes_{i=1}^{k} \Psi_u,$$

where

$$\Psi_u := \mathcal{L}\left(S \mid F_R(R) > u\right).$$

Proof. Since F_R is continuous, the order statistics $R_{n:1}, \ldots, R_{n:n}$ are a.s. different. Hence the permutation π in (13.17) is a.s. unique and the indices $i(n, 1), \ldots, i(n, k)$ are well defined. With f_R denoting the density of the random variable R, the density of the $(k + 1)$-th order statistic $R_{n:k+1}$ is given by

$$f_{R_{n:k+1}}(t) = \frac{n!}{(n - k - 1)!} F_R^{n-k-1}(t) f_R(t) (1 - F_R(t))^k$$

and the common density of the $(k + 1)$ upper order statistics is given by

$$f_{R_{n:1}, \ldots, R_{n:k+1}}(r_1, \ldots, r_k, t)$$
$$= 1\{t < r_k < \ldots < r_1\} \frac{n!}{(n - k - 1)!} F_R^{n-k-1}(t) f_R(t) \prod_{j=1}^{k} f_R(r_j).$$

Thus the conditional density of $R_{n:1}, \ldots, R_{n:k}$ given $R_{n:k+1} = t$ is a.s. equal to

$$f_{R_{n:1}, \ldots, R_{n:k} \mid R_{n:k+1} = t}(r_1, \ldots, r_k)$$
$$= \frac{f_{R_{n:1}, \ldots, R_{n:k+1}}(r_1, \ldots, r_k, t)}{f_{R_{n:k+1}}(t)}$$
$$= 1\{t < r_k < \ldots < r_1\} k! (1 - F_R(t))^{-k} \prod_{j=1}^{k} f_R(r_j),$$

which is the density of the order statistics of k iid random variables with probability distribution $\mathcal{L}(R \mid R > t)$. Since the permutation π is independent from the order statistics and uniformly distributed over the permutation group $\mathcal{S}^k$, the subsample $R_{i(n,1)}, \ldots, R_{i(n,k)}$ is conditionally iid, given $R_{n:k+1} = t$. This property is inherited by the subsample $X_{i(n,1)}, \ldots, X_{i(n,k)}$ and its angular parts $S_{i(n,1)}, \ldots, S_{i(n,k)}$, yielding

$$\mathcal{L}\left(\left(S_{i(n,1)}, \ldots, S_{i(n,k)}\right) \mid R_{n:k+1} = t\right) = \otimes_{i=1}^{k} \mathcal{L}(S \mid R > t).$$

Setting $t := F_R^{-1}(u)$ and rewriting the conditions in an appropriate way completes the proof.   □

Lemma 13.7 implies a representation of $\mu_{n,k} := \mathcal{L}(S_{i(n,1)}, \ldots, S_{i(n,k)})$ as a mixture of product measures.

Corollary 13.8. *If F_R is continuous, then*

$$\mu_{n,k}(A) = \int_0^1 \Psi_u^k(A) \, dP^{U_n}(u) \qquad (13.29)$$

for $A \in \mathcal{B}\left((\Sigma^d)^k\right)$, where

$$\Psi_u^k := \otimes_{i=1}^k \Psi_u, \quad u \in [0, 1].$$

Since $F_R^{-1}(u) \to \infty$ for $u \uparrow 1$, the behaviour of Ψ_u for $u \uparrow 1$ is related to the regular variation of X.

Lemma 13.9. *Suppose that the random variable X is multivariate regularly varying. Then*

$$\Psi_u \overset{w}{\to} \Psi, \quad u \uparrow 1.$$

Proof. The measure Ψ_u is obtained from the measure

$$\mu_u := \mathcal{L}\left(t(u)^{-1}X \mid R > t(u)\right), \quad t(u) := F_R^{-1}(u),$$

by application of the transformation $\tau : x \mapsto \|x\|_1^{-1} x$, i. e.

$$\Psi_u = \mu_u^\tau.$$

The representation of the multivariate regular variation property in (13.5) implies

$$\mu_u \overset{w}{\to} \nu|_{A_1},$$

where $\nu|_{A_1}$ is the restriction of ν to the set $A_1 = \{x \in \mathbb{R}_+^d; \ \|x\|_1 > 1\}$. Hence, the Continuous Mapping Theorem yields

$$\mu_u^\tau \overset{w}{\to} \nu|_{A_1}^\tau = \Psi. \qquad \square$$

The empirical version $\widehat{\gamma}_\xi$ of the extremal risk index has the representation

$$\widehat{\gamma}_\xi = \mathbb{P}_n f_{\xi,\widehat{\alpha}},$$

where $\mathbb{P}_n$ is the empirical measure of $S_{i(n,1)}, \ldots, S_{i(n,k)}$, indexed by elements of the function class

$$\mathcal{F} := \{f_{\xi,\alpha}; \ \alpha \in (0,\infty), \xi \in \Sigma^d\}. \tag{13.30}$$

Corollary 13.8 implies that the empirical measure $\mathbb{P}_n$ is a mixture of empirical measures constructed from iid observations:

$$\mathcal{L}(\mathbb{P}_n) = \int_0^1 \mathcal{L}(\mathbb{P}_{k,\Psi_u}) \, \mathrm{P}^{U_n}(u).$$

Moreover, $\Psi_u \overset{w}{\to} \Psi$ and further $U_n \uparrow 1$ P-a.s. Thus the consistency and the asymptotic normality of the estimator $\mathbb{P}_n f_{\xi,\widehat{\alpha}}$ are related to the uniformity of the

Glivenko–Cantelli and Donsker properties (13.25) and (13.27) of the class $\mathcal{F}$ and the underlying probability measure $\Psi \in \{\Psi_u;\ u \in (0, 1]\}$.

While the uniform Donsker property of $\mathcal{F}$ provides that convergence of empirical processes constructed from iid observations is uniform in the underlying probability measure, the uniform "pre-Gaussian" property allows to extend the convergence of empirical processes to the case when the underlying probability distribution converges for $k \to \infty$. Let $\mathcal{P}$ be a class of probability measures on Σ^d.

The function class $\mathcal{F}$ is "pre-Gaussian uniformly in $\Psi \in \mathcal{P}$" if the following two conditions are satisfied:

$$\sup_{\Psi \in \mathcal{P}} E \,\|\mathbb{G}_\Psi\|_{\ell^\infty(\mathcal{F})} < \infty$$

and   (13.31)

$$\limsup_{\delta \downarrow 0} E \sup_{\substack{\Psi \in \mathcal{P} \\ \varrho_\Psi(f-g)<\delta}} |\mathbb{G}_{k,\Psi}(f) - \mathbb{G}_{k,\Psi}(g)| = 0,$$

where $\varrho_\Psi(f)$ is the seminorm $\|f - \Psi f\|_{\Psi,2}$.

The following lemma states that the class $\mathcal{F}$ is "universally Glivenko–Cantelli", "Donsker and pre-Gaussian", i.e. that these properties are uniform over the class of all probability measures on $(\Sigma^d, \mathcal{B}(\Sigma^d))$.

Lemma 13.10. *The function class $\mathcal{F}$ is universally Glivenko–Cantelli, Donsker and pre-Gaussian.*

Proof. We first note that $\mathcal{F}$ is measurable (i.e. all $f \in \mathcal{F}$ are measurable) and that $\mathcal{F}$ is uniformly bounded by 1:

$$f(s) \le 1, \quad \forall s \in \Sigma^d, \forall f \in \mathcal{F}.$$

Therefore we can take

$$F(s) := 1_{\Sigma^d}(s)$$ (13.32)

as an "envelope function" for $\mathcal{F}$.

According to van der Vaart and Wellner (2000), $\mathcal{F}$ is universally Glivenko–Cantelli if it satisfies the "entropy condition"

$$\sup_{Q \in \mathcal{Q}_n} \log N\left(\varepsilon \|F\|_{Q,1}, \mathcal{F}, L^1(Q)\right) = o(n),$$ (13.33)

where $\mathcal{Q}_n$ denotes the class of all discrete probability measures on Σ^d with atoms of size which are integer multiples of $1/n$. Moreover, $\mathcal{F}$ is universally Donsker and pre-Gaussian if it satisfies the "uniform entropy condition"

$$\int_0^\infty \sup_{Q \in \mathcal{Q}} \sqrt{\log N\left(\varepsilon \|F\|_{Q,2}, \mathcal{F}, L^2(Q)\right)}\, d\varepsilon < \infty,$$ (13.34)

where $\mathcal{Q}$ denotes the class of all discrete probability measures on Σ^d. Here the "covering number" $N(\varepsilon, \mathcal{F}, \|\cdot\|)$ is defined as the minimal number of balls $\{g;\ \|g - f\| < \varepsilon\}$ of radius ε needed to cover the class $\mathcal{F}$. The "entropy" is the logarithm of the covering number. For these results and notions we refer to van der Vaart and Wellner (2000, Theorems 2.8.1 and 2.8.3).

The verification of the entropy conditions in (13.33) and (13.34) can be based on the properties of "Vapnik–Červonenkis (VC)" classes of sets and functions. (For definition and properties see van der Vaart and Wellner (2000, Section 2.6.))

To apply these results to the class $\mathcal{F}$ in (13.30) we note that any $f_{\xi,\alpha} \in \mathcal{F}$ is obtained by composition of a linear and a monotone function:

$$f_{\xi,\alpha} = g_\alpha \circ h_\xi,$$

where $g_\alpha : [0, 1] \to [0, 1]$, $t \mapsto t^\alpha$ and $h_\xi : \Sigma^d \to [0, 1]$, $s \mapsto \xi^\top s$. The function class

$$\mathcal{H} := \{h_\xi; \ \xi \in \Sigma^d\}$$

is a subset of a finite-dimensional vector space of functions. Hence it is a "VC-major" class. Since the functions g_α are monotone, the class $\mathcal{F}$ is also VC-major. Furthermore, the uniform boundedness of $\mathcal{F}$ implies that it is a "VC-hull" class. As a result we obtain from van der Vaart and Wellner (2000) (see Section 2.6, Theorem 2.6.9, Lemmas 2.6.13, 2.6.15 and 2.6.20) that $\mathcal{F}$ satisfies (13.34) and (13.33). $\square$

After these preparations we next give the proof of strong consistency of $\widehat{\gamma}_\xi$ in Theorem 13.4.

Proof of Theorem 13.4: Part (a) Consider the decomposition

$$\widehat{\gamma}_\xi - \gamma_\xi = \left(\widehat{\Psi} f_{\xi,\widehat{\alpha}} - \mathrm{E}\widehat{\Psi} f_{\xi,\widehat{\alpha}}\right) + \left(\mathrm{E}\widehat{\Psi} f_{\xi,\widehat{\alpha}} - \Psi f_{\xi,\widehat{\alpha}}\right) + \left(\Psi f_{\xi,\widehat{\alpha}} - \Psi f_{\xi,\alpha}\right). \tag{13.35}$$

First we show that

$$\Psi f_{\xi,\widehat{\alpha}} - \Psi f_{\xi,\alpha} \to 0 \text{ P-a.s.}$$

uniformly in ξ. Since

$$\left|\Psi f_{\xi,\widehat{\alpha}} - \Psi f_{\xi,\alpha}\right| \leq \left\| f_{\xi,\widehat{\alpha}} - f_{\xi,\alpha} \right\|_\infty$$

for all $\xi \in \Sigma^d$, it suffices to show that

$$\sup_{\xi \in \Sigma^d} \left\| f_{\xi,\widehat{\alpha}} - f_{\xi,\alpha} \right\|_\infty \to 0 \text{ P-a.s.} \tag{13.36}$$

Consider the partial derivative of $f_{\xi,\alpha}$ in α:

$$\frac{\partial}{\partial \alpha} \left(\xi^\top s\right)^\alpha = \left(\xi^\top s\right)^\alpha \log \left(\xi^\top s\right).$$

Since $\xi^\top s$ ranges in $[0, 1]$, we obtain

$$\left| \frac{\partial}{\partial \alpha} f_{\xi,\alpha}(s) \right| \leq \sup_{t \in [0,1]} |t^\alpha \log t| = |t_0^\alpha \log(t_0)|,$$

where $t_0 = \exp(-1/\alpha)$. Due to the strong consistency of $\widehat{\alpha} = \widehat{\alpha}(n)$ we have $\widehat{\alpha} > \alpha/2$ P-a.s. for n exceeding a sufficiently large bound n_0 and therefore

$$\left\| f_{\xi,\widehat{\alpha}} - f_{\xi,\alpha} \right\|_\infty \leq 2 (e \cdot \alpha)^{-1} |\widehat{\alpha} - \alpha| \quad \text{P-a.s.} \quad \forall \xi \in \Sigma^d \qquad (13.37)$$

for $n \geq n_0$. Hence (13.36) follows from the strong consistency of $\widehat{\alpha}$.

Now consider the second term on the right side of (13.35). We have

$$\mathrm{E}\widehat{\Psi} f_{\xi,\widehat{\alpha}} - \Psi f_{\xi,\widehat{\alpha}} = \left(\mathrm{E}\widehat{\Psi} - \Psi\right) f_{\xi,\alpha} + \left(\mathrm{E}\widehat{\Psi} - \Psi\right)\left[f_{\xi,\widehat{\alpha}} - f_{\xi,\alpha}\right].$$

Due to

$$\left|\left(\mathrm{E}\widehat{\Psi} - \Psi\right)\left[f_{\xi,\widehat{\alpha}} - f_{\xi,\alpha}\right]\right| \leq 2\left\| f_{\xi,\widehat{\alpha}} - f_{\xi,\alpha} \right\|_\infty$$

and (13.36) we only need to show that $(\mathrm{E}\widehat{\Psi} - \Psi) f_{\xi,\alpha} \to 0$ uniformly in ξ, which is provided by assumption (13.22).

Finally, we consider the first term on the right side of (13.35). Since $\widehat{\Psi} = \mathbb{P}_n$, the mixture representation (13.29) yields

$$\mathcal{L}\left(\widehat{\Psi} f_{\xi,\widehat{\alpha}} - \mathrm{E}\widehat{\Psi} f_{\xi,\widehat{\alpha}}\right) = \mathcal{L}\left(\mathbb{P}_n f_{\xi,\widehat{\alpha}} - \mathrm{E}\mathbb{P}_n f_{\xi,\widehat{\alpha}}\right)$$

$$= \int_{[0,1]} \mathcal{L}\left(\mathbb{P}_n f_{\xi,\widehat{\alpha}} - \mathrm{E}\mathbb{P}_n f_{\xi,\widehat{\alpha}} \mid U_n = u\right) d\mathrm{P}^{U_n}(u)$$

$$= \int_{[0,1]} \mathcal{L}\left(\left(\mathbb{P}_{k,\Psi_u} - \Psi_u\right) f_{\xi,\widehat{\alpha}}\right) d\mathrm{P}^{U_n}(u). \qquad (13.38)$$

Due to the universal Glivenko–Cantelli property of $\mathcal{F}$ (cf. Lemma 13.10) we have

$$\left(\mathbb{P}_{k,\Psi_u} - \Psi_u\right) f \to 0 \quad \text{P-a.s.}$$

uniformly in Ψ_u and $f \in \mathcal{F}$. Applied to the representation in (13.38), this yields

$$\widehat{\Psi} f_{\xi,\widehat{\alpha}} - \mathrm{E}\widehat{\Psi} f_{\xi,\widehat{\alpha}} \to 0 \quad \text{P-a.s.}$$

uniformly in ξ. Hence all terms in (13.35) vanish uniformly in ξ almost surely and the proof of part (a) is finished.

Part (b) follows along the lines of the proof of part (a). If the assumption (13.22) is dropped, we only need to verify

$$\mathrm{E}\widehat{\Psi} f_{\xi,\alpha} - \Psi f_{\xi,\alpha} \to 0 \qquad (13.39)$$

pointwise in ξ. Recall that the mixture representation (13.29) yields

$$\mathcal{L}(\mathbb{P}_n) = \int_{[0,1]} \mathbb{P}_{k,\Psi_u} \, d\mathrm{P}^{U_n}(u)$$

and that we have $U_n \uparrow 1$ P-a.s. and $\Psi_u \overset{w}{\to} \Psi$ for $u \uparrow 1$. As a result, we obtain the weak convergence

$$\mathbb{P}_n \overset{w}{\to} \Psi.$$

Since $\widehat{\Psi} = \mathbb{P}_n$ and all functions $f_{\xi,\alpha}$ are continuous, we obtain (13.39) pointwise in ξ.   □

Remark 13.11. It was noted in Remark 13.5 that the assumption (13.22) is satisfied for all $\alpha \geq 1$. This is due to the fact that weak convergence of Borel measures on a separable metric space $\mathcal{Z}$ is metrizable by the "bounded Lipschitz metric"

$$d_{BL_1}(L_1, L_2) := \sup_{h \in BL_1} \left| \int h \, dL_1 - \int h \, dL_2 \right|, \qquad (13.40)$$

where BL_1 is the set of all functions $h : \ell^\infty(\mathcal{Z}) \to \mathbb{R}$ that are uniformly bounded by 1 and Lipschitz with factor 1:

$$\sup_{z \in \mathcal{Z}} |h(z)| \leq 1,$$

$$|h(z_1) - h(z_2)| \leq \|z_1 - z_2\|_{\mathcal{Z}}$$

(cf. van der Vaart and Wellner (2000, Chapter 1.12)). It is easy to verify that the function class $\{f_{\xi,\alpha}; \ \xi \in \Sigma^d\}$ is uniformly Lipschitz for any $\alpha \geq 1$. Hence weak convergence $\widehat{\Psi} \overset{w}{\to} \Psi$ implies that (13.22) is satisfied for any $\alpha \geq 1$.   ◊

Remark 13.12 (Extension of the consistency result). Under the assumption $\widehat{\alpha} \to \alpha$ and $k(n) \geq \delta n^q$ for some $q \in (0,1)$, $\delta > 0$, $\widehat{\gamma}_\xi$ has been proved to be strongly consistent uniformly on $\xi \in \Sigma^d$ in Mainik (2012). In fact this consistency result is established there in the loss and gains case for compact portfolio sets $H \subset H^1 = \{x \in \mathbb{R}^d; \ \sum x_i = 1\}$. The argument in that paper is based on applying the Hoeffding inequality to bound $\mathcal{L}(\widehat{\Psi} f_{\xi,\widehat{\alpha}} - E\widehat{\Psi} f_{\xi,\widehat{\alpha}})$ in (13.38) which is made possible by the rate assumption on $k(n)$.   ◊

13.4 Asymptotic Normality of $\widehat{\gamma}_\xi$

In this section our aim is to state conditions which imply asymptotic normality of the empirical extremal risk index $\widehat{\gamma}_\xi$. For the proof which is based on empirical process theory as in Section 13.3 we make use of the following condition.

Condition (A) *At least one of the following assumptions is fulfilled:*

(a) The spectral measure Ψ has no mass on the boundary of Σ^d:

$$\Psi\left(\partial\Sigma^d\right) = 0.$$

(b) The tail index α is not smaller than 1 and an upper bound for α is known:

$$\alpha \in [1,\alpha^*], \quad \alpha^* < \infty.$$

The next theorem states the asymptotic normality result in a process version with index $\xi \in \Sigma^d$ under a second order condition and, under weaker conditions, states pointwise convergence for any ξ.

Theorem 13.13. *Let $X_1,\ldots,X_n$ be iid multivariate regularly varying random variables with tail index $\alpha \in (0,\infty)$ and spectral measure Ψ. Assume that the distribution function F_R of the radial part is continuous and that Condition (A) is satisfied.*

(a) Suppose that the estimator $\widehat{\alpha}$ is asymptotically normal,

$$\sqrt{k}\left(\widehat{\alpha} - \alpha\right) \xrightarrow{w} Y \sim \mathcal{N}\left(\mu_\alpha, \sigma_\alpha^2\right), \tag{13.41}$$

and that there exists a mapping $b \in \ell^\infty(\Sigma^d)$ such that

$$\sqrt{k}\left(\mathrm{E}\widehat{\Psi}\, f_{\xi,\alpha} - \Psi f_{\xi,\alpha}\right) \to b(\xi) \quad in\ \ell^\infty\left(\Sigma^d\right). \tag{13.42}$$

Then

$$\sqrt{k}\left(\widehat{\gamma}_\xi - \gamma_\xi\right) \xrightarrow{w} \mathbb{G}_\Psi f_{\xi,\alpha} + b(\xi) + c_{\xi,\alpha}Y \quad in\ \ell^\infty(\Sigma^d), \tag{13.43}$$

where $\mathbb{G}_\Psi$ is a Brownian bridge on the function class $\{f_{\xi,\alpha};\ \xi \in \Sigma^d\}$ "with time" Ψ, $b(\xi)$ is the asymptotic bias term from (13.42), Y is a Gaussian random variable which is independent from $\mathbb{G}_\Psi$ and distributed according to (13.41), and $c_{\xi,\alpha}$ is given by

$$c_{\xi,\alpha} = \int_{\Sigma^d} \left(\xi^\top s\right)^\alpha \log\left(\xi^\top s\right) \Psi(ds).$$

(b) Suppose that (13.41) is satisfied and that

$$\sqrt{k}\left(\mathrm{E}\Psi f_{\xi_i,\alpha} - \Psi f_{\xi_i,\alpha}\right) \to b(\xi_i) \in \mathbb{R} \tag{13.44}$$

holds for $\xi_1,\ldots,\xi_p \in \Sigma^d$. Then

$$\sqrt{k}\left(\widehat{\gamma}_{\xi_1},\ldots,\widehat{\gamma}_{\xi_p}\right) \xrightarrow{w} \mathcal{N}\left(\mu(\alpha,\xi_1,\ldots,\xi_p), \sigma(\alpha,\xi_1,\ldots,\xi_p)\right) \tag{13.45}$$

for all $\alpha \geq 0$. The expectations $\mu^{(i)}(\alpha, \xi_1, \ldots, \xi_p)$ are given by

$$\mu^{(i)} = b(\xi_i) + c_{\alpha, \xi_i} \mu_\alpha, \quad i = 1, \ldots, p,$$

and the covariances $\sigma_{i,j}(\alpha, \xi_1, \ldots, \xi_p)$ are equal to

$$\Psi\left[\left(f_{\xi_i, \alpha} - \Psi f_{\xi_i, \alpha}\right)\left(f_{\xi_j, \alpha} - \Psi f_{\xi_j, \alpha}\right)\right] + c_{\xi_i, \alpha} c_{\xi_j, \alpha} \sigma_\alpha,$$

where μ_α and σ_α are the mean and the variance of the random variable Y in (13.41).

Remark 13.14. (a) Under appropriate conditions that specify the rate of convergence of the distribution $\mathcal{L}(t^{-1}R|R > t)$ the estimators $\widehat{\alpha}$ of α in (13.19) are asymptotically normal (see de Haan and Ferreira (2006), Davis and Resnick (1984), Drees (1995), Dekkers et al. (1989), Smith (1987) and Drees et al. (2004)).

(b) Condition (13.42) can be understood as a second order condition related to the weak convergence of the angular parts $S_{i(n,1)}, \ldots, S_{i(n,k)}$. Since multivariate regular variation leaves convergence rates completely unspecified, similar second order conditions are necessary for establishing asymptotic normality in regularly varying models.

(c) **Extension of the asymptotic normality result.** The asymptotic normality result in (13.43) has been extended to the gains and loss case in Mainik (2012) for compact classes of portfolios $H \subset H_1$. The assumptions made in this paper include the asymptotic normality assumption (13.41), a version of the second order bias condition in (13.42) and an asymptotic independence condition of $Y_n = \sqrt{k(n)}(\widehat{\alpha} - \alpha)$ and $\mathbb{G}_n = \sqrt{k(n)}(\mathbb{P}_n - \Psi_{U_n})$. Condition (A) is dismissed within this paper.   ◊

For the proof of the central limit theorem in Theorem 13.13 we need some further preparation. The central part in this proof is the weak convergence of the empirical process related to the subsample $S_{i(n,1)}, \ldots, S_{i(n,k)}$:

$$\mathbb{G}_n := \sqrt{k}\left(\mathbb{P}_n - P_n\right), \tag{13.46}$$

where the probability measure P_n is defined as the expectation of $\mathbb{P}_n$

$$P_n f := \mathbb{E}\mathbb{P}_n f. \tag{13.47}$$

With the universal Donsker property and pre-Gaussianity in Lemma 13.10, convergence of $\mathbb{G}_n$ is obtained from the convergence of the conditioning random variable U_n.

Lemma 13.15. *Suppose that Condition (A) is satisfied. Then the empirical process $\mathbb{G}_n$ converges to a Brownian Bridge "with time" Ψ:*

$$\mathbb{G}_n \overset{\text{w}}{\to} \mathbb{G}_\Psi \quad \text{in } \ell^\infty(\mathcal{F}). \tag{13.48}$$

Proof. Since the empirical process $\mathbb{G}_n$ is constructed from the subsample $S_{i(n,1)}, \ldots, S_{i(n,k)}$, the mixture representation in (13.29) of $\mathcal{L}(S_{i(n,1)}, \ldots, S_{i(n,k)})$ implies

$$\mathcal{L}(\mathbb{G}_n) = \int_{[0,1]} \mathcal{L}(\mathbb{G}_{k,\Psi_u}) \, d\mathrm{P}^{U_n}(u). \tag{13.49}$$

Moreover, we already know that $U_n \uparrow 1$ P-a.s. and $\Psi_u \overset{w}{\to} \Psi$ for $u \uparrow 1$.

Consider a sequence u_k in $(0,1)$ such that $u_k \uparrow 1$ for $k \to \infty$ and the empirical processes $\mathbb{G}_{k,\Psi_k}$ with the underlying measure $\Psi_k := \Psi_{u_k}$. As shown in Lemma 13.10, the class $\mathcal{F}$ is universally Donsker and pre-Gaussian. According to Lemma 2.8.7 in van der Vaart and Wellner (2000), the convergence

$$\mathbb{G}_{k,\Psi_k} \overset{w}{\to} \mathbb{G}_\Psi \tag{13.50}$$

holds if the class $\mathcal{F}$ and the sequence Ψ_k satisfy

$$\limsup_{k\to\infty} \Psi_k \left[F^2 \cdot 1\{F \geq \varepsilon\sqrt{k}\} \right] = 0, \quad \forall \varepsilon > 0 \tag{13.51}$$

and

$$\sup_{f,g\in\mathcal{F}} |\varrho_{\Psi_k}(f-g) - \varrho_\Psi(f-g)| \to 0, \tag{13.52}$$

where $\varrho_\Psi(f)$ denotes the seminorm $\|f - \Psi f\|_{\Psi,2} = (\Psi((f-\Psi f)^2))^{1/2} = (\Psi f^1 - (\Psi f)^1)^{1/2}$.

Since the envelope function F of $\mathcal{F}$ is bounded, condition (13.51) is trivial and we only need to verify (13.52). For $h := (f-g)$ we have

$$\left|\varrho_{\Psi_k}^2(h) - \varrho_\Psi^2(h)\right| = \left|\Psi_k h^2 - (\Psi_k h)^2 - \left(\Psi h^2 - (\Psi h)^2\right)\right|$$

$$= \left|(\Psi_k h^2 - \Psi h^2) - (\Psi_k h - \Psi h) \cdot (\Psi_k h + \Psi h)\right|$$

$$\leq \left|\Psi_k h^2 - \Psi h^2\right| + |\Psi_k h - \Psi h| \cdot O(1).$$

Thus, due to $(f-g)^2 \leq 2f^2 + 2g^2$ and $|f-g| \leq |f| + |g|$, it suffices to show that for $k \to \infty$

$$\sup_{f\in\mathcal{F}} \left|\Psi_k f^2 - \Psi f^2\right| \to 0 \quad \text{and} \quad \sup_{f\in\mathcal{F}} |\Psi_k f - \Psi f| \to 0.$$

Since $f \in \mathcal{F}$ implies $f^2 \in \mathcal{F}$, we only need to verify

$$\sup_{f\in\mathcal{F}} |\Psi_k f - \Psi f| \to 0, \quad k \to \infty. \tag{13.53}$$

Consider the sets

$$B_\delta := \left\{ s \in \Sigma^d;\ s_i > \delta \right\}, \quad \delta > 0.$$

If Condition (A) (a) is satisfied, i.e. we have $\Psi(\partial \Sigma^d) = 0$, then for any $\varepsilon > 0$ there exists $\delta > 0$ such that

$$\Psi(\Sigma^d \setminus B_\delta) < \varepsilon/4.$$

Since the number of atoms of Ψ is countable, δ can always be chosen such that

$$\Psi(\partial B_\delta) = 0.$$

Hence, $\Psi_k \xrightarrow{w} \Psi$ implies $\Psi_k(\Sigma^d \setminus B_\delta) \to \Psi(\Sigma^d \setminus B_\delta)$ and therefore

$$|\Psi_k f - \Psi f|$$
$$\leq \left| \Psi_k \left(f \cdot 1_{B_\delta} \right) - \Psi \left(f \cdot 1_{B_\delta} \right) \right| + \left| \Psi_k \left(f \cdot (1 - 1_{B_\delta}) \right) - \Psi \left(f \cdot (1 - 1_{B_\delta}) \right) \right|$$
$$\leq \left| \Psi_k \left(f \cdot 1_{B_\delta} \right) - \left(\Psi f \cdot 1_{B_\delta} \right) \right| + \Psi_k \left(\Sigma^d \setminus B_\delta \right) + \Psi \left(\Sigma^d \setminus B_\delta \right)$$
$$\leq \left| \Psi_k \left(f \cdot 1_{B_\delta} \right) - \Psi \left(f \cdot 1_{B_\delta} \right) \right| + \frac{3}{4} \varepsilon$$

for sufficiently large k. Hence we only need to verify

$$\sup_{f \in \mathcal{F}} \left| \Psi_k \left(f \cdot 1_{B_\delta} \right) - \Psi \left(f \cdot 1_{B_\delta} \right) \right| \to 0, \quad k \to \infty.$$

Due to the metrization of weak convergence by the bounded Lipschitz metric (cf. Remark 13.11) it suffices to show that the function class $\mathcal{F}$ is uniformly Lipschitz on B_δ, i.e. that there exists $K > 0$ such that

$$\forall f \in \mathcal{F}, \forall s^1, s^2 \in B_\delta \text{ it holds that: } \left| f(s^1) - f(s^2) \right| \leq K \left| s^1 - s^2 \right|. \quad (13.54)$$

Since all $f \in \mathcal{F}$ are differentiable on B_δ it suffices to show that the partial derivatives of $f \in \mathcal{F}$ are uniformly bounded on B_δ. We have

$$\frac{\partial}{\partial s_i} f_{\xi,\alpha}(s) = \alpha \left(\xi^\top s \right)^{\alpha-1} \cdot \xi_i.$$

Due to $\xi_i \leq 1$ and $s \in B_\delta$ we obtain

$$\sup_{s \in B_\delta} \left| \frac{\partial}{\partial s_i} f_{\xi,\alpha}(s) \right| \leq \sup_{t \in (\delta, 1-\delta)} \alpha \cdot t^{\alpha-1}.$$

The term on the right side is uniformly bounded for $\alpha \in (0, \infty)$ due to

$$\sup_{\alpha \geq 1} \sup_{t \in (\delta, 1-\delta)} \alpha \cdot t^{\alpha-1} = \sup_{\alpha \geq 1} \alpha (1 - \delta)^{\alpha-1} < \infty.$$

and

$$\sup_{\alpha\in(0,1)} \sup_{t\in(\delta,1-\delta)} \alpha\cdot t^{\alpha-1} = \sup_{\alpha\in(0,1)} \alpha\cdot\delta^{\alpha-1} < \infty.$$

As a consequence, we obtain (13.54), which implies (13.53) and (13.52). Hence we obtain the convergence in (13.50).

If Condition (A) (b) is satisfied, i.e. we have $\alpha \in [1,\alpha^*]$, then the class of index functions f can be reduced to

$$\mathcal{F}_{\alpha^*} := \{f_{\xi,\alpha}; \ \xi \in \Sigma^d, \alpha \in [1,\alpha^*]\}.$$

Thus (13.52) is simplified to

$$\sup_{f,g\in\mathcal{F}_{\alpha^*}} |\varrho_{\Psi_k}(f - g) - \varrho_\Psi(f - g)| \to 0,$$

which can be obtained from the uniform Lipschitz property of $f \in \mathcal{F}_{2\alpha^*}$. Since $f \in \mathcal{F}$ are differentiable on Σ^d, the uniform Lipschitz property follows from

$$\sup_{f\in\mathcal{F}_{2\alpha^*}} \sup_{s\in\Sigma^d} \left|\frac{\partial}{\partial s_i} f(s)\right| = 2\alpha^* < \infty.$$

Hence (13.52) is verified and we obtain (13.5).

The proof is finished by combination of (13.50) with the mixture representation (13.49). It was already mentioned above that weak convergence is metrized by the bounded Lipschitz metric d_{BL_1} (cf. Remark 13.11). Hence it suffices to show that

$$d_{BL_1}(\mathbb{G}_n, \mathbb{G}_\Psi) \to 0.$$

Let $h \in BL_1$. Then the mixture representation (13.49) yields

$$\left|\int h\,d\mathcal{L}(\mathbb{G}_n) - \int h\,d\mathcal{L}(\mathbb{G}_\Psi)\right|$$

$$= \left|\int_{[0,1]} \int h\,d\mathcal{L}(\mathbb{G}_{k,\Psi_u})\,d\mathrm{P}^{U_n}(u) - \int h\,d\mathcal{L}(\mathbb{G}_\Psi)\right|$$

$$= \left|\int_{[0,1]} \left(\int h\,d\mathcal{L}(\mathbb{G}_{k,\Psi_u}) - \int h\,d\mathcal{L}(\mathbb{G}_\Psi)\right) d\mathrm{P}^{U_n}(u)\right|$$

$$\leq \int_{[0,u_0)} \left|\int h\,d\mathcal{L}(\mathbb{G}_{k,\Psi_u}) - \int h\,d\mathcal{L}(\mathbb{G}_\Psi)\right| d\mathrm{P}^{U_n}(u)$$

$$+ \int_{[u_0,1]} \left|\int h\,d\mathcal{L}(\mathbb{G}_{k,\Psi_u}) - \int h\,d\mathcal{L}(\mathbb{G}_\Psi)\right| d\mathrm{P}^{U_n}(u)$$

$$\leq 2\mathrm{P}\{U_n < u_0\} + \sup_{u\geq u_0} d_{BL_1}(\mathbb{G}_\Psi u, \mathbb{G}_\Psi).$$

Given a fixed $\varepsilon > 0$, there exists $u_0 \in (0, 1)$ and $n_0 \in \mathbb{N}$ such that

$$d_{BL_1}(\mathbb{G}_{k,\Psi_u}, \mathbb{G}_\Psi) < \varepsilon/2$$

for $u \geq u_0$ and $n \geq n_0$. Since $U_n \to 1$ P-a.s., the index n_0 can be enlarged (if necessary) so that

$$P(\{U_n < u_0\}) < \frac{\varepsilon}{4}$$

for $n \geq n_0$. Now, for $n \geq n_0$, we obtain

$$d_{BL_1}(\mathbb{G}_n, \mathbb{G}_\Psi) = \sup_{h \in BL_1} \left| \int h \, d\mathcal{L}(\mathbb{G}_n) - \int h \, d\mathcal{L}(\mathbb{G}_\Psi) \right| < \varepsilon,$$

which implies $d_{BL_1}(\mathbb{G}_n, \mathbb{G}_\Psi) \to 0$. $\square$

   With Lemma 13.15 we next prove the asymptotic normality results in Theorem 13.13.

Proof asymptotic normality in Theorem 13.13: Part (a) We need to show that the asymptotic normality (13.41) of $\widehat{\alpha}$ and the second order condition (13.42) yield weak convergence of $\sqrt{k}(\widehat{\gamma}_\xi - \gamma_\xi) = \sqrt{k}(\mathbb{P}_n \widehat{f} - \Psi f)$ to the Gaussian process in (13.43). Consider the decomposition

$$\sqrt{k}\left(\mathbb{P}_n \widehat{f} - \Psi f\right)$$
$$= \sqrt{k}\left(\mathbb{P}_n \widehat{f} - \mathbb{P}_n \widehat{f}\right) + \sqrt{k}\left(\mathbb{P}_n \widehat{f} - \Psi \widehat{f}\right) + \sqrt{k}\left(\Psi \widehat{f} - \Psi f\right)$$
$$= \mathbb{G}_n f_{\xi,\widehat{\alpha}} + \sqrt{k}(\mathbb{P}_n - \Psi) f_{\xi,\widehat{\alpha}} + \Psi \sqrt{k}\left(f_{\xi,\widehat{\alpha}} - f_{\xi,\alpha}\right). \tag{13.55}$$

Recall the arguments for inequality (13.37). Since asymptotic normality of $\widehat{\alpha}$ implies $\widehat{\alpha} \xrightarrow{P} \alpha$, for any $\varepsilon > 0$, there exists $n_0 \in \mathbb{N}$ such that

$$P\left(\left\{\widehat{\alpha} < \alpha/2\right\}\right) < \varepsilon \tag{13.56}$$

for $n \geq n_0$. This yields

$$P\left(\left\{\|f_{\xi,\widehat{\alpha}} - f_{\xi,\alpha}\| > 2(e \cdot \alpha)^{-1} |\widehat{\alpha} - \alpha|\right\}\right) < \varepsilon \tag{13.57}$$

for $n \geq n_0$ and, therefore, $\|f_{\xi,\widehat{\alpha}} - f_{\xi,\alpha}\|_\infty \xrightarrow{P} 0$. Hence Lemma 13.15 implies

$$\left(\mathbb{G}_n f_{\xi,\widehat{\alpha}}\right)_{\xi \in \Sigma^d} \xrightarrow{w} \left(\mathbb{G}_\Psi f_{\xi,\alpha}\right)_{\xi \in \Sigma^d} \quad \text{in } \ell^\infty(\Sigma^d)$$

for all $\alpha > 0$.

Now consider the second term in (13.55). Due to the asymptotic normality of $\widehat{\alpha}$, inequality (13.57) implies

$$\left\| f_{\xi,\widehat{\alpha}} - f_{\xi,\alpha} \right\|_\infty = O_P(1/\sqrt{k})$$

and therefore

$$\sqrt{k}(P_n - \Psi) f_{\xi,\widehat{\alpha}} = \sqrt{k}(P_n - \Psi) \left[f_{\xi,\alpha} + O_P(1/\sqrt{k}) \right].$$

Hence assumption (13.42) yields

$$\sqrt{k}(P_n - \Psi) f_{\xi,\widehat{\alpha}} \xrightarrow{P} b(\xi)$$

uniformly in ξ.

The asymptotic distribution of $\Psi\sqrt{k}\left(\widehat{f} - f\right)$ is obtained from the asymptotic normality (13.41) of $\widehat{\alpha}$. Recall that $f_{\xi,\alpha} = (\xi^\top s)^\alpha$ and $\xi^\top s$ ranges in $[0, 1]$. Taylor expansion yields

$$\left(t^{\widehat{\alpha}} - t^\alpha\right) = (\widehat{\alpha} - \alpha) t^\alpha \log t + \frac{1}{2}(\widehat{\alpha} - \alpha)^2 t^{\alpha^*} \log^2 t$$

with some α^* between α and $\widehat{\alpha}$. The asymptotic normality of $\widehat{\alpha}$ implies (13.56). Since the mappings $t \mapsto t^\alpha \log t$ and $t \mapsto t^\alpha \log^2 t$ are bounded on $[0, 1]$ uniformly in $\alpha > \delta$ for any fixed $\delta > 0$, we obtain

$$\sqrt{k}\left(\widehat{f}(s) - f(s)\right) = \sqrt{k}\left((\xi^\top s)^{\widehat{\alpha}} - (\xi^\top s)^\alpha\right)$$

$$= \sqrt{k}\,(\widehat{\alpha} - \alpha)\,(\xi^\top s)^\alpha \log(\xi^\top s) + \sqrt{k}\,O_P(\widehat{\alpha} - \alpha)^2$$

$$\xrightarrow{w} Z(s) := Y \cdot (\xi^\top s)^\alpha \log(\xi^\top s)$$

as an ℓ^∞-function of $\xi \in \Sigma^d$.

Due to the continuity of the mapping $g \mapsto \Psi g$ for $g \in \ell^\infty(\Sigma^d)$ we can apply the Continuous Mapping Theorem and obtain

$$\Psi\sqrt{k}\left(f_{\xi,\widehat{\alpha}} - f_{\xi,\alpha}\right) \xrightarrow{w} \Psi Z = Y \cdot c_{\xi,\alpha} \quad \text{in } \ell^\infty(\Sigma^d),$$

where

$$c_{\xi,\alpha} := \Psi\left((\xi^\top s)^\alpha \log(\xi^\top s)\right).$$

Furthermore, ΨZ and $\mathbb{G}_\Psi f$ are independent. This follows from the asymptotic independence of the radial parts $R_{n:1}, \ldots, R_{n:k}$ and the angular parts $S_{i(n,1)}, \ldots, S_{i(n,k)}$ of the extreme subsample $X_{i(n,1)}, \ldots, X_{i(n,k)}$. Hence we obtain (13.43).

The result of part (b) concerns convergence of the finite-dimensional distributions in part (a). Replacing the assumption (13.42) by (13.44) affects only the middle term in (13.55). This results from the replacement of the uniform convergence to $b(\xi)$ by pointwise convergence. As a consequence the pointwise asymptotic normality (13.45) follows along the lines of the proof for part (a). $\square$

13.5 Application to Risk Minimization

In this section we consider the application of the the extreme risk index γ_ξ and its estimates to risk minimization. The ordering of large quantiles of the portfolio loss $\xi^\top X$ by γ_ξ as in (13.10) has immediate consequences on "risk measures" such as the "Value-at-Risk" (VaR) and the "Expected Shortfall" (ES). For an asset loss Y with distribution function F_Y the Value-at-Risk at level λ for (typically small) $\lambda \in (0,1)$ is defined by

$$\text{VaR}_\lambda(Y) := F_Y^{-1}(1-\lambda).$$

In comparison to Chapter 3 we use here small parameter values λ for large tails. The Expected Shortfall at level λ is defined in its simplified version by

$$\text{ES}_\lambda(Y) = \mathrm{E}\left[Y|Y > \text{VaR}_\lambda\right].$$

The portfolio ξ^{opt} which minimizes the extreme risk index γ_ξ also minimizes the expected shortfall risk as $\lambda \to 0$. For the proof we make use of the fact that the expected shortfall risk measure can be represented as average value at risk

$$\text{ES}_\lambda(Y) = \frac{1}{\lambda}\int_{1-\lambda}^1 F_Y^{-1}(u)du. \tag{13.58}$$

We consider $\text{VaR}_\lambda(\xi^\top X)$ and $\text{ES}_\lambda(\xi^\top X)$ for $\lambda \downarrow 0$ for a multivariate regularly varying loss vector X with non-negative components. Due to (13.10) we obtain that ξ^{opt} is the optimal portfolio if we want to minimize $\text{VaR}_{1-\lambda}$ with respect to extreme risks, i.e. for $\lambda \downarrow 0$. For $\alpha > 1$ we have

$$\lim_{\lambda \downarrow 0} \frac{\text{ES}_\lambda}{\text{VaR}_\lambda}(\xi^\top X) = \frac{\alpha}{\alpha-1}. \tag{13.59}$$

This asymptotic relation is a consequence of the representation in (13.58) and the uniform convergence theorem for slowly varying functions (cf. Theorem 1.5.2 in Bingham et al. 1987). Consequently, ξ^{opt} also minimizes $\text{ES}_\lambda\left(\xi^\top X\right)$ for $\lambda \downarrow 0$.

The asymptotic result in (13.59) can be generalized to the class of "spectral risk measures" (see Section 7.1.2) which are obtained as weighted averages of loss quantiles:

$$M_\varphi(Y) := \int_0^1 F_Y^{-1}(p)\varphi(p)dp,$$

where the weight function $\varphi : [0, 1] \to \mathbb{R}$ is an "admissible risk spectrum", i.e. φ is non-negative, non-decreasing, and satisfies $\int_0^1 \varphi(p)dp = 1$.

As a consequence of (13.58), ES_λ is a spectral risk measure. Thus (13.59) can be viewed as a limit relation for the rescaled and properly normalized risk spectrum. Analogously, for any admissible risk spectrum $\varphi_1 : [0, 1] \to \mathbb{R}$ the transformations $p \mapsto u := 1 - \lambda(1 - p)$ with $\lambda \in (0, 1)$ induce a family of rescaled admissible risk spectra

$$\varphi_\lambda(u) := \lambda^{-1}\varphi_1\left(1 - \lambda^{-1}(1 - u)\right) \cdot 1_{[1-\lambda,1]}(u).$$

This notation leads to the following generalization of (13.59).

Lemma 13.16. *Let Y be a continuously distributed random variable on $\mathbb{R}_+$ and suppose that Y is regularly varying with tail index $\alpha > 1$. Then*

$$\lim_{\lambda \downarrow 0} \frac{M_{\varphi_\lambda}(Y)}{\mathrm{VaR}_\lambda(Y)} = \int_0^1 (1 - p)^{-1/\alpha}\varphi_1(p)dp.$$

Proof. We have

$$M_{\varphi_\lambda}(Y) = \int_{1-\lambda}^1 F_Y^{-1}(u)\varphi_\lambda(u)du.$$

The substitution $u = 1 - \lambda(1 - p)$ yields

$$M_{\varphi_\lambda}(Y) = \int_0^1 F_Y^{-1}(1 - \lambda(1 - p))\varphi_1(p)dp$$

and we obtain

$$\frac{M_{\varphi_\lambda}(Y)}{\mathrm{VaR}_\lambda} = \int_0^1 \frac{F_Y^{-1}(1 - \lambda(1 - p))}{F_Y^{-1}(1 - \lambda)}\varphi_1(p)dp. \tag{13.60}$$

Since regular variation of Y with index α implies that the quantile function F_Y^{-1} is regularly varying in 1 with index $-1/\alpha$, the uniform convergence theorem yields

$$\frac{F_Y^{-1}(1 - \lambda(1 - p))}{F_Y^{-1}(1 - \lambda)} \to (1 - p)^{-1/\alpha} \quad \text{as } \lambda \downarrow 0$$

uniformly in $u \in [0, 1]$ and the result follows from (13.60). $\square$

As a consequence, we obtain that minimization of $M_{\varphi_\lambda}(\xi^\top X)$ for $\lambda \downarrow 0$ can be reduced to the minimization of the extremal risk index γ_ξ.

Chapter 14
Ordering of Multivariate Risk Models with Respect to Extreme Portfolio Losses

The previous chapter (Chapter 13) is concerned with portfolio optimization w.r.t. extreme portfolio losses within a particular model – assumed to be multivariate regularly varying – and with statistical properties of the estimators of the extremal risk index and the optimal portfolio. In this chapter an ordering called asymptotic portfolio loss ordering (apl) is introduced which is closely connected with the extremal risk index and which allows to compare different stochastic models of risk vectors w.r.t. extreme portfolio losses. In particular worst case and best case dependence structures are determined. Various comparison criteria for the apl-ordering are derived in terms of their spectral measures in the frame of multivariate regularly varying models. Comparison criteria in terms of further stochastic ordering notions are derived. The examples include elliptical distributions and multivariate regularly varying models with Gumbel, Archimedean, and Galambos copulas. The results in this chapter are based on Mainik and Rü (2012).

14.1 Asymptotic Portfolio Loss Ordering

Let X be a "random loss vector" with values in $\mathbb{R}^d$; positive values of the components $X^{(i)}$, $i = 1, \ldots, d$, represent losses and negative values of $X^{(i)}$ represent gains of some risky assets. Following the intuition of diversifying a unit capital over several assets, we restrict the set of portfolios to the unit simplex in $\mathbb{R}_+$:

$$\Sigma^d := \left\{ \xi \in \mathbb{R}_+^d; \ \sum_{i=1}^{d} \xi^{(i)} = 1 \right\}.$$

The portfolio loss resulting from a random vector X and the portfolio ξ is given by the scalar product $\xi^\top X$ of ξ and X.

L. Rüschendorf, *Mathematical Risk Analysis*, Springer Series in Operations Research and Financial Engineering, DOI 10.1007/978-3-642-33590-7_14,
© Springer-Verlag Berlin Heidelberg 2013

Definition 14.1. Let X and Y be d-dimensional random vectors. Then X is called smaller than Y in "asymptotic portfolio loss order",

$$X \preceq_{apl} Y,$$

if

$$\limsup_{t \to \infty} \frac{P\{\xi^T X > t\}}{P\{\xi^T Y > t\}} \leq 1 \quad \forall \xi \in \Sigma^d. \tag{14.1}$$

Here, $\frac{0}{0}$ is defined to be 1.

If $Y \preceq_{apl} Y$ and $Y \preceq_{apl} X$, we write $X \sim_{apl} Y$.

Remark 14.2. (a) Although designed for random vectors, the asymptotic portfolio loss order $\preceq_{apl}$ is also defined for random variables. In this case, the portfolio set has only one element, $\Sigma^1 = \{1\}$. In the framework of multivariate regular variation we will derive sufficient criteria for $X \preceq_{apl} Y$ in terms of componentwise ordering $|X^{(i)}| \preceq_{apl} |Y^{(i)}|$ for all i and a proper ordering of the asymptotic dependence structures of X and Y. An equivalent criterion needs $|X^{(i)}| \sim_{apl} |Y^{(i)}|$, which is equivalent to

$$\lim_{t \to \infty} \frac{P\{|X^{(i)}| > t\}}{P\{|Y^{(i)}| > t\}} = 1. \tag{14.2}$$

This is the asymptotic analogue to the assumption of identical marginal distributions, which is typical in the analysis of dependence structures. The asymptotic nature of $\preceq_{apl}$ does not need more than domination of tails.

(b) It is obvious that $\preceq_{apl}$ is invariant under componentwise rescaling. Let vx denote the componentwise product of $v, x \in \mathbb{R}^d$:

$$vx := (v^{(1)} x^{(1)}, \ldots, v^{(d)} x^{(d)}). \tag{14.3}$$

Then it is easy to see that $X \preceq_{apl} Y$ implies $vX \preceq_{apl} vY$ for all $v \in \mathbb{R}_+^d$. Hence condition (14.1) can be equivalently stated for $\xi \in \mathbb{R}_+^d$.

(c) Obviously $\preceq_{apl}$ is not closed under weak convergence. Consider an exponentially distributed X and $Y_n \stackrel{d}{=} \frac{1}{2}(X + 1\{Z > n\}Z)$ where Z is independent of X and Pareto distributed. Here we have $X \preceq_{apl} Y_n$ for all n, but $Y_n \stackrel{w}{\to} \frac{1}{2}X$. ◊

The ordering statement $X \preceq_{apl} Y$ means that for all portfolios $\xi \in \Sigma^d$ the portfolio loss $\xi^T X$ is asymptotically stochastically smaller than $\xi^T Y$. Thus $\preceq_{apl}$ concerns only the extreme portfolio losses. In consequence, this order relation is weaker than the (usual) stochastic ordering $\preceq_{st}$ of the portfolio losses:

$$\xi^T X \preceq_{st} \xi^T Y \text{ for all } \xi \in \Sigma^d \text{ implies } X \preceq_{apl} Y.$$

Here, for real random variables U, V the "stochastic ordering" $U \preceq_{st} V$ is defined by

$$P\{U > t\} \leq P\{V > t\} \quad \forall t \in \mathbb{R}.$$

Some related, well-known stochastic orderings (cf. Müller and Stoyan (2002), Shaked and Shanthikumar (1994)) are collected in the following list. Recall that $f : \mathbb{R}^d \to \mathbb{R}$ is called "supermodular" if

$$f(x \wedge y) + f(x \vee y) \geq f(x) + f(y) \quad \forall x, y \in \mathbb{R}^d.$$

Definition 14.3. Let X, Y be random vectors in $\mathbb{R}^d$. Then X is said to be smaller than Y in

(a) "(increasing) convex order", $X \preceq_{cx} Y$ ($X \preceq_{icx} Y$), if $\mathbb{E}f(X) \leq \mathbb{E}f(Y)$ for all (increasing) convex functions $f : \mathbb{R}^d \mapsto \mathbb{R}$ such that the expectations exist;

(b) "linear convex order", $X \preceq_{lcx} Y$, if $\xi^\top X \preceq_{cx} \xi^\top Y$ for all $\xi \in \mathbb{R}^d$;

(c) "positive linear convex order", $X \preceq_{plcx} Y$, if $\xi^\top X \preceq_{cx} \xi^\top Y$ for all $\xi \in \mathbb{R}^d_+$;

(d) "supermodular order" $X \preceq_{sm} Y$, if $\mathbb{E}f(X) \leq \mathbb{E}f(Y)$ for all supermodular functions $f : \mathbb{R}^d \to \mathbb{R}$ such that the expectations exist;

(e) "directionally convex order", $X \preceq_{dcx} Y$, if $\mathbb{E}f(X) \leq \mathbb{E}f(Y)$ for all directionally convex, i.e., supermodular and componentwise convex functions $f : \mathbb{R}^d \to \mathbb{R}$ such that the expectations exist.

The stochastic orderings listed in Definition 14.3 are useful for describing the risk induced by larger diffusion (convex functions) as well as the risk induced by positive dependence (supermodular and directionally convex functions). The following implications hold generally for random vectors X, Y in $\mathbb{R}^d$:

(a) $(X \preceq_{sm} Y) \Rightarrow (X \preceq_{dcx} Y) \Rightarrow (X \preceq_{plcx} Y)$,

(b) $(X \preceq_{cx} Y) \Rightarrow (X \preceq_{lcx} Y) \Rightarrow (X \preceq_{plcx} Y)$,

(c) $(X \preceq_{icx} Y) \Rightarrow (X \preceq_{plcx} Y)$.

Remark 14.4. (a) It is obvious that the usual stochastic order $\preceq_{st}$ implies $\preceq_{apl}$ in the univariate case.

(b) By definition

$$\liminf_{t \to \infty} \left(P\{\xi^\top X > t\} / P\{\xi^\top Y > t\} \right) > 1$$

implies

$$\mathbb{E}(\xi^\top X - t_0)_+ > \mathbb{E}(\xi^\top Y - t_0)_+ \text{ for sufficiently large } t_0.$$

Hence, if Y is integrable and the limit in (14.1) exists, then

$$X \preceq_{plcx} Y \text{ implies } X \preceq_{apl} Y.$$

However, $\preceq_{\text{plcx}}$ is typically stronger and more difficult to verify than $\preceq_{\text{apl}}$. In Section 14.2 we derive specific conditions for $\preceq_{\text{apl}}$ in multivariate regularly varying models which are mostly stated in terms of spectral measures. These conditions will be verified in particular model classes in Section 14.4.

(c) In the non-integrable case the implications of $\preceq_{\text{cx}}$ and $\preceq_{\text{sm}}$ concerning $\preceq_{\text{apl}}$ may be different from the integrable case. For instance, by Tchen's comparison result, the comonotonic dependence structure is the worst case with respect to the supermodular ordering $\preceq_{\text{sm}}$ (see Section 6.1). On the other hand, we show that comonotonicity is the best case with respect to $\preceq_{\text{apl}}$ for non-integrable multivariate regularly varying random vectors in $\mathbb{R}_+^d$ (cf. Theorem 14.16 and Corollary 14.17). $\Diamond$

The following result gives a sufficient criterion for $\preceq_{\text{apl}}$ in multiplicative models $X = RV$ with a non-negative scaling random variable R independent of V. We use this result then to establish $\preceq_{\text{apl}}$ in elliptical and l^1-symmetric models.

Lemma 14.5. *Let R_1, $R_2 \geq 0$ be real random variables and let V be a real random variable independent of R_i, $i = 1, 2$. If $R_1 \preceq_{\text{apl}} R_2$ and $V \leq K$ for some constant K, then*

$$R_1 V \preceq_{\text{apl}} R_2 V.$$

Proof. Since $R_1 V \preceq_{\text{apl}} R_2 V$ is trivial for $V \leq 0$, we assume that $P\{V > 0\} > 0$. Hence $V \leq K$ implies for all $t > 0$

$$P\{R_1 V > t\} = \int_{(0,K)} f(t/v) P\{R_2 > t/v\}\, dP^V(v), \qquad (14.4)$$

where $f(z) := P\{R_1 > z\}/P\{R_2 > z\}$. An obvious consequence of (14.4) is the inequality

$$P\{R_1 V > t\} \leq \sup\{f(z) : z > t/K\} P\{R_2 V > t\}.$$

As $R_1 \preceq_{\text{apl}} R_2$ is equivalent to $\limsup_{z \to \infty} f(z) \leq 1$, we obtain

$$\limsup_{t \to \infty} \frac{P\{R_1 V > t\}}{P\{R_2 V > t\}} \leq 1. \qquad \qquad \square$$

Remark 14.6. (a) The extension of Lemma 14.5 to multivariate multiplicative models $X = RV$ with $R \geq 0$ and bounded V in $\mathbb{R}^d$ is straightforward. Considering $\xi^\top X = R \xi^\top V$ for arbitrary $\xi \in \Sigma^d$, we immediately obtain that

$$R_1 \preceq_{\text{apl}} R_2 \text{ implies } R_1 V \preceq_{\text{apl}} R_2 V. \qquad (14.5)$$

In particular, this setting includes "ℓ^1-symmetric" models, where V is uniformly distributed on the unit simplex Σ^d. These models are closely related to Archimedean copulas. Any Archimedean copula is obtained as a copula of a model in product form RV (cf. McNeil and Nešlehová (2009)). This important result gives an intuitive understanding of Archimedean copulas explaining that these models arise from a simple stochastic model.

(b) A precise tail behaviour of products RV_i for $R \geq 0$ and $V_i, i = 1, 2$ independent is known assuming that R is "regularly varying" with "tail index" $\alpha > 0$, i.e.,

$$\lim_{t \to \infty} \frac{P\{R > tx\}}{P\{R > t\}} = x^{-\alpha}, \quad x > 0. \tag{14.6}$$

Then Breiman's Theorem (cf. Resnick (2007, Proposition 7.5)) states: If $\mathbb{E}(V_i)_+^{\alpha+\varepsilon} < \infty$ for $i = 1, 2$ and some $\varepsilon > 0$, then

$$\lim_{t \to \infty} \frac{P\{RV_i > t\}}{P\{R > t\}} = E\left[(V_i)_+^{\alpha}\right].$$

This yields the precise asymptotics:

$$\lim_{t \to \infty} \frac{P\{RV_1 > t\}}{P\{RV_2 > t\}} = \frac{\mathbb{E}\left[(V_1)_+^{\alpha}\right]}{\mathbb{E}\left[(V_2)_+^{\alpha}\right]}. \tag{14.7}$$

$\Diamond$

While in general multivariate models criteria for the apl-ordering are not obvious it is possible to state for the important class of elliptical distributions sufficient conditions implying $\preceq_{\text{apl}}$ directly. This will be the subject of the following part of this section. A random vector $X \in \mathbb{R}^d$ is called "elliptically distributed", if there exist $\mu \in \mathbb{R}^d$ and a $d \times d$ matrix A such that X has a representation of the form

$$X \stackrel{d}{=} \mu + RAU, \tag{14.8}$$

where U is uniformly distributed on the Euclidean unit sphere $S_2^d := \{x \in \mathbb{R}^d; \|x\|_2 = 1\}$, and R is a non-negative random variable independent of U. By definition we have that $E\|X\|_2^2 < \infty \Leftrightarrow ER^2 < \infty$, and in this case

$$\text{Cov}(X) = \text{Var}(R)AA^{\top}.$$

The matrix $C := AA^{\top}$ is unique except for a constant factor and is also called the "generalized covariance matrix" of X. We denote the elliptical distribution constructed according to (14.8) by $\mathcal{E}(\mu, C, F_R)$, where F_R is the distribution of R. If convenient, we will also use the notation $\mathcal{E}(\mu, C, R)$.

A classical stochastic ordering result going back to Anderson (1955) and Fefferman, Jodeit, and Perlman (1972) (cf. Tong (1980, p. 70)) says that "positive semidefinite ordering" of the generalized covariance matrices

$$C_1 \preceq_{\text{psd}} C_2,$$

defined as

$$\forall \xi \in \mathbb{R}^d \quad \xi^\top C_1 \xi \leq \xi^\top C_2 \xi, \tag{14.9}$$

implies symmetric convex ordering if the location parameter μ and the distribution F_R of the radial factor R are fixed:

$$\mathcal{E}(\mu, C_1, F_R) \preceq_{\text{symmcx}} \mathcal{E}(\mu, C_2, F_R). \tag{14.10}$$

Analogously to Definition 14.3, $X \preceq_{\text{symmcx}} Y$ means that $\mathbb{E} f(X) \leq \mathbb{E} f(Y)$ for all symmetric and convex functions f such that the expectations exist.

A further relevant ordering result for elliptical distributed random vectors $X \sim \mathcal{E}(\mu, C, F_R)$ is the following:

Theorem 14.7 (Ordering of distribution functions).   *The multivariate distribution function $F(x) := P\{X^{(1)} \leq x^{(1)}, \dots, X^{(d)} \leq x^{(d)}\}$ of an elliptical distributed random vector X is increasing in $C_{i,j}$ for $i \neq j$, where $C = (C_{i,j})$ is the generalized covariance matrix (cf. Joe (1997, Theorem 2.21)).*

The following result is concerned with the apl-ordering $\preceq_{\text{apl}}$ for elliptical distributions.

Theorem 14.8 (apl-ordering of elliptical distributions).
Let $X \sim \mathcal{E}(\mu_1, C_1, R_1)$ and $Y \sim \mathcal{E}(\mu_2, C_2, R_2)$. Further, suppose that

$$\mu_1 \leq \mu_2, \quad R_1 \preceq_{\text{apl}} R_2,$$

and

$$\xi^\top C_1 \xi \leq \xi^\top C_2 \xi, \quad \forall \xi \in \Sigma^d. \tag{14.11}$$

Then $X \preceq_{\text{apl}} Y$.

Proof. It suffices to show that $\xi^\top X \preceq_{\text{apl}} \xi^\top Y$ for an arbitrary portfolio $\xi \in \Sigma^d$. Without loss of generality we can assume $\mu_1 = \mu_2 = 0$. Then we have for all $\xi \in \Sigma^d$

$$\xi^\top X \overset{d}{=} R_1 a_1 v_1^\top U \quad \text{and} \quad \xi^\top Y \overset{d}{=} R_2 a_2 v_2^\top U \tag{14.12}$$

with $a_i = a_i(\xi) := (\xi^\top C_i \xi)^{1/2}$ and $v_i = v_i(\xi) := a_i^{-1} A_i^\top \xi$. We can assume $a_i > 0$ for $i = 1, 2$, because otherwise (14.11) implies $0 = a_1 \leq a_2$, and $\xi^\top X \preceq_{\text{apl}} \xi^\top Y$ is trivial.

Since the vectors $v_i = v_i(\xi)$ have unit length by construction, the random variables $v_i^\top U$ are orthogonal projections of $U \sim \text{unif}(S_2^d)$ on vectors of unit length. Symmetry arguments yield that the distribution of $v_i^\top U$ is independent of v_i. This yields

$$\xi^\top X \overset{d}{=} a_1 R_1 U^{(1)} \quad \text{and} \quad \xi^\top Y \overset{d}{=} a_2 R_2 U^{(1)}.$$

By assumption we have that $a_1 \leq a_2$ and $R_1 \preceq_{\text{apl}} R_2$. Applying Lemma 14.5 we obtain $\xi^\top X \preceq_{\text{apl}} \xi^\top Y$. $\square$

Remark 14.9. (a) Condition (14.11) is weaker than (14.9). Let for example $-1 <$
$\varrho_1 < \varrho_2 < 1$ and consider covariance matrices

$$C_i := \begin{pmatrix} 1 & \varrho_i \\ \varrho_i & 1 \end{pmatrix}, \quad i = 1, 2. \tag{14.13}$$

Straightforward calculation shows that C_i satisfy (14.11), but not (14.9).
(b) For subexponentially distributed R_i the assumption $\mu_1 \leq \mu_2$ in Theorem 14.8
can be omitted. In this case one has $X + \mu \preceq_{apl} X$ for any fixed $\mu \in \mathbb{R}^d$. $\Diamond$

14.2 Characterization of $\preceq_{apl}$ in Multivariate Regularly Varying Models

This section is concerned with the characterization of the asymptotic portfolio loss
order $\preceq_{apl}$ in the framework of multivariate regular variation. The results in this
section show the influence of the tail index α and of the spectral measure Ψ on
$\preceq_{apl}$. It is shown that $\preceq_{apl}$ corresponds to a family of order relations on the set of
canonical spectral measures and that these order relations are closely related to the
extreme risk index γ_ξ.

The main result of this section is stated in Theorem 14.15, which gives
sufficient and equivalent criteria for $X \preceq_{apl} Y$ in terms of componentwise ordering
$X^{(i)} \preceq_{apl} Y^{(i)}$ for $i = 1, \dots, d$ and ordering of canonical spectral measures.
Applying the sufficient criterion, we characterize the dependence structures that
yield the best and the worst possible diversification effects for random vectors in $\mathbb{R}^d_+$
(cf. Theorem 14.16 and Corollary 14.17). The equivalent criterion for $\preceq_{apl}$ stated in
Theorem 14.15 allows to derive ordering of canonical spectral measures. This can
be used to transfer $\preceq_{apl}$ from particular models where this ordering is known to
other models with the same canonical spectral measures. We apply this result to the
elliptical distributions.

14.2.1 Multivariate Regular Variation

In this subsection we give a brief account of the basic notions of multivariate regular
variation ($\mathcal{MRV}$), the spectral measure and its canonical standardization. In the
univariate case, the notion of regular variation can be defined separately for the
lower and the upper tail of a random variable via (14.6). In the following we write
$X \in \mathcal{RV}$ or $X \in \mathcal{RV}_{-\alpha}$.

The Notion of Multivariate Regular Variation A random vector X with values in $\mathbb{R}^d$ is called "multivariate regularly varying" if there exist a sequence $a_n \to \infty$ and a (non-zero) Radon measure ν on the Borel σ-field $\mathcal{B}([-\infty, \infty]^d \setminus \{0\})$ such that $\nu([-\infty, \infty]^d \setminus \mathbb{R}^d) = 0$ and, as $n \to \infty$,

$$n\mathrm{P}^{a_n^{-1}X} \xrightarrow{v} \nu \text{ on } \mathcal{B}([-\infty, \infty]^d \setminus \{0\}), \qquad (14.14)$$

where $\xrightarrow{v}$ denotes the "vague convergence" of Radon measures and $\mathrm{P}^{a_n^{-1}X}$ is the probability distribution of $a_n^{-1}X$.

Many popular risk models are $\mathcal{MRV}$. In particular, according to Hult and Lindskog (2002), multivariate regular variation of an elliptical distribution $\mathcal{E}(\mu, C, R)$ is equivalent to the regular variation of the radial factor R. Other examples can be obtained by endowing regularly varying margins $X^{(i)}$ with an appropriate copula as shown in Proposition 14.26. For an account of the notion of multivariate regular variation, vague convergence, and the Borel σ-fields on the punctured space $[-\infty, \infty]^d \setminus \{0\}$ we refer to Resnick (2007, Section 6.1).

It is well known that the limit measure ν in (14.14) is unique except for a constant factor. The limit measure ν has a singularity at the origin in the sense that $\nu((-\varepsilon, \varepsilon)^d) = \infty$ for any $\varepsilon > 0$, and exhibits the "scaling property"

$$\nu(tA) = t^{-\alpha}\nu(A) \qquad (14.15)$$

for all sets $A \in \mathcal{B}([-\infty, \infty]^d \setminus \{0\})$ that are bounded away from 0. The scaling index $\alpha \in (0, \infty)$ is unique for each X and is called the "tail index" of X.

It is also well known that (14.15) implies $\|X\| \in \mathcal{RV}_{-\alpha}$ for any norm $\|\cdot\|$ on $\mathbb{R}^d$. The tail index α is inherited from X. Moreover, the normalizing sequence a_n can be chosen as

$$a_n := F_{\|X\|}^{-1}(1 - 1/n),$$

where $F_{\|X\|}^{-1}$ is the quantile function of $\|X\|$. With this choice the limit measure ν in (14.14) is normalized on the set $A_{\|\cdot\|} := \{x \in \mathbb{R}^d; \|x\| > 1\}$ by

$$\nu(A_{\|\cdot\|}) = 1. \qquad (14.16)$$

Thus, after normalizing ν by (14.16), the scaling relation (14.15) yields an equivalent form of the $\mathcal{MRV}$ condition (14.14) in terms of weak convergence:

$$\mathcal{L}\{t^{-1}X \mid \|X\| > t\} \xrightarrow{w} \nu|_{A_{\|\cdot\|}} \text{ on } \mathcal{B}(A_{\|\cdot\|}) \qquad (14.17)$$

for $t \to \infty$, where $\nu|_{A_{\|\cdot\|}}(B)$ is the restriction of ν to $\mathcal{B}(A_{\|\cdot\|})$.

In addition to (14.14) we assume throughout this section that the limit measure ν is non-degenerate in the following sense:

$$\nu\left(\{x \in \mathbb{R}^d; |x^{(i)}| > 1\}\right) > 0, \quad i = 1, \ldots, d. \qquad (14.18)$$

This assumption ensures that all asset losses $X^{(i)}$ are relevant for the extremes of the portfolio loss $\xi^{\top} X$. If (14.18) is satisfied in the upper tail region, i.e., if

$$v\left(\left\{x \in \mathbb{R}^d;\ x^{(i)} > 1\right\}\right) > 0, \quad i = 1, \ldots, d,$$

then v also characterizes the asymptotic distribution of the componentwise maxima $M_n := (M^{(1)}, \ldots, M^{(d)})$ with $M^{(i)} := \max\left\{X_1^{(i)}, \ldots, X_n^{(i)}\right\}$. By the "central limit theorem for maxima" it holds that for $x \in (0, \infty]^d$:

$$\mathrm{P}\{a_n^{-1} M_n \in [-\infty, x]\} \xrightarrow{w} \exp\left(-v\left([-\infty, \infty]^d \setminus [-\infty, x]\right)\right). \tag{14.19}$$

Therefore v is called the "exponent measure". For more details concerning the asymptotic distributions of componentwise maxima we refer to Resnick (1987, Chapter 5) and de Haan and Ferreira (2006, Chapter 6).

Scaling and Spectral Measure

An important consequence of the scaling property (14.15) is the product representation of v in polar coordinates

$$(r, s) := \tau(x) := (\|x\|, \|x\|^{-1} x)$$

with respect to an arbitrary norm $\|\cdot\|$ on $\mathbb{R}^d$. The induced measure $v^{\tau} := v \circ \tau^{-1}$ satisfies

$$v^{\tau} = c \cdot \varrho_{\alpha} \otimes \Psi \tag{14.20}$$

with the constant factor $c = v(A_{\|\cdot\|}) > 0$. Here the measure ϱ_{α} on $(0, \infty]$ is defined by

$$\varrho_{\alpha}((x, \infty]) := x^{-\alpha}, \quad x \in (0, \infty],$$

and Ψ is a probability measure on the unit sphere $S^d_{\|\cdot\|}$ with respect to $\|\cdot\|$,

$$S^d_{\|\cdot\|} := \left\{s \in \mathbb{R}^d;\ \|s\| = 1\right\}.$$

The measure Ψ is called the "spectral measure" of v or of X. In the special case of $\mathbb{R}^d_+$-valued random vectors X it may be convenient to reduce the domain of Ψ to $S^d_{\|\cdot\|} \cap \mathbb{R}^d_+$. In the following we will write $X \in \mathcal{MRV}_{-\alpha, \Psi}$ for multivariate regular variation of X with tail index α and spectral measure Ψ.

Although the domain of the spectral measure Ψ depends on the norm $\|\cdot\|$ used for the polar coordinates, the representation (14.20) is norm-independent in the following sense: If (14.20) holds for some norm $\|\cdot\|$, then it also holds for any other norm $\|\cdot\|_*$ that is equivalent to $\|\cdot\|$. The tail index α is the same, the factor c equals $v(A_{\|\cdot\|_*})$, and the spectral measure Ψ_* on the unit sphere S^d_* corresponding to $\|\cdot\|_*$ is obtained from Ψ by the following transformation:

$$\Psi_* = \Psi^T, \quad T(s) := \|s\|_*^{-1} s.$$

In the following we will use the sum norm $\|\cdot\|_1$, so that the spectral measures will be defined on $\mathcal{B}(S_1^d)$. Without loss of generality, ν will be normalized by (14.16), which yields $c = 1$ in (14.20). Thus by (14.17) Ψ can be understood as the asymptotic distribution of excess directions on S_1^d. This interpretation is a relevant hint for the modelling with multivariate regular varying models.

$\mathcal{MRV}$, portfolios, and extreme risk index

A useful property of the $\mathcal{MRV}$ notion is that $\mathcal{MRV}$ of the loss vector X is strongly related to the univariate regular variation of linear combinations $\xi^T X$. It is shown in Basrak et al. (2002) that:

$$X \in \mathcal{MRV} \text{ implies existence of } \xi_0 \in \mathbb{R}^d \text{ such that } \xi_0^T X \in \mathcal{RV}_{-\alpha}.$$

Further for any portfolio vector ξ holds:

$$\lim_{t \to \infty} \frac{P\{\xi^T X > t\}}{P\{\xi_0^T X > t\}} = c(\xi, \xi_0) \in [0, \infty). \tag{14.21}$$

In consequence any portfolio loss $\xi^T X$ is regularly varying with tail index α or is asymptotically negligible compared with $\xi_0^T X$, i.e. the limit in (14.21) is zero.

Moreover, for $\mathbb{R}_+^d$-valued random vectors X the following converse implication is true

$$\xi_0^T X \in \mathcal{RV}_{-\alpha} \text{ implies } X \in \mathcal{MRV}. \tag{14.22}$$

This theorem is due to Basrak et al. (2002) and Boman and Lindskog (2009).

Under the assumption that $X \in \mathcal{MRV}$ the "extreme risk index" $\gamma_\xi = \gamma_\xi(X)$ is defined in Chapter 13 as

$$\gamma_\xi(X) = \lim_{t \to \infty} \frac{P\{\xi^T X > t\}}{P\{\|X\|_1 > t\}}. \tag{14.23}$$

In Chapter 13 the random vector X is restricted to $\mathbb{R}_+^d$ and the portfolio vector ξ is restricted to Σ^d. The general case with X in $\mathbb{R}^d$ and possible short positions, i.e., $\xi \in H_1 := \{x \in \mathbb{R}^d; \sum_{i=1}^d x^{(i)} = 1\}$, is considered in Mainik (2010). Normalizing the exponent measure ν by (14.16) and using $\|\cdot\|_1$ there, one obtains that

$$\gamma_\xi(X) = \nu\left(\{x \in \mathbb{R}^d; \xi^T x > 1\}\right). \tag{14.24}$$

Rewriting this representation in terms of the spectral measure Ψ and the tail index α yields

$$\gamma_\xi = \int_{S_1^d} (\xi^T s)_+^\alpha \, d\Psi(s), \quad \forall \xi \in H_1.$$

Denoting the integrand by $f_{\xi,\alpha}$, we will write as in Chapter 13 this representation as $\gamma_\xi = \Psi f_{\xi,\alpha}$ for all $\xi \in \mathbb{R}^d$. The study of diversification effects achievable with d assets and short positions can be reduced to considering $\xi \in H_1$. The definition of the portfolio loss order $\preceq_{\mathrm{apl}}$ excludes unbounded short positions, and restricts ξ to Σ^d. For the ordering of portfolio risks for $\xi \in H_1 \setminus \Sigma^d$ we refer to Mainik (2010, pp. 49–50).

The extreme risk index $\gamma_\xi(X)$ allows to compare the risks of different portfolios within one risk model since

$$\lim_{t\to\infty} \frac{P\{\xi_1^T X > t\}}{P\{\xi_2^T X > t\}} = \frac{\gamma_{\xi_1}(X)}{\gamma_{\xi_2}(X)}.$$

Thus, by construction, ordering of the extreme risk index γ_ξ is closely related to the asymptotic portfolio loss order $\preceq_{\mathrm{apl}}$.

Since the extreme index γ_ξ is designed for the comparison of different portfolio risks within one model, it cannot be directly applied to the comparison of different models. The problem is the standardization by $P\{\|X\|_1 > t\}$ in (14.23). Given $\mathcal{MRV}$ of X and Y, (14.23) yields

$$\limsup_{t\to\infty} \frac{P\{\xi^T X > t\}}{P\{\xi^T Y > t\}} = \frac{\gamma_\xi(X)}{\gamma_\xi(Y)} \limsup_{t\to\infty} \frac{P\{\|X\|_1 > t\}}{P\{\|Y\|_1 > t\}}. \tag{14.25}$$

The lim sup term on the right-hand side depends on the spectral measures of X and Y. Thus the ratio $\gamma_\xi(X)/\gamma_\xi(Y)$ does not carry all information needed to compare portfolio risk asymptotics across models.

In the following we describe the relation between $\mathcal{MRV}$, spectral measures, and $\preceq_{\mathrm{apl}}$. The $\preceq_{\mathrm{apl}}$ ordering can be easily verified for $\mathcal{MRV}$ random vectors with different tail indices and non-degenerate portfolio losses.

Proposition 14.10. *Let X and Y be $\mathcal{MRV}$ random vectors in $\mathbb{R}^d$ and let the extremal risk index $\gamma_\xi(Y) > 0$ for all $\xi \in \Sigma^d$.*

(a) If

$$\lim_{t\to\infty} \frac{P\{\|X\|_1 > t\}}{P\{\|Y\|_1 > t\}} = 0, \tag{14.26}$$

then $X \preceq_{\mathrm{apl}} Y$.
(b) If $\alpha_X > \alpha_Y$, then $X \preceq_{\mathrm{apl}} Y$.

Proof. (a) This is an immediate consequence of (14.26) and (14.25).
(b) Recall that $X \in \mathcal{MRV}_{-\alpha_X, \Psi_X}$ implies $\|X\|_1 \in \mathcal{RV}_{-\alpha_X}$ and similarly $|Y|_1 \in \mathcal{RV}_{-\alpha_Y}$. The condition $\alpha_X > \alpha_Y$ implies (14.26) and by part (14.10) we obtain $X \preceq_{\mathrm{apl}} Y$. □

Thus the primary setting for studying the influence of dependence structures on the ordering of extreme portfolio losses is the case of random variables X and Y with equal tail indices:

$$\alpha_X = \alpha_Y =: \alpha.$$

The Canonical Spectral Measure

Another technical issue arises from the invariance of $\preceq_{\mathrm{apl}}$ under componentwise rescalings. Since the spectral measure Ψ does not exhibit this property, ordering of spectral measures needs additional normalization of margins that makes it consistent with $\preceq_{\mathrm{apl}}$. To solve these problems, we use an alternative representation of γ_ξ in terms of the so-called canonical spectral measure Ψ^*, which has standardized marginal weights.

This representation is closely related to the asymptotic risk aggregation coefficient discussed in Barbe et al. (2006). The link between canonical spectral measures and extreme value copulas allows to transfer ordering results for copulas into the $\preceq_{\mathrm{apl}}$ setting. These results are described in Section 14.3.

In the framework of multivariate regular variation, asymptotic dependence in the tail region is characterized by the spectral measure Ψ and also by its canonical version Ψ^*. The "canonical exponent measure" ν^* of X is obtained from the exponent measure ν as

$$\nu^* = \nu \circ T = \nu^T$$

with the transformation $T : \mathbb{R}^d \to \mathbb{R}^d$ defined by

$$T(x) := \left(T_\alpha \left(\nu(B_1) \cdot x^{(1)} \right), \ldots, T_\alpha \left(\nu(B_d) \cdot x^{(d)} \right) \right), \qquad (14.27)$$

where

$$T_\alpha(t) := \left(t_+^{1/\alpha} - t_-^{1/\alpha} \right) \text{ and } B_i := \{ x \in \mathbb{R}^d ; \ |x^{(i)}| > 1 \}. \qquad (14.28)$$

It is easy to see that $\nu^*(B_i) = 1$ for all i. Furthermore, ν^* exhibits the scaling property

$$\nu^*(tA) = t^{-1} \nu^*(A), \quad t > 0,$$

and, analogously to (14.20), has a product structure in polar coordinates:

$$\nu^* \circ T^{-1} = \varrho_1 \otimes \Psi^*. \qquad (14.29)$$

The measure Ψ^* is the "canonical spectral measure" of X. The normalization $\nu^*(B_i) = 1$ is equivalent to

$$\forall i = 1, \ldots, d \quad \int_{S_1^d} |s^{(i)}| \, d\Psi^*(s) = 1. \qquad (14.30)$$

This standardization of marginal weights entails invariance of both ν^* and Ψ^* under componentwise rescaling. That is, for any $X \in \mathcal{MRV}_{-\alpha, \Psi_X}$ and any $w \in (0, \infty)^d$ we have

$$\Psi^*_{wX} = \Psi^*_X.$$

The product wX is understood componentwise, as in (14.3).

Depending on the application, both Ψ and Ψ^* have their advantages. The canonical standardization of exponent and spectral measures has no equivalent in the general setting, where the components $X^{(i)}$ may have different tail indices. However, in the special case of $\mathcal{MRV}$ models it is rather natural to define the spectral measure according to (14.20). The spectral measure Ψ has an interpretation as the asymptotic distribution of the loss directions on the unit sphere. For random vectors in $\mathbb{R}_+^d$, there are several marginal standardizations that lead to the canonical spectral measure Ψ^*, such as the standardization to Pareto(1) or to Fréchet(1) margins (cf. Klüppelberg and Resnick (2008)). The transformation (14.27) extends the corresponding transformation of spectral measures to the $\mathbb{R}^d$ case. After the standardization to Ψ^*, this information can only be recovered if the tail index α and the marginal component weights $\Psi(B_i)$ are known.

14.2.2   Ordering of Canonical Spectral Measures

As $\preceq_{apl}$ is invariant under componentwise rescalings, the canonical spectral measure Ψ^* is more suitable for the characterization of $\preceq_{apl}$ than Ψ. The following result provides a representation of the extreme risk index γ_ξ in terms of Ψ^*. Derived from the representation $\gamma_\xi = \Psi f_{\xi, \alpha}$, it is valid for $\xi \in \mathbb{R}^d \setminus \{0\}$. Studying $\preceq_{apl}$, we will only need $\xi \in \Sigma^d$.

Proposition 14.11. *Let $X \in \mathcal{MRV}_{-\alpha, \Psi}$. If X satisfies the non-degeneracy condition (14.18), then*

$$\gamma_\xi(X) = \int_{S_1^d} g_{\xi, \alpha}(\nu s) \, d\Psi^*(s), \qquad (14.31)$$

where Ψ^* is the canonical spectral measure of X, the rescaling vector $\nu = (\nu^{(1)}, \dots, \nu^{(d)})$ is defined by

$$\nu^{(i)} := (\gamma_{e_i}(X) + \gamma_{-e_i}(X)),$$

and the function $g_{\xi, \alpha} : \mathbb{R}^d \to \mathbb{R}$ is given by

$$g_{\xi, \alpha}(x) := \left(\sum_{i=1}^d \xi^{(i)} \cdot ((x^{(i)})_+^{1/\alpha} - (x^{(i)})_-^{1/\alpha}) \right)_+^\alpha.$$

Proof. Combining (14.24) with the definition of v^*, we obtain that

$$
\begin{aligned}
\gamma_\xi(X) &= v(\{x \in \mathbb{R}^d;\ \xi^\top x > 1\}) \\
&= v^*\{x \in \mathbb{R}^d;\ \xi^\top T(x) > 1\} \\
&= \int_{S_1^d} \int_{(0,\infty)} 1\{\xi^\top T(rs) > 1\}\, d\varrho_1(r)\, d\Psi^*(s).
\end{aligned}
\tag{14.32}
$$

The definition of $T_\alpha(t)$ in (14.28) implies the scaling property $T_\alpha(rt) = r^{1/\alpha} T_\alpha(t)$ for $r > 0$ and $t \in \mathbb{R}$. Consequently, (14.27) yields

$$
T(rx) = r^{1/\alpha} T(x)
\tag{14.33}
$$

for $r > 0$ and $x \in \mathbb{R}^d$. Applying (14.33)–(14.32), we obtain

$$
\begin{aligned}
\gamma_\xi(X) &= \int_{S_1^d} \int_{(0,\infty)} 1\{\xi^\top T(s) > 0\} 1\{r > (\xi^\top T(s))^{-\alpha}\}\, d\varrho_1(r)\, d\Psi^*(s) \\
&= \int_{S_1^d} 1\{\xi^\top T(s) > 0\}(\xi^\top T(s))^\alpha\, d\Psi^*(s) \\
&= \int_{S_1^d} (\xi^\top T(s))_+^\alpha\, d\Psi^*(s).
\end{aligned}
$$

Finally, for the sets B_i defined in (14.28) it is easy to see that

$$
v(B_i) = \gamma_{e_i}(X) + \gamma_{-e_i}(X) = v^{(i)}.
$$

Hence

$$
(\xi^\top T(s))_+^\alpha = \left(\sum_{i=1}^d \xi^{(i)} \cdot \left(T_\alpha(v^{(i)} s^{(i)}) \right) \right)_+^\alpha = g_{\xi,\alpha}(vs). \qquad \square
$$

As already mentioned above, $\preceq_{\mathrm{apl}}$ and Ψ^* are invariant under rescaling of components. This suggests to start the analysis of the relation between $\preceq_{\mathrm{apl}}$ and canonical spectral measures in a standardized case. A particularly convenient rescaling of the components $X^{(i)}$ is one that makes all marginal weights $v^{(i)} = \gamma_{e_i}(X) + \gamma_{-e_i}(X)$ in Proposition 14.11 standardized by

$$
\lim_{t\to\infty} \frac{P\{|X^{(i)}| > t\}}{P\{|X^{(j)}| > t\}} = 1, \quad \forall i, j \in \{1,\ldots,d\}.
\tag{14.34}
$$

This condition will be referred to as the "balanced tails condition" and is equivalent to $|X^{(i)}| \sim_{\mathrm{apl}} |X^{(j)}|$ for all i, j. The balanced tails condition is the asymptotic counterpart to the assumption of equal margins. It is a special case of tail equivalence cf. Resnick (1987, Section 1.5), where the ratio on the left side in (14.34) is assumed to converge to some $c \in (0, \infty)$. The following result shows that balanced tails significantly simplify the representation (14.31). This will allow us to characterize $\preceq_{apl}$ in terms of ordered canonical spectral measures.

Proposition 14.12. *Let $X \in \mathcal{MRV}_{-\alpha, \Psi}$.*

(a) *If X has balanced tails in the sense of (14.34), then*

$$\frac{\gamma_\xi(X)}{\gamma_{e_1}(X) + \gamma_{-e_1}(X)} = \int_{S_1^d} g_{\xi,\alpha}(s)\, d\Psi^*(s) =: \Psi^* g_{\xi,\alpha}. \qquad (14.35)$$

(b) *The non-degeneracy condition (14.18) is equivalent to the existence of a vector $w \in (0, \infty)^d$ such that wX has balanced tails.*

(c) *The extreme risk index γ_ξ of the rescaled vector wX obtained in part (b) satisfies*

$$\frac{\gamma_\xi(wX)}{\gamma_{e_1}(wX) + \gamma_{-e_1}(wX)} = \Psi_X^* g_{\xi,\alpha}. \qquad (14.36)$$

Proof. (a) Consider the integrand $g_{\xi,\alpha}(vs)$ in the representation (14.31):

$$g_{\xi,\alpha}(vs) = \left(\sum_{i=1}^d \xi^{(i)} \cdot \left((v^{(i)} s^{(i)})_+^{1/\alpha} - (v^{(i)} s^{(i)})_-^{1/\alpha} \right) \right)_+^\alpha.$$

The balanced tails condition (14.34) implies that X is non-degenerate in the sense of (14.18). Furthermore, all weights $v^{(i)}$ in (14.31) are equal:

$$1 = \lim_{t \to \infty} \frac{P\{|X^{(i)}| > t\}/P\{\|X\|_1 > t\}}{P\{|X^{(j)}| > t\}/P\{\|X\|_1 > t\}} = \frac{\gamma_{e_i}(X) + \gamma_{-e_i}(X)}{\gamma_{e_j}(X) + \gamma_{-e_j}(X)}$$

$$= \frac{v^{(i)}}{v^{(j)}}, \quad i, j \in \{1, \dots, d\}.$$

Hence $g_{\xi,\alpha}(vs) = v^{(1)} g_{\xi,\alpha}(s)$ and (14.31) yields

$$\gamma_\xi(X) = (\gamma_{e_1}(X) + \gamma_{-e_1}(X)) \Psi^* g_{\xi,\alpha}.$$

(b) Suppose that X satisfies (14.18). Then the sets B_i defined in (14.28) satisfy $v(B_i) > 0$ for $i = 1, \dots, d$. Hence $|X^{(i)}| \in \mathcal{RV}_{-\alpha}$ for all i, and denoting $w^{(i)} := (v(B_i))^{-1/\alpha}$, we obtain that

$$\lim_{t \to \infty} \frac{P\{w^{(i)}|X^{(i)}| > t\}}{P\{\|X\|_1 > t\}} = \lim_{t \to \infty} \left(\frac{P\{|X^{(i)}| > t/w^{(i)}\}}{P\{|X^{(i)}| > t\}} \cdot \frac{P\{|X^{(i)}| > t\}}{P\{\|X\|_1 > t\}} \right)$$

$$= (w^{(i)})^\alpha \cdot v(B_i)$$

$$= 1$$

for $i = 1, \ldots, d$. That is, for any $i, j \in \{1, \ldots, d\}$ we have

$$\lim_{t \to \infty} \frac{\mathrm{P}\{|w^{(i)}X^{(i)}| > t\}}{\mathrm{P}\{|w^{(j)}X^{(j)}| > t\}} = 1.$$

To prove the inverse implication, suppose that $Z := wX$ has balanced tails for some $w \in (0, \infty)^d$. Then the exponent measure v of X satisfies

$$\frac{v(B_i)}{v(B_1)} = \lim_{t \to \infty} \frac{\mathrm{P}\{|Z^{(i)}| > w^{(i)}t\}}{\mathrm{P}\{|Z^{(1)}| > w^{(1)}t\}}$$

$$= \lim_{t \to \infty} \left(\frac{\mathrm{P}\{|Z^{(i)}| > w^{(i)}t\}}{\mathrm{P}\{|Z^{(i)}| > t\}} \cdot \frac{\mathrm{P}\{|Z^{(1)}| > t\}}{\mathrm{P}\{|Z^{(1)}| > w^{(1)}t\}} \cdot \frac{\mathrm{P}\{|Z^{(i)}| > t\}}{\mathrm{P}\{|Z^{(1)}| > t\}} \right)$$

$$= \left(\frac{w^{(i)}}{w^{(1)}} \right)^{-\alpha} \in (0, \infty), \quad i \in \{1, \ldots, d\}.$$

As $X \in \mathcal{MRV}$ implies $v(B_j) > 0$ for at least one index $j \in \{1, \ldots, d\}$, this yields $v(B_i) > 0$ for all i.

(c) This is an immediate consequence of (a) and the invariance of canonical spectral measures under componentwise rescaling. □

Thus, in the balanced tails setting, representation (14.35) says that the ordering of the normalized extreme risk indices $\gamma_\xi / (\gamma_{e_1} + \gamma_{-e_1})$ uniformly in $\xi \in \Sigma^d$ can be considered as an "integral order relation" for canonical spectral measures with respect to the function class

$$\mathcal{G}_\alpha := \{g_{\xi,\alpha}; \ \xi \in \Sigma^d\}.$$

This ordering of asymptotic dependence structures will be shown to be the key to the characterization of $\preceq_{\mathrm{apl}}$ in $\mathcal{MRV}$ models and suggests the following definition.

Definition 14.13 (Ordering of canonical spectral measures).   Let Ψ^* and Φ^* be canonical spectral measures on S_1^d and let $\alpha > 0$. Then the order relation $\Psi^* \preceq_{\mathcal{G}_\alpha} \Phi^*$ is defined by

$$\Psi^* g \leq \Phi^* g, \quad \forall g \in \mathcal{G}_\alpha.$$

Remark 14.14. By definition $g_{\xi,\alpha}(s) = \xi^\top s$ for $\alpha = 1$ and $\xi \in \Sigma^d$. Hence, if a canonical spectral measure Ψ^* is concentrated on Σ^d, then the standardization (14.30) implies that $\Psi^* g_{\xi,\alpha} = 1$ for all $\xi \in \Sigma^d$. Thus, $\preceq_\mathcal{G}$ is indifferent to spectral measures Ψ^* and Φ^* on Σ^d:

$$\Psi^* \preceq_{\mathcal{G}_1} \Phi^* \quad \text{and} \quad \Phi^* \preceq_{\mathcal{G}_1} \Psi^*.$$

In terms of the asymptotic portfolio loss, this means that the extreme risk index γ_ξ is linear in $\xi \in \Sigma^d$ for $\alpha = 1$ and spectral measures on Σ^d (cf. Lemma 13.2). ◊

The following central theorem provides sufficient and equivalent criteria for $\preceq_{\mathrm{apl}}$ in terms of ordered margins and ordered asymptotic dependence structures. It allows to reduce the $\preceq_{\mathrm{apl}}$ ordering for random vectors to the $\preceq_{\mathrm{apl}}$ ordering for components and to the $\preceq_{\mathcal{G}_\alpha}$ ordering for the canonical spectral measures.

Theorem 14.15 (Ordering of canonical spectral measures and $\preceq_{\mathrm{apl}}$). *Let $X \in \mathcal{MRV}_{-\alpha,\Psi_X}$ and $Y \in \mathcal{MRV}_{-\alpha,\Psi_Y}$. Further, suppose that X and Y satisfy the balanced tails condition* (14.34).

(a) If $|X^{(1)}| \preceq_{\mathrm{apl}} |Y^{(1)}|$, then $\Psi_X^* \preceq_{\mathcal{G}_\alpha} \Psi_Y^*$ implies $X \preceq_{\mathrm{apl}} Y$.
(b) If $|X^{(1)}| \sim_{\mathrm{apl}} |Y^{(1)}|$, then $\Psi_X^* \preceq_{\mathcal{G}_\alpha} \Psi_Y^*$ is equivalent to $X \preceq_{\mathrm{apl}} Y$.

Due to the link between the canonical spectral measures and the copulas of multivariate extreme value distributions, Theorem 14.15 can be used to derive $\preceq_{\mathrm{apl}}$ from corresponding copula ordering results. A detailed discussion of this connection will be given in Section 14.3, and some applications will be given in Section 14.4.

An interesting consequence of characterizing $\preceq_{\mathrm{apl}}$ in terms of spectral measures is that one can derive $\preceq_{\mathcal{G}_\alpha}$ ordering from $\preceq_{\mathrm{apl}}$ ordering in particular models and then conclude $\preceq_{\mathrm{apl}}$ ordering in further models with the same canonical spectral measures and appropriate margins. Further, given explicit representations of spectral measures or their canonical versions, one can verify $\preceq_{\mathrm{apl}}$ numerically by testing $\preceq_{\mathcal{G}_\alpha}$ on a sufficiently fine grid for the integrands $g_{\xi,\alpha}$.

Proof of Theorem 14.15: (a) As X has balanced tails, Proposition 14.12(a) yields

$$\lim_{t \to \infty} \frac{\mathrm{P}\{\xi^\top X > t\}}{\mathrm{P}\{|X^{(1)}| > t\}} = \lim_{t \to \infty} \left(\frac{\mathrm{P}\{\xi^\top X > t\}}{\mathrm{P}\{\|X\|_1 > t\}} \cdot \frac{\mathrm{P}\{\|X\|_1 > t\}}{\mathrm{P}\{|X^{(1)}| > t\}} \right)$$

$$= \frac{\gamma_\xi(X)}{\gamma_{e_1}(X) + \gamma_{-e_1}(X)}$$

$$= \Psi_X^* g_{\xi,\alpha},$$

and, analogously, $\lim_{t \to \infty} \left(\mathrm{P}\{\xi^\top Y > t\}/\mathrm{P}\{|Y^{(1)}| > t\} \right) = \Psi_Y^* g_{\xi,\alpha}$.

Therefore, $\Psi_X^* \preceq_{\mathcal{G}_\alpha} \Psi_Y^*$ and $|X^{(i)}| \preceq_{\mathrm{apl}} |Y^{(i)}|$ imply that

$$\limsup_{t \to \infty} \frac{\mathrm{P}\{\xi^\top X > t\}}{\mathrm{P}\{\xi^\top Y > t\}}$$

$$= \limsup_{t \to \infty} \left(\frac{\mathrm{P}\{\xi^\top X > t\}}{\mathrm{P}\{|X^{(1)}| > t\}} \cdot \frac{\mathrm{P}\{|Y^{(1)}| > t\}}{\mathrm{P}\{\xi^\top Y > t\}} \cdot \frac{\mathrm{P}\{|X^{(1)}| > t\}}{\mathrm{P}\{|Y^{(1)}| > t\}} \right)$$

$$= \frac{\Psi_X^* g_{\xi,\alpha}}{\Psi_Y^* g_{\xi,\alpha}} \cdot \limsup_{t \to \infty} \frac{\mathrm{P}\{|X^{(1)}| > t\}}{\mathrm{P}\{|Y^{(1)}| > t\}}$$

$$\leq 1. \tag{14.37}$$

(b) As $|X^{(1)}| \sim_{\mathrm{apl}} |Y^{(1)}|$ by assumption, (14.2) and (14.37) yield

$$\frac{\Psi_X^* g_{\xi,\alpha}}{\Psi_Y^* g_{\xi,\alpha}} = \limsup_{t\to\infty} \frac{P\{\xi^\top X > t\}}{P\{\xi^\top Y > t\}}.$$

Thus $X \preceq_{\mathrm{apl}} Y$ implies $\Psi_X^* \preceq_{\mathcal{G}_\alpha} \Psi_Y^*$. $\qquad\square$

The following result is a consequence of Theorem 14.15. It answers the question which dependence structures correspond to the best and the worst possible diversification effects for $\mathcal{MRV}$ random vectors in $\mathbb{R}_+^d$. This case is particularly relevant to applications in insurance and operational risk. By Theorem 14.15 it suffices to find the upper and the lower elements with respect to the $\preceq_{\mathcal{G}_\alpha}$-ordering in the set of canonical spectral measures on Σ^d. As expected for $\alpha > 1$ the best diversification effects are obtained in case of asymptotic independence, i.e., the $\preceq_{\mathcal{G}_\alpha}$-minimal element is given by

$$\Psi_0^* := \sum_{i=1}^d \delta_{e_i}, \tag{14.38}$$

whereas the worst diversification effects are obtained in case of the asymptotic comonotonicity, represented by

$$\Psi_1^* := d \cdot \delta_{(1/d,\dots,1/d)}. \tag{14.39}$$

The ordering result for $\alpha < 1$ however is reciprocal. The independence case is here the worst case while the asymptotic comonotone case is the best. In Chapter 13 are given some simulations to this effect.

Theorem 14.16. *Let Ψ^* be a canonical spectral measure on Σ^d and let Ψ_0^* and Ψ_1^* be the canonical spectral measures in (14.38) and (14.39). Then*

(a) $\Psi_0^* \preceq_{\mathcal{G}_\alpha} \Psi^* \preceq_{\mathcal{G}_\alpha} \Psi_1^*$ *for $\alpha \geq 1$;*
(b) $\Psi_1^* \preceq_{\mathcal{G}_\alpha} \Psi^* \preceq_{\mathcal{G}_\alpha} \Psi_0^*$ *for $\alpha \in (0,1]$.*

Proof. Let $X \in \mathcal{MRV}_{-\alpha,\Psi}$. Without loss of generality we assume that X satisfies the balanced tails condition (14.34). Then, according to (14.35), we have

$$\Psi^* g_{\xi,\alpha} = \frac{\gamma_\xi(X)}{\gamma_{e_1}(X)}. \tag{14.40}$$

In particular, this implies $\Psi^* g_{e_i,\alpha} = 1$ for $i = 1,\dots,d$. Furthermore, the mapping $\xi \mapsto \gamma_\xi$ is by Lemma 13.2 convex for $\alpha \geq 1$, and this property is inherited by the mapping $\xi \mapsto \Psi^* g_{\xi,\alpha}$. Thus for $\alpha \geq 1$ we have $\Psi^* g_{\xi,\alpha} \leq 1 = \Psi_1^* g_{\xi,\alpha}$ for all $\xi \in \Sigma^d$, i.e. $\Psi^* \preceq_{\mathcal{G}_\alpha} \Psi_1^*$ for $\alpha \geq 1$.

To complete the proof of part (a), note that the normalization of canonical spectral measures yields

$$\Psi_0^* g_{\xi,\alpha} = \sum_{i=1}^d (\xi^{(i)})^\alpha = \int_{\Sigma^d} \sum_{i=1}^d (\xi^{(i)})^\alpha s^{(i)} \, \Psi^*(ds), \quad \forall \xi \in \Sigma^d. \tag{14.41}$$

Comparing the integrand on the right side of (14.41) with the function $g_{\xi,\alpha}(s) = (\xi^T s^{1/\alpha})^\alpha$, we see that

$$\sum_{i=1}^{d} (\xi^{(i)})^\alpha s^{(i)} = g_{\xi,\alpha}(s) \cdot \sum_{i=1}^{d} z_i^\alpha$$

with

$$z_i := \frac{\xi^{(i)} \cdot (s^{(i)})^{1/\alpha}}{\xi^T s^{1/\alpha}}.$$

Thus it suffices to demonstrate that $\sum_{i=1}^{d} z_i^\alpha \leq 1$, which however follows from $z_i \in [0,1]$, $z_i^\alpha \leq z_i$ for $\alpha \geq 1$, and $\sum_{i=1}^{d} z_i = 1$.

The ordering result for $\alpha \in (0,1]$ stated in (b) follows from the concavity of the mapping $\xi \mapsto \Psi^* g_{\xi,\alpha}$ and the inequality $z_i^\alpha \geq z_i$.   □

Due to Theorem 14.15, an analogue of the foregoing spectral ordering result for $\preceq_{\mathrm{apl}}$ is straightforward.

Corollary 14.17 (Diversification and dependence). *Let X be $\mathcal{MRV}$ in $\mathbb{R}_+^d$ with tail index $\alpha \in (0,\infty)$ and identically distributed margins $X^{(i)} \sim F$, $i = 1,\ldots,d$. Further, let Y be a random vector with independent margins $Y^{(i)} \sim F$, and let Z be a random vector with totally dependent margins $Z^{(i)} = Z^{(1)}$ P-a.s. and $Z^{(1)} \sim F$. Then*

(a) $Y \preceq_{\mathrm{apl}} X \preceq_{\mathrm{apl}} Z$ for $\alpha \geq 1$;
(b) $Z \preceq_{\mathrm{apl}} X \preceq_{\mathrm{apl}} Y$ for $\alpha \in (0,1]$.

Remark 14.18. (a) Related results for the case of equally weighted portfolios can be found in Alink et al. (2004, 2005), Nešlehová et al. (2006b), Barbe et al. (2006), and Embrechts et al. (2009a,b).

(b) The strict assumptions of Corollary 14.17 can be weakened. The independence of $Y^{(i)}$ and the total dependence of $Z^{(i)}$ are needed only in the tail region. That is, it suffices for Y and Z to be $\mathcal{MRV}$ with canonical spectral measures Ψ_0^* and Ψ_1^*, respectively. Furthermore, the assumption of identically distributed margins can be replaced by mutual tail domination i.e., $X^{(i)} \sim_{\mathrm{apl}} Y^{(i)} \sim_{\mathrm{apl}} Z^{(i)}$ for all i. Finally, the non-negativity of $X^{(i)}$, $Y^{(i)}$, and $Z^{(i)}$ is needed only in the asymptotic sense. The ordering results remain true if the canonical spectral measure Ψ_X^* of X satisfies $\Psi_X^*(S_1^d \setminus \Sigma^d) = 0$.   ◊

We next give an application of the interplay of the ordering canonical spectral measures with apl-ordering in the case of elliptical distributions.

Combining Theorem 14.15 with Theorem 14.8 for elliptical distributions, one obtains an ordering result for the canonical spectral measures of multivariate regularly varying elliptical distributions. This implies $X \preceq_{\mathrm{apl}} Y$ for all $X, Y \in \mathcal{MRV}$ with tail index α, balanced tails, ordered margins $|X^{(1)}| \preceq_{\mathrm{apl}} |Y^{(1)}|$, and canonical spectral measures $\Psi_X^* = \Psi^*(\alpha, C_1)$, $\Psi_Y^* = \Psi^*(\alpha, C_2)$ with C_1, C_2

satisfying (14.11). In particular, this includes models of the type $X = X' + Z$, where X' is elliptically distributed and the distortion term Z is asymptotically negligible for large $\|X\|$.

The notation $\Psi^* = \Psi^*(\alpha, C)$ is justified by the fact that spectral measures of elliptical distributions depend only on the tail index α and the generalized covariance matrix C. An explicit representation of spectral densities for bivariate elliptical distributions is obtained by Hult and Lindskog (2002). Alternative representations that are valid for all dimensions $d \geq 2$ are given in Mainik (2010, Lemma 2.8).

Proposition 14.19. *Let C_1 and C_2 be d-dimensional positive definite matrices with equal diagonal elements $C_{1;i,i} = C_{2;i,i}$. If C_1 and C_2 satisfy (14.11), then*

$$\forall \alpha > 0 \quad \Psi^*(\alpha, C_1) \preceq_{\mathcal{G}_\alpha} \Psi^*(\alpha, C_2).$$

Proof. Fix $\alpha \in (0, \infty)$ and consider random vectors

$$X \overset{d}{=} RA_1 U, \quad Y \overset{d}{=} RA_2 U,$$

where $C_i = A_i A_i^\top$ for $i = 1, 2$, and a non-negative random variable $R \in \mathcal{RV}_{-\alpha}$, independent of U.

Theorem 14.8 yields $X \preceq_{\text{apl}} Y$. Further invariance of $\preceq_{\text{apl}}$ under componentwise rescaling implies $wX \preceq_{\text{apl}} wY$ for $w = (w^{(1)}, \dots, w^{(d)})$ with

$$w^{(i)} := C_{1;i,i}^{-1/2} = C_{2;i,i}^{-1/2}, \quad i = 1, \dots, d.$$

Moreover, by arguments as used for the proof of (14.12), one obtains

$$w^{(i)} X^{(i)} \overset{d}{=} w^{(j)} Y^{(j)}, \quad i, j \in \{1, \dots, d\}.$$

Hence the random vectors wX and wY satisfy the balanced tails condition (14.34), whereas their components are mutually ordered with respect to $\preceq_{\text{apl}}$. Finally, Theorem 14.15(b) and invariance of canonical spectral measures under componentwise rescalings yield

$$\Psi^*(\alpha, C_1) = \Psi^*_{wX} \preceq_{\mathcal{G}_\alpha} \Psi^*_{wY} = \Psi^*(\alpha, C_2). \qquad \square$$

14.2.3 Unbalanced Tails

In this final part of Section 14.2 we give an extension of Theorem 14.15 to random vectors with unbalanced tails and a special ordering result for $\alpha = 1$.

If the tails of $Y \in \mathcal{MRV}$ are non-degenerate in the sense of (14.18), then Proposition 14.12(b) gives us a weight vector $w \in (0, \infty)^d$ such that wY has balanced tails. However, the tails of wX are not necessarily balanced. Thus the rescaling invariance of $\preceq_{\mathrm{apl}}$ allows to reduce the general case only to the ordering of a balanced Y and a potentially unbalanced X.

To establish $X \preceq_{\mathrm{apl}} Y$ in this case, it would suffice to find a $v \in (0, \infty)^d$ such that $vX \preceq_{\mathrm{apl}} Y$ and

$$X \preceq_{\mathrm{apl}} vX \quad \text{or} \quad Y \preceq_{\mathrm{apl}} vY. \tag{14.42}$$

If X or Y is restricted to $\mathbb{R}_+^d$, then (14.42) is trivially satisfied for any $v \in [1, \infty)^d$. Indeed, $X \in \mathbb{R}_+^d$ implies $vX - X \in \mathbb{R}_+^d$ and hence $X \preceq_{\mathrm{apl}} vX$. In this case, the natural choice of v for applying Theorem 14.15 to vX and Y is with the smallest $v \in [1, \infty)^d$ that makes the tails of X balanced. That is,

$$v := \left(\min\{u^{(1)}, \ldots, u^{(d)}\} \right)^{-1} u, \tag{14.43}$$

where $u \in (0, \infty)^d$ is such that uX has balanced tails. This yields the following conclusion from Theorem 14.15.

Corollary 14.20. *Let $X \in \mathcal{MRV}_{-\alpha, \Psi_X}$ satisfy equation (14.18) and let $Y \in \mathcal{MRV}_{-\alpha, \Psi_Y}$ have balanced tails.*

(a) If there exists $v \in (0, \infty)^d$ such that vX has balanced tails, $|v^{(1)} X^{(1)}| \preceq_{\mathrm{apl}} |Y^{(1)}|$, and (14.42) is satisfied, then $\Psi_X^* \preceq_{\mathcal{G}_\alpha} \Psi_Y^*$ implies $X \preceq_{\mathrm{apl}} Y$.

(b) If X or Y assumes values in $\mathbb{R}_+^d$ only, and v defined in (14.43) satisfies $|v^{(1)} X^{(1)}| \preceq_{\mathrm{apl}} |Y^{(1)}|$, then $\Psi_X^* \preceq_{\mathcal{G}_\alpha} \Psi_Y^*$ implies $X \preceq_{\mathrm{apl}} Y$.

Remark 14.21. In the general setting without restriction to $\mathbb{R}_+^d$, there is no simple recipe for v like (14.43). The problem is that $v \in [1, \infty)^d$ does not imply $X \preceq_{\mathrm{apl}} vX$ in general. The following counterexample is similar to Embrechts et al. (2009a, Theorem 6.3). It illustrates the problems arising from negative dependence. Consider $X := (Z, -\frac{1}{2}Z)$ where Z is Student-t distributed with $\alpha > 0$ degrees of freedom. The construction rule (14.43) would give $v = (1, 2)$, but $X \preceq_{\mathrm{apl}} vX$ does not hold. We need $\xi^\top X \preceq_{\mathrm{apl}} \xi^\top (vX)$ for all $\xi \in \Sigma^2$. As $\xi^{(2)} = 1 - \xi^{(1)}$, it is easy to see that

$$\xi^\top X \stackrel{d}{=} \left| \frac{3}{2} \xi^{(1)} - \frac{1}{2} \right| Z \quad \text{and} \quad \xi^\top (vX) \stackrel{d}{=} \left| 2 \xi^{(1)} - 1 \right| Z.$$

Hence $X \preceq_{\mathrm{apl}} vX$ requires $|\frac{3}{2} \xi^{(1)} - \frac{1}{2}| \leq |2 \xi^{(1)} - 1|$ for all $\xi^{(1)} \in [0, 1]$, which is obviously not true. The random vectors X and $Y := vX$ satisfy all conditions of Corollary 14.20(b) except the restriction on $\mathbb{R}_+^d$. Although $\Psi_Y^* = \Psi_{vX}^* = \Psi_X^*$, and hence $\Psi_Y^* \preceq_{\mathcal{G}_\alpha} \Psi_X^*$ for all $\alpha > 0$, we do not have $X \preceq_{\mathrm{apl}} Y$. This suggests that in the most general case $\preceq_{\mathrm{apl}}$ cannot be reduced to $\preceq_{\mathrm{apl}}$ for the components and the ordering of canonical spectral measures with respect to $\preceq_{\mathcal{G}_\alpha}$. Furthermore, the choice of v that gives (14.43) seems to involve the non-canonical spectral measures Ψ_X and Ψ_Y.   ◊

The final result of this section is based on the indifference of $\preceq_{\mathcal{G}_\alpha}$ for $\alpha = 1$ mentioned in Remark 14.14. This special property of spectral measures on Σ^d allows to reduce $\preceq_{apl}$ to the ordering of components. It cannot be extended to spectral measures on S_1^d as shown in Mainik (2010, Lemma 5.22).

Lemma 14.22. *Let X and Y be $\mathcal{MRV}$ in $\mathbb{R}_+^d$ with tail index $\alpha = 1$. Further, suppose that Y satisfies (14.18) and that $X^{(i)} \preceq_{apl} Y^{(i)}$ for $i = 1, \ldots, d$. Then $X \preceq_{apl} Y$.*

14.3 Relations to the Convex and Supermodular Order

In this section we consider the relations between $\preceq_{apl}$-ordering and some other well-known ordering notions under the $\mathcal{MRV}$ assumption. Applying the relationship between $\preceq_{apl}$ and the ordering of canonical spectral measures by $\preceq_{\mathcal{G}_\alpha}$ established in Theorem 14.15, we obtain criteria for $\preceq_{apl}$ in terms of the dependence orders $\preceq_{sm}, \preceq_{dcx}$ and convexity orders $\preceq_{cx}, \preceq_{icx}, \preceq_{plcx}$. Theorem 14.23 entails a collection of sufficient criteria for $\preceq_{apl}$ in terms of convex and supermodular order relations. In particular, we see the inversion of diversification effects for $\alpha < 1$. An application to copula based models is given in Proposition 14.26.

The effect of $\preceq_{sm}$ on the ordering of canonical spectral measures is applied in Embrechts et al. (2009b) to the ordering of risks for the portfolio vector $\xi = (1, \ldots, 1)$ in a specific family of $\mathcal{MRV}$ models with identically distributed, non-negative margins $X^{(i)}$ (cf. Example 14.28 in Section 14.4). The canonical spectral measure of that model corresponds to the Galambos copula.

The arguments in that paper apply in a similar way to general $\mathcal{MRV}$ random vectors in $\mathbb{R}^d$ with balanced tails and tail index $\alpha \neq 1$. The case $\alpha = 1$ is excluded for two reasons. First, this case is partly trivial due to the indifference of $\preceq_{\mathcal{G}_\alpha}$ for spectral measures on Σ^d (cf. Remark 14.14 and Lemma 14.22). Second, Karamata's theorem used in the proof of the integrable case $\alpha > 1$ does not yield the desired result for random variables with tail index $\alpha = 1$.

The result of Theorem 14.23 is stated in terms of the $\preceq_{\mathcal{G}_\alpha}$-ordering, and the conditions are formulated directly in terms of expectations of the portfolio losses $\xi^\top X$. This emphasizes that strong order relations such as $\preceq_{sm}$ or $\preceq_{cx}$ are far from necessary for $\preceq_{apl}$. Furthermore, strong ordering results need not be available in the particular model of interest. However, in some cases, one can switch to some idealized model that has the same canonical spectral measure as the original one and exhibits additional properties that imply the $\preceq_{\mathcal{G}_\alpha}$-ordering. Then one may obtain $\preceq_{apl}$ in the original model from Theorem 14.15.

Theorem 14.23. *Let X and Y be $\mathcal{MRV}$ in $\mathbb{R}^d$ with identical tail index $\alpha \neq 1$. Further, assume that X and Y satisfy the balanced tails condition (14.34).*

(a) If $\alpha > 1$,

$$\limsup_{t \to \infty} \frac{P\{|X^{(1)}| > t\}}{P\{|Y^{(1)}| > t\}} = 1, \tag{14.44}$$

and

$$\limsup_{u \to \infty} \frac{\mathbb{E}\big[(\xi^\top X - u)_+\big]}{\mathbb{E}\big[(\xi^\top Y - u)_+\big]} \le 1, \quad \forall \xi \in \Sigma^d \tag{14.45}$$

then $\Psi_X^* \preceq_{\mathcal{G}_\alpha} \Psi_Y^*$.
(b) If $\alpha < 1$, $|X^{(1)}| \sim_{\mathrm{apl}} |Y^{(1)}|$, and

$$\limsup_{u \to \infty} \frac{\mathbb{E}\big[(\xi^\top X)_+ \wedge u\big]}{\mathbb{E}\big[(\xi^\top Y)_+ \wedge u\big]} \ge 1, \quad \forall \xi \in \Sigma^d \tag{14.46}$$

then $\Psi_Y^* \preceq_{\mathcal{G}_\alpha} \Psi_X^*$.

The proof of this result will be given after some remarks and conclusions. Part (a) generalizes the intuitively obvious arguments presented in Remark 14.4(b) to the case when the limit in (14.2) does not exist. Part (b) treats the case $\mathbb{E}(\xi^\top X)_+ = \infty$, which is not covered by Remark 14.4(b).

The relation between $\preceq_{\mathcal{G}_\alpha}$ and $\preceq_{\mathrm{apl}}$ established in Theorem 14.15 immediately yields the following corollary.

Corollary 14.24. *(a) If random vectors X and Y satisfy the conditions of Theorem 14.23(a), then $X \preceq_{\mathrm{apl}} Y$;*
(a) If X and Y satisfy the conditions of Theorem 14.23(b), then $Y \preceq_{\mathrm{apl}} X$.

The conditions (14.45) and (14.46) are asymptotic forms of the increasing convex ordering $\xi^\top X \preceq_{\mathrm{icx}} \xi^\top Y$ and the decreasing convex ordering $(\xi^\top X)_+ \preceq_{\mathrm{decx}} (\xi^\top Y)_+$, respectively. Some comments are given in the following remark.

Remark 14.25. (a) The following criteria are sufficient for (14.45) and (14.46) to hold:

(i) $(\xi^\top X)_+ \preceq_{\mathrm{cx}} (\xi^\top Y)_+$ for all $\xi \in \Sigma^d$;
(ii) X and Y are restricted to $\mathbb{R}_+^d$ and $X \preceq Y$ with $\preceq$ denoting either $\preceq_{\mathrm{plcx}}$, $\preceq_{\mathrm{lcx}}$, $\preceq_{\mathrm{cx}}$, $\preceq_{\mathrm{dcx}}$, or $\preceq_{\mathrm{sm}}$.

(b) Additionally, condition (14.45) follows from $X \preceq Y$ with $\preceq$ denoting either $\preceq_{\mathrm{plcx}}$, $\preceq_{\mathrm{lcx}}$, $\preceq_{\mathrm{cx}}$, $\preceq_{\mathrm{dcx}}$, or $\preceq_{\mathrm{sm}}$.
(c) The final comment to Theorem 14.23 concerns convex ordering of non-integrable random variables and diversification for $\alpha < 1$. The occurrence of negative diversification effects for $\alpha < 1$ demonstrates that the implications of convex ordering are essentially different for integrable and non-integrable random variables. If a random variable Z in $\mathbb{R}$ satisfies $\mathbb{E}[Z_+] = \mathbb{E}[Z_-] = \infty$, then the only integrable convex functions of Z are the constant ones thus convex ordering does not make sense. Moreover, if Z is restricted to $\mathbb{R}_+$ and $\mathbb{E}Z = \infty$, then any integrable convex function of Z is necessarily non-increasing. ◇

Proof of Theorem 14.23: (a) Denote $h_u(t) := (t - u)_+$. Then, for $u > 0$, we have

$$\frac{1}{u}\mathbb{E}h_u(\xi^\top X) = \int_{(1,\infty)} P\{\xi^\top X > tu\}\, dt$$

and, as a consequence,

$$\frac{u^{-1}\mathbb{E}h_u(\xi^\top X)}{P\{|X^{(1)}| > u\}} = \frac{P\{\xi^\top X > u\}}{P\{|X^{(1)}| > u\}} \int_{(1,\infty)} \frac{P\{\xi^\top X > tu\}}{P\{\xi^\top X > u\}}\, dt.$$

Proposition 14.12(a) implies

$$\lim_{u\to\infty} \frac{P\{\xi^\top X > u\}}{P\{|X^{(1)}| > u\}} = \frac{\gamma_\xi(X)}{\gamma_{e_1}(X) + \gamma_{-e_1}(X)} = \Psi_X^* g_{\xi,\alpha} \tag{14.47}$$

and Karamata's Theorem (see Bingham et al. (1987, Theorem 1.5.11(ii))) yields

$$\lim_{u\to\infty} \int_{(1,\infty)} \frac{P\{\xi^\top X > tu\}}{P\{\xi^\top X > u\}}\, dt = \int_{(1,\infty)} t^{-\alpha}\, dt = \frac{1}{\alpha - 1}.$$

As a result one obtains

$$\lim_{u\to\infty} \frac{u^{-1}\mathbb{E}h_u(\xi^\top X)}{P\{|X^{(1)}| > u\}} = \frac{1}{\alpha - 1}\Psi_X^* g_{\xi,\alpha}$$

and, analogously,

$$\lim_{u\to\infty} \frac{u^{-1}\mathbb{E}h_u(\xi^\top Y)}{P\{|Y^{(1)}| > u\}} = \frac{1}{\alpha - 1}\Psi_Y^* g_{\xi,\alpha}.$$

Hence conditions (14.45) and (14.44) yield

$$1 \geq \limsup_{u\to\infty} \frac{u^{-1}\mathbb{E}h_u(\xi^\top X)}{u^{-1}\mathbb{E}h_u(\xi^\top Y)}$$

$$= \limsup_{u\to\infty} \left(\frac{u^{-1}\mathbb{E}h_u(\xi^\top X)}{P\{|X^{(1)}| > u\}} \cdot \frac{P\{|Y^{(1)}| > u\}}{u^{-1}\mathbb{E}h_u(\xi^\top Y)} \cdot \frac{P\{|X^{(1)}| > u\}}{P\{|Y^{(1)}| > u\}} \right)$$

$$= \frac{\Psi_X^* g_{\xi,\alpha}}{\Psi_Y^* g_{\xi,\alpha}}$$

for all $\xi \in \Sigma^d$. This is exactly $\Psi_X^* \preceq_{\mathcal{G}_\alpha} \Psi_Y^*$.

(b) Assume that $\Psi_Y^* \preceq_{\mathcal{G}_\alpha} \Psi_X^*$ is not satisfied, i.e., there exists $\xi \in \Sigma^d$ such that

$$\varepsilon := \Psi_Y^* g_{\xi,\alpha} - \Psi_X^* g_{\xi,\alpha} > 0. \tag{14.48}$$

Denoting

$$\delta := \frac{\varepsilon}{2(\Psi_X^* g_{\xi,\alpha} + \varepsilon)}, \tag{14.49}$$

we have that $\delta \in (0, 1/2]$, and condition (14.46) implies that there exists an increasing, non-negative sequence $u_n \to \infty$ such that

$$\mathbb{E}\left[(\xi^T X)_+ \wedge u_n\right] \geq (1 - \delta)\, \mathbb{E}\left[(\xi^T Y)_+ \wedge u_n\right], \quad \forall n \in \mathbb{N}. \tag{14.50}$$

Furthermore, it is easy to see that any random variable Z in $\mathbb{R}_+$ satisfies

$$\mathbb{E}\left[Z \wedge u_{n+m}\right] = \mathbb{E}\left[Z \wedge u_n\right] + \int_{(u_n, u_{n+m})} \mathrm{P}\{Z > t\}\, dt, \quad \forall n, m \in \mathbb{N}.$$

Hence, replacing u_n by u_{n+m} in (14.50), we obtain that

$$\mathbb{E}\left[(\xi^T X)_+ \wedge u_n\right] - (1 - \delta)\, \mathbb{E}\left[(\xi^T Y)_+ \wedge u_n\right] \geq I(n, m), \quad \forall n, m \in \mathbb{N}, \tag{14.51}$$

where

$$\begin{aligned}
I(n, m) &:= \int_{(u_n, u_{n+m})} \left((1 - \delta)\mathrm{P}\{\xi^T Y > t\} - \mathrm{P}\{\xi^T X > t\}\right) dt \\
&= \int_{(u_n, u_{n+m})} \varphi(t) \cdot \mathrm{P}\{|X^{(1)}| > t\}\, dt
\end{aligned}$$

with

$$\varphi(t) := \frac{(1 - \delta)\mathrm{P}\{\xi^T Y > t\} - \mathrm{P}\{\xi^T X > t\}}{\mathrm{P}\{|X^{(1)}| > t\}}.$$

According to (14.2), $|X^{(1)}| \sim_{\mathrm{apl}} |Y^{(1)}|$ is equivalent to

$$\lim_{t \to \infty} \frac{\mathrm{P}\{|X^{(1)}| > t\}}{\mathrm{P}\{|Y^{(1)}| > t\}} = 1.$$

Hence, applying (14.47) to X and Y, we obtain that

$$\begin{aligned}
\varphi(t) &= (1 - \delta)\frac{\mathrm{P}\{\xi^T Y > t\}}{\mathrm{P}\{|Y^{(1)}| > t\}} \cdot \frac{\mathrm{P}\{|Y^{(1)}| > t\}}{\mathrm{P}\{|X^{(1)}| > t\}} - \frac{\mathrm{P}\{\xi^T X > t\}}{\mathrm{P}\{|X^{(1)}| > t\}} \\
&\to (1 - \delta)\Psi_Y^* g_{\xi,\alpha} - \Psi_X^* g_{\xi,\alpha}, \quad t \to \infty.
\end{aligned}$$

Furthermore, (14.48) and (14.49) yield

$$(1 - \delta)\Psi_Y^* g_{\xi,\alpha} - \Psi_X^* g_{\xi,\alpha} = (1 - \delta)(\Psi_Y^* g_{\xi,\alpha} - \Psi_X^* g_{\xi,\alpha}) - \delta\Psi_X^* g_{\xi,\alpha} = \frac{\varepsilon}{2}.$$

This implies

$$I(n,m) \geq \frac{\varepsilon}{4} \int_{(u_n, u_{n+m})} \mathrm{P}\{|X^{(1)}| > t\} \, dt \tag{14.52}$$

for sufficiently large n and any $m \in \mathbb{N}$. As $|X^{(1)}| \in \mathcal{RV}_{-\alpha}$ with $\alpha < 1$ implies $\mathbb{E}|X^{(1)}| = \infty$, the integral on the right side of (14.52) tends to infinity for $m \to \infty$:

$$\lim_{m \to \infty} \int_{(u_n, u_{n+m})} \mathrm{P}\{|X^{(1)}| > t\} \, dt = \infty, \quad \forall n > 0.$$

Hence, choosing m sufficiently large, one can achieve $I(n,m) > c$ for any $c \in \mathbb{R}$. In particular, n and m can be chosen such that

$$I(n,m) > \mathbb{E}\left[(\xi^\top X)_+ \wedge u_n\right] - (1-\delta)\,\mathbb{E}\left[(\xi^\top Y)_+ \wedge u_n\right],$$

which contradicts (14.51). Thus (14.48) cannot be true and we necessarily have $\Psi_Y^* \preceq_{\mathcal{G}_\alpha} \Psi_X^*$. $\qquad\qquad \square$

Now we return to the ordering criterion in terms of the supermodular order $\preceq_{\mathrm{sm}}$ stated in Remark 14.25. The invariance of $\preceq_{\mathrm{sm}}$ under non-decreasing component transformations allows to transfer these criteria to copula models. Furthermore, since we are interested in the ordering of the asymptotic dependence structures represented by the canonical spectral measures, Ψ_1^* and Ψ_2^*, we can take any copulas that yield Ψ_1^* and Ψ_2^* as asymptotic dependence structures.

A natural choice of such copulas is given by the "extreme value copulas", defined as the copulas of "simple max-stable distributions" corresponding to Ψ_i^*, i.e., the distributions

$$G_i^*(x) := \exp(-v_i^*(-[\infty, x]^c)), \quad x \in \mathbb{R}_+^d \tag{14.53}$$

where v_i^* is the canonical exponent associated with Ψ_i^* via (14.29). For further details on max-stable distributions we refer to Resnick (1987, Section 5.4). Since extreme value copulas and canonical spectral measures can be regarded as alternative parametrisations of the same asymptotic dependence structures, we can start with supermodular ordering of copulas and derive $\preceq_{\mathcal{G}_\alpha}$ for the corresponding canonical spectral measures. This is particularly useful as it yields $\preceq_{\mathrm{apl}}$ for models with involved spectral measures, but not necessarily identical copulas.

Proposition 14.26. *Let Ψ_1^* and Ψ_2^* be canonical spectral measures on Σ^d. Further, for $i = 1, 2$, let C_i denote the copula of the simple max-stable distribution G_i^* induced by Ψ_i^* according to (14.53) and (14.29). Then $C_1 \preceq_{\mathrm{sm}} C_2$ implies*

(a) $\Psi_1^* \preceq_{\mathcal{G}_\alpha} \Psi_2^*$ *for $\alpha \in (1, \infty)$;*
(b) $\Psi_2^* \preceq_{\mathcal{G}_\alpha} \Psi_1^*$ *for $\alpha \in (0, 1)$.*

Proof. Let v_i^* denote the canonical exponent measures corresponding to Ψ_i^* and G_i^*. It is easy to see that the transformed measures

$$v_{\alpha,i} := v_i^* \circ T^{-1}, \quad i = 1, 2,$$

with $\alpha > 0$ and $T(x) := \left((x^{(i)})^{1/\alpha}, \ldots, (x^{(d)})^{1/\alpha}\right)$ have the scaling property described in (14.15). Hence the transformed distributions

$$G_{\alpha,i}(x) := G_i^* \circ T^{-1}(x) = \exp\left(-\nu_{\alpha,i}([0,x]^c)\right)$$

are max-stable with exponent measures $\nu_{\alpha,i}$.

It is well known that max-stable distributions with identical heavy-tailed margins are $\mathcal{MRV}$ (cf. Resnick (1987, Section 5.4.2)). Moreover, the limit measure ν in the $\mathcal{MRV}$ condition can be chosen equal to the exponential measure associated with the property of max-stability. Consequently, the probability distributions $G_{\alpha,i}$ for $i = 1, 2$ and $\alpha > 0$ are $\mathcal{MRV}$ with tail index α and canonical spectral measures Ψ_i^*.

It is easy to see that $X \sim G_{\alpha,1}$ and $Y \sim G_{\alpha,2}$ have identical margins:

$$X^{(i)} \overset{d}{=} Y^{(j)}, \quad i, j \in \{1, \ldots, d\}.$$

Moreover, due to the invariance of $\preceq_{sm}$ under non-decreasing marginal transformations, $C_1 \preceq_{sm} C_2$ implies

$$G_{\alpha,1} \preceq_{sm} G_{\alpha,2}$$

for all $\alpha > 0$. Thus an application of the ordering criteria from Remark 14.25 to $X \sim G_{\alpha,1}$ and $Y \sim G_{\alpha,2}$ completes the proof. $\square$

14.4   Examples of apl-Ordering

In this final section we illustrate the results established in Sections 14.2 and 14.3 by a series of examples with parametric models. Examples 14.27 and 14.28 demonstrate application of Proposition 14.26 to copula based models and the inversion of diversification effects for random vectors in $\mathbb{R}_+^d$ with tail index $\alpha < 1$. The fact that inverse diversification effects do not necessarily appear in the general case with gains and losses is demonstrated by multivariate Student-t distributions in Example 14.29.

Example 14.27. The family of Gumbel copulas is given by

$$C_\vartheta(u) := \exp\left(-\left(\sum_{i=1}^d \left(-\log u^{(i)}\right)^\vartheta\right)^{1/\vartheta}\right), \quad \vartheta \in [1, \infty). \tag{14.54}$$

Gumbel copulas are extreme value copulas, i.e., they are copulas of simple max-stable distributions. According to Wei and Hu (2002), Gumbel copulas with dependence parameter $\vartheta \in [1, \infty)$ are ordered by $\preceq_{sm}$:

$$\vartheta_1 \leq \vartheta_2 \Rightarrow C_{\vartheta_1} \preceq_{sm} C_{\vartheta_2}, \quad \forall \vartheta_1, \vartheta_2 \in [1, \infty).$$

Consequently, Proposition 14.26 applies to the family of canonical spectral measures Ψ_ϑ^* corresponding to the Gumbel copulas C_ϑ. Thus

$$1 \le \vartheta_1 \le \vartheta_2 < \infty \text{ implies } \Psi_{\vartheta_1}^* \preceq_{\mathcal{G}_\alpha} \Psi_{\vartheta_2}^* \text{ for } \alpha > 1. \qquad (14.55)$$

The diversification effects become inverse for $\alpha \in (0, 1)$. In this case one has

$$\vartheta_1 < \vartheta_2 \text{ implies } \Psi_{\vartheta_2}^* \preceq_{\mathcal{G}_\alpha} \Psi_{\vartheta_1}^*. \qquad (14.56)$$

Applying Theorem 14.15, one obtains ordering with respect to $\preceq_{\mathrm{apl}}$ for random vectors X and Y in $\mathbb{R}_+^d$ that are $\mathcal{MRV}$ with canonical spectral measures of Gumbel type and have balanced tails ordered by $\preceq_{\mathrm{apl}}$. In particular, this is the case if X and Y have identical regularly varying marginal distributions and Archimedean copulas that satisfy an appropriate regularity condition (cf. Genest and Rivest (1989), Barbe et al. (2006)). This condition is stated in terms of the generator φ of the Archimedean copula

$$C(u) = \varphi^{-1}\left(\sum_{i=1}^d \varphi(u^{(i)})\right).$$

If the function $t \mapsto \varphi(1 - t)$ is regularly varying at 0 with index ϑ for $\vartheta > 1$, then the corresponding extreme value copula is the Gumbel copula with parameter ϑ. Moreover, $\mathcal{MRV}$ random vectors with Archimedean copulas can only induce extreme value copulas of Gumbel type (cf. Genest and Rivest (1989)).

Figure 14.1 illustrates the resulting diversification effects in the bivariate case, including indifference to portfolio diversification for $\alpha = 1$ and the inversion of diversification effects occurring when α crosses this critical value. The graphics show the function $\xi^{(1)} \mapsto \Psi_\vartheta^* g_{\xi,\alpha}$ for selected values of ϑ and α. Due to $X \in \mathbb{R}_+^d$, the representation $\Psi_\vartheta^* g_{\xi,\alpha} = \gamma_\xi/(\gamma_{e_1} + \gamma_{-e_1})$ simplifies to $\Psi_\vartheta^* g_{\xi,\alpha} = \gamma_\xi/\gamma_{e_1}$ and therefore

$$\Psi_\vartheta^* g_{e_1,\alpha} = \Psi_\vartheta^* g_{e_2,\alpha} = 1. \qquad \Diamond$$

Example 14.28. Another family of extreme value copulas that are ordered by $\preceq_{\mathrm{sm}}$ is the family of "Galambos copulas" with parameter $\vartheta \in (0, \infty)$:

$$C_\vartheta(u) := \exp\left(\sum_{I \subset \{1,\dots,d\}} (-1)^{|I|}\left(\sum_{i \in I}\left(-\log u^{(i)}\right)^{-\vartheta}\right)^{-1/\vartheta}\right). \qquad (14.57)$$

According to Wei and Hu (2002), $\vartheta_1 \le \vartheta_2$ implies $C_{\vartheta_1} \preceq_{\mathrm{sm}} C_{\vartheta_2}$. Thus Proposition 14.26 yields ordering of the corresponding canonical spectral measures Ψ_ϑ^* with respect to $\preceq_{\mathcal{G}_\alpha}$:

$$\vartheta_1 \le \vartheta_2 \text{ implies } \Psi_{\vartheta_1}^* \preceq_{\mathcal{G}_\alpha} \Psi_{\vartheta_2}^* \text{ for } \alpha > 1 \text{ and } \Psi_{\vartheta_2}^* \preceq_{\mathcal{G}_\alpha} \Psi_{\vartheta_1}^* \text{ for } \alpha \in (0, 1). \quad (14.58)$$

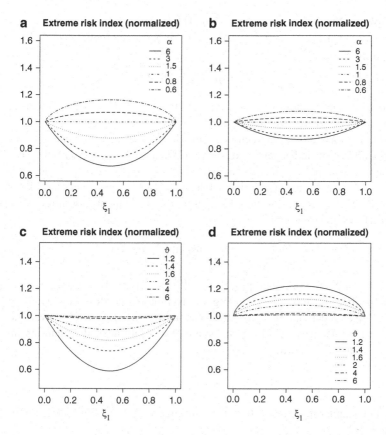

Figure 14.1 Bivariate Gumbel copulas: Diversification effects represented by functions $\xi^{(1)} \mapsto \Psi_\vartheta^* g_{\xi,\alpha}$ for selected values of ϑ and α. (a) Varying α for $\vartheta = 1.4$. (b) Varying α for $\vartheta = 2$. (c) Varying ϑ for $\alpha = 3$. (d) Varying ϑ for $\alpha = 0.6$ (Published as Figure 5.1 in *Statistics & Risk Modeling*, 29: 73–105, 2012)

Galambos copulas correspond to the canonical exponent measures of random vectors X in $\mathbb{R}_+^d$ with identically distributed regularly varying margins $X^{(i)}$ and dependence structure of $-X$ given by an Archimedean copula. If the Archimedean generator φ is regularly varying at 0 with index $-\vartheta$ for $\vartheta > 0$, then the extreme value copula of X is the Galambos copula with parameter ϑ (cf. Barbe et al. (2006)). Models of this type were discussed in studies of aggregation effects for extreme risks (cf. Alink et al. (2004, 2005), Nešlehová et al. (2006b), Barbe et al. (2006), Embrechts et al. (2009a,b)).   ◊

The final example illustrates results established in Proposition 14.19 and Theorem 14.8. In particular, it shows that elliptical distributions do not exhibit inverse diversification effects for any tail index α.

Example 14.29. A random vector X in $\mathbb{R}^d$ is multivariate Student-t distributed with location parameter μ, $\alpha > 0$ degrees of freedom, and (generalized) covariance matrix C, if

$$X \stackrel{d}{=} \mu + \sqrt{\alpha/W}\, Z$$

with independent random variables $Z \sim \mathcal{N}(0, C)$ and $W \sim \chi_\alpha^2$. It is well known that X is elliptically distributed and $\mathcal{MRV}$ with tail index α.

Let the bivariate random vectors X and Y be Student-t distributed with $\alpha > 0$ degrees of freedom and generalized covariance matrices $C_1 = C(\varrho_1)$ and $C_2 = C(\varrho_2)$, respectively, where $C(\varrho_i)$ is defined in (14.2). If $\varrho_1 < \varrho_2$, then, according to Remark 14.9(a), C_1 and C_2 satisfy condition (14.11), and Theorem 14.8 yields $X \preceq_{\text{apl}} Y$. Moreover, Proposition 14.19 implies an identical ordering of diversification effects in the sense that

$$\Psi_{\alpha,\varrho_1}^* \preceq_{\mathcal{G}_\alpha} \Psi_{\alpha,\varrho_2}^* \text{ for all } \alpha \in (0, \infty).   \tag{14.59}$$

Figure 14.2 shows functions $\xi^{(1)} \mapsto \Psi_{\alpha,\varrho}^* g_{\xi,\alpha}$ for selected parameter values ϱ and α that illustrate the ordering of asymptotic portfolio losses by ϱ and positive diversification effects for all α. The indifference to portfolio diversification for $\alpha = 1$ is also absent. Moreover, symmetry of elliptical distributions implies $\gamma_{-e_1} = \gamma_{e_1}$ and, as a result,

$$\Psi_{\alpha,\varrho}^* g_{e_1,\alpha} = \Psi_{\alpha,\varrho}^* g_{e_2,\alpha} = 1/2.$$

Thus the standardization of the plots in Figure 14.2 is different from that in Figure 14.1.

$\Diamond$

Remark 14.30. Experience of several examples seems to suggest that the diversification coefficient $\Psi^* g_{\xi,\alpha}$ is decreasing in α. This means that risk diversification is stronger for lighter component tails than for heavier ones. The influence of the tail index α on the risk coefficient aggregation is different from that. The asymptotic risk aggregation coefficient

$$q_d := \lim_{t \to \infty} \frac{P\{X^{(1)} + \ldots + X^{(d)} > t\}}{P\{X^{(1)} > t\}}   \tag{14.60}$$

studied in Alink et al. (2004, 2005), Nešlehová et al. (2006b), Barbe et al. (2006), and Embrechts et al. (2009a,b) is known to be increasing in α when the loss components $X^{(i)}$ are non-negative (cf. Barbe et al. (2006)). It is easy to see that for non-negative $X^{(i)}$

$$q_d = \lim_{t \to \infty} \frac{P\{\|X\|_1 > t\}}{P\{X^{(1)} > t\}} = \frac{1}{\gamma_{e_1}}.   \tag{14.61}$$

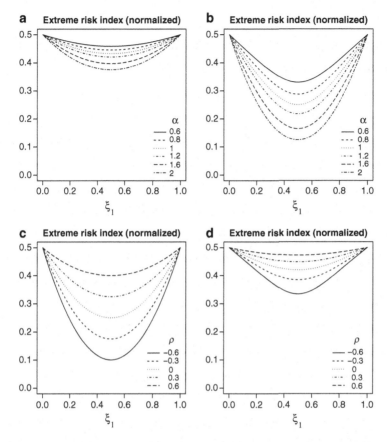

Figure 14.2 Bivariate elliptical distributions with generalized covariance matrices defined in (14.2): Diversification effects represented by functions $\xi^{(1)} \mapsto \Psi^*_{\alpha,\varrho} g_{\xi,\alpha}$ for selected values of ϱ and α. (**a**) Varying α for $\varrho = 0.5$. (**b**) Varying α for $\varrho = -0.5$. (**c**) Varying ϱ for $\alpha = 2$. (**d**) Varying ϱ for $\alpha = 0.5$ (Published as Figure 5.2 in *Statistics & Risk Modeling*, 29: 73–105, 2012)

Moreover, if X takes values in $\mathbb{R}^d_+$ and has balanced tails, then Proposition 14.12(a) yields

$$q_d = \lim_{t \to \infty} \frac{P\{\eta^\top X > d^{-1} t\}}{P\{X^{(1)} > t\}} = d^\alpha \frac{\gamma_\eta}{\gamma_{e_1}} = d^\alpha \Psi^* g_{\eta,\alpha}, \qquad (14.62)$$

where $\eta := (1/d, \ldots, 1/d)$ is the equally weighted portfolio.

Thus q_d is a product of the factor d^α, which is increasing in α, and the diversification coefficient $\Psi^* g_{\eta,\alpha}$. The coefficients $\Psi^* g_{\xi,\alpha}$ with $\xi \in \Sigma^d$ are decreasing in all examples considered here. This means that the aggregation and the diversification of risks are influenced by the tail index α in different ways. One can easily see that the extreme risk index γ_ξ is decreasing in α for $\xi \in \Sigma^d$. However, this result cannot be extended to $\Psi^* g_{\xi,\alpha}$ directly since $\Psi^* g_{\xi,\alpha}$ is related to γ_ξ by the normalizations (14.35) and (14.36). ◊

References

B. Acciaio, Optimal risk sharing with non-monotone monetary functionals. Finance Stoch. **11**, 267–289 (2007)

B. Acciaio, G. Svindland, Optimal risk sharing with different reference probabilities. Insur. Math. Econ. **44**, 426–433 (2009)

C. Acerbi, Spectral measures of risk: a coherent representation of subjective risk aversion. J. Bank. Finance **26**, 1505–1518 (2002)

G. Alberti, On the structure of singular sets of convex functions. Calc. Var. Partial Differ. Equ. **2**, 17–27 (1994)

S. Alink, M. Löwe, M.V. Wüthrich, Diversification of aggregate dependent risks. Insur. Math. Econ. **35**, 77–95 (2004)

S. Alink, M. Löwe, M.V. Wüthrich, Analysis of the expected shortfall of aggregate dependent risks. ASTIN Bull. **35**, 25–43 (2005)

C.D. Aliprantis, K.C. Border, *Infinite Dimensional Analysis: A Hitchhiker's Guide* (Springer, Berlin/New York, 1994)

T.W. Anderson, The integral of a symmetric unimodal function over a symmetric convex set and some probability inequalities. Proc Am. Math. Soc. **6**, 170–176 (1955)

E. Arjas, T. Lehtonen, Approximating many server queues by means of single server queues. Math. Oper. Res. **3**, 205–223 (1978)

B.C. Arnold, *Majorization and the Lorenz Order: A Brief Introduction*. Volume 43 of Lecture Notes in Statistics (Springer, Berlin, 1987)

P. Artzner, F. Delbaen, J.-M. Eber, D. Heath, Coherent measures of risk. Math. Finance **9**, 203–228 (1999)

J.-P. Aubin, *Optima and Equilibria: An Introduction to Nonlinear Analysis*. Number 140 in Graduate Texts in Mathematics (Springer, Berlin/New York, 1993)

F. Aurenhammer, F. Hoffmann, B. Aronov, Minkowski-type theorems and least-squares clustering. Algorithmica **20**, 61–76 (1998)

F. Baccelli, A.M. Makowski, Multidimensional stochastic ordering and associated random variables. Oper. Res. **37**, 478–487 (1989)

G. Balkema, P. Embrechts, *High Risk Scenarios and Extremes*. Zurich Lectures in Advanced Mathematics (European Mathematical Society EMS, Zürich, 2007)

P. Barbe, A.-L. Fougères, C. Genest, On the tail behavior of sums of dependent risks. ASTIN Bull. **36**, 361–373 (2006)

V. Barbu, T. Precupanu, *Convexity and Optimization in Banach Spaces*. Mathematics and Its Applications (East European Series), vol. 10, (D. Reidel, Dordrecht/Boston/Lancaster, 2nd rev. and extended edn., 1986)

386   References

R.E. Barlow, F. Proschan, *Statistical Theory of Reliability and Life Testing*. International Series in Decision Processes (Holt, Rinehart and Winston, New York, 1975)

P. Barrieu, N. El Karoui, Optimal derivatives design under dynamic risk measures, in *Mathematics of Finance*, ed. by G. Yin et al. Volume 351 of Contemporary Mathematics. Summer Research Conference on Mathematics on Finance, June 22–26, 2003, in Snowbird, Utah (AMS, Providence RI, 2004)

P. Barrieu, N. El Karoui, Inf-convolution of risk measures and optimal risk transfer. Finance Stoch. **9**, 269–298 (2005)

P. Barrieu, G. Scandolo, General Pareto optimal allocations and applications to multi-period risk. ASTIN Bull. **38**, 105–126 (2008)

B. Basrak, R.A. Davis, T. Mikosch, A characterization of multivariate regular variation. Ann. Appl. Probab. **12**, 908–920 (2002)

N. Bäuerle, Inequalities for stochastic models via supermodular orderings. Commun. Stat. **13**, 181–201 (1997)

N. Bäuerle, A. Müller, Modeling and comparing dependencies in multivariate risk portfolios. ASTIN Bull. **28**, 59–76 (1998)

T. Bedford, R. Cooke, *Probabilistic Risk Analysis: Foundations and Methods* (Cambridge University Press, Cambridge/New York, 2001)

T. Bedford, R.M. Cooke, Vines – a new graphical model for dependent random variables. Ann. Stat. **30**, 1031–1068 (2002)

M. Beiglböck, W. Schachermayer, Duality for Borel measurable cost functions. Trans. AMS **363**, 4203–4224 (2011)

M. Beiglböck, M. Goldstern, G. Maresch, W. Schachermayer, Optimal and better transport plans. J. Funct. Anal. **256**, 1907–1927 (2009)

D. Belomestny, V. Krätschmer, Central limit theorems for law-invariant coherent risk measures. J. Appl. Probab. **49**, 1–21 (2012)

R. Bergmann, Some classes of semi-ordering relations for random vectors and their use for comparing covariances. Math. Nachr. **82**, 103–114 (1978)

C. Bernard, P.P. Boyle, Explicit representation of cost-efficient strategies (2010), Available at SSRN: http://ssrn.com/abstract=1695543, Oct. 2010

C. Bernard, M. Maj, S. Vanduffel, Improving the design of financial products in a multidimensional Black–Scholes market. North Am. Actuar. J. **14**, 77–96 (2011)

C. Bernard, P.P. Boyle, S. Vanduffel, Explicit representation of cost-efficient strategies (2012). Available at SSRN: http://ssrn.com/abstract=1561272, March 2012

S. Biagini, M. Frittelli, On continuity properties and dual representation of convex and monotone functionals on Fréchet lattices. Working paper (2006)

N.H. Bingham, C.M. Goldie, J.L. Teugels, *Regular Variation*. Volume 27 of Encyclopedia of Mathematics and Its Applications (Cambridge University Press, Cambridge, 1987)

T. Björk, *Arbitrage Theory in Continuous Time*, 2nd edn. (Oxford University Press, Oxford, 2004)

J. Boman, F. Lindskog, Support theorems for the Radon transform and Cramér-Wold theorems. J. Theor. Probab. **22**, 683–710 (2009)

K. Borch, Reciprocal reinsurance treaties. ASTIN Bull. **1**, 170–191 (1960a)

K. Borch, The safety loading of reinsurance premiums. Skand. Aktuarietidsskr. **1**, 163–184 (1960b)

K. Borch, Equilibrium in a reinsurance market. Econometrica **30**, 424–444 (1962)

K. Borch, General equilibrium in the economic of uncertainty, in *Risk and Uncertainty*, vol. 30, ed. by K.H. Borch, J. Mossin (Macmillan, London, 1968), pp. 247–264

R.I. Boţ, G. Wanka, A weaker regularity condition for subdifferential calculus and fenchel duality in infinite dimensional spaces. Nonlinear Anal. **64**, 2787–2804 (2006)

Y. Brenier, Polar factorization and monotone rearrangement of vector-valued functions. Commun. Pure Appl. Math. **44**, 375–417 (1991)

H. Brezis, *Analyse Fonctionnelle, Théorie et Application* (Dunod, Paris, 1999)

H. Bühlmann, *Mathematical Methods in Risk Theory*. Volume 172 of Die Grundlehren der mathematischen Wissenschaften (Springer, Berlin/Heidelberg/New York, 1970)

H. Bühlmann, W.S. Jewell, Optimal risk exchanges. ASTIN Bull. **10**, 243–263 (1979)

R.S. Burachik, V. Jeyakumar, A dual condition for the convex subdifferential sum formula with applications. J. Convex Anal. **12**, 279–290 (2005)

C. Burgert, L. Rüschendorf, Allocation of risks and equilibrium in markets with finitely many traders. Insur. Math. Econ. **42**, 177–188, (2005/2008). Preprint (2005)

C. Burgert, L. Rüschendorf, Consistent risk measures for portfolio vectors. Insur. Math. Econ. **38**, 289–297 (2006a)

C. Burgert, L. Rüschendorf, On the optimal risk allocation problem. Stat. Decis. **24**, 153–171 (2006b)

S. Cambanis, G. Simons, W. Stout, Inequalities for $Ek(X, Y)$ when the marginals are fixed. Z. Wahrscheinlichkeitstheorie Verw. Geb. **36**, 285–294 (1976)

G. Carlier, R.A. Dana, Core of convex distortions of a probability. J. Econ. Theory **113**, 199–222 (2003)

G. Carlier, R.-A. Dana, Rearrangement inequalities in non-convex insurance models. J. Math. Econ. **41**, 483–503 (2005)

G. Carlier, R.-A. Dana, A. Galichon, Pareto efficiency for the concave order and multivariate comonotonicity. Working paper, available under (2011), http://arxiv.org/pdf/0912.0509v2

A. Chateauneuf, R.-A. Dana, J.-M. Tallon, Optimal risk-sharing rules and equilibria with Choquet-expected-utility. J. Math. Econ. **34**, 191–214 (2000)

P. Cheridito, T. Li, Risk measures on Orlitz hearts. Math. Finance **19**, 189–214 (2009)

P. Cheridito, F. Delbaen, M. Kupper, Dynamic monetary risk measures for bounded discrete-time processes. Electron. J. Probab. **11** (Paper No. 3), 57–106 (2006)

A.S. Cherny, Weighted VR and its properties. Finance Stoch. **10**, 367–393 (2006)

K.C. Cheung, C.J. Sung, S.C.P. Yam, S.P. Yung, Risk minimizing reinsurance for multivariate risks. J. Risk Insur. (2010). http://www.sta.cuhk.edu.hk/scpy/Preprints/Joseph%20Sung/Risk-MinimizingReinsurance_final.pdf

K.C. Cheung, C.J. Sung, S.C.P. Yam, S.P. Yung, Optimal reinsurance under general law-invariant risk measures. Scand. Actuarial J. (2011). http://dx.doi.org/10.1080/03461238.2011.636880

E. Chevallier, H.H. Müller, Risk allocation in capital markets: Portfolio insurance, tactical asset allocation and collar strategies. ASTIN Bull. **24**, 5–18 (1994)

K.-M. Chong, Some extensions of a theorem of Hardy, Littlewood and Pólya and their applications. Can. J. Math. **26**, 1321–1340 (1974)

K.-M. Chong, N.M. Rice, *Equimeasurable Rearrangements of Functions*. Volume 28 of Queen's Papers in Pure and Applied Mathematics (Queen's University, Kingston, 1971)

T.C. Christofides, E. Vaggelatou, A connection between supermodular ordering and positive/negative association. J. Multivar. Anal. **88**, 138–151 (2004)

J.A. Cuesta-Albertos, A. Tuero-Diaz, A characterization for the solution of the Monge-Kantorovich mass transference problem. Stat. Probab. Lett. **16**, 147–152 (1993)

J.A. Cuesta-Albertos, L. Rüschendorf, A. Tuero-Diaz, Optimal coupling of multivariate distributions and stochastic processes. J. Multivar. Anal. **46**, 335–361 (1993)

J.A. Cuesta-Albertos, C. Matrán, S.T. Rachev, L. Rüschendorf, Mass transport ation problems in probability theory. Math. Sci. **21**, 37–72 (1996)

J.A. Cuesta-Albertos, C. Matrán, A. Tuero-Diaz, On the monotonicity of optimal transportation plans. J. Math. Anal. Appl. **215**, 86–94 (1997)

C. Czado, *Pair Copula Constructions of Multivariate Copulas*. Volume 198 of Lecture Notes in Statistics (Springer, New York, 2010), pp. 93–109

G. Dall'Aglio, Sugli estremi dei momenti delle funzioni di ripartizione doppie. Ann. Scuola Normale Superiore Di Pisa, Cl. Sci. **3**, 33–74 (1956)

G. Dall'Aglio, Fréchet classes and compatibility of distribution functions, in *Symposia Mathematica*, vol. 9 (AMS, Providence RI, 1972), pp. 131–150

R.-A. Dana, A representation result for concave Schur concave functions. Math. Finance **14**, 613–634 (2005)

R.-A. Dana, M. Jeanblanc, *Financial Markets in Continuous Time* (Springer, Berlin/New York, 2003)

R.-A. Dana, I.I. Meilijson, Modelling agents' preferences in complete markets by second order stochastic dominance (2003). Working paper 0333, available at http://www.ceremade.dauphine. fr/preprints/cmd/2003-33.pdf, Cahiers du CEREMADE (2003)

R.-A. Dana, M. Scarsini, Optimal risk sharing with background risk. J. Econ. Theory **133**, 152–176 (2007)

J. Daníelsson, B.N. Jorgensen, M. Sarma, C.G. de Vries, Sub-additivity re-examined: the case for Value-at-Risk. EURANDOM Reports (2005). TU Eindhoven Report-no 2005-006, http://www. eurandom.tue.nl/reports/2005/006-report.pdf

J. Daníelsson, B.N. Jorgensen, M. Sarma, C.G. de Vries, Comparing downside risk measures for heavy tailed distributions. Econ. Lett. **92**, 202–208 (2006)

R. Davis, S. Resnick, Tail estimates motivated by extreme value theory. Ann. Stat. **12**, 1467–1487 (1984)

P.W. Day, Rearrangement inequalities. Can. J. Math. **24**, 930–943 (1972)

L. de Haan, A. Ferreira, *Extreme Value Theory.* Springer Series in Operations Research and Financial Engineering (Springer, New York, 2006)

L. de Haan, S.I. Resnick, Limit theory for multivariate sample extremes. Z. Wahrscheinlichkeitstheorie Verw. Geb. **40**, 317–337 (1977)

P. Deheuvels, La fonction de dépendance empirique et ses propriétés. Un test non paramètrique d'independance. Bulletin de la Classe des Sciences, V. Série, Académie Royale de Belgique **65**, 274–292 (1979)

P. Deheuvels, A Kolmogorov–Smirnov type test for independence and multivariate samples. Rev. Roum. Math. Pures Appl. **26**, 213–226 (1981)

A.L.M. Dekkers, J.H.J. Einmahl, L. de Haan, A moment estimator for the index of an extreme-value distribution. Ann. Stat. **17**, 1833–1855 (1989)

F. Delbaen, *Coherent Risk Measures.* Cattedra Galileiana (Scuola Normale Superiore, Pisa, 2000)

F. Delbaen, Coherent risk measures on general probability spaces, in *Advances in Finance and Stochastics: Essays in Honor of Dieter Sondermann*, ed. by K. Sandmann, P. Schönbucher (Springer, Berlin/New York, 2002), pp. 1–37

M. Denault, Coherent allocation of risk capital. J. Risk **4**, 1–34 (2001)

D. Denneberg, Distorted probabilities and insurance premiums. Math. Oper. Res. **63**, 3–5 (1990)

D. Denneberg, *Non-Additive Measure and Integral* (Kluwer, Dordrecht/Boston, 1997). 1st edn., 1994, 2nd edn., 1997

M. Denuit, C. Genest, E. Marceau, Stochastic bounds on sums of dependent risks. Insur. Math. Econ. **25**, 85–104 (1999)

M. Denuit, J. Dhaene, C. Ribas, Does positive dependence between individual risks increase stop-loss premiums? Insur. Math. Econ. **28**, 305–308 (2001)

M. Denuit, J. Dhaene, M. Goovaerts, R. Kaas, *Actuarial Theory for Dependent Risks: Measures, Orders and Models* (Wiley, Hoboken, 2005)

M. Denuit, J. Dhaene, M. Goovaerts, R. Kaas, R. Laeren, Risk measurement with equivalent utility principles. Stat. Decis. **24**, 1–26 (2006)

O. Deprez, H.U. Gerber, On convex principles of premium calculation. Insur. Math. Econ. **4**, 179–189 (1985)

J. Dhaene, M. Denuit, M.J. Goovaerts, R. Kaas, D. Vyncke, The concept of comonotonicity in actuarial science and finance: applications. Insur. Math. Econ. **31**, 133–161 (2002)

D. Dowson, B. Landau, The Frechet distance between multi-variate normal distributions. J. Multivar. Anal. **12**, 450–455 (1982)

H. Drees, Refined Pickands estimators of the extreme value index. Ann. Stat. **23**, 2059–2080 (1995)

H. Drees, A. Ferreira, L. de Haan, On maximum likelihood estimation of the extreme value index. Ann. Appl. Probab. **14**, 1179–1201 (2004)

W. Du Mouchel, The Pareto-optimality of an n-company reinsurance treaty. Skand. Aktuarietidsskr. **51**, 165–170 (1968)

D. Duffie, Martingales, arbitrage, and portfolio choice, in *First European Congress of Mathematics (ECM), Paris, France, July 6–10, 1992.* Volume II: Invited Lectures (Birkhäuser, Basel, 1994), pp. 3–21

D. Duffie, *Dynamic Asset Pricing Theory* (Princeton University Press, Princeton, 1st edn., 1992, 2nd edn., 2001)

N. Dunford, J.T. Schwartz, *Linear Operators. Part I: General Theory*, (Wiley, New York, 1st edn., 1958, 2nd edn., 1967)

N. Dunford, J.T. Schwartz, *Linear Operators. Part II: Spectral Theory. Self Adjoint Operators in Hilbert Space* (Wiley, New York, 1963)

F. Durante, C. Sempi, *Copula Theory: An Introduction*. Volume 198 of Lecture Notes in Statistics (Springer, Berlin/Heidelberg, 2010), pp. 3–31

P. Dybvig, Distributional analysis of portfolio choice. J. Bus. **61**, 369–393 (1988)

J.H.J. Einmahl, L. de Haan, X. Huang, Estimating a multidimensional extreme-value distribution. J. Multivar. Anal. **47**, 35–47 (1993)

J.H.J. Einmahl, L. de Haan, V.I. Piterbarg, Nonparametric estimation of the spectral measure of an extreme value distribution. Ann. Stat. **29**, 1401–1423 (2001)

I. Ekeland, W. Schachermayer, Law invariant risk measures on $L^{\infty}(\mathbb{R}^d)$. Stat. Risk Model. **28**, 195–225 (2011)

I. Ekeland, R. Teman, *Analyse Convexe et Problèmes Variationnels* (Dunod, Paris, 1974)

I. Ekeland, A. Galichon, M. Henry, Comonotonic measures of multivariate risks. Math. Finance **20**, 1–24 (2010)

P. Embrechts, G. Puccetti, Aggregating risk capital, with an application to operational risk. Geneva Risk Insur. Rev. **31**(2), 71–90 (2006a)

P. Embrechts, G. Puccetti, Bounds for functions of dependent risks. Finance Stoch. **10**, 341–352 (2006b)

P. Embrechts, G. Puccetti, Bounds for the sum of dependent risks having overlapping marginals. J. Multivar. Anal. **101**, 177–190 (2010)

P. Embrechts, A. Höing, A. Juri, Using copulae to bound the value-at-risk for functions of dependent risks. Finance Stoch. **7**, 145–167 (2003)

P. Embrechts, D.D. Lambrigger, M.V. Wüthrich, Multivariate extremes and the aggregation of dependent risks: examples and counter-examples. Extremes **12**, 107–127 (2009a)

P. Embrechts, J. Nešlehová, M.V. Wüthrich, Additivity properties for Value-at-Risk under Archimedean dependence and heavy-tailedness. Insurance **44**, 164–169 (2009b)

P. Embrechts, G. Puccetti, L. Rüschendorf, *Model Uncertainty and VaR Aggregation* (University of Freiburg, Freiburg, 2012). Preprint

J.D. Esary, F. Proschan, D.W. Walkup, Association of random variables, with applications. Ann. Math. Stat. **38**, 1466–1474 (1967)

M. Falk, J. Hüsler, R.-D. Reiss, *Laws of Small numbers: Extremes and Rare Events*. Volume 23 of DMV Seminar (Birkhäuser, Basel, 1994)

K. Fan, G.G. Lorentz, An integral inequality. Am. Math. Mon. **61**, 626–631 (1954)

C. Fefferman, M. Jodeit, JR, M.D. Perlman, A spherical surface measure inequality for convex sets. Proc. Am. Math. Soc. **33**, 114–119 (1972)

D. Filipović, M. Kupper, Optimal capital and risk transfers for group diversification. Math. Finance **18**, 55–76 (2008)

D. Filipović, G. Svindland, Optimal capital and risk allocations for law- and cash-invariant convex functions. Finance Stoch. **12**, 423–439 (2008)

D. Filipović, G. Svindland, The canonical model space for law-invariant convex risk measures is L^1. Math. Finance **22**, 585–589 (2012)

H. Föllmer, A. Schied, *Stochastic Finance. An Introduction in Discrete Time*. Volume 27 of Studies in Mathematics (de Gruyter, New York, 2011, 1st edn., 2002, 2nd edn., 2004)

M.J. Frank, R.B. Nelsen, B. Schweizer. Best-possible bounds for the distribution of a sum – a problem of Kolmogorov. Probab. Theory Relat. Fields **74**, 199–211 (1987)

M. Fréchet, Généralisations du théorème des probabilités totales. Fundam. Math. **25**, 379–387 (1935)

M. Fréchet, Les probabilités associées à un système d'événements compatibles et dépendants; I. Événements en nombre fini fixe. volume 859 of Actual. sci. industr. (Hermann & Cie, Paris)

M. Fréchet, Sur les tableaux de corrélation dont les marges sont données. Annales de l'Université de Lyon, Section A, Series 3 **14**, 53–77 (1951)

M. Frittelli, E. Rosazza Gianin, Putting order in risk measures. J. Bank. Finance **26**, 1473–1486 (2002)

M. Frittelli, E. Rosazza Gianin, Law invariant convex risk measures. in *Advances in Mathematical Economics*, vol. 7, ed. by S. Kusuoka et al. (Springer, Tokyo, 2005), pp. 33–46

M. Frittelli, G. Scandolo, Risk measure and capital requirements for processes. Math. Finance **16**, 589–612 (2006)

B. Fuchssteiner, W. Lusky, *Convex Cones*. Volume 56 of North-Holland Mathematics Studies (North-Holland, 1981)

N. Gaffke, L. Rüschendorf, On a class of extremal problems in statistics. Math. Operationsforschung Stat. **12**, 123–135 (1981)

N. Gaffke, L. Rüschendorf, On the existence of probability measures with given marginals. Stat. Decis. **2**, 163–174 (1984)

J. Galambos, *The Asymptotic Theory of Extreme Order Statistics* (Wiley, New York, 1978)

W. Gangbo, R.J. McCann, The geometry of optimal transportation. Acta Math **177**, 113–161 (1996)

W. Gangbo, A. Święch, Optimal maps for the multidimensional Monge–Kantorovich problem. Commun. Pure Appl. Math. **51**, 23–45 (1998)

C. Genest, L.-P. Rivest, A characterization of Gumbel's family of extreme value distributions. Stat. Probab. Lett. **8**, 207–211 (1989)

H.U. Gerber, Pareto-optimal risk exchanges and related decision problems. ASTIN Bull. **10**, 25–33 (1978)

H.U. Gerber, *An Introduction to Mathematical Risk Theory*. Number 8 in S. S. Huebner Foundation Monograph Series (S. S. Huebner Foundation for Insurance Education, Wharton School, University of Pennsylvania, Philadelphia, 1979)

H.U. Gerber, Credibility for Esscher premiums. Mitt. Ver. Schweiz. Versicherungsmath. **1980**, 307–312 (1980)

T. Goll, L. Rüschendorf, Minimax and minimal distance martingale measures and their relationship to portfolio optimization. Finance Stoch. **5**, 557–581 (2001)

M. Goovaerts, F. de Vylder, J. Haezendonck, *Insurance Premiums. Theory and Applications* (North-Holland, Amsterdam/New York, 1984)

I.H. Haff, K. Aas, A. Frigessi, On the simplified pair-copula construction – simply useful or too simplistic? J. Multivar. Anal. **101**, 1296–1310 (2010)

T. Hailperin, Best possible inequalities for the probability of a logical function of events. Am. Math. Mon. **72**, 343–359 (1965)

G.H. Hardy, J.E. Littlewood, G. Pólya, Some simple inequalities satisfied by convex functions. Messenger **58**, 145–152 (1929)

G.H. Hardy, J.E. Littlewood, G. Pólya, *Inequalities*, 2nd edn. (Cambridge University Press, Cambridge, 1952)

H.A. Hauksson, M.M. Dacorogna, T. Domenig, U.A. Müller, G. Samorodnitsky, Multivariate extremes, aggregation and risk estimation. *SSRN eLibrary* (2000). http://ssrn.com/paper= 254392

H. He, N.D. Pearson, Consumption and portfolio policies with incomplete markets and short-sale constraints: the finite-dimensional case. Math. Finance **1**(3), 1–10 (1991a)

H. He, N.D. Pearson, Consumption and portfolio policies with incomplete markets and short-sale constraints: the infinite dimensional case. J. Econ. Theory **54**, 259–304 (1991b)

D. Heath, H. Ku, Pareto equilibria with coherent measures of risk. Math. Finance **14**, 163–172 (2004)

B.M. Hill, A simple general approach to inference about the tail of a distribution. Ann. Stat. **3**, 1163–1174 (1975)

W. Hoeffding, Maßstabinvariante Korrelationstheorie. PhD thesis, *Schr. Math. Inst. u. Inst. Angew. Math. Univ. Berlin* 5, 181–233 (1940) and Berlin: Dissertation (1940)

H. Hult, F. Lindskog, Multivariate extremes, aggregation and dependence in elliptical distributions. Adv. Appl. Probab. **34**, 587–608 (2002)

H. Hult, F. Lindskog, Regular variation for measures on metric spaces. Publ. Inst. Math. (Beograd) (N.S.) **80**(94), 121–140 (2006)

D. Hunter, An upper bound for the probability of a union. J. Appl. Probab. **13**, 597–603 (1976)

T.P. Hutchinson, C.D. Lai, *Continuous Bivariate Distributions, Emphasising Applications* (Rumsby Scientific Publishing, Adelaide, 1990)

A. Inoue, On the worst conditional expectation. J. Math. Anal. Appl. **286**, 237–247 (2003)

P. Jaworski, F. Durante, W.K. Härdle, T. Rychlik (eds.), *Copula Theory and Its Applications. Proceedings of the Workshop held in Warsaw, Poland, 25–26 September 2009*. Number 198 in Lecture Notes in Statistics (Springer, Berlin/Heidelberg, 2010)

H. Joe, *Multivariate Models and Dependence Concepts*. Volume 73 of Monographs on Statistics and Applied Probability (Chapman & Hall, London, 1997)

K. Jogdeo, An inequality for a strict unimodal function with an application to a characterization of independence. Sankhyā A **32**, 405–410 (1970)

N.L. Johnson, S. Kotz, On some generalized Farlie–Gumbel–Morgenstern distributions. Commun. Stat. **4**, 415–427 (1975)

E. Jouini, H. Kallal, Efficient trading strategies in the presence of market frictions. Rev. Finance Stud. **14**, 343–369 (2001)

E. Jouini, M. Meddeb, N. Touzi, Vector-valued coherent risk measures. Finance Stoch. **8**, 531–552 (2004)

E. Jouini, W. Schachermayer, N. Touzi, Law invariant risk measures have the Fatou property, in *Advances in Mathematical Economics*, vol. 9, ed. by S. Kusuoka et al. (Springer, Tokyo, 2006), pp. 49–71

E. Jouini, W. Schachermayer, N. Touzi, Optimal risk sharing for law invariant monetary utility functions. Math. Finance **18**, 269–292 (2008)

R. Kaas, M. Goovaerts, J. Dhaene, M. Denuit, *Modern Actuarial Risk Theory* (Kluwer, Boston, 2001)

M. Kaina, Darstellung und Stetigkeitseigenschaften von Risikomaßen. Diplomarbeit, Universität Freiburg (2007)

M. Kaina, L. Rüschendorf. On convex risk measures on L^p-spaces. Math. Methods Oper. Res. **69**, 475–495 (2009)

J. Kallsen, Optimal portfolios for exponential Lévy processes. Math. Methods Oper. Res. **51**, 357–374 (2000)

J. Kallsen, Utility-based derivative pricing in incomplete markets, in *Mathematical Finance – Bachelier Congress 2000*, ed. by H. Geman, D. Madan, S. Pliska, T. Vorst (Springer, Berlin/New York, 2002), pp. 313–338

G. Kalmykov, On the partial ordering of one-dimensional Markov processes. Theory Probab. Appl. **7**, 456–459 (1962)

L.V. Kantorovich, On the translocation of masses. C. R. (Dokl.) Acad. Sci. URSS, n. Ser. **37**, 199–201 (1942)

L.V. Kantorovich, On a problem of Monge. Uspekhi Mat. Nauk. **3**, 225–226 (1948). In Russian

L.V. Kantorovich, G.P. Akilov, *Functional Analysis*. Transl. from the Russian by H. L. Silcock, 2nd edn. (Pergamon Press, Oxford, 1982)

L.V. Kantorovich, G.S. Rubinstein, On a function space in certain extremal problems. Dokl. Akad. Nauk. USSR **115**, 1058–1061 (1957)

J. Karamata, Sur une inégalité relative aux fonctions convexes. Publ. Math. Univ. Belgrade **1**, 145–148 (1932)

S. Karlin, A. Novikoff, Generalized convex inequalities. Pac. J. Math. **13**, 1251–1279 (1963)

S. Karlin, Y. Rinott, Classes of orderings of measures and related correlation inequalities. I. Multivariate totally positive distributions. J. Multivar. Anal. **10**, 467–498 (1980)

H.G. Kellerer, Maßtheoretische Marginalprobleme. Math. Ann. **153**, 168–198 (1964)

H.G. Kellerer, Duality theorems and probability metrics, in *Proceedings 7th Brasov Conference (Bucuresti, Romania, 1982)* (VNU Science Press, Utrecht, 1984a), pp. 211–220

H.G. Kellerer, Duality theorems for marginal problems. Z. Wahrscheinlichkeitstheorie Verw. Geb. **67**, 399–432 (1984b)

H.G. Kellerer, Measure theoretic versions of linear programming. Math. Z. **198**, 367–400 (1988)

S. Kiesel, L. Rüschendorf, Characterization of optimal risk allocations for convex risk functionals. Stat. Decis. **26**, 303–319 (2008)

S. Kiesel, L. Rüschendorf, On optimal allocation of risk vectors. Insur. Math. Econ. **47**, 167–175 (2010)

G. Kimeldorf, A. Sampson, One-parameter families of bivariate distributions with fixed marginals. Commun. Stat. **4**, 293–301 (1975a)

G. Kimeldorf, A. Sampson, Uniform representations of bivariate distributions. Commun. Stat. **4**, 617–627 (1975b)

C. Klüppelberg, S.I. Resnick, The Pareto copula, aggregation of risks, and the emperor's socks. J. Appl. Probab. **45**, 67–84 (2008)

M. Knott, C.S. Smith, On the optimal mapping of distributions. J. Optim. Theory Appl. **43**, 39–49 (1984)

M. Knott, C.S. Smith, Note on the optimal transportation of distributions. J. Optim. Theory Appl. **52**, 323–329 (1987)

M. Knott, C.S. Smith, On a generalization of cyclic monotonicity and distances among random vectors. Linear Algebr. Appl. **199**, 363–371 (1994)

D. Kurowicka, R.M. Cooke, *Uncertainty Analysis with High Dimensional Dependence Modelling* (Wiley, Hoboken, 2006)

S. Kusuoka, On law invariant coherent risk measures, in *Advances in Mathematical Economics*, vol. 3, ed. by S. Kusuoka et al. (Springer, Tokyo, 2001), pp. 83–95

T.L. Lai, M. Robbins, Maximally dependent random variables. Proc. Nat. Acad. Sci. USA **73**, 286–288 (1976)

T.L. Lai, H. Robbins, A class of dependent random variables and their maxima. Z. Wahrscheinlichkeitstheorie Verw. Geb. **42**, 89–111 (1978)

M. Landsberger, I.I. Meilijson, Co-monotone allocations, Bickel–Lehmann dispersion and the Arrow–Pratt measure of risk aversion. Ann. Oper. Res. **52**, 97–106 (1994)

E.L. Lehmann, Some concepts of dependence. Ann. Math. Stat. **37**, 1137–1153 (1966)

H.E. Leland, Who should buy portfolio insurance? J. Finance **35**, 581–596 (1980)

J. Lemaire, Exchange de Risques Entre Assureurs et Théorie des Jeux. ASTIN Bull. **9**, 155–179 (1977)

J. Lembcke, Gemeinsame Urbilder endlich additiver Inhalte. (Common inverse images of finitely additive contests). Math. Ann. **198**, 239–258 (1972)

J. Lembcke, Reguläre Maße mit einer gegebenen Familie von Bildmaßen. Sitzungsber. Bayr. Akad. Wiss. 61–115 (1976)

V.L. Levin, On the problem of mass transfer. Soviet Math. Dokl. **16** 1349–1353 (1975)

V.L. Levin. The Monge–Kantorovich problem on mass transfer, in *Methods of Functional Analysis in Mathematical Economics* (Nauka, Moscow, 1978), pp. 23–55

V.L. Levin, The problem of mass transfer in a topological space, and probability measures having given marginal measures on the product of two spaces. Sov. Math. Dokl. **29**, 638–643 (1984)

V.L. Levin, A.A. Milyutin, The mass transfer problem with discontinuous cost function and a mass setting for the problem of duality of convex extremum problems. Trans. Russ. Math. Surv. **34**, 1–78 (1979)

F. Liese, I. Vajda, *Convex Statistical Distances* (Teubner, Leipzig, 1987)

A.W. Lo, The three P's of total risk management. Finance Anal. J. **55**, 13–26 (1999)

G.G. Lorentz, An inequality for rearrangements. Am. Math. Mon. **60**, 176–179 (1953)

M. Ludkovski, L. Rüschendorf, On comonotonicity of Pareto optimal risk sharing. Stat. Probab. Lett. **78**, 1181–1188 (2008)

H. Luschgy, W. Thomsen, Extreme points in the Hahn–Banach–Kantorovich setting. Pac. J. Math. **105**, 387–398 (1983)

W.A.J. Luxemburg, Rearrangement invariant Banach function spaces. Queen's Papers Pure Appl. Math. **10**, 83–144 (1967)

G. Mainik, On asymptotic diversification effects for heavy-tailed risks. PhD thesis, University of Freiburg (2010)

G. Mainik, Estimating asymptotic dependence functionals in multivariate regularly varying models (ETH Zurich 2012). Lithuanian Mathematical Journal, July 2012, Volume 52, Issue 3, pp. 259–281

G. Mainik, L. Rüschendorf, On optimal portfolio diversification with respect to extreme risks. Finance Stoch. **14**, 593–623 (2011)

G. Mainik, L. Rüschendorf, Ordering of multivariate risk models with respect to extreme portfolio losses. Stat. Risk Model. **29**, 73–105 (2012)

G.D. Makarov, Estimates for the distribution function of a sum of two random variables when the marginal distributions are fixed. Theory Probab. Appl. **26**, 803–806 (1981)

Y. Malevergne, D. Sornette, *Extreme Financial Risks: From Dependence to Risk Management* (Springer, Berlin/Heidelberg/New York, 2006)

K.V. Mardia, Multivariate Pareto distributions. Ann. Math. Stat. **33**, 1008–1015 (1962)

D.D. Mari, S. Kotz, *Correlation and Dependence* (Imperial College Press, London 2001)

H.M. Markowitz, J. Finance **VII**, 77–91 (1952)

H.M. Markowitz, Foundations of portfolio theory. J. Finance **46**, 469–477 (1991)

A.W. Marshall, I. Olkin, *Inequalities: Theory of Majorization and Its Applications*. Number 143 in Mathematics in Science and Engineering (Academic, New York, 1979)

A.W. Marshall, I. Olkin, B.C. Arnold, *Inequalities: Theory of Majorization and Its Applications*. Springer Series in Statistics, 2nd edn. (Springer, New York/Dordrecht/Heidelberg/London, 2011)

W. Maurer, Bivalent trees and forests or upper bounds for the probability of a union revisited. Discret. Appl. Math. **6**, 157–171 (1983)

A.J. McNeil, J. Nešlehová, Multivariate Archimedean copulas, d-monotone functions and l_1-norm symmetric distributions. Ann. Stat. **37**, 3059–3097 (2009)

A.J. McNeil, J. Nešlehová, From Archimedean to Liouville copulas. J. Multivar. Anal. **101**, 1772–1790 (2010)

A.J. McNeil, R. Frey, P. Embrechts, *Quantitative Risk Management. Concepts, Techniques, and Tools*. Princeton Series in Finance (Princeton University Press, Princeton, 2005a)

A.J. McNeil, R. Frey, P. Embrechts, *Quantitative Risk Management. Concepts, Techniques, and Tools* (Princeton University Press, Princeton, 2005b)

L.E. Meester, J.G. Shanthikumar, Stochastic convexity on general space. Math. Oper. Res. **24**, 472–494 (1999)

I. Meilijson, A. Nadas, Convex majorization with an application to the length of critical paths. J. Appl. Probab. **16**, 671–677 (1979)

R.C. Merton, Optimum consumption and portfolio rules in a continuous-time model. J. Econ. Theory **3**, 373–413 (1971)

P.A. Meyer, *Probability and Potentials* (Blaisdell Publishing Company, a Division of Ginn and Company, Waltham, 1966)

Y. Miyahara, Minimal entropy martingale measures of jump type price processes in incomplete assets markets. Asia-Pac. Finance Mark. **6**, 97–113 (1999)

G. Monge, Mémoire sur la théorie des déblais et des remblais, in *Histoire de l'Académie Royale des Sciences de Paris, avec les Mémoires de Mathématiques et de Physique, pour la même Année, Tirés des Registres de cette Académie* (Paris, Mémoires de l'Académie Royale, 1781), pp. 666–704

D.S. Moore, M.C. Spruill, Unified large-sample theory of general chi-squared statistics for tests of fit. Ann. Stat. **3**, 599–616 (1975)

M. Moscadelli, The modelling of operational risk: experience with the analysis of the data collected by the Basel Committee (2004). Available at http://ideas.repec.org/p/bdi/wptemi/td_517_04.html, July 2004

K. Mosler, M. Scarsini, Some theory of stochastic dominance, in *Stochastic Orders and Decision Under Risk: Papers from the International Workshop held in Hamburg, Germany, May 16–20, 1989*. Volume 19 of Lecture Notes (Institute of Mathematical Statistics, Hayward, 1991a), pp. 261–284

K. Mosler, M. Scarsini (eds.), *Stochastic Orders and Decision Under Risk: Papers from the International Workshop held in Hamburg, May 16–20, 1989*. Volume 19 of Lecture Notes (Institute of Mathematical Statistics, Hayward, 1991b)

R. Moynihan, B. Schweizer, A. Sklar, Inequalities among operations on probability distribution functions. Tagung Oberwolfach 1976, ISNM **41**, 133–149 (1978)

A. Müller, M. Scarsini, Stochastic comparison of random vectors with a common copula. Math. Oper. Res. **26**, 723–740 (2001)

A. Müller, D. Stoyan, *Comparison Methods for Stochastic Models and Risks* (Wiley, Chichester, 2002)

D.W. Müller, *Statistische Entscheidungstheorie*. Lecture Notes (University of Göttingen, Göttingen, 1971)

Y. Nakano, Efficient hedging with coherent risk measure. J. Math. Anal. Appl. **293**, 345–354 (2004)

R.B. Nelsen, *An Introduction to Copulas. Properties and Applications*. Volume 139 of Lecture Notes in Statistics (Springer, New York, 1st edn., 1999, 2nd edn., 2006)

J. Nešlehová, P. Embrechts, V. Chavez-Demoulin, Infinite-mean models and the LDA for operational risk. J. Oper. Risk **1**, 3–25 (2006a)

J. Nešlehová, P. Embrechts, V. Chavez-Demoulin, Infinite-mean models and the LDA for operational risk. J. Oper. Risk **1**, 3–25 (2006b)

S.Y. Novak, Inference on heavy tails from dependent data. Siberian Adv. Math. **12**, 73–96 (2002)

G.L. O'Brien, The comparison method for stochastic processes. Ann. Probab. **3**, 80–88 (1975)

K.R. Parthasarathy, *Probability Measures on Metric Spaces*. Probability and Mathematical Statistics. A Series of Monographs and Textbooks (Academic, New York, 1967)

G.C. Pflug, W. Römisch, *Modeling, Measuring and Managing Risk* (World Scientific, Hackensack, 2007)

J. Pickands III, Statistical inference using extreme order statistics. Ann. Stat. **3**, 119–131 (1975)

J.I. Pickands, Multivariate extreme value distributions. Bull. Int. Stat. Inst. **49**, 859–878 (1981)

V. Piterbarg, Spread options, Farka's lemma and linear programming. Risk.net, Sept. (2011)

L.D. Pitt, Positively correlated normal variables are associated. Ann. Probab. **10**, 496–499 (1982)

B.L.S. Prakasa Rao, *Asymptotic Theory of Statistical Inference* (Wiley, New York, 1987)

G. Puccetti, L. Rüschendorf, Sharp bounds for sums of dependent risks. Journal of Applied Probability, to appear (2011)

G. Puccetti, L. Rüschendorf, Bounds for joint portfolios of dependent risks. Stat. Risk Model. **29**, 107–132 (2012a)

G. Puccetti, L. Rüschendorf, Computation of sharp bounds on the distribution of a function of dependent risks. J. Comput. Appl. Math. **236**, 1833–1840 (2012b)

G. Puccetti, M. Scarsini, Multivariate comonotonicity. J. Multivar. Anal. **101**, 291–304 (2010)

G. Puccetti, B. Wang, R. Wang, Advances in complete mixability. J. Appl. Probab. **49**(2), 430–440 (2012)

S.T. Rachev, The Monge–Kantorovich mass transference problem and its stochastic applications. Theory Probab. Appl. **29**, 647–676 (1985)

S.T. Rachev, *Probability Metrics and the Stability of Stochastic Models* (Wiley, Chichester, 1991)

S.T. Rachev, L. Rüschendorf, A transformation property of minimal metrics. Theory Probab. Appl. **35**, 110–117 (1990)

S.T. Rachev, L. Rüschendorf, Recent results in the theory of probability metrics. Stat. Decis. **9**, 327–373 (1991)

S.T. Rachev, L. Rüschendorf, Solution of some transportation problems with relaxed and additional constraints. SIAM Control Optim. **32**, 673–689 (1994)

S.T. Rachev, L. Rüschendorf, *Mass Transportation Problems. Vol. I: Theory*. Probability and Its Applications (Springer, New York/Heidelberg/Berlin, 1998a)

S.T. Rachev, L. Rüschendorf, *Mass Transportation Problems. Vol. II: Applications*. Probability and Its Applications (Springer, New York/Heidelberg/Berlin, 1998b)

S. T. Rachev, C. Menn, F. Fabozzi, *Fat-Tailed and Skewed Asset Return Distributions* (Wiley, Hoboken, 2005)

D. Ramachandran, L. Rüschendorf, A general duality theorem for marginal problems. Probab. Theory Relat. Fields **101**, 311–319 (1995)

D. Ramachandran, L. Rüschendorf, Duality theorems for assignments with upper bounds, in *Distributions with Given Marginals and Moment Problems. Proceedings of the 1996 Conference, Prague, Czech Republic*, ed. by V. Beneš et al. (Kluwer Academic, Boston, 1997), pp. 283–290

D. Ramachandran, L. Rüschendorf, On the validity of the Monge–Kantorovich duality theorem. Theory Probab. Appl. **45**, 350–356 (2000)

D. Ramachandran, L. Rüschendorf, Assignment models for constrained marginals and restricted markets, in *Distributions with Given Marginals and Statistical Modelling*, ed. by C.M. Cuadras, J. Fortiana, J.A. Rodriguez-Lallena (Kluwer Academic, Dordrecht/Boston, 2002), pp. 195–210

K. Rao, M. Rao, *Theory of Charges. A Study of Finitely Additive Measures* (Academic Press, London, 1983)

S.I. Resnick, *Extreme Values, Regular Variation, and Point Processes*. Volume 4 of Applied Probability. A Series of the Applied Probability Trust (Springer, New York, 1987)

S.I. Resnick, On the foundations of multivariate heavy-tail analysis. J. Appl. Probab. **41A**, 191–212 (2004).

S.I. Resnick, *Heavy-Tail Phenomena*. Springer Series in Operations Research and Financial Engineering (Springer, New York, 2007)

T. Rheinländer, Optimal martingale measures and their applications in mathematical finance. Dissertation, Dissertation Technische Universität Berlin, (1999)

R.T. Rockafellar, *Conjugate Duality and Optimization* (SIAM, Philadelphia, 1974)

R.T. Rockafellar, S. Uryasev, Optimization of conditional value-at-risk. J. Risk **2**(3), 21–41 (2000)

R.T. Rockafellar, S. Uryasev, M. Zabarankin, Optimality conditions in portfolio analysis with general deviation measures. Math. Program. **108**, 515–540 (2006)

V.A. Rohlin, On the fundamental ideas of measure theory. Am. Math. Soc. Transl. **1952**(71), 55p (1952)

H. Rootzén, N. Tajvidi, Multivariate generalized Pareto distributions. Bernoulli **12**, 917–930 (2006)

M. Rosenblatt, Remarks on a multivariate transformation. Ann. Math. Stat. **23**, 470–472 (1952)

M. Rubinstein, An aggregation theorem for securities markets. J. Finance Econ. **1**, 225–244 (1974)

B. Rüger, Scharfe untere und obere Schranken für die Wahrscheinlichkeit der Realisation von *k* unter *n* Ereignissen. Metrika **28**, 71–77 (1981)

L. Rüschendorf, Verteilungskonvergenz in ϕ-mischenden Prozessen mit Anwendungen für Order- und Rangstatistiken. PhD thesis, University of Hamburg (1974)

L. Rüschendorf, Asymptotic distributions of multivariate rank order statistics. Ann. Stat. **4**, 912–923 (1976)

L. Rüschendorf, *Vergleich von Zufallsvariablen bzgl. integralinduzierter Halbordnungen*. Habilitationsschrift, Mathematisch-Naturwissenschaftliche Fakultät der Rheinisch-Westfälischen Technischen Hochschule Aachen. 133 S. (1979)

L. Rüschendorf, Inequalities for the expectation of Δ-monotone functions. Z. Wahrscheinlichkeitstheorie Verw. Geb. **54**, 341–354 (1980)

L. Rüschendorf, Characterization of dependence concepts for the normal distribution. Ann. Inst. Stat. Math. **33**, 347–359 (1981a)

L. Rüschendorf, Ordering of distributions and rearrangement of functions. Ann. Probab. **9**, 276–283 (1981b)

L. Rüschendorf, Sharpness of Fréchet-bounds. Z. Wahrscheinlichkeitstheorie Verw. Geb. **57**, 293–302 (1981c)

L. Rüschendorf, Stochastically ordered distributions and monotonicity of the OC-function of sequential probability ratio tests. Mathematische Operationsforschung und Statistik Series Statistics **12**, 327–338 (1981d)

L. Rüschendorf, Weak association of random variables. J. Multivar. Anal. **11**, 448–451 (1981e)

L. Rüschendorf, Random variables with maximum sums. Adv. Appl. Probab. **14**, 623–632 (1982)

L. Rüschendorf, Solution of a statistical optimization problem by rearrangement methods. Metrika **30**, 55–62 (1983)

L. Rüschendorf, On the minimum discrimination information theorem. Stat. Decis. **1** (Suppl. issue), 263–283 (1984)

L. Rüschendorf, The Wasserstein distance and approximation theorems. Z. Wahrscheinlichkeitstheorie Verw. Geb. **70**, 117–129 (1985)

L. Rüschendorf, Monotonicity and unbiasedness of tests via a.s. constructions. Statistics **17**, 221–230 (1986)

L. Rüschendorf, Bounds for distributions with multivariate marginals, in *Stochastic Order and Decision under Risk*, vol. 19, ed. by K. Mosler, M. Scarsini (IMS Lecture Notes, Hayward, 1991a), pp. 285–310

L. Rüschendorf, Fréchet-bounds and their applications, in *Advances in Probability Measure with Given Marginals (Rome, Italy, 1990)*. Volume 67 of Math. Appl., ed. by G. Dall'Aglio, S. Kotz, G. Salinetti (Kluwer, 1991b), pp. 151–187

L. Rüschendorf, Convergence of the iterative proportional fitting procedure. Ann. Stat. **23**, 1160–1174 (1995a)

L. Rüschendorf, Optimal solutions of multivariate coupling problems. Appl. Math. **23**, 325–338 (1995b)

L. Rüschendorf, Developments on Fréchet bounds, in *Proceedings of Distributions with Fixed Marginals and Related Topics*, vol. 28. IMS Lecture Notes Monograph Series (IMS, Hayward, 1996), pp. 273–296

L. Rüschendorf, Wasserstein metric, in *Encyclopedia of Mathematics*, ed. by M. Hazewinkel (Springer, Dordrecht/Boston/London, 2001)

L. Rüschendorf, Comparison of multivariate risks and positive dependence. J. Appl. Probab. **41**, 391–406 (2004)

L. Rüschendorf, Stochastic ordering of risks, influence of dependence and a.s. constructions, in *Advances on Models, Characterizations and Applications*, ed. by N. Balakrishnan, I.G. Bairamov, O.L. Gebizlioglu (Chapman & Hall/CRC, Boca Raton, 2005), pp. 19–56

L. Rüschendorf, Law invariant convex risk measures for portfolio vectors. Stat. Decis. **24**, 97–108 (2006)

L. Rüschendorf, Monge–Kantorovich transportation problem and optimal couplings. Jahresbericht der DMV **109**, 113–137 (2007)

L. Rüschendorf, On the distributional transform, Sklar's thorem, and the empirical copula process. J. Stat. Plan. Infer. **139**, 3921–3927 (2009)

L. Rüschendorf, Risk measures for portfolio vectors and allocation of risks, in *Risk Assessment. Decisions in Banking and Insurance*, ed. by G. Bol, S.T. Rachev, R. Würth (Physica-Verlag, Heidelberg, 2010), pp. 133–164

L. Rüschendorf, Worst case portfolio vectors and diversification effects. Finance Stoch. **16**, 155–175 (2012a). Preprint 2009

L. Rüschendorf, Risk bounds, worst case dependence, and optimal claims and contracts, in *Proceedings of AFMathConf 2012: Actuarial and Financial Mathematics Conference* (Koninklijke Vlaamse Academie van Belgie voor Wetenschappen en Kunsten, Brussels, 2012b), pp. 23–36

L. Rüschendorf, V. de Valk, On regression representations of stochastic processes. Stoch. Process. Appl. **46**, 183–198 (1993)

L. Rüschendorf, S.T. Rachev, A characterization of random variables with minimum L^2-distance. J. Multivar. Anal. **32**, 48–54 (1990)

L. Rüschendorf, W. Thomsen, Note on the Schrödinger equation and I-projections. Stat. Probab. Lett. **17**, 369–375 (1993)

L. Rüschendorf, W. Thomsen, Closedness of sum spaces and the generalized Schrödinger problem. Theory Probab. Appl. **42**, 483–494 (1997)

L. Rüschendorf, L. Uckelmann, On optimal multivariate couplings, in *Distributions With Given Marginals and Moment Problems, Proceedings of the 1996 Conference, Prague, Czech Republic*, ed. by V. Beneš et al. (Kluwer Academic, Dordrecht, 1997), pp. 261–273

L. Rüschendorf, L. Uckelmann, Numerical and analytical results for the transportation problem of Monge–Kantorovich. Metrika. Int. J. Theor. Appl. Stat. **51**, 245–258 (2000)

L. Rüschendorf, L. Uckelmann, On the n-coupling problem. J. Multivar. Anal. **81**, 242–258 (2002)

A. Ruszczyński, A. Shapiro, Optimization of risk measures, in *Probabilistic and Randomize Methods for Design Under Uncertainty*, ed. by G. Calafiore et al. (Springer, London, New York, 2006), pp. 119–1157

J.V. Ryff, Orbits of L^1 functions under doubly stochastic transformations. Trans. Am. Math. Soc. **117**, 92–100 (1965)

G. Scandolo, Models of capital requirements in static and dynamic settings. Econ. Notes **33**, 415–435 (2004)

A. Schied, On the Neyman–Pearson problem for law-invariant risk measures and robust utility functionals. Ann. Appl. Probab. **14**, 1398–1423 (2004)

A. Schied, Optimal investments for robust utility functionals in complete market models. Math. Oper. Res. **30**, 750–764 (2005)

F. Schmid, R. Schmidt, T. Blumentritt, S. Gaisser, M. Ruppert, *Copula-Based Measures of Multivariate Association*. Volume 198 of Lecture Notes in Statistics (Springer, Heidelberg/Dordrecht/London/New York, 2010), pp. 209–236

R. Schmidt, U. Stadtmüller, Nonparametric estimation of tail dependence. Scand. J. Stat. **33**, 307–335 (2006)

B. Schweizer, Thirty years of copulas, in *Advances in Probability Distributions with Given Marginals*. Number 67 in Mathematics and Its Applications (Kluwer Academic, Dordrecht/Boston, 1991), pp. 13–50

H.L. Seal, *Stochastic Theory of Risk Business* (Wiley, New York, 1969)

M. Shaked, J.G. Shanthikumar, *Stochastic Orders and Their Applications*. Probability and Mathematical Statistics (Academic, Boston, 1994)

M. Shaked, J.G. Shanthikumar, Supermodular stochastic orders and positive dependence of random vectors. J. Multivar. Anal. **61**, 86–101 (1997)

M. Shaked, Y.L. Tong, Some partial orderings of exchangeable random variables by positive dependence. J. Multivar. Anal. **17**, 333–349 (1985)

A. Sklar, Fonctions de répartition à n dimensions et leurs marges. Publ. Inst. Stat. Univ. Paris **8**, 229–231 (1959)

A. Sklar, Random variables, joint distribution functions, and copulas. Kybernetika **9**, 449–460 (1973)

A.V. Skorohod, On a representation of random variables. Theory Probab. Appl. **21**, 645–648 (1976)

R.L. Smith, Estimating tails of probability distributions. Ann. Stat. **15**, 1174–1207 (1987)

C. Smith, M. Knott, On Hoeffding–Fréchet bounds and cyclic monotone relations. J. Multivar. Anal. **40**, 328–334 (1992)

D. Stoyan, Halbordnungsrelationen für Verteilungsgesetze. Math. Nachr. **52**, 315–331 (1972)

D. Stoyan, *Qualitative Eigenschaften und Abschätzungen stochastischer Modelle* (Oldenbourg, München, 1977)

V. Strassen, The existence of probability measures with given marginals. Ann. Math. Stat. **36**, 423–439 (1965)

P. Tankov, Improved Fréchet bounds and model-free pricing of multi-asset options. J. Appl. Probab. **48**, 389–403 (2011)

D. Tasche, Expected shortfall and beyond. J. Bank. Finance **26**, 1519–1533 (2002)

A.H. Tchen, Inequalities for distributions with given marginals. Ann. Probab. **8**, 814–827 (1980)

Y.L. Tong, *Probability Inequalities in Multivariate Distributions*. Probability and Mathematical Statistics (Academic, New York, 1980)

L. Uckelmann, Optimal couplings between one-dimensional distributions, in *Distributions With Given Marginals and Moment Problems, Proceedings of the 1996 Conference, Prague, Czech Republic*, ed. by V. Beneš et al. (Kluwer Academic, Dordrecht, 1997), pp. 275–281

A.W. van der Vaart, J.A. Wellner, *Weak Convergence and Empirical Processes*. Springer Series in Statistics (Springer, New York, 1st edn., 1996, corrected 2nd printing edn., 2000)

L.N. Vasershtein, Markov processes over denumerable products of spaces describing large system of automata. Problemy Peredači Informacii **5**, 64–72 (1969)

A.F. Veinott, Optimal policy in a dynamic, single product, nonstationary inventory model with several demand classes. Oper. Res. **13**, 761–778 (1965)

C. Villani, *Topics in Optimal Transportation*. Number 58 in Graduate Studies in Mathematics (American Mathematical Society, Providence RI, 2003)

C. Villani, *Optimal Transport, Old and New* (Springer, Berlin, 2009)

A. Vorobev, Consistent families of measures and their extensions. Theory Probab. Appl. **7**, 147–163 (1962)

B. Wang, R. Wang, The complete mixability and convex minimization problems with monotone marginal densities. J. Multivar. Anal. **102**, 1344–1360 (2011)

B. Wang, L. Peng, J. Yang, Bounds for the sum of dependent risks and worst Value-at-Risk with monotone marginal densities. Preprint (2011)

S. Wang, Premium calculation by transforming the layer premium density. ASTIN Bull. **26**, 71–92 (1996)

S.S. Wang, V.R. Yang, H.H. Panjer, Axiomatic characterization of insurance prices. Insur. Math. Econ. **21**, 173–183 (1997)

W. Warmuth, Marginal-Fréchet-bounds for multidimensional distribution functions. Statistics **19**, 283–294 (1988)

I. Weber, Komonotone Verbesserung und optimale Risikoallokation. Diplomarbeit, University of Freiburg (2008)

G. Wei, T. Hu, Supermodular dependence ordering on a class of multivariate copulas. Stat. Probab. Lett. **57**, 375–385 (2002)

G. Wei, H.B. Fang, K.T. Fang, The dependence of patterns of random variables – elementary algebraic and geometrical properties of copulas. Technical Report. No. 190, Dept. of Math. (Hong Kong Baptist University, 1998)

M. Whitt, Bivariate distributions with given marginals. Ann. Stat. **4**, 1280–1289 (1976)

W. Whitt, Stochastic comparisons for non-Markov processes. Math. Oper. Res. **11**, 608–618 (1986)

R.C. Williamson, T. Downs, Probabilistic arithmetic. I. Numerical methods for calculating convolutions and dependency bounds. Int. J. Approx. Reason **4**, 89–158 (1990)

M.V. Wüthrich, Asymptotic Value-at-Risk estimates for sums of dependent random variables. ASTIN Bull. **33**, 75–92 (2003)

C. Zălinescu. *Convex Analysis in General Vector Spaces* (World Scientific, River Edge, 2002)

List of Symbols

a.s.	almost surely
$X \sim F$	X has distribution function F
$X \uparrow_{\text{st}} Y$	stochastic increasing
$X = Y[P]$	X equal to Y P almost surely
CI	conditional increasing
PID	positive orthant dependence
PSMD	positive supermodular dependence
SIS	strong intersection condition
X^c	comonotone vector
WAS	weakly associated in sequence
$f(Q\|P)$	divergence distance
ℓ_2	minimal ℓ_2-metric
$\mathbb{Q}_{d,p}$	positive dual space of L_d^p
γ_ξ	extremal risk index
$ES_\alpha(X)$	expected shortfall
$L_d^p(P)$	product of L^p spaces
$\widehat{\varrho}(X)$	infimal convolution
$SR_p(X)$	shortfall risk
$\text{VaR}_\lambda(X)$	value at risk
$\partial f(x)$	subgradient of f
core(D)	core of D
dom f	domain of f
MRV	multivariate regular varying

L. Rüschendorf, *Mathematical Risk Analysis*, Springer Series in Operations Research and Financial Engineering, DOI 10.1007/978-3-642-33590-7,
© Springer-Verlag Berlin Heidelberg 2013

$\mathcal{F}(F_1, \ldots, F_n)$ Fréchet class
$\mathcal{M}_{\mathcal{E}}$ generalized Fréchet class
$\mathcal{M}(P_1, \ldots, P_n)$ Fréchet class

$\mathcal{E}(\mu, \Sigma, F_R)$ elliptical distribution
$F(Y, V)$ distributional transform
P^h distribution of h
τ_F multivariate quantile transform
$U(0, 1)$ uniform distribution on $(0, 1)$

$\leq_c$ concordance order
$\leq_{ccx}$ componentwise convex order
$\leq_{cx}$ convex order
$\leq_\Delta$ ordering by Δ-monotone functions
$\leq_{dcx}$ directionally convex order
$\leq_{\mathcal{F}}$ integral induced order
$\leq_{icx}$ increasing convex order
$\leq_S$ Schur order
$\leq_{sm}$ supermodular order
$\leq_{st}$ stochastic order
$\leq_{uo}$ upper orthant order

$\preceq_{apl}$ asymptotic portfolio loss order
$\preceq_{psd}$ positive semidefinite order

Index

vine, D-, 18
vine representation, 20

WCAP$_\Psi$. *See* Worst case average portfolio (WCAP$_\psi$)
WCP$_\Psi$. *See* Worst case portfolio (WCP$_\psi$)
weak association, 116
weak association in sequence, 121
weak majorization order, 58
weakly associated vector, 121
weakly conditional increasing in sequence order, 121
weighted minimal convolution, 271, 287
weighting scheme bounds, 111
well posed, 285

worst case average joint portfolio, 210
worst case average portfolio (WCAP$_\psi$), 208
worst case dependence, 319
worst case diversification effect, 212
worst case pairs, 75
worst case portfolio, 208
worst case portfolio (WCP$_\psi$), 208, 291
worst case reinsurance problem, 318
worst case scenario, 209
    measure, 154, 196, 209, 290
worst case solutions, 76
worst case total risk, 211

zero utility premiums, 151

Printed in the United States
By Bookmasters